Supramolecular Chemistry at Surfaces

Monographs in Supramolecular Chemistry

Series Editors:
Professor Philip Gale, *University of Southampton, UK*
Professor Jonathan Steed, *Durham University, UK*

Titles in this Series:
1: Cyclophanes
2: Calixarenes
3: Crown Ethers and Cryptands
4: Container Molecules and Their Guests
5: Membranes and Molecular Assemblies: The Synkinetic Approach
6: Calixarenes Revisited
7: Self-assembly in Supramolecular Systems
8: Anion Receptor Chemistry
9: Boronic Acids in Saccharide Recognition
10: Calixarenes: An Introduction, 2nd Edition
11: Polymeric and Self Assembled Hydrogels: From Fundamental Understanding to Applications
12: Molecular Logic-based Computation
13: Supramolecular Systems in Biomedical Fields
14: Synthetic Receptors for Biomolecules: Design Principles and Applications
15: Polyrotaxane and Slide-Ring Materials
16: Boron: Sensing, Synthesis and Supramolecular Self-Assembly
17: Porous Polymers: Design, Synthesis and Applications
18: Pillararenes
19: Supramolecular Chemistry at Surfaces

How to obtain future titles on publication:
A standing order plan is available for this series. A standing order will bring delivery of each new volume immediately on publication.

For further information please contact:
Book Sales Department, Royal Society of Chemistry, Thomas Graham House, Science Park, Milton Road, Cambridge, CB4 0WF, UK
Telephone: +44 (0)1223 420066, Fax: +44 (0)1223 420247
Email: booksales@rsc.org
Visit our website at http://www.rsc.org/Shop/Books/

Supramolecular Chemistry at Surfaces

David B. Amabilino
School of Chemistry, The University of Nottingham, UK
Email: david.amabilino@nottingham.ac.uk

ROYAL SOCIETY OF CHEMISTRY

THE QUEEN'S AWARDS
FOR ENTERPRISE:
INTERNATIONAL TRADE
2013

Monographs in Supramolecular Chemistry No. 19

Print ISBN: 978-1-84973-952-8
PDF eISBN: 978-1-78262-216-1
EPUB eISBN: 978-1-78262-818-7
ISSN: 1368-8642

A catalogue record for this book is available from the British Library

Published by The Royal Society of Chemistry,
Thomas Graham House, Science Park, Milton Road,
Cambridge CB4 0WF, UK

Registered Charity Number 207890

Visit our website at www.rsc.org/books

Printed in the United Kingdom by CPI Group (UK) Ltd, Croydon, CR0 4YY, UK

Preface

for Lluïsa, Adrià, Brian and Mònica

The idea for this text came during a Master level lecture course that I was participating in at the *Universitat de Barcelona* on the Bottom-up Growth of Materials on Surfaces, especially focussing on molecular materials. I felt that in the nanoscience area, the vast body of work on mono- and multilayers had very good specific reviews but there was no overview of the area as a whole, showing the requirements or needs for any particular kind of assembly route and the advantages and disadvantages of the various strategies for surface assembly, or indeed the opportunities that existed should one route have been tried and not reached expectations. Where was one to turn for alternatives? This monograph modestly aims to help address this space and provide an (albeit highly personal and far from comprehensive) overview of supramolecular chemistry at surfaces.

The text aims to introduce the main concepts involved in the growth of layers of molecules on surfaces, and show the breadth of work that has been done, and the excitement that this growing area generates. It does not attempt to be an exhaustive treatment of supramolecular chemistry at surfaces, and I apologise in advance to those kind enough to read the work but feel their important research under-represented. It was simply impossible to give a totally comprehensive coverage of all the fields that are discussed here. The treatment is broad and is the author's view of the main principle concepts across strongly and weakly interacting systems at interfaces.

I have to thank all the people, mentors, students and colleagues I have worked with during my scientific career, in every aspect from my PhD through my postdocs up to the present time, and especially those who have carried out research into monolayers in my group in the Materials Science Institute of Barcelona (ICMAB-CSIC)—Elba Gomar-Nadal,

Monographs in Supramolecular Chemistry No. 19
Supramolecular Chemistry at Surfaces
By David B. Amabilino

Published by the Royal Society of Chemistry, www.rsc.org

Josep Puigmartí-Luis, Patrizia Iavicoli, Mathieu Gonidec, Wojciech Saletra, Magali Lingenfelder, Ángela Bejarano, Arántzazu González Campo, Cristina Oliveras and Marta Riba—because I've learned a lot with you. Also, I'm indebted to my collaborators and dear friends Steven De Feyter, Rasmita Raval, Roberto Lazzaroni, Mathieu Linares, Hans Elemans, Ángel Pérez del Pino, Ismael Díez-Pérez and Carmen Ocal who have taught me an enormous amount, as well as to people in their groups, André Gesquiere who taught me how to drive an STM, Mohamed Abdel-Mottaleb, Sam Haq, Wael Mamdouh, Shengbin Lei, Mendel In't Veld, Andrea Minoia and Hong Xu. To my present group and colleagues, Neil Champness, Peter Beton and Rob Jones, I thank them for their patience and kindness in answering the often stupid questions I posed to them!

Scientists like what they do too much (or at least they should do), but the strength from our loved ones is vital, and the most important thing; to my family I dedicate this book. To my wife Lluïsa, who is my soul mate and also professional confidante, you've helped me more than you know, and to Adrià, Brian and Mònica for making me the proudest Dad in the world.

Abbreviations

AES	Auger electron spectroscopy
AFM	atomic force microscopy
APTES	(3-aminopropyl)triethoxysilane
BAM	Brewster angle microscopy
CV	cyclic voltammetry
CyDs	cyclodextrins
DFT	density functional theory
FESEM	field emission scanning electron microscopy
GIXRD	grazing incidence X-ray reflectivity
HOMO	highest occupied molecular orbital
HOPG	highly oriented pyrolytic graphite
HWE	hot wall epitaxy
IRRAS	IR reflection absorption spectroscopy
ITO	indium tin oxide
LB	Langmuir–Blodgett
LbL	layer-by-layer
LEED	low energy electron diffraction
LUMO	lowest unoccupied molecular orbital
MBE	molecular beam epitaxy
MOF	metal–organic framework
μCP	microcontact printing
ML	monolayer
OMBE	organic molecular beam epitaxy
NEXAFS	near-edge X-ray absorption fine structure spectroscopy
PDMS	poly(dimethylsiloxane)
PEM	polyelectrolyte multilayer
PM-IRRAS	polarization modulation-infrared reflection-adsorption spectroscopy
QCM	quartz crystal microbalance

Monographs in Supramolecular Chemistry No. 19
Supramolecular Chemistry at Surfaces
By David B. Amabilino

Published by the Royal Society of Chemistry, www.rsc.org

RAIRS	reflection-absorption infrared spectroscopy
SAM	self-assembled monolayer
SEM	scanning electron microscopy
SFGS	sum frequency generation vibrational spectroscopy
STM	scanning tunnelling microscopy
STS	scanning tunnelling spectroscopy
SURMOF	surface grown metal–organic frameworks
TOF-SIMS	time-of-flight secondary ion mass spectrometry
UHV	ultrahigh vacuum
XPS	X-ray photoelectron spectroscopy
XRD	X-ray diffraction

Contents

Monographs in Supramolecular Chemistry No. 19
Supramolecular Chemistry at Surfaces
By David B. Amabilino

Published by the Royal Society of Chemistry, www.rsc.org

CHAPTER 1

Surfaces for Supramolecular Systems

1.1 Introduction

The non-covalent bond is omnipresent in molecular systems at surfaces and the frontiers between materials in general.[1–5] This environment is distinct to either bulk materials or solution-borne aggregates, although as we shall see, the principles of supramolecular chemistry can be applied to great effect at surfaces. A great deal can be learned about supramolecular objects by studying them at surfaces where they can be probed with powerful techniques that allow precise information to be garnered about their structural and physical characteristics. Apart from the purely scientific fascination with these systems, opportunities are evident for their use in electronic devices including sensors, or for their optical or mechanical properties. While many of the phenomena apply to nanoparticulate systems with solid cores where supramolecular chemistry from the surface is extremely versatile,[6–11] the most precise way to study these systems is on flat surfaces where molecules are attached at more or less specific locations. This kind of system is the subject of this book.[†]

The supramolecular chemistry in these environments can be complex, because, as is shown graphically in Figure 1.1, the molecules that form an eventual layer on a bulk material have interactions between themselves and also with the surface of the bulk material and whatever medium is over them. We must therefore consider conscientiously this kind of situation in

---

[†]The frontiers between two bulk phases in colloidal systems are not covered here; the interested reader is recommended the following text for treatment of these equally fascinating areas: H.-J. Butt, K. Graf and M. Kappl, *Physics and Chemistry of Interfaces, second revised and enlarged edition*, Wiley-VCH, Weinheim, 2006.

Monographs in Supramolecular Chemistry No. 19
Supramolecular Chemistry at Surfaces
By David B. Amabilino

Published by the Royal Society of Chemistry, www.rsc.org

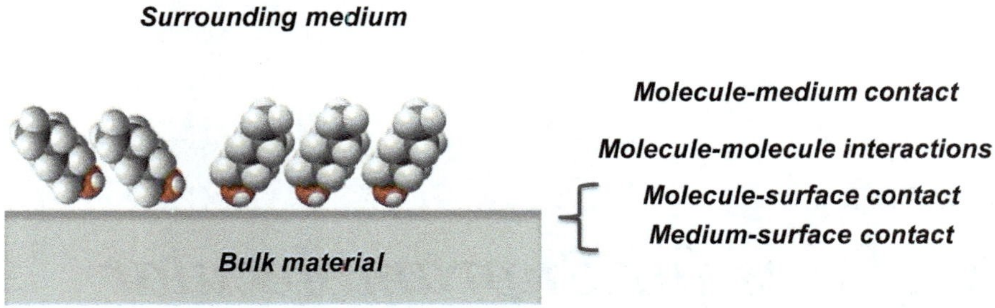

Figure 1.1 Molecules at surfaces and the frontiers between the components of the system.

all its guises as a system, a whole rather than a sum of parts. The chemical characteristics of all components of the system can influence dramatically the relative importance of any particular non-covalent interaction. There is also the case where the surrounding medium contains other molecules capable of interaction with the layer of surface-bound molecules. It is desired to provide the reader with an understanding of surfaces and how supramolecular systems behave at them, and how one can characterise them. It will be shown why these systems are of interest from a fundamental point of view, but also in terms of materials with practical applications. In this sense, opportunities for surface-grown supramolecular systems will be given.

Surfaces surround us, literally, and play defining roles in the properties of all kinds of materials (Figure 1.2), and have done for millennia.[12] In natural systems, the interactions and reactions between molecules take place very often in heterogeneous media: at surfaces, or more properly at *interfaces*, the area between two continuous phases. The self-assembly of lipids and macromolecules at interfaces is widespread and intriguing.[13] Indeed, it was in biological systems where scientists initially saw the importance of supramolecular chemistry and mentioned it in these terms in the context of collagen self-assembly.[14] A recent book discusses in detail the fascinating topic of biointerfaces.[15] All the frontiers between the combinations of solids, gases, and liquids are involved in important chemical processes, from the capture of oxygen by lungs or gills when animals breathe[16,17] to the deposition of inorganic materials in processes such as bone formation. Supramolecular chemistry is of vital importance in the functioning of these systems. In man-made materials, "non-stick" surfaces in cookware have revolutionised the way we prepare food and the remarkable advances in adhesives have made fixing broken objects easy. Surfactants, natural and artificial, are perhaps the most frequently observed system where a molecule interacts through non-covalent bonding with itself and a surface to make bubbles or clean dirt.

The physical or chemical examples given so far refer principally to surfaces that are the contact point between two continuous materials (phases). The surface is the outermost layer of a material (Figure 1.3). However, this plane

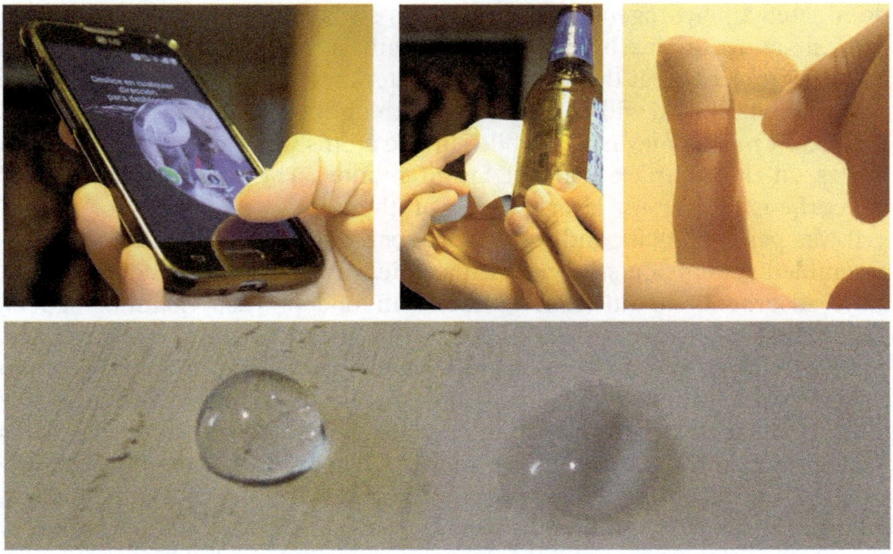

Figure 1.2 Household interfaces in action; from top left, our fingers with the screens of mobile tactile devices for function and casing for grip and support, a label on a glass bottle held in place by an adhesive, a quite different sticky situation for a plaster that must adhere to damp fingers, and the wetting of paper by a drop of water (right) or a wax crayon coated area (left) where the drop stays intact and round.

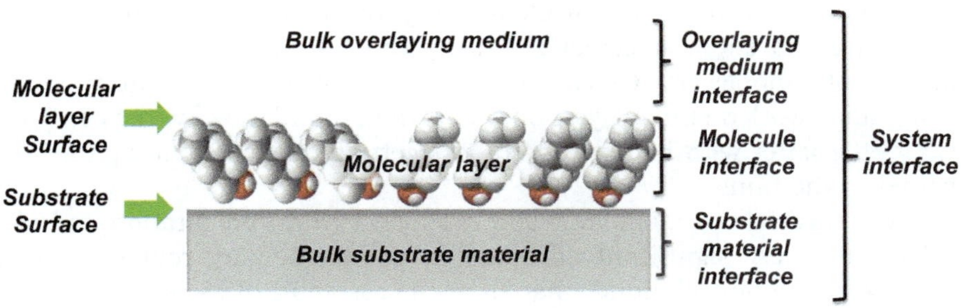

Figure 1.3 The concepts of surface and interface, of each material and the system.

and even the region immediately below the surface of a material is not representative of the bulk, especially in the case of liquid systems or metals, for example. When considering phenomena such as the reflection light from solid materials or the properties of conductors or magnets near their surface, it is far more appropriate to speak of the interface between two bulk phases. The term implies that there is a region near the surface that is influenced by it in a structural and/or property sense. The properties of materials near their contact point with a different phase will change compared with the bulk. The same is true when considering liquids, water perhaps being the best known case,[18] where the short range ordered structure dominated by hydrogen bonds makes its surface tension high in contact with air and explains how it covers certain surfaces. The mobility of the atoms or molecules at the surface of a material will generally be different to that in the bulk and will depend on the strength of the interaction between them. Soft metals may have mobile surface atoms while the strong bonds in oxides on their surface will not favour motion. These characteristics will affect the nature of any layer forming on them.

It was perhaps Allara and Nuzzo that sparked the present interest in self-assembled systems on surfaces with a series of papers on the adsorption of alkyl adsorbents to different surfaces. In one of their seminal papers, "Spontaneously Organized Molecular Assemblies. 1. Formation, Dynamics, and Physical Properties of *n*-Alkanoic Acids Adsorbed from Solution on an Oxidized Aluminum Surface" published in the first issue of the then new journal *Langmuir*,[19] they showed how long alkyl chain carboxylic acids formed layers spontaneously from solution. They also presented a scholarly perspective concerning the previous work on adsorption at surfaces. Indeed, in 1946 (in the first volume of *Journal of Colloid Science*!), the formation of monolayer films from non-polar liquids was established.[20] The authors of that work noted then that "the greater the chain length of the molecules the more condensed and rigid the film will be", a hypothesis that has truly stood the test of time, as we shall see. Remarkably, they also found that for platinum and Pyrex "while weight concentrations of only 10^{-7} M were required for primary aliphatic amines and monocarboxylic acids, roughly 1000 times more was needed for the aliphatic alcohols, esters and ketones and for cholesterol". This pioneering work already gave us the evidence that specific molecular and supramolecular chemistry can be used to control the formation of monolayers. Of course, all this work was surely inspired by Langmuir's work on monolayers on water[21] and multilayers by transfer from it,[22,23] the latter with his co-workers Blodgett and Schaefer that give their names to the films.

These methods of monolayer and multilayer (film) preparation and the subsequent very significant advances in complementary routes make a toolbox of unique techniques that allow the assembly of molecules on surfaces. Some of the main methods are represented in Figure 1.4, and they are the ones that will appear throughout this text.

Building on the shoulders of the pioneering scientists working on adsorption at surfaces and those of the founders of supramolecular chemistry,[24–26] the development of supramolecular systems at surfaces has been spectacular in recent years. It is a truly interdisciplinary area with multiple potential and actual applications across a huge range of areas. One need only consider the following achievements observed at the molecular level to appreciate this: precise binding of molecules in host–guest systems at interfaces;[27] observation of reactivity of single molecules on surfaces;[28] the proof of spontaneous symmetry breaking on a surface;[29] the measurement of conductivity through single molecules,[30] and the self-assembly of single molecule magnets on surfaces.[31,32]

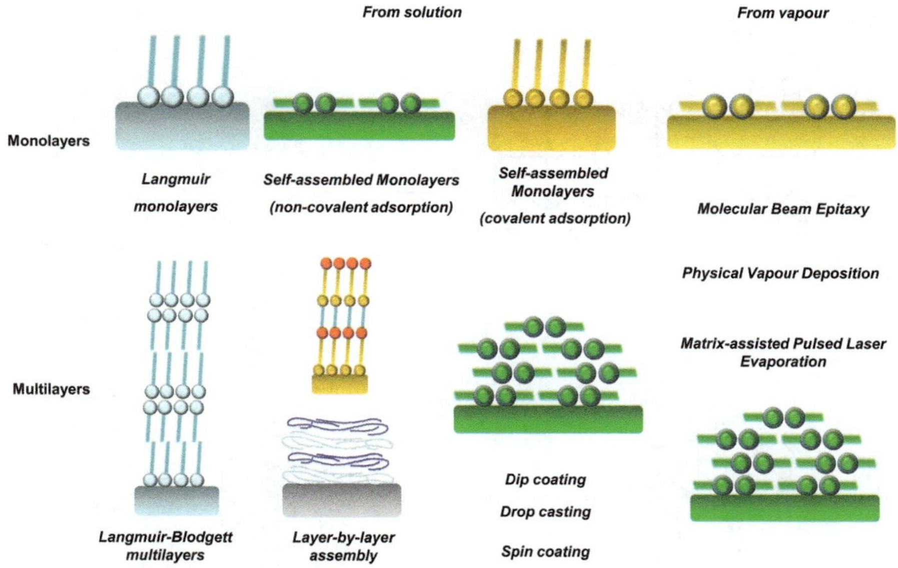

Figure 1.4 Some routes to the formation of layers of molecules on surfaces and cartoon representations of the films they produce.

As specific landmarks in the development of supramolecular systems on surfaces (Figure 1.5), the observation of binding and formation of organised domains of a complex formed between a biotin-derived surfactant and streptavidin is paradigmatic of the way host–guest interactions can be used to assemble materials.[33] This biological recognition motif forms remarkably well defined monolayers at the surface of water. In a more purely synthetic example, the proof that resorcin[4]arenes adsorbed on a gold surface through thioether groups showed a remarkable selectivity and sensitivity (down to nanograms) for tetrachloroethene in a quartz microbalance oscillator (Figure 1.5).[34] This example shows how a host–guest system from solution can be turned into a device that does not require solution processing at all. Surface immobilisation is an excellent route for applying developed host–guest systems in a way where solution processing is not necessary and where the limits of detection of spectroscopy are avoided.

The observation of extended surface chirality when the enantiomers of tartaric acid are adsorbed on copper was remarkable.[35] It showed how modern imaging techniques at metal surfaces can give precise structural information—in this case, the chirality in extended areas—of molecules on surfaces to show their supramolecular organisation as well as proving the potential to probe heterogeneous catalysis in this kind of adsorbed system. The final example here is that of another biologically inspired system—duplex DNA—that when confined in a monolayer is capable of polarising electrons, an important feat for molecules within the context of spintronics.[36]

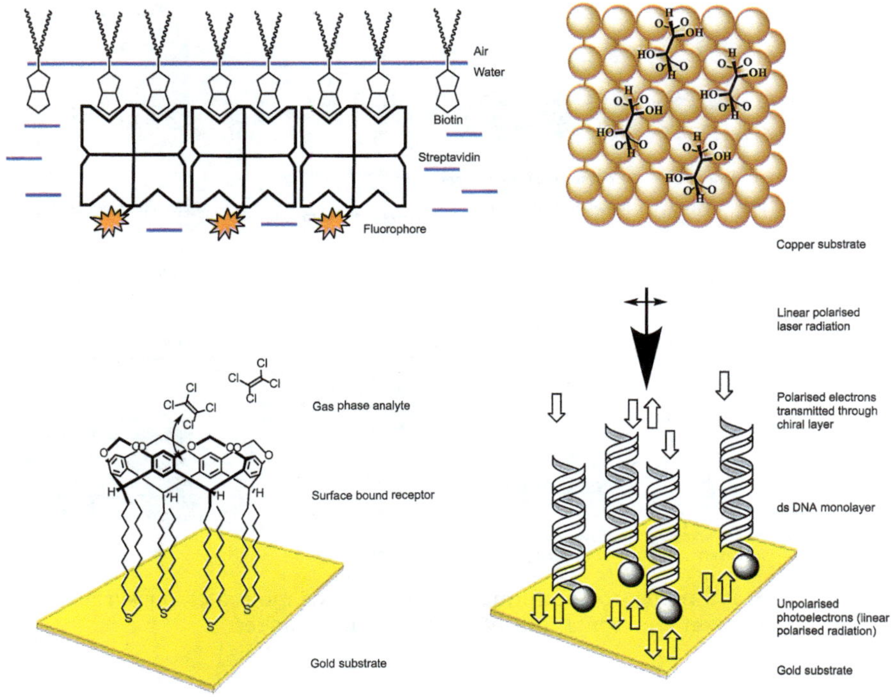

Figure 1.5 Four examples of supramolecular systems on surfaces that were important breakthroughs for science. Top left, the binding of a biotin derived surfactant to a fluorescently-labelled streptavidin at the air–water interface; bottom left, the selective detection of tetrachloroethene by a self-assembled monolayer formed by a resorcinarene derivative; top right, the expression of structural chirality in monolayers of chemisorbed tartrate on metallic copper; and bottom right, the observation of spin-selective passage of electrons through a monolayer of double-stranded DNA.

The study of self-assembly on surfaces has led to an area referred to as "two-dimensional supramolecular chemistry",[37,38] focused mainly on the plane parallel to the initial surface, which shares the feedback of information with crystal engineering that many supramolecular systems do. The difference now is that very precise information can be determined on surfaces, perhaps leading to insight that crystal science can sometimes not achieve and revealing the beauty and complexity of superstructures and complexes.[39] Merely by way of example, the observation of benzoic acid derivatives on surfaces reveals that the structures that they form can be very similar to those formed in the solid state. The dicarboxylic acid compound shown in Figure 1.6 forms a structure on the surface of graphite very similar to planes within the bulk crystals of the compound.[40] Chains of hydrogen bonded acids run parallel to one another and the alkyl chains attached to the phenyl ring interdigitate to give a crystalline packing. Apart from single-component systems whose structure can be studied precisely at surfaces,

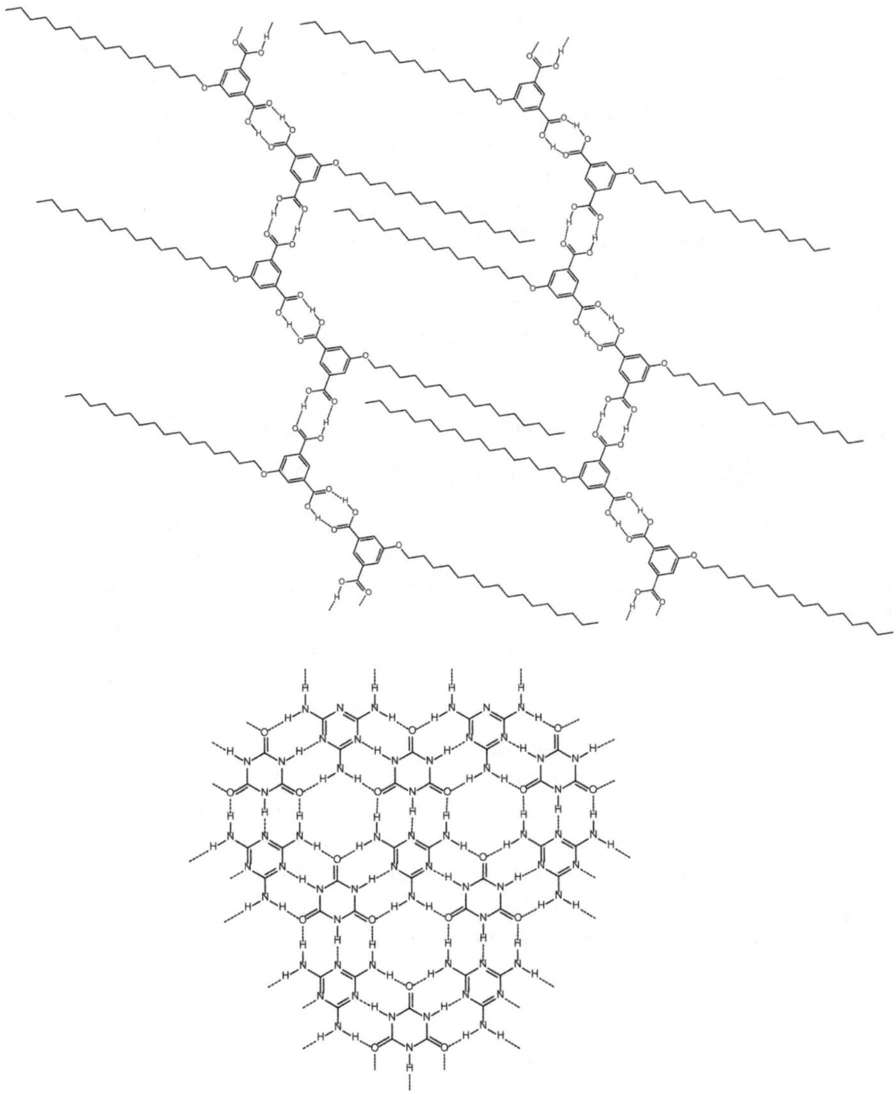

Figure 1.6 Drawings of ordered supramolecular structures observed both in three-dimensional solids and on certain surfaces. The top example shows chains of a phthalic acid derivative where the pendant long alkyl chains interdigitate and interact through van der Waals interactions. Below, the complex formed between melamine and cyanuric acid.

ordered multiple-component systems are also formed readily. The classic example from supramolecular chemistry and crystal engineering[41] of the complex formed between melamine and cyanuric acid (Figure 1.6) forms the same bimolecular network on a surface.[42] While not all adsorbates show identical two-dimensional structure to the solid state, there are factors in the molecular and supramolecular design that can be optimised to result in remarkable similarities: pyrene-2,7-dicarboxylic acid and the tetrahydro-analogue crystallise in three and two dimensions with remarkable similarity, while biphenyl derivatives do not.[43] This effect was ascribed to the high planarity of the former. From these examples of strict assembly, it is possible to develop systems that show molecular recognition at interfaces,[44] as we shall see in forthcoming chapters.

Whether supramolecular structures on surfaces mimic what they would do in bulk or do something different depends in grand measure on the nature of assembly and surface. The foundation for the supramolecular "buildings" included here are all important. Let us then first consider the nature of surfaces and how they interact with molecules.

1.2 The Nature of Surfaces, Their Interactions with Molecules and Growth of Layers

Surfaces of materials are generally fascinating places to look at chemistry: the energy of the surface atoms or molecules is higher than in the bulk, because all the theoretically possible interactions that would give them their lowest energy state are not achievable.[45] Apart from their accessibility, surfaces are the part of a material where reactions and interactions take place first because of this higher energy. Their interaction with an incoming molecule lowers the energy of the system as a whole.

When an atom or molecule adheres at the interface between two bulk phases, the process is termed adsorption. The attached chemical is termed the adsorbate (while before it is attached, it is termed adsorpt or adsorptive, Figure 1.7) and the bulk phase it adheres to most strongly is the adsorbent. The amount of adsorpt that ends up attached to the adsorbent in the most simple case is a function of the concentration or vapour pressure of the adsorpt and the temperature of the system.

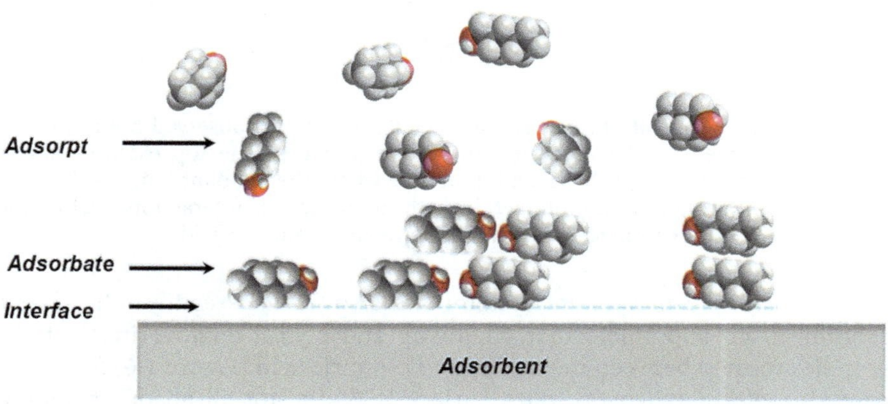

Figure 1.7 A schematic view of adsorption and the definition of the components in the system.

The mobility of a molecule on a surface and the interactions it will be able to enter into with other like, or unlike, molecules will depend largely on the nature of the surface–molecule interaction. The wetting of surfaces by molecules, as in macroscopic systems, is determined by the balance between adhesive forces (between the adsorbent and the adsorbate) and the cohesive forces that tend to make the adsorpt aggregate homogeneously and not stick to the interface.

Surfaces are never perfect, therefore, it is important to consider the nature of a substrate's surface before the addition of any material to it as it will have an influence on the type of adherence that takes place. For all surfaces, be they liquid or solid, chemical impurities can influence dramatically the local surface energy. These impurities can affect the structural and therefore chemical and physical properties,[46] for bad or good: doping (introduction of cations or anions with chemical "impurities" sometimes through oxidation or reduction) can be essential for the optimisation of certain electrical[47] and thermal[48] properties and can affect beneficial growth modes.[49] It should also be noted that the purity of any adsorpt is similarly very influential in the self-assembly of surface-based structures, and on eventual properties.[50]

For solid surfaces (that are never atomically flat on a micron scale), common defects are step edges, where the plateau of the surface rises or lowers by one atomic or molecular plane (Figure 1.8). The sizes of the plateaus can vary tremendously. Pits, trenches or simple vacancies can occur, just as adatoms and molecules and islands comprised of them can be present. The defects can attract adsorbates first, rather like the seats in a train where everyone sits down first, the seats correspond to the most favourable absorption sites in this simile. All of the seats are filled until no more are available. People will then sit on the floor, lean or stand as there are no predominantly favoured sites: the flat space fills up. Back to the molecular world, it is frequent to observe molecules and their assemblies aggregate first at step edges.

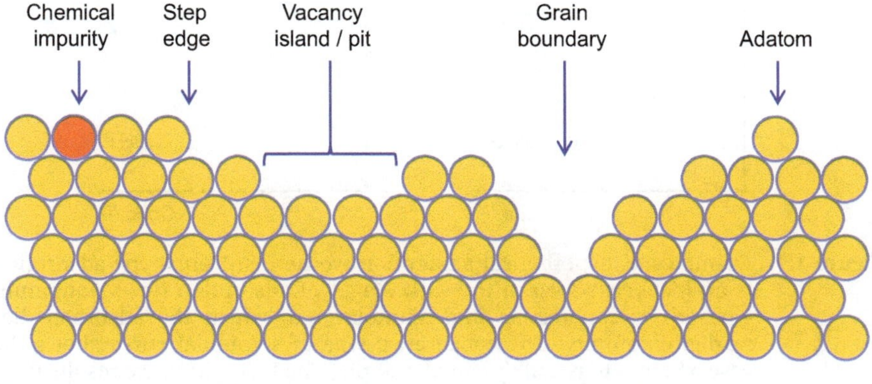

Figure 1.8 A lateral view representation of an adsorbent showing some common surface defects on otherwise atomically flat surfaces, where the density of defects is grossly exaggerated.

The reader will find references to "the" solid–liquid interface, but in fact this term is not precise because there are a multitude of solid–liquid interfaces, all with their own particular characteristics. It is obvious that the interface between water and fresh teflon is completely different to that between water and hydrophilic glass, yet both are solid–liquid interfaces.

The consecutive adsorption of molecules on surfaces can occur in several ways that can be distinguished from one another by their adsorption isotherms. The isotherms in Figure 1.9 show some of the contrasting behaviours that can occur, where the interfacial excess (Γ in mol m^{-2})—the amount of adsorbate in touch with the surface—is plotted against the concentration of adsorbate (either the pressure, in the case of a gas, or the molarity of a solute in a liquid). The linear relationship between interfacial excess and concentration, corresponding to Henry's rule, implies a proportional increase in adsorbate molecules on the surface with the surface pressure or concentration. Although this ideal kind of behaviour might be possible through isotropic interactions of a single material being grown from the vapour state, in the context of the systems described in this text, it is more likely only to be observed in the initial phases of adsorption of layers when the interfacial excess is very low. It can be regarded as the first term in many of the isotherms.

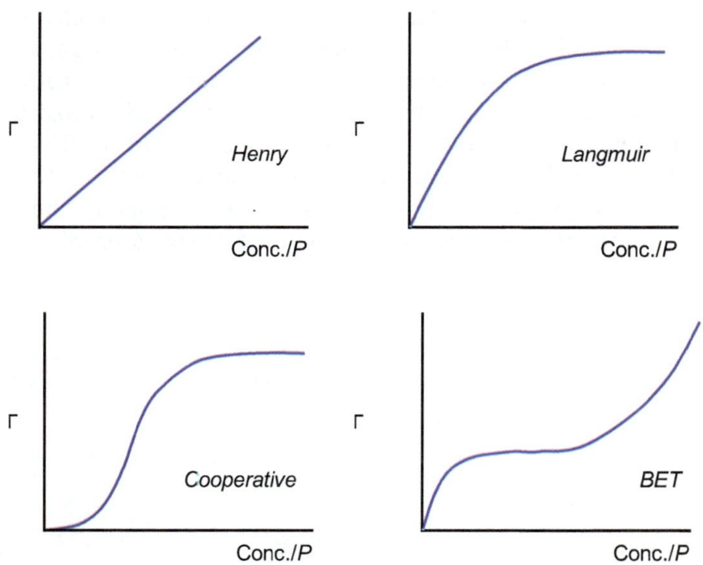

Figure 1.9 Examples of hypothetical adsorption isotherms: Henry-type adsorption is an ideal case seen in the early stages of adsorption (low). Langmuir-type adsorption results in a monolayer to which the adsorbate has negligible affinity. The cooperative type of sigmoidal curve shows the case where self-assembly can play a role. The BET-type curve results from the adsorption of a monolayer that subsequently has a weak affinity for further adsorbate growth, which does happen at higher concentrations or pressures.

The Langmuir isotherm shows a principally linear increase of surface excess at low concentration but then flattens and reaches a constant value despite any increase in concentration or pressure of the adsorpt. The molecule does not interact with itself: indeed, the Langmuir equation[51] shown below assumes that there are no interactions between adsorbate molecules, or between adsorpts and adsorbates. In this equation, θ is the fraction of adsorbate on the surface compared to a complete monolayer, α is the Langmuir constant and P the pressure.

$$\theta = \frac{\alpha P}{1 + \alpha P}$$

This equation was derived starting with the assumption that there is an equilibrium between adsorbate and adsorpt with occupied and non-occupied sites on the adsorbent, as shown in the following equation:

$$S^* + P \rightleftarrows SP$$

which would remind many supramolecular chemists of any host–guest equilibrium in homogeneous media. Indeed, the Langmuir equation is analogous to the Hill equation used in biochemical systems in order to evaluate cooperative binding. Once all binding sites are occupied, no more material can enter in this model, and the Langmuir-type curve effectively corresponds to a self-limiting growth scenario. This situation is observed often for solution-based systems on solid substrates.

At this point, it is also important to consider the relative amounts of each component in the system. The stoichiometry of interactions between adsorbate and adsorbent accessible surface is usually quantified in terms of the *coverage*. More properly, this is termed the fractional coverage (θ as shown in the Langmuir equation above), and corresponds to the ration of N_s (the number of surface adsorption sites occupied by adsorbate) and N (the total number of adsorption sites on the adsorbent). A complete monolayer is often referred to as 1.0 ML ($\theta = 1$ for monolayer) coverage, corresponding to the plateau in the Langmuir isotherm. This situation corresponds to a particular molecular orientation with respect to the surface. Supposing half of the material with respect to the full monolayer were adsorbed: that would correspond to 0.5 ML coverage. A bilayer would correspond to 2.0 ML, and so on. The coverage is a key and useful concept in defining the formation of monolayers, and can be used to quantify the efficiency of layer formation.

The formation of layers can involve a very significant interaction between the adsorbate molecules, and in this case a cooperative binding curve (Figure 1.9) results for the formation of a monolayer. This type of curve is typical for systems where "all or nothing" adsorption takes place because of strong non-covalent bonds in-between the adsorbed molecules themselves as well as a significant interaction between their assembly and the substrate surface. For this kind of curve, after a monolayer is formed, no more adsorpt is incorporated into the layer because the interaction between the molecule in the layer and that in the bulk overlayer medium is very weak.

Should a complete monolayer be formed through relatively strong adsorbent–adsorpt interactions (full monolayer coverage) but the resulting adsorbate is capable of forming relatively weaker but significant interactions with like molecules in the overlayer, the BET-type isotherm (after Brunauer, Emmett and Teller) will result. The layer will get increasingly thick as the adsorpt pressure or concentration in the overlayer medium is increased, until the isotherm becomes essentially that of the adsorpt condensing on itself as a bulk material. The effect of the interface is lost as the number of layers increases.

The different kind of adsorption curves can also be explained at the molecular level from a qualitative structural viewpoint. The phenomena that are manifested in the different growth models (Figure 1.10) of island (Volmer–Weber), layer by layer (Frank–Van der Merwe) or layer followed by island (Stranski–Krastanov) correspond to different kinds of adsorption isotherms. In the full layer growth model (Frank–Van der Merwe), one would expect an isotherm similar to that of the Langmuir isotherm. In the cartoon in Figure 1.10, this would correspond to a full monolayer as in the second drawing from the top. Should additional layers be formed specifically on top, at sufficiently high concentrations or pressures, subsequent Langmuir-type isotherms would come about. It would not be expected that this type of behaviour would proceed beyond many layers unless very specific and strong interactions were at play, though we shall see examples of this in this book.

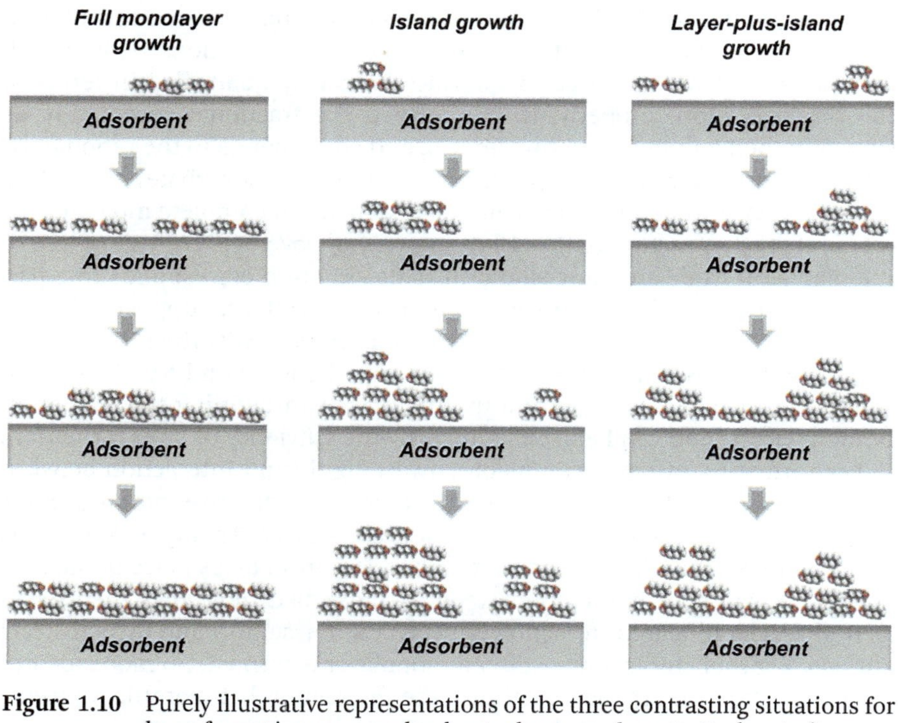

Figure 1.10 Purely illustrative representations of the three contrasting situations for layer formation on an adsorbent when an adsorpt attaches to it.

Should the interaction between adsorpt and adsorbent be of lower order to the interactions between adsorbate and adsorpt, islands of material will grow on the initial surface, as indicated in Figure 1.10 in the middle column. It will be appreciated that the number of interactions between adsorbate molecules themselves increases dramatically more than the interactions between adsorbate and adsorbent. Therefore, we would expect Henry-type behaviour with a negative deviation at very low surface coverages.

A combination of the two previous contrasting behaviours is at play in the Stranski–Krastanov type of film growth: the surface has a strong interaction with the adsorbate in the first layer, influencing the manner in which it organises itself in the single molecule thick film. This film interacts with adsorpt so that second layers form, but the interaction of this almost bulk-like second layer is not sufficient to make it adopt the packing of the molecules in contact with the adsorbent. Therefore, islands form on the monolayer with a packing similar to the bulk material.

The preceding discussion demonstrates that the result of adsorption of a molecule onto a surface and its subsequent organisation when more of the same molecule deposits on the surface will depend on the relative strengths of interactions between all the components of the system: surface, adsorpt, and overlayer medium. Indeed, the relative orientation of the adsorbate on the surface will be determined by the strength of interaction between the two in one orientation or another and the strength of interaction between the adsorbate molecules themselves. A strong adsorbate–adsorbate interaction (compared with adsorbate–adsorbent) can lead to a perpendicular arrangement of the molecules (Figure 1.11, top). If the surface–adsorbate interaction dominates, then the molecules will lay coplanar to the surface. The situation is complicated further in solution, where the adsorpt is solvated prior to adsorption and likewise the surface (Figure 1.11, bottom). Thus, the relative interaction of the medium with both components plays a determining role in the formation of any layer.[52]

Up to this point, equilibriums have been inferred, and, therefore, clearly we are dealing with situations where molecules are weakly bound—physisorbed—to the adsorbent. Under these conditions, the system reaches equilibrium rapidly, because the molecules diffuse with ease over the surface, and the desorption energy is low, in the region of 20–90 kJ mol^{-1}. Therefore, the molecules can be in a state of flux within the plane of the surface and can even de-adsorb and re-adsorb. On the other hand, when chemical bonds are formed (more or less reversibly), the energy associated with de-sorption is often considered to be in the range 100–400 kJ mol^{-1}, so molecules are much less mobile at room temperature and below. These cases will be discussed individually in the forthcoming sections, but some consideration is important at this point.

The chemical bonding between adsorbate and adsorbent determine many of the characteristics of the systems, and broadly they can be grouped into two classes: physisorbed and chemisorbed. In the following sections, these two classes of monolayers will be discussed in turn, but the general

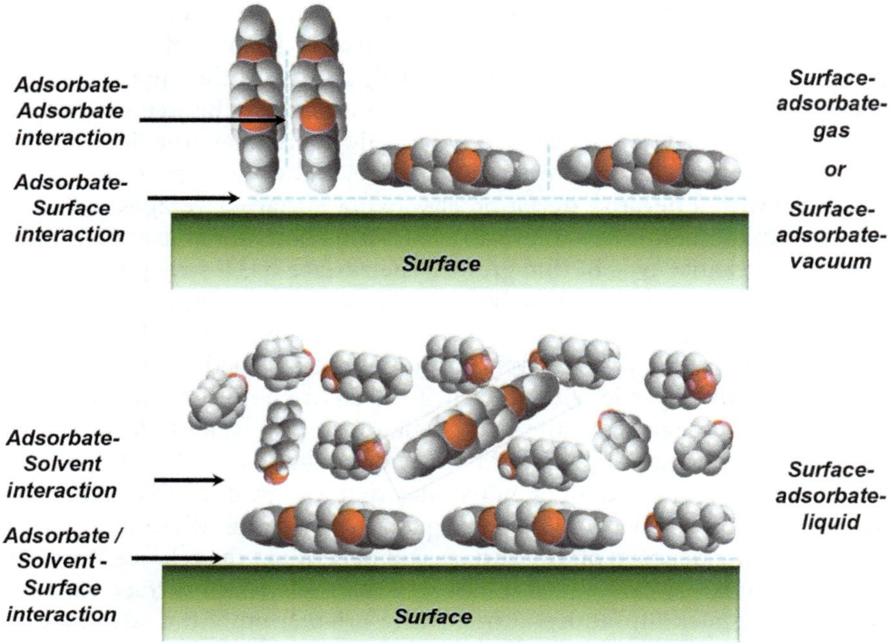

Figure 1.11 The interactions and relative orientations of molecules on surfaces in a gas or vacuum (top) or in liquid solution (bottom).

characteristics are summarised in Figure 1.12. Broadly speaking, chemisorbed systems are robust and have the organisation of the molecules determined by specific bonds between adsorpt and adsorbent. While this type of layer is referred to as "self-assembled", it is in fact physisorbed systems that are much closer to what a supramolecular scientist might consider as such. The weaker interactions between adsorbent and adsorpt mean that structures that are generated are much closer to equilibrium. To emphasise this point, consider that the sizes of the domains (the areas with a certain type and orientation of structure) are usually much bigger in physisorbed systems. There is clearly a very thin line between what one might consider a chemisorbed or physisorbed species. For the purposes of this monograph, each system has been classified according to the reversibility of the adsorbate–adsorbent bond. If it is clearly reversible, the system will be considered physisorbed, despite the nature of the adsorbent.‡

There is an often quite heated debate as to what exactly is meant when referring to as a "self-assembled monolayer" (a SAM). Here, the term is used in its broadest sense, and no distinction shall be made into systems that are very close to or conversely far from equilibrium and dominated by kinetic factors, such as reactivity or mass transport. The variation of the relative

‡One might consider that a metal adsorbent always forms chemisorbed systems with organic adsorpts because of the strength of the bonds between them, but many of these systems are dynamic with no specific metal–adsorbate bonds, and the adsorbate can be desorbed intact.

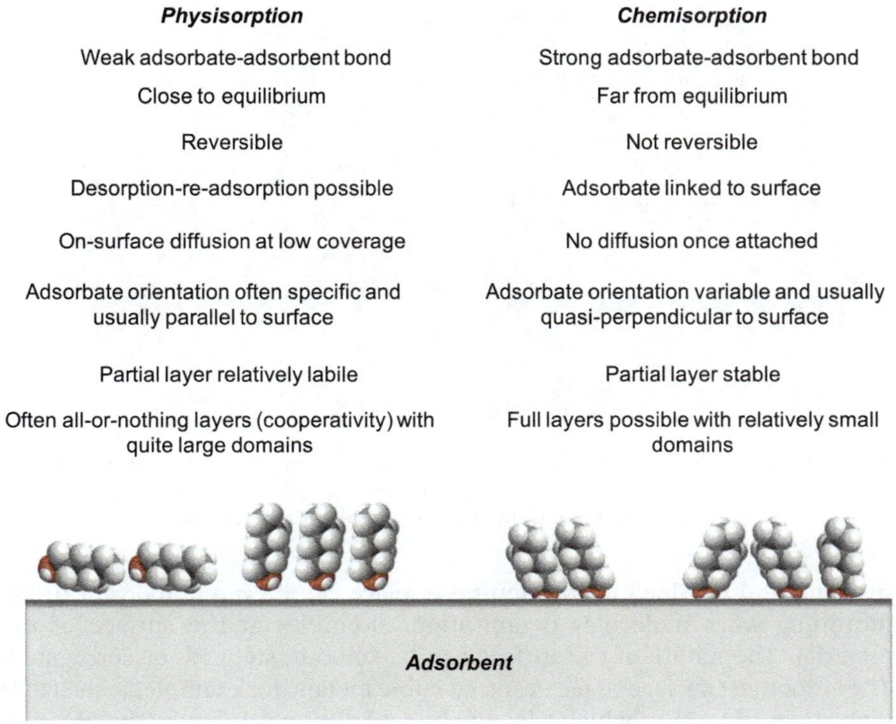

Figure 1.12 Some general characteristics of physisorbed and chemisorbed monolayers.

dominance of kinetic and thermodynamic factors can lead to a variety of structural outcomes. These can be controlled through temperature, adsorpt concentration and medium control.[53] While it is still complex to predict the behaviour that molecules have when they self-assemble at surfaces, it is important to be aware that, just as in solution, the assembly can be influenced by modulating these parameters for both physisorbed and chemisorbed systems.

The nature of the substrate—monocrystalline, polycrystalline or amorphous—also has a determining influence on the organisation of matter adsorbing at its surface. One might assume that order begets order, and disorder begets chaos, yet this extrapolation is not general. As shall be seen throughout this book, it is often true that crystalline layers of materials can be formed on monocrystalline substrates, but disordered layers are also frequent. A poorly-organised surface can lead to well-ordered overlayers if there is no strong interaction between the two materials. It is true that defects in surfaces can be translated into monolayers.

Crystalline surfaces with long range order are often the ideal place to look precisely at supramolecular phenomena because they can be extremely well defined which leads to huge simplifications when determining the structure of adsorbed molecules at the surface. The crystal face that is exposed to the

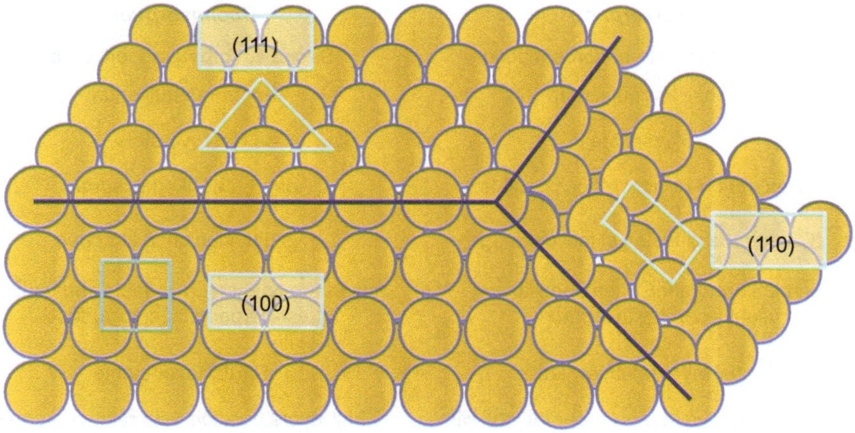

Figure 1.13 A schematic drawing of atoms in a face-centred cubic crystal exposing (111), (110) and (100) faces, with the unit cells of each surface indicated, and the dark lines indicating the edges of the crystal.

medium and to which adsorption takes place can be very important in determining what molecular organisation assembles at the surface of the material. The nature of the surface can be smooth, stepped, or corrugated. The smoothest surface of face-centred cubic metals, for example, is the (111) plane (Figure 1.13), of which a lot will be said during this book. The notation (111) refers to the Miller indices that indicate how the plane crosses the main axes of the crystal. Because it is the smoothest surface, it is also the most stable, since all surface atoms are surrounded in the plane. It has trigonal symmetry. Figure 1.13 shows this plane in a hypothetical object that also exposes the (100) surface (that has square symmetry and can be seen as a simple translation to a plane above the main cubic plane) and the (110) plane that is corrugated and has a rectangular unit cell with rows of atoms running along one direction. It is useful to remember these structures, as they are by far the most common in studies of self-assembled monolayers on metals. Apart from these very simple cases that suffice for our purposes, here, the surfaces of crystals can contain varying amounts of kink sites and atomic steps are defined by the Miller planes. A useful tool for visualising the different surfaces of crystals can be found at: http://surfexp.fhi-berlin.mpg.de/.

When a second material is deposited on the crystalline face of an adsorbent, the adsorbate can have its packing dictated by the surface. If the crystal structure is the same (as it can be in the case of certain metals deposited on other metals) then the growth is said to be *pseudomorphic*. In this case, the lattice parameters and crystalline orientation of adsorbate and adsorbent are the same. Molecules cannot take on the structure of metals, for example, but their packing can be influenced, with preferred orientations and distances between molecules different from the bulk. In this case, we can talk of *epitaxial growth*. The name epitaxy is derived from *epi* and *taxis* in Greek, meaning "above" and "in ordered manner", respectively.

The adsorbent forces the adsorbate to adopt a structure that is not its preferred one as a pure material. A monolayer of a molecular material on a crystalline substrate is said to be *commensurate* when epitaxial growth leads to formation of a multilayer on a crystalline substrate where the new surface takes on the registry of the "template" surface. If the lattice positions of adsorbent do not match the adsorbate but the latter still has a preferred orientation, epitaxial growth takes place, but the nature is incommensurate. For all the examples in this book, we are dealing with *heteroepitaxy* where the deposited material and substrate are different.

The concepts of epitaxy and commensurate growth can apply to molecular layers formed under a variety of conditions, from ultrahigh vacuum (UHV) through different kinds of deposition—notably molecular beam epitaxy (MBE) where chemicals are sublimed onto the adsorbate—to solution state adsorption and electrochemically-grown films. The origins of epitaxial growth lie in the formation of inorganic and metallic films of compounds. Precise information on the growth modes of organic thin films was first developed using the MBE technique.[54]

To understand commensurate structures in terms of molecules on surfaces, a couple of examples are called for. The first that has been chosen is that of *N*,*N*′-dimethylperylene-3,4,9,10-bis(dicarboximide) (DM-PBDCI) when adsorbed onto different crystal faces of metallic silver.[55] This case also shows how the surface to which an adsorpt attaches can influence dramatically the structural outcome (and as a consequence the eventual properties). Consider the layer of the DM-PBDCI molecules on an Ag (110) surface, shown in cartoon form in Figure 1.14 at the top. It is immediately obvious that the long molecular axis is perfectly perpendicular to the silver atom rows of this corrugated surface. Furthermore, the oxygen atoms of the adsorbate are located over silver atoms in the rows, as are the central bonds of the naphthalene residues. One could draw a unit cell on these molecules that would coincide with a multiple of the unit cell of the silver surface exactly, and therefore the structure is commensurate. The structural outcome of adsorption onto the Ag (111) crystal face is quite different: inspection of the lower cartoon in Figure 1.14 will make clear that the symmetry of the adsorbate packing and its unit cell cannot be brought to coincide or put into register with the metal underneath, and therefore an incommensurate structure is formed.

More than one commensurate structure can be formed by a given molecule on a given adsorbent. A fine example is that of coronene on gold (111).[56] When this compound was physisorbed from heptanoic acid at relatively high concentration (8×10^{-6} M), a regular dense hexagonal packing of the compound was observed, with a separation between the centres of the molecules of approximately 1.19 nm (Figure 1.15). This distance is in accord with a dense hexagonal packing matching the symmetry of the adsorbent, with the molecules spaced every fifth gold atom in each symmetry direction. Lowering the concentration to 3×10^{-6} M resulted in a more open structure (described as epitaxial) with the molecules located every sixth gold atom (with a distance of 1.45 nm between the centres of the molecules), a structure

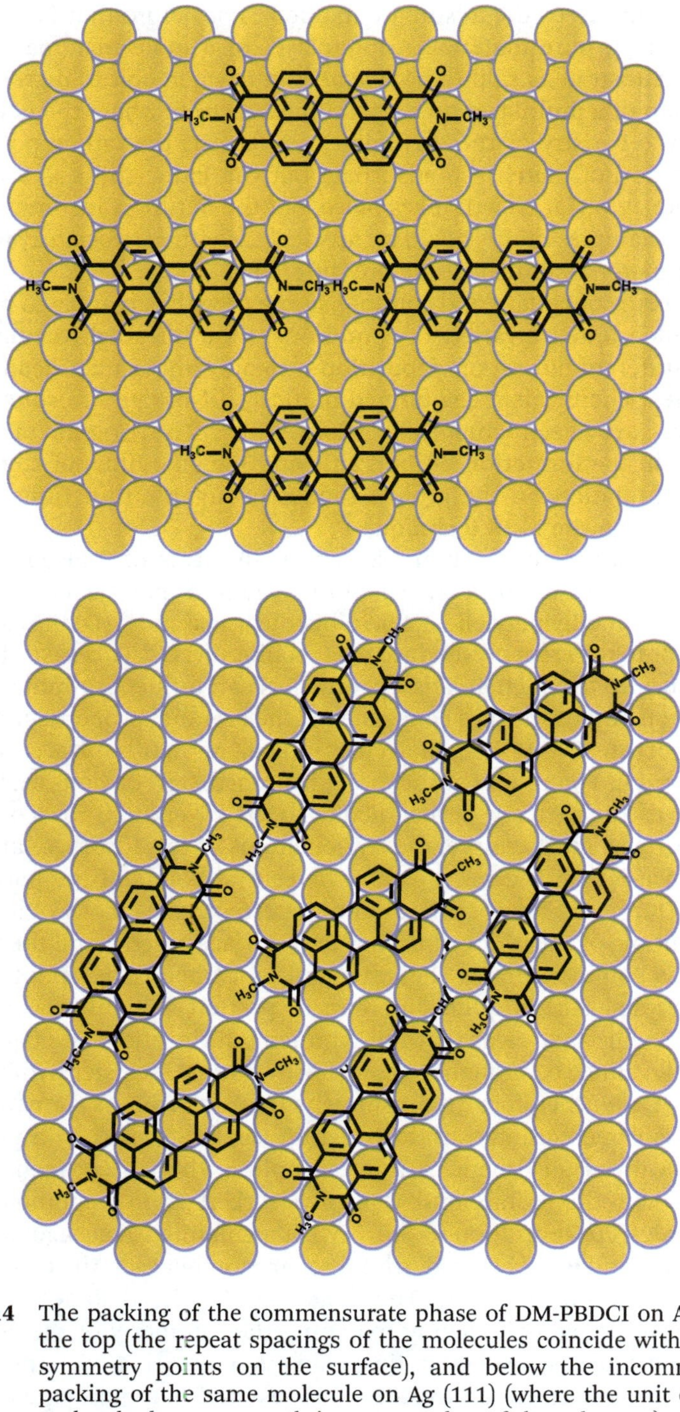

Figure 1.14 The packing of the commensurate phase of DM-PBDCI on Ag (110) at the top (the repeat spacings of the molecules coincide with the same symmetry points on the surface), and below the incommensurate packing of the same molecule on Ag (111) (where the unit cell of the molecule does not match in any way that of the substrate).

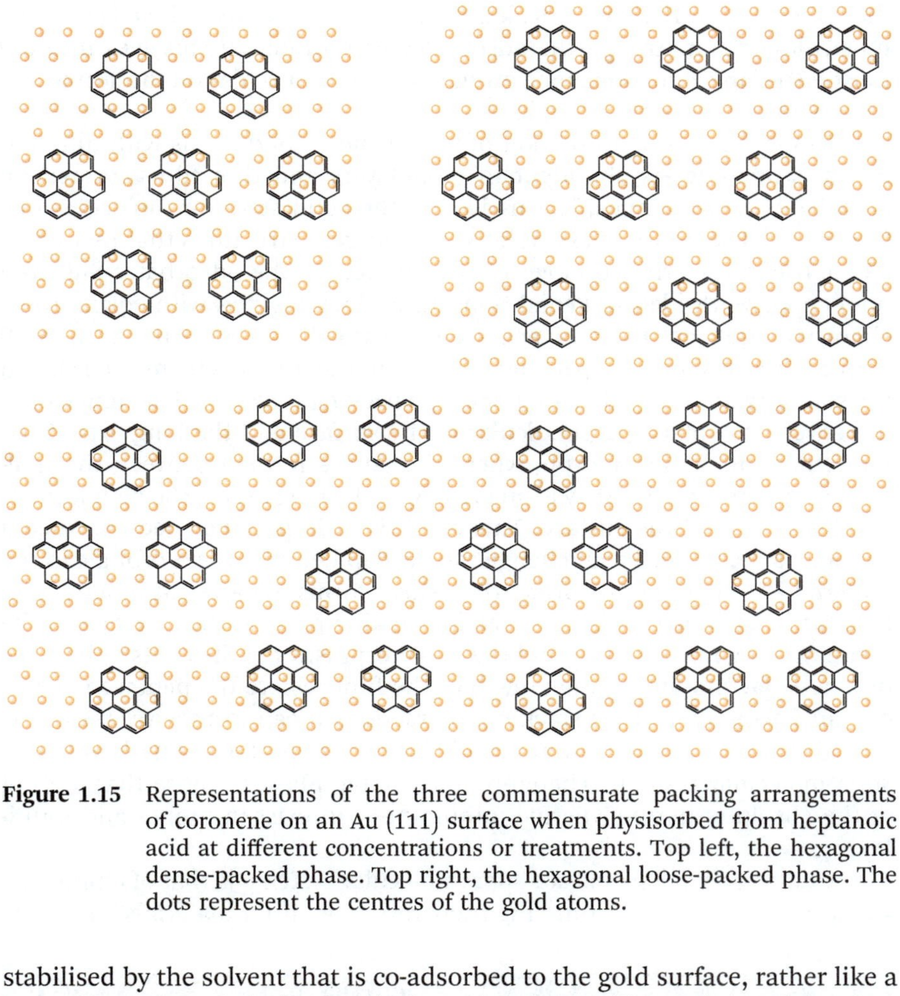

Figure 1.15 Representations of the three commensurate packing arrangements of coronene on an Au (111) surface when physisorbed from heptanoic acid at different concentrations or treatments. Top left, the hexagonal dense-packed phase. Top right, the hexagonal loose-packed phase. The dots represent the centres of the gold atoms.

stabilised by the solvent that is co-adsorbed to the gold surface, rather like a crystalline solvate in bulk crystals that might be familiar to the reader. When the low density phase was exposed to additional solvent (or sometimes when more dilute solutions were used), a third structure was formed—also hexagonal in line with the adsorbent symmetry—but with three molecules per unit cell, as shown schematically in Figure 1.15 below the other two structures. The difference in energy between the two structures was assigned to surface–solvent and coronene–adsorbed solvent interactions, since the coronene–gold interactions must be practically identical in the two phases.

The reason that molecules can have a preference for the formation of commensurate epitaxial monolayers is the gain in potential energy that the system has when the adsorbates are located at preferred adsorption sites on the surface.[57] It is clear that any particular surface will have a potential energy map that effectively acts as a template for the adsorbed molecule to locate itself in, a local energy well. Provided that the interaction between molecules is not sufficient to overcome this energy—usually the case when the molecules lay flat and provided the interaction between the edges of the molecule are not strong—the adsorbates will arrange themselves into the most favourable situations for a given density of molecules on the surface, as we saw in the previous example.

Polymorphism—the formation of two or more solid forms with different crystal structures—is abundant in monolayers and thin films formed on adsorbents by molecules, just as it is in the solid state for molecular materials.[58–61] The previous examples show different structures that may come about because of the coverage of the surface by the adsorbate. But even under apparently identical conditions, thin films can grow with polymorphic character. Epitaxial films of pentacene 30 nm thick grown on a Cu (110) surface show two types of packing in different experiments (Figure 1.16).[62] In the standing up form, a herringbone-type packing is seen with an angle of 3° in-between the plane perpendicular to the surface and the long axis of the molecules. In the polymorph, where the angles formed between the molecules and the lamellae they form are larger, the long axis of the pentacene is parallel to that of the surface. This result is of importance because such a subtle change in angle can change the overlap between the π orbitals in the material and therefore its properties. These forms are a result of a delicate balance in the non-covalent interactions taking place at the interface of the metal and organic systems, and the polymorphism manifests itself because of kinetic factors during the growth of the films where the pentacene has a certain flexibility in the π stacking interactions it can form with itself. It is a good example of how a surface can influence the packing of molecules in a supramolecular fashion. Therefore, metastable phases—those that are not at the absolute thermodynamic minimum—can exist in mono- and multilayer growth.

A recurring theme in surface supramolecular systems is that of chirality,[63] which is of interest regarding the preparation of chiral materials[64] and has

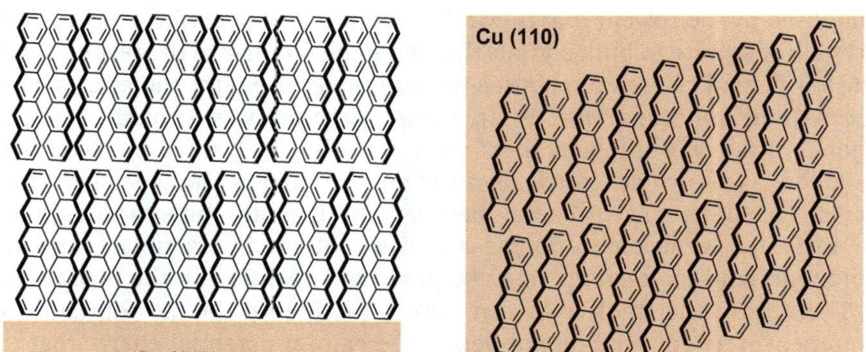

Figure 1.16 Representations of the polymorphic forms of pentacene on a Cu (110) surface. On the left is the standing up conformation viewed parallel to the copper surface and on the right is the packing parallel to the surface viewed from the direction perpendicular to the plane.

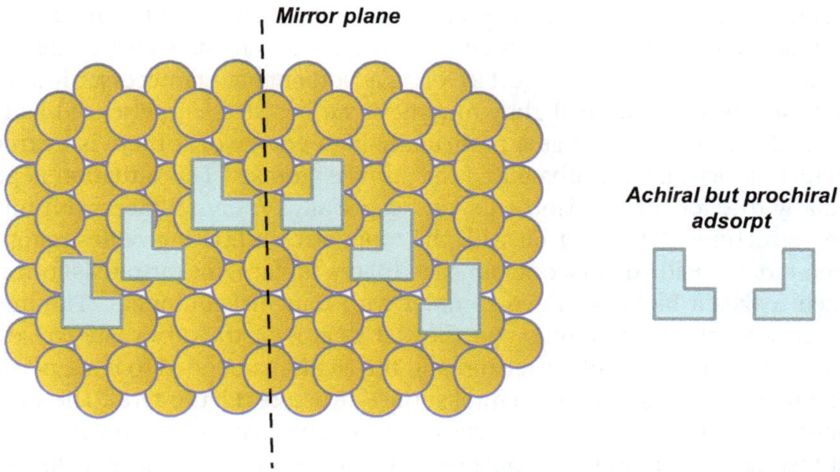

Figure 1.17 Cartoons showing the formation of chiral lines of prochiral objects (that can appear as an L or an r on the surface) on a (110) crystal face, and the mirror plane that relates them.

even been inferred in the origins of natural homochirality.[65] Even achiral molecules can adsorb to achiral adsorbents to give chiral structures, with the proviso that the adsorpt is prochiral. The origin of this emergence of chirality is the angles that are formed between the molecules and the symmetry axes of the underlying surface (Figure 1.17). In the case of chiral molecules, this leads to an intrinsic preference for one chirality or the other, while for achiral molecules, the local chiral chance preference can be propagated through lines (as in Figure 1.17) or to complete domains. Single molecules of cysteine adsorbed onto a gold (110) face show the kind of three-point interaction that can give the L or r shape shown in Figure 1.17 at the single molecule level.[66] There is a hierarchy of levels in the formation of chiral monolayers,[67,68] from the single molecules that are chiral at the surface or can form diastereomeric conformations if intrinsically chiral, to the domains, to the surface as a whole. The only way at present to have a homochiral surface is through the use of intrinsically chiral molecules.

The reason that chirality has a special significance in surface chemistry in general is the reduced symmetry operations available when compared with three-dimensional crystals.[69,70] Molecules restricted to a plane have considerably fewer possible packing arrangements than in three-dimensional crystals. There are fewer 2D space groups: 17 compared with 230 for 3D, but the proportion is similar. Five of the 2D space groups are chiral when laid on a surface (while in 3D there are 65 chiral space groups), that constitute about 29 and 28%, respectively.[71] But the real key is the symmetry operations available for packing on a surface, because an adsorbate on a surface cannot be related to another by inversion symmetry. Similarly, the glide plane parallel to the surface is not likely. The nature of the surface and molecular orientation provide more specific restrictions.[72] This fact was demonstrated in dramatic fashion in the construction and analysis of a particular subset of monolayers that showed a dramatic preference for the *p*2 space group.[73]

The atoms on a crystalline surface of an adsorbent are not necessarily totally immobile, but can move slightly to remove surface stress experienced by the atoms if the crystal face has a particularly high energy or if there is a strong interaction between adsorbate and the surface atoms.[74] This situation is particularly true for metal adsorbents, and especially for crystal faces with more open structures. Following on with the topic of chirality to give an example, when tartaric acid adsorbs onto an Ni (110) crystal face, bis-tartrate is formed, which makes a highly strained complex with four nickel atoms.[75] The interaction between the Ni atoms and the OH groups of the adsorbate causes a distortion of the molecular skeleton and the Ni atoms bonded to it are twisted and pulled apart resulting in a chiral surface structure (Figure 1.18). The strong bonds between adsorbate and surface atoms can often cause a restructuring, not just locally, but over the whole surface if an ordered monolayer is formed.

Should crystalline monolayers of adsorbate be formed, the size of the crystalline domains is an important characteristic of the overall surface. The minimum in energy, assuming an optimal packing match between adsorbent and adsorbate, will comprise a single crystalline surface domain. This situation has not been reached in molecular systems: domains of differing sizes are seen, and therefore the surface is polycrystalline, irrespective of the nature of the adsorbent mostly. Just as bulk crystals can undergo Ostwald ripening, so do "two-dimensional crystals". The phase boundaries are minimised in order to minimise the energy of the system. This phenomenon was shown in alkyl derivatives of anthraquinone and octathiophene adsorbed on graphite from solution,[76] where domains with edges tens of nanometres were seen to disappear at the expense of larger domains on a timescale of less than a minute! Subsequently, this effect has been shown to be general, though the timescale depends on the temperature and the strength of the molecule–surface interaction.

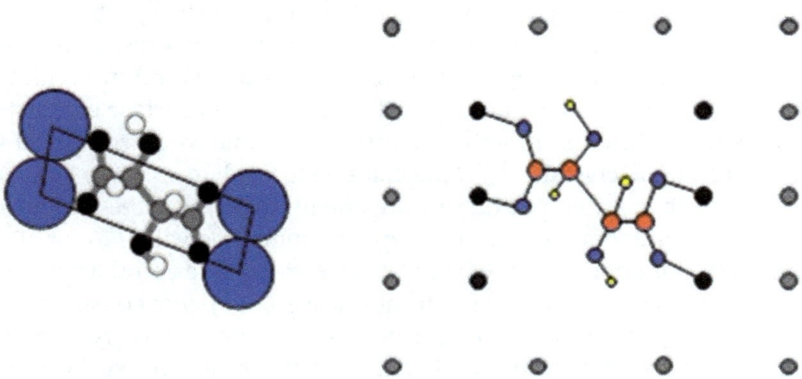

Figure 1.18 The adsorption configuration of (*S*,*S*)-bis-tartrate on Ni (110). The cartoon on the right shows the restructuring of the metal surface atoms to adapt to the strong coordination bonds.

For polycrystalline surfaces—where areas of crystalline material, each of them oriented in a random manner, make up the surface—the adsorbate can follow the same kinds of adsorption as for crystalline materials. In the case of epitaxial growth, the adsorbent crystallites in each area will dominate the packing, especially at low temperatures. However, for incommensurate films or those where epitaxy is lacking totally, the crystal domains can be bigger. Returning to pentacene on copper (which was discussed above, see Figure 1.16), for the polycrystalline metal surface, the nature of the epitaxy and adsorbate crystallite size depend on the temperature at which the surface is held.[77] At higher temperatures (333 K), crystallites of the organic molecule are over 100 nm long, but at the same time, polymorphism is again evident. At 77 K, long range order is not present in the film. As we have discussed for Ostwald ripening, heating (during deposition or post-deposition annealing) can influence greatly the quality of self-assembled monolayers, both in terms of adsorbate and adsorbent. This has been demonstrated, for example, in the case of monolayers of thiols on gold where the surface was annealed to give the flat metal[78] or in the preparation of silane layers on silicon oxide on silicon where increasing the temperature during deposition increases greatly coverage and molecular order.[79]

An emerging area of great potential is that of what has been termed organic–organic heteroepitaxy or "weak epitaxial growth"[80] (WEG) of organic materials on organic surfaces (that are themselves surface-supported, Figure 1.19). Layers of iron porphyrins on a phthalocyanine[81] proved the idea that matching of packing types of aromatic compounds could lead to heterostructures.[82] This kind of organic–organic heterostructure is of interest for the preparation of materials with a variety of properties.[83,§] Some examples of supramolecular influence in this kind of directed growth will be

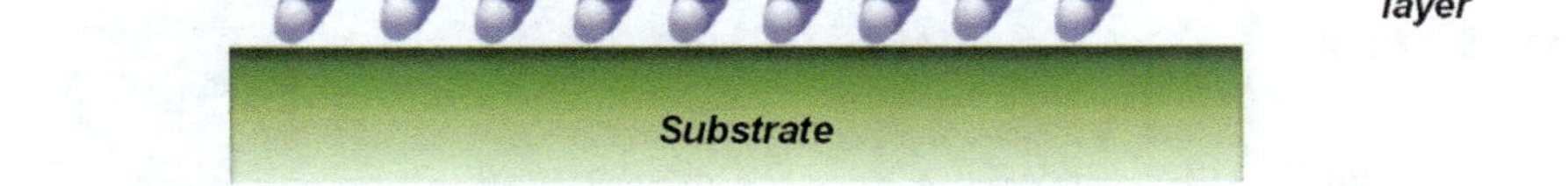

Figure 1.19 The principle of organic–organic heteroepitaxy or weak epitaxial growth. The purely hypothetical molecules in the first layer on the substrate are ordered and induce epitaxy in the (equally hypothetical) second layer, and subsequent ones.

§While the idea of organic–organic heteroepitaxy and heterostructures is not general, one could extend this idea to molecule–molecule heteroepitaxy and heterostructures to include all kinds of coordination molecules, organometallics and clusters.

shown in Chapter 7, although the use of non-covalent bonds to direct specifically and selectively is underexploited at present.

Amorphous adsorbents—such as glasses or amorphous silica—will usually result in the growth of poorly structured layers of adsorbate when compared with flat crystalline substrates. As far as applications of films go, however, that characteristic can be irrelevant: consider the coatings of glass, silica support materials for chromatography or treated textiles. Many of these rely on chemisorbed molecules that have a function as the new surface of the material. However, if the interaction between the adsorbent and the adsorbate is weak, the non-covalent interaction between molecules on the surface becomes dominant in their organisation. Thus, the π stacking between large flat aromatic molecules can give rise to highly-ordered self-assembled structures, as is the case with certain materials that aggregate on the surface of polymers.[84] A similar situation arises when the substrate is a liquid, as is the case for Langmuir layers of molecules on water, which will be discussed in Section 1.5. Provided the interactions of supramolecular synthons with the amorphous substrate are weak, the self-association can lead to very ordered systems where the surface merely acts to orient the sample, something absent in three dimensions and an opportunity for the organisation of molecules.

All of the cases of surface-based supramolecular systems can be subject to analysis using phase diagrams, where the layers can change their structure as a function of two principle factors when adsorbate composition is held: temperature and coverage. Returning to the case of tartaric acid on Ni (110), the adsorption phase diagram is shown in Figure 1.20.[85] At low temperature, the molecules are physisorbed to the surface; increasing the amount of compound on the surface by evaporation merely makes multilayers.

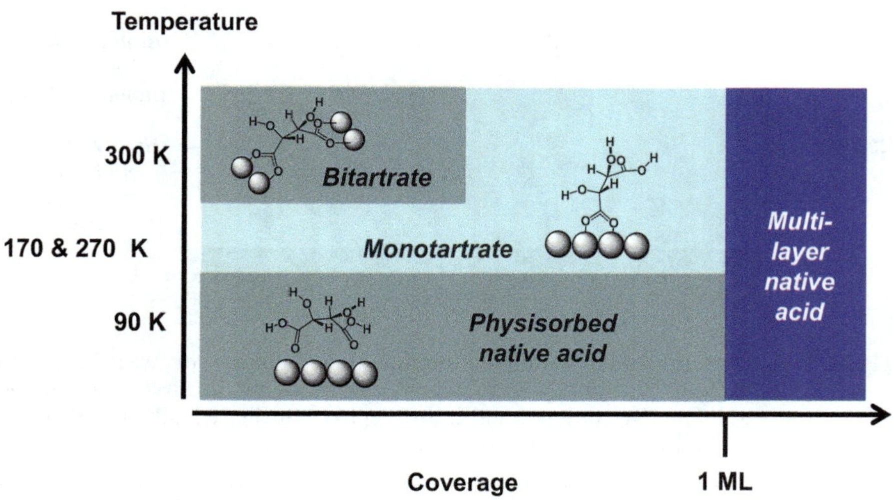

Figure 1.20 An adsorption phase diagram as a function of temperature and coverage for (*R*,*R*)-tartaric acid deposited onto a Ni (110) face in vacuum.

At 170 K and above, however, a chemical reaction takes place; hydrogen is lost and the tartrate is formed. The monotartrate is formed only between 170 K and 270 K up to monolayer coverage, and also at 300 K when coverage is high. However, at less than half monolayer coverage, the molecule has sufficient space that the bistartrate form is generated. Thus, structure and bonding are controlled by temperature and coverage.

In the case of adsorption from a solution, the coverage is affected by concentration, and obviously the solvent will play a role in the nature of the layer that is formed. An interesting case in this regard is that of a mixture of triacontane and triacontanol adsorbed onto graphite from different solvents.[86] Firstly, regarding the phase diagram, it was observed that the two adsorbates form separate domains and have clearly different adsorption energies, and a difference of only 0.5 kcal mol^{-1} is sufficient to generate a 2 : 1 bias. The solvent is important: phenyloctane allows formation of the alcohol layer predominantly, while incorporation of a different alcohol into the solvent makes adsorption of triacontanol less favourable because it is solvated more favourably. Evidently, changing the proportion of the adsorbates will influence the surface coverage, and it is clear from the results on the system that solution composition does not correspond to monolayer composition when the adsorption energies are different. Another interesting case of solvent effect is that of monolayers of 3,5-di(4-*n*-tetradecyloxyphenyl)pyrazole on graphite. When pure toluene was used as a solvent, rod-shaped molecules were observed with an achiral structure.[87] On the other hand, the packing of the molecules was chiral when a 1 : 3 toluene–chloroform mixture was used. More cases shall appear in Chapter 4, although the prediction of how solvent will affect the structure of monolayers remains difficult, and empirical testing is usually required to obtain optimum conditions for monolayer formation.

Supramolecular approaches to assembly on adsorbents very often use neutral components assembled on formally neutral surfaces. However, many surfaces have formal charges, particularly where ionic species are involved and where water is present in the medium, charging can become very important. While a full description is beyond the scope of this text, the importance of charged interfaces is enormous and the science rich; the reader is encouraged to explore more specific texts related to these matters.[88] When oxides are in contact with aqueous media, their surface is often anionic because of protonation of water. This feature presents an opportunity for the assembly of cationic species on the surface of oxides, as we shall see (particularly in Chapter 6). Mica is another widely used surface that presents a charged surface, where sheets of aluminosilicate are held together by cations in the mineral that is easily cleaved. In an appropriate medium, the cations can dissociate away from the surface leaving a net negative charge locally. Also, charges can be introduced by binding ions at interfaces as well as during electrochemical experiments. Evidently, these polar surfaces cause an electric dipole that attracts oppositely charged ions. In order for adsorption to take place, it is clear that an interchange of counter ions must take place.

Formally, the layer of ions linked directly to the bulk material is referred to as the Stern layer, which is regarded as immobile. Counter ions are attracted to the surface by the electric field, but are mobile, and are referred to as the diffuse layer. Together, these areas comprise the electric double layer. This layer can be replaced, taking advantage of strong interactions between surface- and solution-borne counter ions that are drawn to the surface because of their multiply-charged nature or specific bonds. The former corresponds to the layer-by-layer formation of polyelectrolytes and the latter to chemisorption at interfaces.

In the following sections, the general characteristics of physisorption, chemisorption, and assembly at liquid interfaces shall be discussed, to provide the reader with an initial understanding of the characteristics of each class of system.

1.3 Physisorption

When molecules bind reversibly to an adsorbent—and can diffuse freely over it without the formation of significantly strong bonds—the adsorbate is considered to be physisorbed. The general characteristics have been compared and contrasted with chemisorption in Figure 1.12. The energy of sublimation (typically considered to be approximately 20–90 kJ mol^{-1}) of a molecule adsorbed onto the surface in question is of the order of that required for the breaking of typical non-covalent interactions. Particularly, van der Waals interactions are very important in determining whether molecules will physisorb to a surface or not. This reversible attachment of molecules to a surface—which is a general phenomenon—can be made specific through the use of non-covalent interactions between adsorbate molecules. Classic examples are alkanes and their derivatives adsorbed onto graphite-like materials. Here, the methylene groups of the aliphatic component form van der Waals interactions (of the [C–H$\cdots\pi$] type) with the surface where there is near perfect registry for a commensurate epitaxial layer.[89] The all *trans* conformation of the alkanes is preferred, with each alternate methylene hydrogen atom interacting with neighbouring six membered rings in the graphitic surface (Figure 1.21). Typically, the alkanes are surrounded by their like neighbours generating close packed lamellar-like structures. Submonolayer coverages do not habitually show ordered material (except at very low temperatures) because diffusion takes place. Clearly, the larger the molecule, the more interactions that can take place, and the adsorption enthalpy increases while the entropic importance of liberating any adsorbed solvent or gas molecules at the surface is considerable. This strategy can be exploited to adhere individual molecules to a surface in solution. The tetrathiafulvalene derivative in Figure 1.21 adsorbs onto graphite from solution to give not only close packed layers but also individual molecules adsorbed onto the surface.[90] The adsorption is favoured by the 72 methylene and methyl groups that anchor the central unit parallel to the graphite surface.

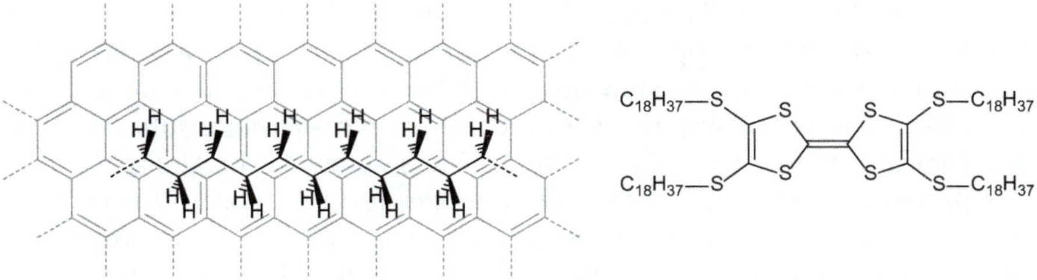

Figure 1.21 The quasi-commensurate arrangement of all-*trans* alkyl chains on graphite showing the hydrogen atoms perfectly located for [C–H···π] interactions, and the chemical structure of a tetrathiafulvalene derivative with four octadecyl chains that allow adsorption as an apparently isolated molecule on graphite interfacing with a solution.

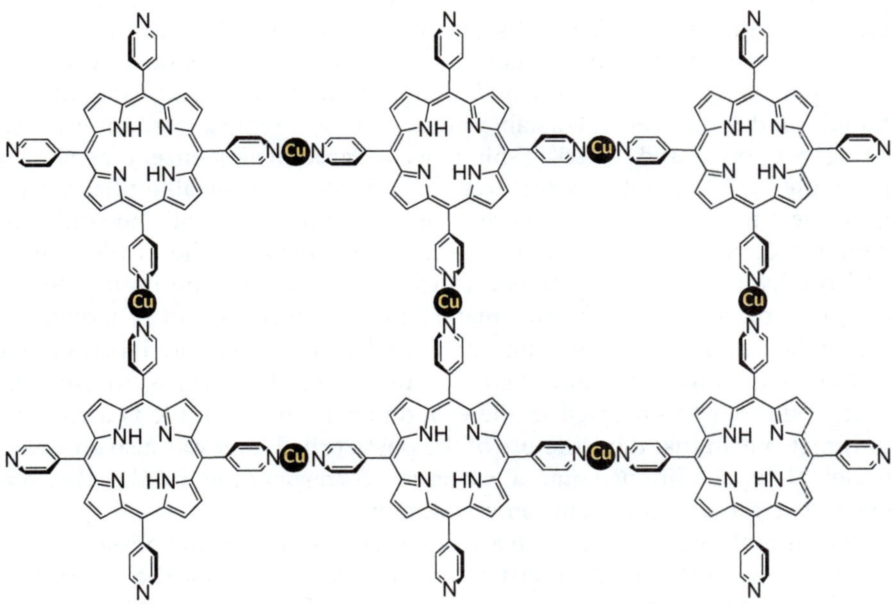

Figure 1.22 Part of the two-dimensional metal–organic framework nanocrystals formed by a porphyrin with metallic copper on the (111) surface of gold (these sheets lie parallel to the interface).

The mobility of molecules in physisorbed systems can be very high, especially at low coverages (below a complete layer) but even when complete monolayers are formed. This diffusion is important in the growth of ordered regions of molecules. A beautiful example of this phenomenon is the formation of two-dimensional supramolecular assemblies of metal–organic frameworks based on porphyrins and copper atoms (Figure 1.22) on a gold (111) surface.[91] This study, performed depositing the components under ultrahigh vacuum conditions onto a surface, showed that by varying the deposition rate and the temperature of the gold and measuring the sizes of the supramolecular grids, the molecular diffusion barrier could be derived. The formation and disappearance of very small aggregates was observed at temperatures below 373 K, showing that diffusion is possible, characteristic of a truly physisorbed system even though it involves a metal substrate, and that Ostwald ripening takes place, for clusters below a critical size.

A particularly elegant example demonstrating molecular diffusion on a surface showed how C_{60} molecules forming a non-covalent chain around the edge of a silver metal island fluctuate in their position over time,[92] and emphasised the dynamic nature that molecule–metal contacts can have. The amplitudes of the fluctuations in this system are largest in the frequency range of 1.3–105 Hz in the modes resulting from Ag motion and 0.01–0.08 Hz arising from C_{60} movement.

Just as in the crystallisation of molecular materials in bulk, metastable phases can be formed in self-assembled monolayers, where they are most clearly identified as polymorphs or solvates, as seen for the case of the polymorphs of pentacene earlier. In physisorbed monolayers, where in principle molecular diffusion and reorganisation are facile especially at liquid–solid interfaces, metastable phases present themselves, an example being the compound 3,5-tris(4′-biphenyl-4″-carbonitrile)benzene as shown in Figure 1.23.[93] The two structures correspond to metastable phases that upon heating convert into a more dense structure where all the voids are compacted, and all of the inferred interactions between the nitrile groups and the hydrogen atoms attached to the aromatic rings are optimised. In fact, it is probable that the two metastable structures are in fact solvates, where the heptanoic and nonanoic acid used as the solvent are incorporated in the voids, thus explaining the different nature of the voids, whose contents could not be observed in the experiments. This example is a particularly nice one in that the structure of the physisorbed layer was also observed under UHV conditions, and a structure corresponding to the thermodynamic structure from solution was obtained.

Porous materials in bulk form are potentially useful for the physisorption of molecules to the internal surfaces of the materials from the gas phase, and is of special relevance today with the hive of activity related to porous materials derived from molecules, be they based on coordination chemistry[94–97] or on covalent frameworks.[98,99] Physisorbed monolayers at interfaces can also form network-like structures with spaces in the layers. Jürgen-Hinrich Fuhrhop championed the idea of “yoctowells”[100] with volumes of 1–10 yoctolitres (1 yL) (10^{-24} L or 1 nm^3), and other groups have demonstrated the formation of frameworks by physisorption of disc-shaped molecules, as shown for the formally C_3 symmetric compound in Figure 1.24.[101] At the interface between graphite and 1,2,4-trichlorobenzene, this molecule, like others with long alkyl chains, favours interdigitation of the oligomethylene fragments thanks to their alignment and quasi-commensurate layering on graphite and the dispersion interactions between them. The open structure only covers the surface after heating to 353 K (at a concentration of 2.87×10^{-5} M), after which a highly ordered layer with very

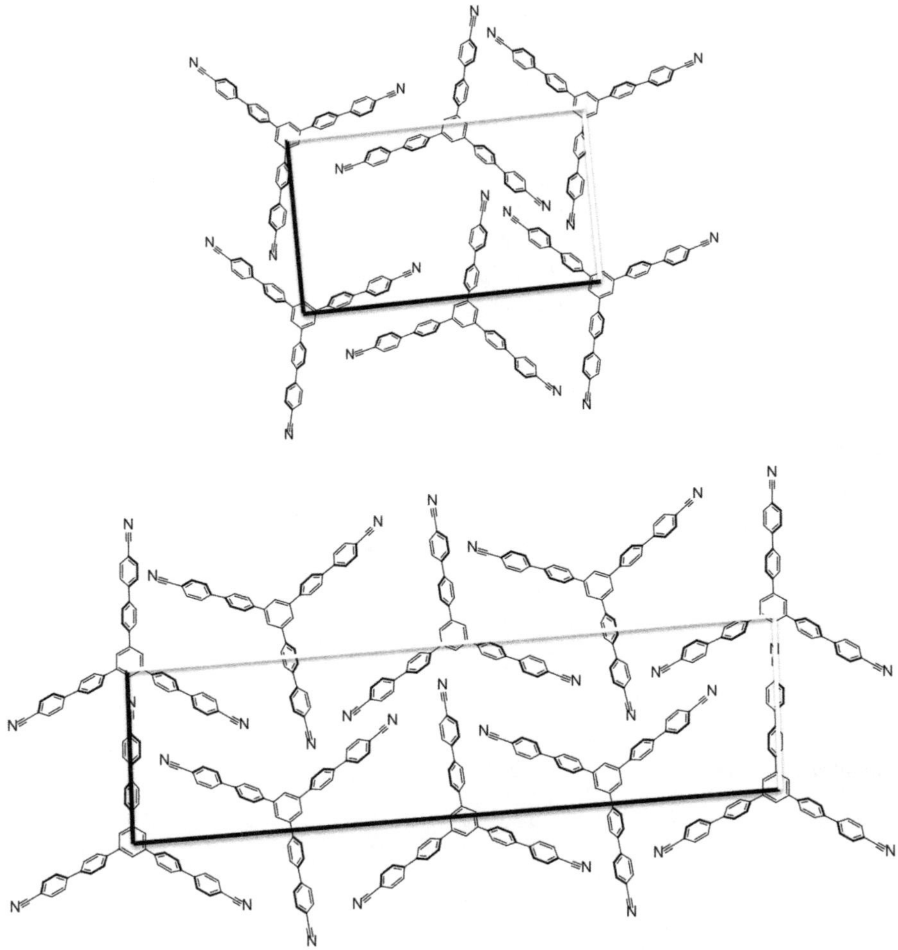

Figure 1.23 The two-dimensional metastable phases of the compound shown at the liquid–graphite interface, with the unit cells of the structures indicated. The top structure corresponds to that from heptanoic acid, and the bottom one to nonanoic acid.

few domain boundaries exists. The spaces in the layer are occupied by solvent molecules. At room temperature, the vast majority of the monolayer comprises the lamellar structure that is densely packed. Interestingly, the lamellar phase does not have all the alkyl chains adsorbed to the graphite, and they are presumed to be solvated above the layer, a not uncommon situation in this kind of physisorbed system where dense packing can be achieved without part of the molecule. The inter-conversion between the two phases can be performed reversibly by raising and lowering the temperature, showing a true thermodynamic process. An analysis of the changes of coverage with concentration and temperature showed the importance of both entropy and enthalpy in determining the relative coverage, and pointed to a

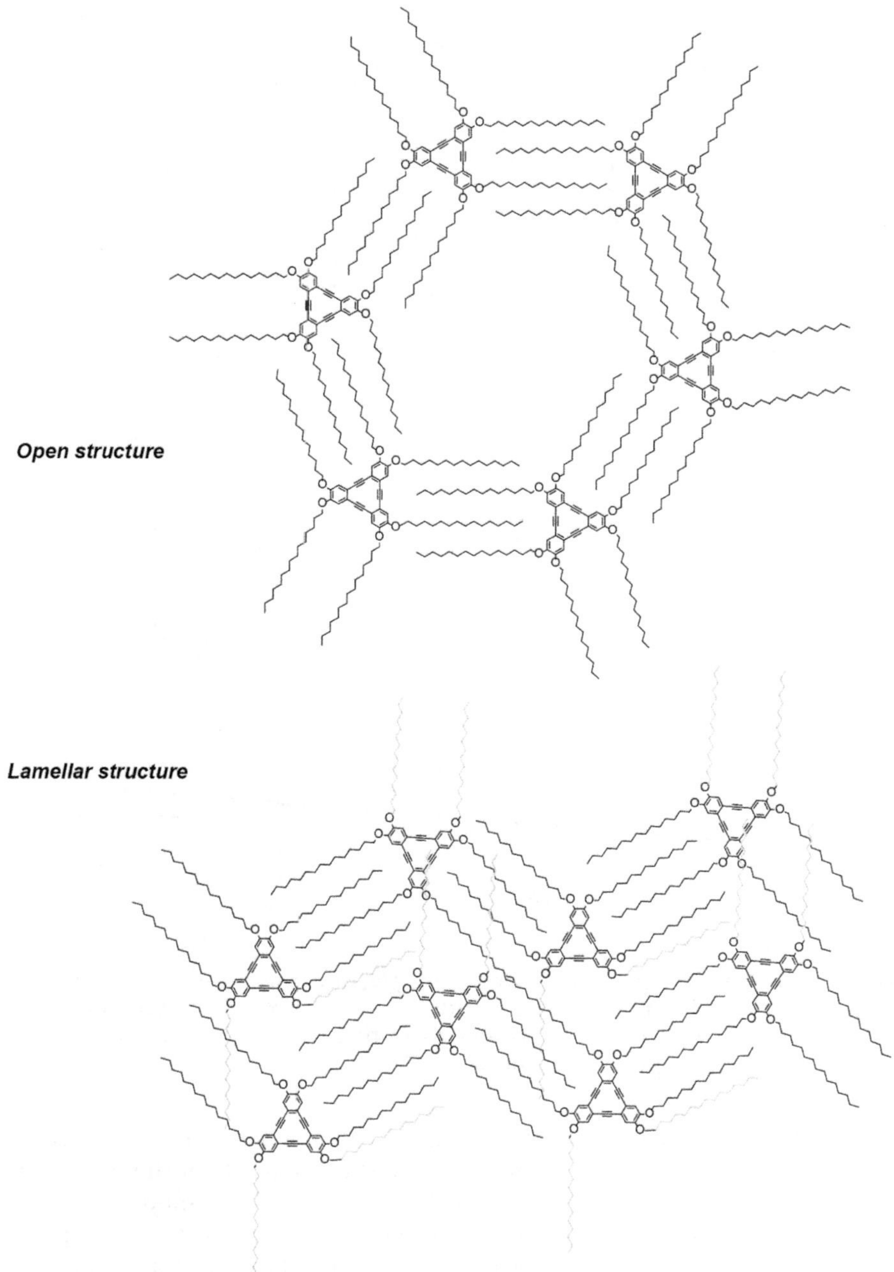

Figure 1.24 Representations of two phases formed by the C_3 symmetric molecule on graphite at its interface with 1,2,4-trichlorobenzene. Higher temperatures favour the porous phase while the lamellar structure is predominant at room temperature. Note that in the lamellar form, the grey alkyl chains represent those not adsorbed to the surface, and probably solvated above the layer.

significant entropy gain upon desorption of the adsorbate thanks to the degrees of conformational freedom it gains upon dissolution. At a lower concentration (2.30×10^{-4} M), the equilibrium is shifted toward the porous phase, that occupied more than half of the interface (while at higher concentration, less than 10% was present), showing how the nature of the supramolecular can be influenced by parameters commonly used in studies of self-assembly in solution.

Reversibility and therefore the thermodynamic effects associated with it are the hallmark of physisorbed layers. While metastable systems are formed, annealing can interconvert surface-based structures. When there is a very strong bond in-between adsorbate and adsorbent, the situation is quite different.

1.4 Chemisorption

Specific strong bonds between a surface and an adsorbed molecule lead to robust chemisorbed structures that are usually very different to physisorbed systems both structurally and chemically. On the one hand, chemisorption tends to produce structures which are to a large extent imperfect because the outcomes of adsorption are determined kinetically, predominantly. Once a strong bond between surface atoms and an adsorpt is formed, there is generally little movement of one with respect to the other. On the other hand, these monolayers are far more robust than physisorbed layers precisely because strong bonds unite adsorbent and adsorbate. Because of the very localised bond that links the two, and its geometric restrictions, the orientation of the molecule can be in any direction, allowed by the conformations of the bonds in the rest of the adsorbate molecule. At low coverages, most adsorbates would tend to interact weakly with the surface in a non-covalent manner away from the point of strong bond formation. In other words, the part of the adsorbate away from the point of covalent attachment is "physisorbed", meaning it can be mobile, although it can't escape to the surroundings, it is anchored. At low temperatures, this part of the molecule may be slumped next to the connection point. But at high coverages, the molecules would tend to align perpendicular to the plane of the surface of the adsorbent so that the maximum number of strong bonds can be made by the surface atoms of the adsorbent.

During the chemisorption process, the first step is usually considered to be physisorption where the molecule adheres to the surface but is able to diffuse in the plane to a degree determined by the substrate temperature (Figure 1.25).[102] This situation is clearly witnessed in the case of molecules adsorbed onto surfaces from the vapour state under vacuum conditions. A paradigmatic case is that of alkanethiols on gold, where the compounds physisorb (they have a vaporisation enthalpy 20% greater than their bulk values) with an enthalpy of the order of 6.1 kJ mol^{-1} per methylene unit, with a physisorption enthalpy of 33 kJ mol^{-1} from the thiol group.[103] The chemisorption enthalpy is 126 kJ mol^{-1} regardless of chain length

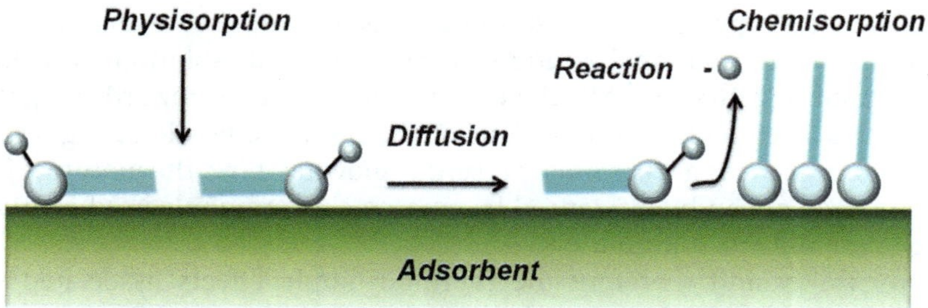

Figure 1.25 The processes leading to chemisorption.

and activation energy is approximately 29 kJ mol^{-1}. Therefore, for chain lengths longer than fourteen carbon atoms, the physisorption enthalpy is higher than the chemisorption enthalpy. Thus, detailed studies of these processes paint a precise picture of how molecules adhere to surfaces.

The magnitude of interaction locally is determined by the strength of the chemical bond between adsorbent and adsorbate. The bigger the molecule gets, provided there is one anchoring point, there is little influence on the local interaction energy between adsorbent and adsorbate, as was mentioned for the case of thiols on gold. Non-specific van der Waals interactions will increase the overall adsorption enthalpy for larger molecules though. There will also clearly be kinetic effects on the growth of the layer, because of the slower diffusion of the molecules.

The most frequent cases of chemisorbed molecules are those involving the bonding of organic molecules to metals, inorganic semiconductors or their oxides, and they are the basis of the development of a variety of self-assembled monolayers[104] with a gamut of processing methods.[105] The bonding is typically fully covalent, between oxygen or carbon atoms of the adsorbate and the surface atoms most commonly, or is essentially covalent through coordination between heteroatoms and inorganic surface atoms. Two examples are shown in Figure 1.26, the first layers of thiols on gold and the second siloxanes on silicon oxide. The latter are particularly common because of the ease of preparing this oxide on silicon wafers. Self-assembled monolayers are often distinguished from systems that are grafted covalently—through carbon–carbon bond formation for instance (Figure 1.26)—because in SAMs, non-covalent interactions influence film structure greatly. Yet both involve an adsorption and a covalent bond formation between substrate and "chemisorpt", the molecule that attaches to the surface. Grafting organic compounds to organic materials, organic layers or inorganic materials[106,107] (Figure 1.26) normally implies a limited influence of non-covalent interactions during the film formation. Grafting is a term used especially when polymers are formed on surfaces.[108,109] That said, "self-assembled layer" and "grafted layer" are sometimes used interchangeably, but all agree the films are chemisorbed! Whether non-covalent interactions play a large role in the adsorption or not, the surfaces that result can clearly be of interest from a supramolecular perspective.

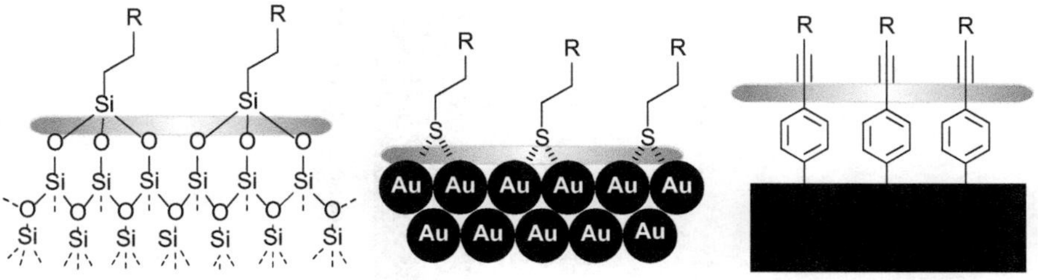

Figure 1.26 Chemisorbed monolayers of organic compounds to the surfaces of oxides (silicon oxide is shown), metals (for thiolates on gold) and organic surfaces where the term "grafting" is used frequently. The grey shaded region indicates the covalent bonds that are created during chemisorption.

The tremendous advantage of chemisorbed layers, when compared to physisorbed layers especially, is their robustness. The stability of the films provided by the chemical bond between adsorbate and adsorbent often makes the films cleanable, manipulable in air, and they can be taken between different analytical characterisation methods without a need for vacuum transfer techniques. The chemical and thermodynamic stability also makes chemisorbed layers more suitable for incorporation into useful products. In addition, the surfaces can be built using covalent or non-covalent assembly. An early example of this approach that has been exploited widely is the chemisorption of a siloxane bearing a terminal amino group in its "tail" (the now commercially available 3-aminopropyltriethoxysilane) to a layer of silicon oxide on silicon (Figure 1.27).[110] This work showed many interesting features of the formation of the layers, and came up with an ingenious way to avoid the formation of particulate silicates that can arise in solution treatments: the substrates were placed in an inert atmosphere with refluxing toluene or xylene solutions of the siloxane, that gave a smooth monolayer thanks to the low vapour pressure of the adsorpt in this mixture. Furthermore, the reaction of the amine group in the resulting monolayer was transformed using a number of reagents, building up grafted levels on top of the initial one. This possibility is available because of the stability of the bonds between the surface and initial adsorpt.

Of course, chemical bond forming reactions on the surface of silica and other hydroxyl terminated supports including different varieties of polymers, natural and not, are responsible for the application of certain particles for biological separation and studies,[111,112] chromatography of different types,[113] solid phase synthesis[114–116] in chemistry, and the modification of surfaces for improved adhesion.[117] An early example of the "insolubilisation" of a catalyst is the functionalisation of glass beads with 3-aminopropyltriethoxysilane. Reaction with 4-nitrobenzoic acid and formation of the diazonium salt gave a surface that was reacted with an alkaline phosphatase enzyme that provided a material that maintained the catalytic

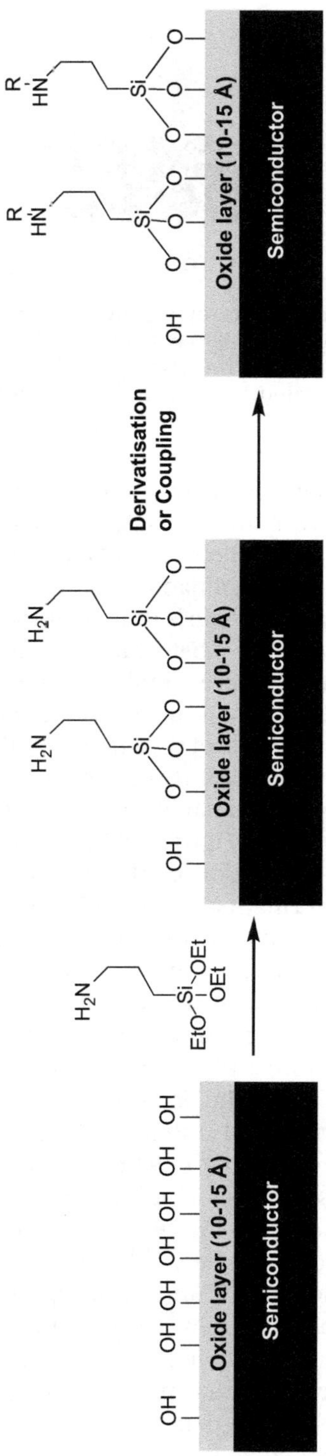

Figure 1.27 The chemical steps leading to the functionalisation of semiconductor surfaces by chemisorption of a siloxane followed by covalent modification.

activity of the biological compound.[118] A nice example of the supramolecular immobilisation of a functional unit is the binding of iron porphyrins to silica gel particles functionalised with a siloxane bearing an imidazole group: the resulting material is reversibly bound oxygen.[119] Carbon monoxide was also proven to be bound. The surface configuration of the iron porphyrin prevents the dimerisation of the system that leads to decomposition in solution in an atmosphere containing oxygen, demonstrating the special properties of the interface.

Often, the function of surface-modified solids requires that the reactive groups are separated from the substrate so that they end up in a solvated situation—for example in the use of polymeric supports for solid phase synthesis—and are therefore immobilised but in a liquid environment. Similarly, the grafting of polymers on surfaces is a noteworthy activity[120,121] but only certain examples of special relevance to supramolecular surface effects will be given in this text. Rather, emphasis shall be given to systems where the proximity to the interface provides an insight and application that are not necessarily evident in solution.

A nice example of the incorporation of supramolecular ideas into functional self-assembled monolayers is the preparation of a caesium sensor on gold-coated microcantilevers.[122] The monolayer formed on the gold was made by co-depositing 1,3-alternate 25,27-bis(11-mercapto-1-undecanoxy)-26,28-calix[4]benzocrown-6 (the selective ion binding component) and decane-1-thiol (Figure 1.28). The alkyl thiol was added to ensure a dense monolayer and so as not to interfere with the binding site of the crown ether. Microcantilevers can be used generally for the detection of analytes because they undergo bending—that can be detected optically—because of the binding of molecules or ions to the layer that generates in-plane stress.[123,124] In this case, deflection of the device is caused by the binding of the caesium ion in solution, and a deflection is clearly seen at 10^{-10} M, three orders of

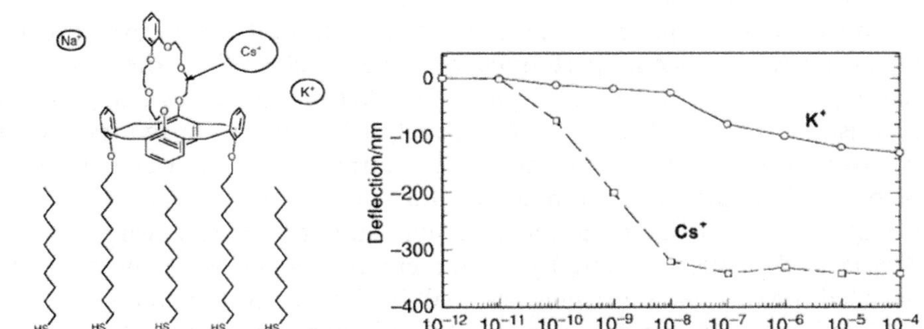

Figure 1.28 A self-assembled monolayer that senses caesium selectively, as shown by the deflection of the cantilever to which it is attached.
Reproduced from ref. 125 with permission from the Royal Society of Chemistry.

magnitude before potassium ions are detected clearly. While the achievement is remarkable, and the approach quite universal,[125,126] it is generally true that these studies are focussed on the functionality of the device, and in the sensor area by and large the precise nature of the monolayer is not studied in depth. After all, the application matters. However, notice that in Figure 1.28 the thiol is still intact on the gold, something that today would be ascertained precisely using analytical techniques to ensure film structure. And the structure under different processing conditions can lead to different properties: durability, and so on. That said, the detection of nerve agents using the complexation of copper(II) to a cysteine monolayer on gold to detect dimethyl methyl phosphonate from a concentration of 10^{-15} M shows the power of the technology.[127] It is hard to imagine a solution-based recognition method that could provide this sensitivity, showing the huge potential of interfaces for sensing and based on the robustness of chemisorbed monolayers.

1.5 Assembly at the Surface of Liquids

The spontaneous formation of supramolecular systems at the interface between liquids and other media is perhaps the most intuitive and general assembly method for molecules. The overriding requirement of the molecular components is that they are poorly soluble in the liquid(s). The main driving force is the assembly of the surfactant—called so because it affects the surface tension of the liquid—at the interface between the two media because of the ability of *part* of the molecule to interact strongly with at least one of them. If the surfactant is applied to water, the most common case with this kind of film, it can be done in a solvent that is lost through evaporation from the interface into the bulk phase, usually to the air that is over the water. A drop of soap on an oily water surface causes an instantaneous change in the appearance of the liquid–air interface. Therefore, the assembly of molecules at the surface of water should have been noticed by the majority of the people reading this text! Indeed, it was in experiments on oil on water systems performed by Agnes Pockels[128,129] then by Lord Rayleigh[130] and Henri Devaux[131] that were later recognised by Langmuir in his historic paper in 1917.[132] In this paper, it is pointed out that castor oil begins to have an appreciable effect on the surface tension of water "when it is in sufficient quantity to form a layer 1.3×10^{-7} cm thick". That is, 1.3 nm, the thickness of a single molecule, as pointed out by Devaux who also observed the formation of monolayers on mercury. In many ways, these results have led the way to nanoscience, as shall be detailed in Chapter 3.

Liquid surfaces present a completely different environment for non-covalent interactions when compared with solid substrates.[133] Apart from the huge mobility of the surface atoms compared with other substrates, an adsorbate can enter into the liquid—affecting directly its surface tension—and can be solvated by it. Thus, the interface region is potentially wider than for an adsorbate–solid interaction and the number of interactions are greater, if not very strong or directional. The reason that water is so effective for the formation of layers of organic molecules is the hydrophobic effect coupled with the lower density of organic materials that means they can be manipulated at the water surface (they do not sink). The formation of layers at the air–water interface has also been studied most widely because of the tools available to "see" formation of films as well as the possibility of transferring these assemblies onto solid surfaces from the liquid, mainly based on simple optical methods (see Chapter 3). From its beginnings where the formation of layers could be studied in relative detail for molecule-thick films, the study of molecular systems at the surface of water has implications ranging from the origin of life to new materials for nanotechnology.[134]

The apparatus that today we call the Langmuir trough is the tool that allows precise measurement of the formation of molecule-thick layers on water.[135] The original drawing of the equipment is shown in Figure 1.29. Habitually, the layer-forming molecule is deposited in a suitable solvent on the dust-free surface of water (the dust would cause imperfections in the film) with the two barriers far apart, and the solvent is allowed to evaporate. The mobile barrier (A in Figure 1.29) is moved so that the surface area available to the sample is reduced, until a monolayer film is formed.

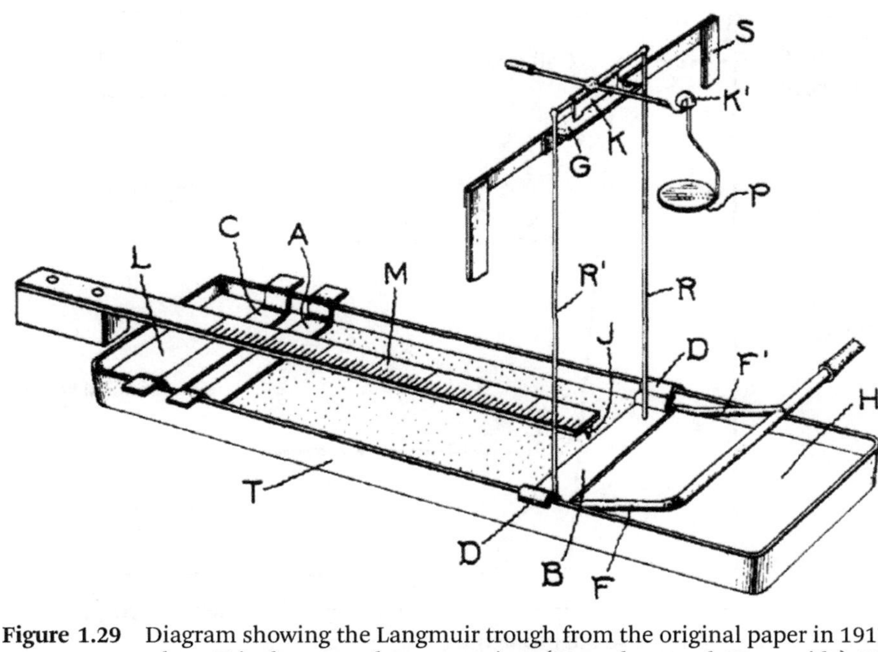

Figure 1.29 Diagram showing the Langmuir trough from the original paper in 1917, where T is the enamel tray container (60 cm long and 15 cm wide), K a knife edge lying on a glass plate G, attached to a support (S). Glass rods (R) attached to the knife edge extend down and pass through holes in a paraffin-coated paper strip (B) that is on the surface of the water. Treated paper strip A is a mobile barrier.
Reprinted with permission from ref. 135. Copyright 1917 American Chemical Society.

A most informative *in situ* measurement of these films is that of the surface pressure in the system as the interfacial area is reduced by adjusting the mobile barrier in the Langmuir trough. It provides a direct measure of the phase in which the surfactant exists, either "bi-dimensional gas", liquid or solid (Figure 1.30). When the barriers are far apart and provided a sufficiently dilute sample is applied, the surfactant molecules are free to diffuse around the surface and if they do not interact significantly with one another may do so. The layer is very easy to compress and has all the characteristics of a gas in this regard. If the molecules do interact with one another, the surface may be occupied by islands of molecules, but will have the characteristics of a two-dimensional gas when the surface pressure isotherm is recorded. Once a value of area per molecule is reached that obliges the molecules (or islands of them) to come into contact, the resistance of the film to pressure increase heightens, and the film enters a liquid-like state. A bi-dimensional solid is reached when the surface area per molecule approaches the cross section of the part of the compound in contact with the water's surface. Evidently, these phase transformations depend on the temperature of the system, and distinctive phase behaviour is seen for

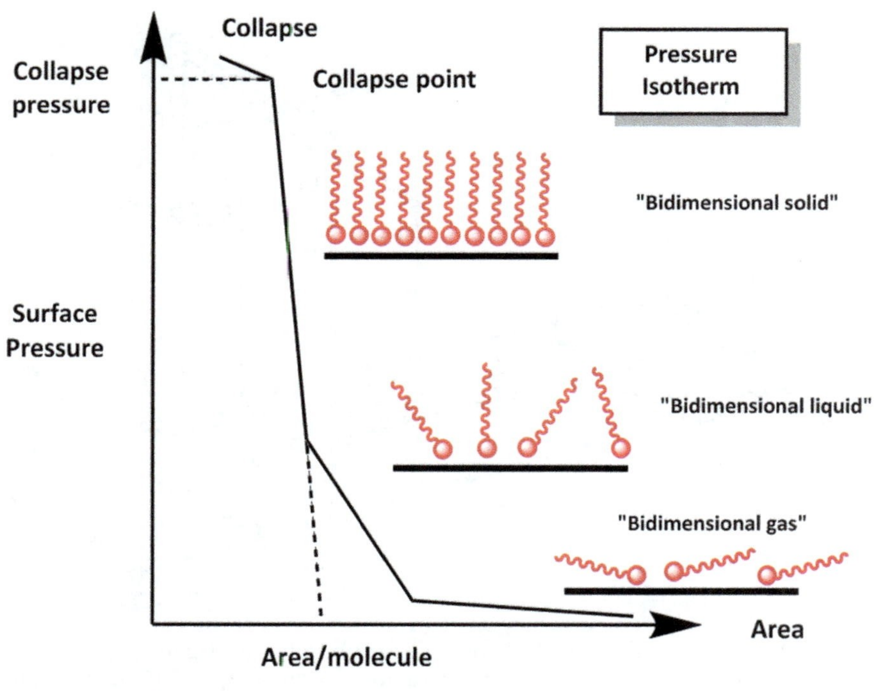

Figure 1.30 The surface pressure isotherm of the type obtained with a Langmuir trough. At low surface pressure, a "bi-dimensional gas" is present, and as the barrier in the trough is used to condense the surface layer, a bi-dimensional liquid is formed before a bi-dimensional solid is generated. If too much surface pressure is applied, the monolayer can become a disordered multilayer after collapse.

different compounds.[136] Compression of this film is now hard, as it is for a bulk solid. If too much pressure is applied, the films break, and they are said to collapse, with the formation of uneven multilayers.

This way of generating ordered monolayers is quite different to those where molecules adhere to an adsorbent. In the present case, a physical force—the compression of the barrier—obliges the molecules to come into close contact, whether they have an inherent tendency to do so or not. They are forced to have a certain orientation, determined by the part of the molecule that is in contact with the water's surface and the shape of the rest of the molecule. An anisotropic arrangement is favoured by the mechanical compression, with a uniformity that is hard to achieve using self-assembly alone.

Having monolayers on the surface of water is scientifically of great interest because the surface pressure can be used to probe interactions of the monolayer-forming material with water-bound hosts or guests. From a materials and technological viewpoint, transferring the monolayers onto a solid substrate to make use of the properties of the thin film is attractive. The Langmuir–Blodgett technique—named after its inventors—is quite unique in the area of supramolecular surface systems in that the monolayers that are generated can be transferred onto solid surfaces. This coating is performed by vertical immersion or emersion of a solid substrate into or out of the Langmuir trough while applying constant pressure to the surface, and thereby favouring the integrity of the layer. The general principle of Langmuir–Blodgett (LB) film formation is shown in Figure 1.31 along with the horizontal lift-off technique, also known as the Langmuir–Schaefer

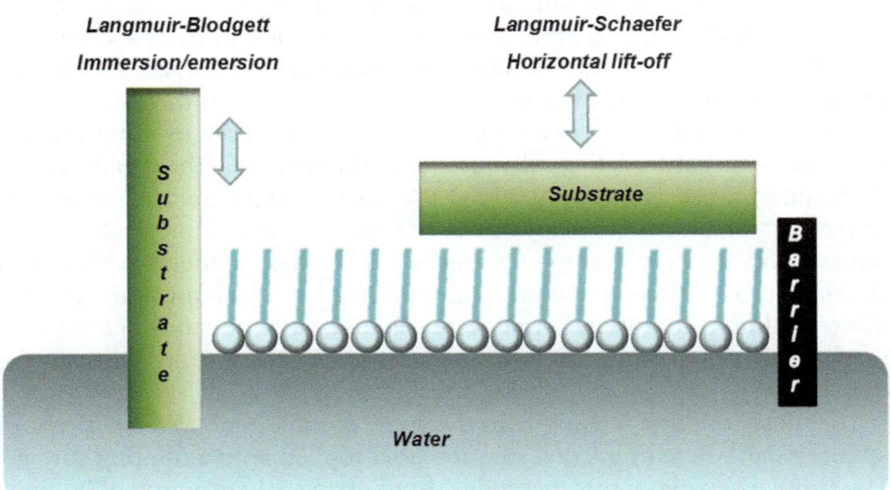

Figure 1.31 A cartoon showing the methods used to transfer Langmuir layers onto solid substrates, by dipping the solid substrate into the trough (the Langmuir–Blodgett method) or by contacting the solid parallel to the air–water interface (Langmuir–Schaefer, or horizontal lift-off method).

method,[137] where the solid substrate is lowered onto the monolayer and contacts at the monolayer–air interface and is subsequently removed.

The transfer of each monolayer onto a substrate is of interest for a number of reasons, not least because each layer, in principle, could be made of a different material and with an orientation that could be determined by the transfer method used. The number of layers adsorbed, in principle, is limited only by one's patience! One of the areas of research where the LB films made a great impact was in the preparation of synthetic metals, that is, molecular systems that display conducting or magnetic properties.[138]

1.6 Self-assembly Under Kinetic Control Upon Deposition or Evaporative Self-assembly

The deposition of organic materials onto surfaces is performed very often by application of a solution of the compound to the potential adsorbent followed by evaporation of the solvent, simply because it is in principle an easy screening method of self-assembly on a surface and because it may not require any special equipment for performing the deposition. During this process of solvent evaporation, it is clear that the point where the system is on the solvent–solute–adsorbent phase diagram will change: the solution becomes more concentrated and generally aggregation and/or adsorption are favoured. The rate at which this process takes place will clearly influence the aggregation process that is far from equilibrium.[139] This change in the point on the phase diagram location has been used widely for the preparation of nanostructures from bulk solution/suspension[140] where control over the phase is more precise, but has also been exploited for the preparation of spectacular surface-based films. The formation of patterns can have a mainly physical origin,[141] but the influence of non-covalent interactions can frequently hold sway in determining the morphology of thin films.

It should be noted that in a drop on a surface, there are two interfaces to consider, that between the substrate (and potential adsorbent) and the liquid and that between the liquid and the surroundings. Gold nanoparticles can undergo kind of kinetically-driven self-assembly at the interface between toluene and air when a drop of a solution of the nanoparticles is placed on silicon nitride.[142] The assembly of the particles at the liquid–air interface under optimum conditions—that were achieved by a full study of the phase diagram of the system—leads to highly homogeneous thin films of the material.

The most common technique in the study of molecular systems is that of spin casting, where drops of a solution are dropped onto a rotating surface. The simpler version of this method is drop casting, where a drop of liquid is placed on a surface and allowed to evaporate slowly. While the latter does not generally lead to such uniform films, it can reveal remarkable effects. For example, the crystallinity of polymers can be higher under drop casting conditions.[143] The evaporation of a solution containing molecules capable of mutual recognition leads to the formation of distinctive patterns as a result of simultaneous solvent flow in a sessile drop[144] during vapour formation and self-assembly of the solute.[145] Flow can either be capillary, or differences in surface tension of an evaporating liquid can give rise to Marangoni flow.[146] These localised variations in surface tension can be caused by differential evaporation giving compositional variation, temperature differences, surface charges and/or compression or dilatation of the adsorbed films. Frequently, this situation on a surface gives rise to spinodal dewetting (a particular case of spinodal decomposition) where a single thermodynamically stable phase is taken far from equilibrium to give two (or more) separate phases, and for thin films this means the formation of curved agglomerations of material of a particular shape.[147] While many examples exist for polymer patterns[148] in which non-covalent bonding is not considered to have a dominant role over the physical phenomenon of drying where a solute is deposited when supersaturation is reached, the examples in this text will show how supramolecular systems can form ordered structures on surfaces and where non-covalent interactions play a determining role in the structural outcome.

A beautiful case that demonstrates the huge potential of evaporation-induced self-assembly on surfaces is that of the C_3 symmetric tris-porphyrin compound shown in Figure 1.32.[149] Ordered fibres of the material are observed over millimetres thanks to the combined effects of self-assembly and dewetting from the surface. This molecule self-assembles to give stacked columns where the hydrogen bonds between the amide groups near the centre of the structure result in a helical fibre the width of the compound, approximately 4.5 nm. When a chloroform solution of the compound is allowed to evaporate quickly on a mica surface, fibres a single molecule thick are formed on the mica whose long axis is parallel to the evaporating solvent front. These single molecule thick but up to millimetre long fibres are spaced remarkably regularly, approximately 700 nm apart. The reason for this phenomenon is believed to be a result of spinodal dewetting, the separation of the fibre phase from the surroundings. There are areas observed where a more classical wavy cylindrical pattern expected for non-specific dewetting was observed, but these are by far the minority case. When evaporation was slower, much thicker surface-bound fibres were observed that are believed to be formed as a result of the pinning of the evaporating solvent front, and fibres perpendicular to this line are observed that are 55 nm wide and clearly consist of bundles of the supramolecular fibres.

1.7 Supramolecular Design for Adsorpts

Supramolecular chemistry creates its object, as Jean-Marie Lehn famously stated and he has developed this idea to remarkable lengths.[150] The Nobel Laureate himself has shown that whether that object is created or not depends on the conditions in which the self-assembly processes take place. The characteristics of surfaces, of particular kinds of growth and

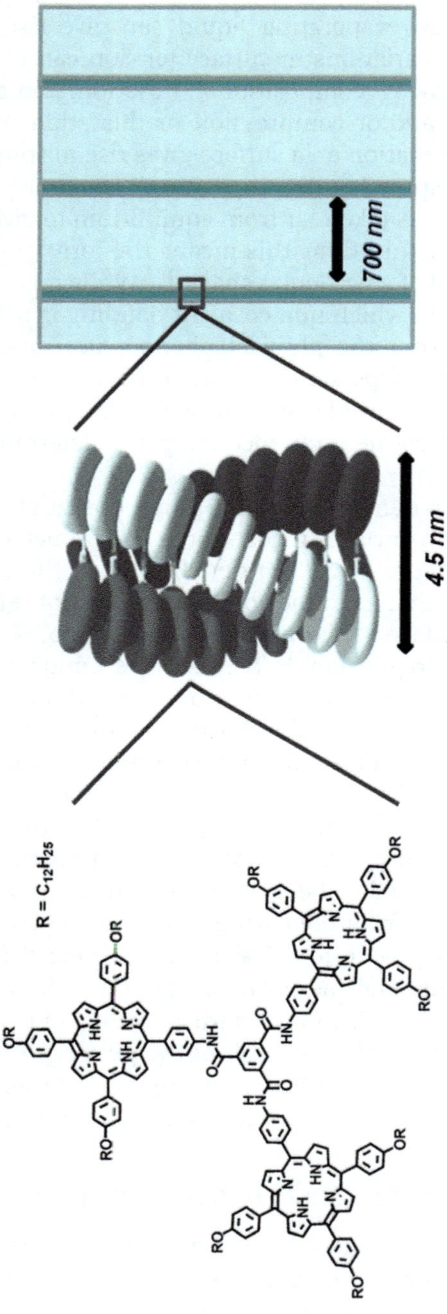

Figure 1.32 A tris-porphyrin that self-assembles through hydrogen bonds to give helical stacks in solution that upon deposition from chloroform and evaporation of the solvent on mica, self-organise to give equidistant (approximately 700 nm) parallel lines of the material.

interactions between adsorpts and surfaces have been outlined in the previous sections. They are an important foundation for the subsequent chapters, and knowledge of supramolecular science has been assumed. The particular question that the supramolecular scientist reading this book might have asked themselves is "what do I need to build into a supramolecular synthon to make it useful for surface assembly"? Naturally, it depends on the surface: one size does not fit all. In addition, the deposition technique can impose limitations on the choice of structure: vacuum deposition techniques generally require thermally-stable compounds and solution versions compounds with a high affinity for the adsorbent. Nonetheless, some general features can be identified.

In approximate historical order, for layers on water, a molecule with a calamitic (rod-like) shape is generally a sure bet for well-behaved film formation, and it is clear that a polar group of a comparable cross section to the remainder of the molecule will give rise to effective packing. The polar moiety is important, but interactions of this group with ions in the subphase and hydrogen bonding units in the calamitic section can help order in monolayers, aiding the formation of crystalline aliphatic regions.[151,152] A strict rod-like shape is not essential, but deviations from this form would generate voids in a monolayer, and this feature favours curved superstructures as in the formation of micelles and vesicles. It is precisely the kind of molecule that form these water-dispersed aggregates that can form interesting superstructures at the interface of air and water. The layers are generated because they are deposited on the surface of water rather than being dispersed in it. The polar head group does not have to be charged, and even relative hydrophilicity can lead to ordered layers on water where, by and large, molecules are oriented quasi-perpendicularly to the interface.

The formation of ordered layers by chemisorption bears remarkable parallels to the generation of Langmuir layers. In the latter, the strong interaction between the polar head group and the water determines orientation, and pressure in the trough helps the alkyl chains orient. In chemisorbed layers on solid substrates, it is the geometry of the chemical bond that helps determine in which direction the molecule "points" and high coverage will give an alignment perpendicular to the interface, an identical orientation to Langmuir layers. The chemistry is often quite specific, thiolates on gold being a good example, and pairing of reactive adsorpts and adsorbents is necessary to ensure layer formation in the same way that functional groups in any kind of compound are made to react with others. In a similar vein, functional groups may be incorporated into adsorbates such that they are reacted subsequently to give layer-by-layer assembly. An additional important factor in the formation of chemisorbed layers is that the adsorbates can be chosen such that specific crystal faces are exposed, meaning that not only is the chemistry selective, but the distances between surface atoms can influence packing of the adsorbates.

Long-chain aliphatic compounds also favour adsorption through purely physical attraction onto hydrophobic surfaces (graphite being a good example) from solution. In this physisorbed state, the molecules lie coplanar with the interface in order to maximise the number of covalent bonds. In this case, the linear chains of alkyl chains pack—rather like the chains in membranes—so that either interdigitated lamellae form, or that the more familiar lamellae double the molecular length with abutting chains are generated. The relative size of head group and chains influence whether interdigitated or abutted domains exist; the larger the head group, the more probable is interdigitation. This feature can be exploited for the formation of porous superstructures (see Chapter 4). On the other hand, specific non-covalent interactions, such as hydrogen bonds, can influence the monolayer structure more readily than in chemisorbed structures, because the layer structure is determined thermodynamically, mainly.

Apart from the functional groups present in a molecule, the overall shape has great significance in all monolayers and films. It has been found that amorphous layers of linear- or planar-shaped molecules are oriented co-planar to the substrate and that the longer the molecule, the greater the anisotropy of the films.[153,154] Subtle effects can give rise to large differences in packing, such as constitutional isomers[155–157] or the apparently small differences that might arise because of the use of odd or even numbers of carbon atoms in a chain, the odd–even effect. This phenomenon applies to bulk systems[158–160] as well as to monolayers.[161]

The functionality of the surface assemblies is of course important, and for this reason, the core of many adsorpts contains π-functional units. Tetrapyrrolic macrocycles—porphyrins and phthalocyanines mainly—are particularly popular for constructing supramolecular architectures on surfaces[162] because of their multiple potential uses (from catalysis, to molecular recognition and optoelectronics[163]) as well as their large dimensions and multiple chemical functionality possibilities that make them attractive for studies by different types of microscopy.[164] Therefore, by and large, the design of artificial surfactants takes into account the following main aspects:

- Functional groups for specific adsorpt–adsorbent interactions;
- Non-covalent interactions to aid ordered surface assembly;
- Molecular shape to ensure appropriate packing in "two dimensions";
- Functional moieties to perform particular tasks.

1.8 Actual and Potential Applications of Surface-based Supramolecular Systems

In our everyday activity, we encounter, consciously or not, treated interfaces that are present to improve the material performance of a product. Corrosion inhibitors protect metals from atmospheric chemical attack, glass coatings improve the performance regarding wetting or glare, heterogeneous catalysts clean up effluent gases, and sensors detect chemicals or radiation, usually at an interface. Consider our touch devices that employ multiple interfaces; from the moment our finger touches a screen, to the information reaching a processor until the signal eventually comes back to our screen, every component is linked through an interface between materials. In that case, they are usually inorganic, though the device is held together with an organic adhesive. The bonding together of surfaces is undoubtedly dependent on chemisorption and polymerisation,[165–167] but supramolecular chemistry can play a role.[168] For instance, it has been demonstrated that biofouling (the adhesion of biological systems to surfaces) can be reversed using polymeric coatings that undergo a phase change.[169]

In biological systems, interfaces are responsible for a multitude of processes. Supramolecular effects in biochemistry are dominant here. Synthetic supramolecular chemistry has opportunities to learn from these biological systems, and also to provide sensors and medicines that can provide therapeutically important benefits for humankind. The use of host–guest interactions at these interfaces provides stimuli-responsive systems.[170] Adhesives are also fascinating in the biological area, where the challenge of this function in aqueous systems has been solved by marine organisms through adsorption to surfaces by phosphorylated serines and hydroxylated tyrosines.[171]

A remarkable panel review paper published in 1987[172] identified promising areas of interest for interface science and the chemical control of these areas to provide improved sensors, catalysts, mechanical resistance, electronic performance, and so on. Perhaps out of view at that point were responsive systems involving chemical reactions or conformational changes.[173] Yet, in general, it is likely that complete films—pure or composite—will provide materials of interest for applications. The formation of these ultrathin films of molecules is potentially interesting from a number of viewpoints,[174] and the traits afforded by specific supramolecular chemistry provide unique characteristics.

The coating of surfaces to protect them from getting damaged through chemical reactions with environmental agents or simply from dirt is an industrially important challenge. The passivation of the surface of metals is particularly important. For example, metallic copper—whose characteristic green colour after oxidation decorates innumerable buildings around the globe—can be protected by the formation of self-assembled monolayers. Surfactants with both sulfur and nitrogen head groups are excellent candidates for this purpose. Benzotriazole was identified for preventing the tarnishing of copper in the middle of the last century.[175,176] A recent example is the adsorption of 3-methyl-5-octadecylsulfanyl-[1,2,4]triazole-4-ylamine, which was identified as preventing dissolution of copper in either acidic or saline solutions by the formation of a chelate complex of the adsorbate with Cu(I) at the interface.[177]

The coating of calcium carbonate is also important, as it can be used as a filler material[178] and is also a viable route to old building conservation, as demonstrated for York Minster's stone blocks.[179] In this latter example, as in the case of superhydrophobic calcium carbonate powder synthesised in one step from sodium carbonate and calcium chloride in the presence of surfactants,[180] both chemisorbed siloxyl and carboxylate ligands were employed. As a filler material in polymers, coating the calcium carbonate with stearate affects both adhesion and tensile strength of the composites,[181] while the nature of organic phosphonic acids also increases the lipophilicity of the particles.[182] The formation of the layer on the calcium carbonate is not necessarily straightforward; stearate forms a monolayer on nanoparticulate form of the salt that rearranges on heating,[183] and this kind of restructuring in materials would benefit greatly from precise views of the supramolecular chemistry taking place at the interface between the components of the system.

Working the other way around, monolayers can template the formation of minerals. Aragonite is a metastable form of calcium carbonate, and its formation can be templated by a film of 5-hexadecyloxyisophthalic acid on supersaturated calcium bicarbonate.[184] The aragonite needles grew away from the interface after their nucleation there. The process needed no additional nucleating agent. This type of control over crystal growth at the interface of water, monolayer and air has also proven effective for organic materials: glycine was crystallised at the water–monolayer–air interface when long-chain enantiomeric α-amino acids were used as the amphiphile.[185,186] The concept of pairing recognition sites there was also used for the crystallisation of hydroxybenzoic acid under monolayers of an amphiphilic derivative of the same compound.[187] On the other hand, chemisorbed SAMs on gold have been used to template the formation of the diverse forms of calcium carbonate,[188,189] as well as organic crystals.[190] The templated formation of crystals is an area of significance from the point of view of promotion of crystallisation, as in biomineralisation, as well as in the prevention of it, thinking of maladies where crystallisation of solid fibres or stones is a problem. The supramolecular control of these effects can be studied with great precision using well defined interfaces.

The preparation of devices that rely on optoelectronic effects frequently requires the control of interfaces between the materials in the systems. The promise of organic electronics means that the formation of SAMs and the grafting of conjugated polymers onto both organic and inorganic substrates (adsorbents) is gaining increasing importance.[191] The reason that chemisorption is so important in this case is two-fold: to ensure the robustness of the interfaces and to control the orientation and relative electronic energy levels of the often disparate materials. While the formation of covalent bonds undoubtedly leads to robustness, as for example in films comprising zinc porphyrins and 4,4-difluoro-4-bora-3*a*,4*a*-diaza-*s*-indacene (BODIPY) derivatives aimed at light-harvesting applications,[192] precise assembly of chromophores requires the use of non-covalent bonds, and thus supramolecular chemistry is bound to play a role in generating high performance materials. A much longer term goal for the area is to find applications for responsive systems, an area that will be touched on in the final chapter of this monograph.

Superhydrophobic (repelling water), superoleophobic (repelling oil),[193] and superamphiphobic (repelling both water and apolar solvents)[194] surfaces are a particularly important family of materials. They have potential application in self-cleaning surfaces for a variety of solutions, anti-freezing properties at the surface, or anti-bacterial growth and corrosion resistance. This property is a combination of both the surface chemistry and the intrinsic topographical roughness of the surface, as we shall see. In particular, polymer membrane fouling—the attachment of unwanted macromolecules, salts, colloids or particles that reduce flow through it—is a problem that can be addressed by surface modification of the bulk membrane material.[195] Here, the formation of hydrophilic or charged surfaces can be beneficial, although precise surface studies would still reveal fundamental phenomena that are at present not totally established, such as the exact relationship between hydrophilicity and anti-fouling performance.

In the subsequent chapters, different kinds of surface-based supramolecular systems and their varied applications will be discussed. But before considering these cases, it is helpful to understand the tools that can be employed to characterise them. We need many kinds of "eye" to "see" what is happening at surfaces. It is extremely rare that one technique alone can provide us with a complete picture of what takes place during the self-assembly of molecules at surfaces or their interactions with the surrounding medium. In many cases, the design of the supramolecular systems must take into account the inclusion of probe moieties in the component structure to be able to track the assembly at the interfaces. An appreciation of what we can see with which technique is presented in the following chapter.

References

1. B. H. Huisman, F. C. J. M. van Veggel and D. N. Reinhoudt, *Pure Appl. Chem.*, 1998, **70**, 1985.
2. E. Gomar-Nadal, J. Puigmartí-Luis and D. B. Amabilino, *Chem. Soc. Rev.*, 2008, **37**, 490.
3. N. Lin, S. Stepanow, M. Ruben and J. V. Barth, *Top. Curr. Chem.*, 2009, **287**, 1.
4. J. A. A. W. Elemans, S. Lei and S. De Feyter, *Angew. Chem., Int. Ed.*, 2009, **48**, 7298.
5. A. G. Slater, L. M. A. Perdigao, P. H. Beton and N. R. Champness, *Acc. Chem. Res.*, 2014, **47**, 3417.
6. M. C. Daniel and D. Astruc, *Chem. Rev.*, 2004, **104**, 293.
7. R. Klajn, J. F. Stoddart and B. A. Grzybowski, *Chem. Soc. Rev.*, 2010, **39**, 220.
8. C. Knorowski and A. Travesset, *Curr. Opin. Solid State Mater. Sci.*, 2011, **15**, 262.
9. L. Xu, W. Ma, L. Wang, C. Xu, H. Kuang and N. A. Kotov, *Chem. Soc. Rev.*, 2013, **42**, 3114.

10. S. Kinge, M. Crego-Calama and D. N. Reinhoudt, *ChemPhysChem*, 2008, **9**, 20.
11. S. Mann, *Nat. Mater.*, 2009, **8**, 781.
12. J. E. Greene, *Appl. Phys. Rev.*, 2015, **2**, 011101.
13. K. S. Birdi, *Self-Assembly Monolayer Structures of Lipids and Macromolecules at Interfaces*, Springer, 2002, ISBN: 978-0-306-46099-9.
14. R. R. Bruns, *J. Cell Biol.*, 1976, **68**, 521.
15. *Biointerfaces: Where Material Meets Biology*, ed. D. Hutmacher and W. Chrzanowski, RSC Publishing, 2015.
16. C. Alonso, T. Alig, J. Yoon, F. Bringezu, H. Warriner and J. A. Zasadzinski, *Biophys. J.*, 2004, **87**, 4188.
17. E. Parra and J. Perez-Gil, *Chem. Phys. Lipids*, 2015, **185**, 153.
18. E. A. Vogler, *Adv. Colloid Interface Sci.*, 1998, **74**, 69.
19. D. L. Allara and R. G. Nuzzo, *Langmuir*, 1985, **1**, 45.
20. W. C. Bigelow, D. L. Pickett and W. A. Zisman, *J. Colloid Sci.*, 1946, **1**, 513.
21. I. Langmuir, *J. Am. Chem. Soc.*, 1917, **39**, 1848.
22. K. B. Blodgett, *J. Am. Chem. Soc.*, 1935, **57**, 1007.
23. I. Langmuir and V. J. Schaefer, *J. Am. Chem. Soc.*, 1938, **60**, 1351.
24. D. J. Cram and J. M. Cram, *Science*, 1974, **183**, 803.
25. C. J. Pedersen, *Angew. Chem., Int. Ed.*, 1988, **27**, 1021.
26. J. M. Lehn, *Angew. Chem., Int. Ed.*, 1988, **27**, 89.
27. M. J. W. Ludden, D. N. Reinhoudt and J. Huskens, *Chem. Soc. Rev.*, 2006, **35**, 1122.
28. S.-W. Hla, L. Bartels, G. Meyer and K.-H. Rieder, *Phys. Rev. Lett.*, 2000, **85**, 2777.
29. R. Viswanathan, J. A. Zasadzinski and D. K. Schwartz, *Nature*, 1994, **368**, 440.
30. L. Sun, Y. A. Diaz-Fernandez, T. A. Gschneidtner, F. Westerlund, S. Lara-Avila and K. Moth-Poulsen, *Chem. Soc. Rev.*, 2014, **43**, 7378.
31. A. Saywell, G. Magnano, C. J. Satterley, L. M. A. Perdigao, A. J. Britton, N. Taleb, M. C. Gimenez-Lopez, N. R. Champness, J. N. O'Shea and P. H. Beton, *Nat. Commun.*, 2010, **1**, 75.
32. M. Gonidec, R. Biagi, V. Corradini, F. Moro, V. De Renzi, U. del Pennino, D. Summa, L. Muccioli, C. Zannoni, D. B. Amabilino and J. Veciana, *J. Am. Chem. Soc.*, 2011, **133**, 6603.
33. R. Blankenburg, P. Meller, H. Ringsdorf and C. Salesse, *Biochemistry*, 1989, **28**, 8214.
34. K. D. Schierbaum, T. Weiss, E. U. Thoden van Veizen, J. F. J. Engbersen, D. N. Reinhoudt and W. Gopel, *Science*, 1994, **265**, 1413.
35. M. Ortega Lorenzo, C. J. Baddeley, C. Muryn and R. Raval, *Nature*, 2000, **404**, 376.
36. B. Gohler, V. Hamelbeck, T. Z. Markus, M. Kettner, G. F. Hanne, Z. Vager, R. Naaman and H. Zacharias, *Science*, 2011, **331**, 894.
37. A. G. Slater (nee Phillips), P. H. Beton and N. R. Champness, *Chem. Sci.*, 2011, **2**, 1440.

38. A. G. Phillips, P. H. Beton and N. R. Champness, in *Supramolecular Chemistry: From Molecules to Nanomaterials*, ed. P. A. Gale and J. W. Steed, Nanotechnology, J. Wiley & Sons, 2012, vol. 8, pp. 3625–3646.
39. K. S. Mali, J. Adisoejoso, E. Ghijsens, I. de Cat and S. De Feyter, *Acc. Chem. Res.*, 2012, **45**, 1309.
40. K. Eichhorst-Gerner, A. Stabel, G. Moessner, D. Declerq, S. Valiyaveettil, V. Enkelmann, K. Mullen and J. P. Rabe, *Angew. Chem., Int. Ed.*, 1996, **35**, 1492.
41. T. J. Prior, J. A. Armstrong, D. M. Benoit and K. L. Marshall, *CrystEngComm*, 2013, **15**, 5838.
42. L. M. A. Perdigao, N. R. Champness and P. H. Beeton, *Chem. Commun.*, 2006, 538.
43. H. Dang, T. Maris, J.-H. Yi, F. Rosei, A. Nanci and J. D. Wuest, *Langmuir*, 2007, **23**, 11980.
44. D. Bonifazi, S. Mohnani and A. Llanes-Pallas, *Chem. - Eur. J.*, 2009, **15**, 7004.
45. C. Q. Sun, *Prog. Solid State Chem.*, 2007, **35**, 1.
46. A. B. H. Hamouda, T. J. Stasevich, A. Pimpinelli and T. L. Einstein, *J. Phys.: Condens. Matter*, 2009, **21**, 084215.
47. J. H. C. S. He, C. S. Lao, L. J. Chen, D. Davidovic and Z. L. Wang, *J. Am. Chem. Soc.*, 2005, **127**, 16376; J. E. Yoo, J. L. Cross, T. L. Bucholz, K. S. Lee, M. P. Espe and Y.-L. Loo, *J. Mater. Chem.*, 2007, **17**, 1268; D. C. Wei, Y. Q. Liu, Y. Wang, H. L. Zhang, L. P. Huang and G. Yu, *Nano Lett.*, 2009, **9**, 1752.
48. J. R. Sootsman, D. Y. Chung, Y. Duck and M. G. Kanatzidis, *Angew. Chem., Int. Ed.*, 2009, **48**, 8616; G. Zhang and B. Li, *Nanoscale*, 2010, **2**, 1058.
49. H. A. Vandervegt, H. M. Vanpinxerten, M. Lohmeier, E. Vlieg and J. M. C. Thornton, *Phys. Rev. Lett.*, 1992, **68**, 3335.
50. B. R. Conrad, E. Gomar-Nadal, W. G. Cullen, A. Pimpinelli, T. L. Einstein and E. D. Williams, *Phys. Rev. B: Condens. Matter Mater. Phys.*, 2008, **77**, 205328.
51. I. Langmuir, *J. Am. Chem. Soc.*, 1918, **40**, 1361.
52. D. B. Amabilino, S. De Feyter, R. Lazzaroni, E. Gomar-Nadal, J. Veciana, C. Rovira, M. M. Abdel-Mottaleb, W. Mamdouh, P. Iavicoli, K. Psychogyiopoulou, M. Linares, A. Minoia, H. Xu and J. Puigmartí-Luis, *J. Phys. Condens. Matter*, 2008, **20**, 184003.
53. R. Gutzler, L. Cardenas and F. Rosei, *Chem. Sci.*, 2011, **2**, 2290.
54. A. Koma, *Prog. Cryst. Growth Charact.*, 1995, **30**, 129.
55. K. Glökler, C. Seidel, A. Soukopp, M. Sokolowski, E. Umbach, M. Böhringer, R. Berndt and W.-D. Schneider, *Surf. Sci.*, 1998, **405**, 1.
56. A. Jahanbekam, S. Vorpahl, U. Mazur and K. W. Hipps, *J. Phys. Chem. C*, 2013, **117**, 2914.
57. C. Wagner, R. Forker and T. Fritz, *J. Phys. Chem. Lett.*, 2012, **3**, 419.
58. J. Bernstein, R. J. Davey and J. O. Henck, *Angew. Chem., Int. Ed.*, 1999, **38**, 3440.

59. J. P. Brog, C. L. Chanez, A. Crochet and K. M. Fromm, *RSC Adv.*, 2013, **3**, 16905.
60. S. L. Price, *Chem. Soc. Rev.*, 2014, **43**, 2098.
61. A. J. Cruz-Cabeza and J. Bernstein, *Chem. Rev.*, 2014, **114**, 2170.
62. T. Djuric, T. Ules, H.-G. Flesch, H. Plank, Q. Shen, C. Teichert, R. Resel and M. G. Ramsey, *Cryst. Growth Des.*, 2011, **11**, 1015.
63. *Chirality at the Nanoscale*, ed. D. B. Amabilino, Wiley-VCH, 2009.
64. D. B. Amabilino and J. Veciana, *Top. Curr. Chem.*, 2006, **265**, 253.
65. A. González-Campo and D. B. Amabilino, *Top. Curr. Chem.*, 2013, **333**, 109.
66. A. Kuhnle, T. R. Linderoth, B. Hammer and F. Besenbacher, *Nature*, 2002, **415**, 891.
67. S. M. Barlow and R. Raval, *Surf. Sci. Rep.*, 2003, **50**, 201.
68. K.-H. Ernst, *Top. Curr. Chem.*, 2006, **265**, 209.
69. M. V. Stewart and E. M. Arnett, *Top. Stereochem.*, 1982, **13**, 195.
70. I. Kuzmenko, I. Weissbuch, E. Gurovich, L. Leiserowitz and M. Lahav, *Chirality*, 1998, **10**, 415.
71. L. Pérez-García and D. B. Amabilino, *Chem. Soc. Rev.*, 2007, **36**, 941.
72. R. Raval, *Chem. Soc. Rev.*, 2009, **38**, 707.
73. K. E. Plass, A. L. Grzesiak and A. J. Matzger, *Acc. Chem. Res.*, 2007, **40**, 287.
74. G. A. Somorjai and M. A. Van Hove, *Prog. Surf. Sci.*, 1989, **30**, 201.
75. V. Humblot, S. Haq, C. Muryn, W. A. Hofer and R. Raval, *J. Am. Chem. Soc.*, 2002, **124**, 503.
76. A. Stabel, R. Heinz, F. C. De Schryver and J. P. Rabe, *J. Phys. Chem.*, 1995, **99**, 505.
77. M. Oehzelt, R. Resel, C. Suess, R. Friedlein and W. R. Salaneck, *J. Chem. Phys.*, 2006, **124**, 054711.
78. Y.-K. Hong, H. Yu, T. G. Lee, N. Lee, J. H. Bahng, N. W. Song, W. Chegal, H. K. Shon and J.-Y. Koo, *Chem. Phys.*, 2014, **428**, 105.
79. Y. Gong, M. C. P. Wang, X. Zhang, H. Wai Ng and B. D. Gates, *Langmuir*, 2012, **28**, 11790.
80. J. Yang and D. Yan, *Chem. Soc. Rev.*, 2009, **38**, 2634.
81. S. Isoda, I. Kubo, A. Hoshino, N. Asaka, H. Kurata and T. Kobayashi, *J. Cryst. Growth*, 1991, **115**, 388.
82. A. Hoshino, S. Isoda and T. Kobayashi, *J. Cryst. Growth*, 1991, **115**, 826.
83. A. Hinderhofer and F. Schreiber, *ChemPhysChem*, 2012, **13**, 628.
84. M. Wiatrowski, E. Dobruchowska, W. Maniukiewicz, U. Pietsch, J. Kowalski, Z. Szamel and J. Ulanski, *Thin Solid Films*, 2010, **518**, 2266.
85. V. Humblot, S. Haq and R. Raval, *J. Catal.*, 2004, **228**, 130.
86. B. Venkataraman, J. J. Breen and G. W. Flynn, *J. Phys. Chem.*, 1995, **99**, 6608.
87. C.-J. Li, Q.-D. Zeng, C. Wang, L.-J. Wan, S.-L. Xu, C.-R. Wang and C.-L. Bai, *J. Phys. Chem. B*, 2003, **107**, 747.
88. *Electrical Phenomena at Interfaces and Biointerfaces*, ed. H. Ohshima, Wiley, 2012.

89. J. P. Rabe and S. Buchholz, *Science*, 1991, **253**, 424.
90. M. M. S. Abdel-Mottaleb, E. Gomar-Nadal, S. De Feyter, M. Zdanowska, J. Veciana, C. Rovira, D. B. Amabilino and F. C. De Schryver, *Nano Lett.*, 2003, **3**, 1375.
91. Y. Li and N. Lin, *Phys. Rev. B: Condens. Matter Mater. Phys.*, 2011, **84**, 125418.
92. C. Tao, T. J. Stasevich, W. G. Cullen, T. L. Einstein and E. D. Williams, *Nano Lett.*, 2007, 7, 1495.
93. T. Sirtl, W. Song, G. Eder, S. Neogi, M. Schmittel, W. M. Heckl and M. Lackinger, *ACS Nano*, 2013, 7, 6711.
94. S. R. Batten and R. Robson, *Angew. Chem., Int. Ed.*, 1998, **37**, 1460.
95. G. Ferey, *Chem. Soc. Rev.*, 2008, **37**, 191.
96. J.-R. Li, R. J. Kuppler and H. C. Zhou, *Chem. Soc. Rev.*, 2009, **38**, 1477.
97. X. Lin, N. R. Champness and M. Schroeder, *Top. Curr. Chem.*, 2010, **293**, 35.
98. X. Feng, X. Ding and D. Jiang, *Chem. Soc. Rev.*, 2012, **41**, 6010.
99. S.-Y. Ding and W. Wei, *Chem. Soc. Rev.*, 2013, **42**, 548.
100. T. Wang, S. Bhosale, S. Bhosale, G. Li and J.-H. Fuhrhop, *Acc. Chem. Res.*, 2006, **39**, 498.
101. M. O. Blunt, J. Adisoejoso, K. Tahara, K. Katayama, M. Van der Auweraer, Y. Tobe and S. De Feyter, *J. Am. Chem. Soc.*, 2013, **135**, 12068.
102. A. Bilic, J. R. Reimers and N. S. Hush, *J. Chem. Phys.*, 2005, **122**, 094708.
103. D. J. Lavrich, S. M. Wetterer, S. L. Bernasek and G. Scoles, *J. Phys. Chem. B*, 1998, **102**, 3456.
104. A. Ulman, *Chem. Rev.*, 1996, **96**, 1533.
105. F. Schreiber, *Prog. Surf. Sci.*, 2000, **65**, 2000.
106. G. Lecayon, Y. Bouizem and C. Le Gressus, *Chem. Phys. Lett.*, 1982, **91**, 506.
107. Q. Liu, J. R. de Wijn, K. de Groot and C. A. van Blitterswijk, *Biomaterials*, 1998, **19**, 1067.
108. D. Roy, M. Semsarilar, J. T. Guthrie and S. Perrier, *Chem. Soc. Rev.*, 2009, **38**, 2046.
109. J. Haccoun, C. Vautrin-Ul, A. Chaussé and A. Adenier, *Prog. Polym. Sci.*, 2008, **63**, 18.
110. I. Haller, *J. Am. Chem. Soc.*, 1978, **100**, 8050.
111. H. H. Weetall, *Nature*, 1969, **223**, 959.
112. P. J. Robinson, P. Dunnill and M. D. Lilly, *Biochim. Biophys. Acta*, 1971, **242**, 659.
113. H. A. Claessens and M. A. van Straten, *J. Chromatogr. A*, 2004, **1060**, 23.
114. J. W. Labadie, *Curr. Opin. Chem. Biol.*, 1998, **2**, 346.
115. C. Blackburn, *Biopolymers*, 1998, **47**, 311.
116. F. García-Martín and F. Albericio, *Chim. Oggi–Chemistry Today*, 2008, **26**, S29.
117. K. L. Mittal and D. F. O'Kane, *J. Adhes.*, 1976, **8**, 93.
118. H. H. Weetal, *Nature*, 1969, **223**, 959.

119. O. Leal, D. L. Anderson, R. G. Bowman, F. Basolo and R. L. Burwell, Jr., *J. Am. Chem. Soc.*, 1975, **97**, 5125.
120. B. Zhao and W. J. Brittain, *Prog. Polym. Sci.*, 2000, **25**, 677.
121. K. Kato, E. Uchida, E.-T. Kang, Y. Uyama and Y. Ikada, *Prog. Polym. Sci.*, 2003, **28**, 209.
122. H.-F. Ji, E. Finot, R. Dabestani, T. Thundat, G. M. Brown and P. F. Britt, *Chem. Commun.*, 2000, 457.
123. H.-F. Ji and T. Thundat, *Biosens. Bioelectron.*, 2002, **17**, 337.
124. J. J. Headrick, M. J. Sepaniak, N. V. Lavrik and P. G. Datskos, *Ultramicroscopy*, 2003, **97**, 417.
125. N. V. Lavrik, C. A. Tipple, M. J. Sepaniak and P. G. Datskos, *Biomed. Microdevices*, 2001, **3**, 35.
126. C. Ziegler, *Anal. Bioanal. Chem.*, 2004, **379**, 946.
127. Y. Yang, H.-F. Ji and T. Thundat, *J. Am. Chem. Soc.*, 2003, **125**, 1124.
128. A. Pockels, *Nature*, 1891, **43**, 437.
129. A. Sella, *Chem. World*, 2015, **11**(issue 6, June), 36.
130. Lord Rayleigh, *Philos. Mag.*, 1899, **48**, 321.
131. H. Devaux, *Annu. Rep. Smithsonian Institute*, 1913, 261.
132. I. Langmuir, *J. Am. Chem. Soc.*, 1917, **39**, 1848.
133. G. L. Gaines, Jr., *Insoluble Monolayers at Liquid-Gas Interfaces*, Interscience, New York, 1966.
134. K. Ariga and J. P. Hill, *Chem. Rec.*, 2011, **11**, 199.
135. I. Langmuir, *J. Am. Chem. Soc.*, 1917, **39**, 1848.
136. D. G. Dervichian, *J. Chem. Phys.*, 1939, 7, 931.
137. I. Langmuir and V. J. Schaefer, *J. Am. Chem. Soc.*, 1938, **60**, 1351.
138. D. R. Talham, *Chem. Rev.*, 2004, **104**, 5479.
139. S. Fabiano and B. Pignataro, *Chem. Soc. Rev.*, 2012, **41**, 6859.
140. C. J. Brinker, Y. Lu, A. Sellinger and H. Fan, *Adv. Mater.*, 1999, **11**, 579.
141. M. C. Cross and P. C. Hohenberg, *Rev. Mod. Phys.*, 1993, **65**, 851.
142. T. P. Bigioni, X. M. Lin, T. T. Nguyen, E. I. Corwin, T. A. Witten and H. M. Jaeger, *Nat. Mater.*, 2006, **5**, 265.
143. T. S. Shabi, S. Grigorian, M. Brinkmann, U. Pietsch, N. Koenen, N. Kayunkid and U. Scherf, *J. Appl. Polym. Sci.*, 2012, **125**, 2335.
144. H. Y. Erbil, *Adv. Colloid Interface Sci.*, 2012, **170**, 67–86.
145. W. Han and Z. Lin, *Angew. Chem., Int. Ed.*, 2012, **51**, 1534.
146. H. Hu and R. G. Larson, *J. Phys. Chem. B*, 2006, **110**, 7090.
147. S. Herminghaus, K. Jacobs, K. Mecke, J. Bischof, A. Fery, M. Ibn-Elhaj and S. Schlagowski, *Science*, 1998, **282**, 916.
148. R. Xie, A. Karim, J. F. Douglas, C. C. Han and R. A. Weiss, *Phys. Rev. Lett.*, 1998, **81**, 1251.
149. R. van Hameren, P. Schön, A. M. van Buul, J. Hoogboom, S. V. Lazarenko, J. W. Gerritsen, H. Engelkamp, P. C. M. Christianen, H. A. Heus, J. C. Maan, T. Rasing, S. Speller, A. E. Rowan, J. A. A. W. Elemans and R. J. M. Nolte, *Science*, 2006, **314**, 1433.
150. J. M. Lehn, *C. R. Chim.*, 2011, **14**, 348.

151. A. Ge, Q. Peng, H. L. Wu, H. Liu, Y. Tong, T. Nishida, N. Yoshida, K. Suzuki, T. Sakai, M. Osawa and S. Ye, *Langmuir*, 2013, **29**, 14411.
152. R. S. Clegg and J. E. Hutchison, *Langmuir*, 1996, **12**, 5239.
153. D. Yokoyama, A. Sakaguchi, M. Suzuki and C. Adachi, *Appl. Phys. Lett.*, 2008, **93**, 173302.
154. D. Yokoyama, A. Sakaguchi, M. Suzuki and C. Adachi, *Org. Electron.*, 2009, **10**, 127.
155. E. Gomar-Nadal, M. M. S. Abdel-Mottaleb, S. De Feyter, J. Veciana, C. Rovira, D. B. Amabilino and F. C. De Schryver, *Chem. Commun.*, 2003, 906.
156. J. Puigmartí-Luis, A. Minoia, A. Pérez del Pino, G. Ujaque, C. Rovira, A. Lledòs, R. Lazzaroni and D. B. Amabilino, *Chem. – Eur. J.*, 2006, **12**, 9161.
157. P. Iavicoli, M. Simón-Sorbed and D. B. Amabilino, *New J. Chem.*, 2009, **33**, 358.
158. S. Marčelja, *J. Chem. Phys.*, 1974, **60**, 3599.
159. A. Blumstein and O. Thomas, *Macromolecules*, 1982, **15**, 1264.
160. E. Ramos-Lermo, B. M. W. Langeveld-Voss, R. A. J. Janssen and E. W. Meijer, *Chem. Commun.*, 1999, 791.
161. F. Tao and S. L. Bernasek, *Chem. Rev.*, 2007, **107**, 1408.
162. S. Mohnania and D. Bonifazi, *Coord. Chem. Rev.*, 2010, **254**, 2342.
163. M. Jurow, A. E. Schuckman, J. D. Batteas and C. M. Drain, *Coord. Chem. Rev.*, 2010, **254**, 2297.
164. J. Otsuki, *Coord. Chem. Rev.*, 2010, **254**, 2311.
165. C. E. Hoyle, T. Y. Lee and T. Roper, *J. Polym. Sci., Part A: Polym. Chem.*, 2004, **42**, 5301.
166. *Strucural Adhesives: Chemistry and Technology*, ed. S. R. Hartshorn, Plenum Press, New York, 1986.
167. A. J. Kinloch, *Adhesion and Adhesives*, Chapman and Hall, 1987.
168. Y. Liu, L. Mu, B. H. Liu and K. L. Kong, *Chem. – Eur. J.*, 2005, **11**, 2622.
169. L. K. Ista, V. H. Pérez-Luna and G. P. López, *Appl. Environ. Microbiol.*, 1999, **65**, 1603.
170. H. Yang, B. Yuan, X. Zhang and O. A. Scherman, *Acc. Chem. Res.*, 2014, **47**, 2106.
171. R. J. Stewart, T. C. Ransom and V. Hlady, *J. Polym. Sci., Part B: Polym. Phys.*, 2011, **49**, 757.
172. J. D. Swalen, D. L. Allara, J. D. Andrade, E. A. Chandross, S. Garoff, J. Israelachvili, T. J. McCarthy, R. Murray, R. F. Pease, J. F. Rabolt, K. J. Wynne and H. Yu, *Langmuir*, 1987, **3**, 932.
173. L. Grill, *J. Phys.: Condens. Matter*, 2008, **20**, 053001.
174. A. Ulman, *An Introduction to Ultrathin Organic Films: From Langmuir–Blodgett to Self-Assembly*, Academic Press, Boston, 1991.
175. Procter and Gamble, Ltd. Compositions for Inhibiting Metal Tarnish, British Patent 625339, 09 December 1947.
176. M. Finšgar and I. Milošev, *Corros. Sci.*, 2010, **52**, 2737.

177. B. V. A. Rao, K. C. Kumar, M. Y. Iqbal and B. Sreedhar, *Ind. J. Chem. Technol.*, 2009, **16**, 25.
178. E. Fekete, B. Pukánszky, A. Tóth and I. Bertóti, *J. Colloid Interface Sci.*, 1990, **135**, 200.
179. R. A. Walker, K. Wilson, A. F. Lee, J. Woodford, V. H. Grassian, J. Baltrusaitis, G. Rubasinghege, G. Cibin and A. Dent, *Sci. Rep.*, 2012, **2**, 880.
180. C. Wang, C. Piao, X. Zhai, F. N. Hickman and J. Li, *Powder Technol.*, 2010, **200**, 84.
181. J. Moczo, E. Fekete and B. Pukanszky, *J. Adhes.*, 2002, **78**, 861–875.
182. W. El Malti, D. Laurencin, G. Guerrero, M. E. Smith and P. H. Mutin, *J. Mater. Chem.*, 2012, **22**, 1212.
183. X. T. Shi, I. Bertóti, B. Pukánszky, R. Rosa and A. Lazzeri, *J. Colloid Interface Sci.*, 2011, **362**, 67.
184. A. L. Litvin, S. Valiyaveettil, D. L. Kaplan and S. Mann, *Adv. Mater.*, 1997, **9**, 124.
185. E. M. Landau, M. Levanon, L. Leiserowitz, M. Lahav and J. Sagiv, *Nature*, 1985, **318**, 353.
186. E. M. Landau, S. G. Wolf, M. Levanon, L. Leiserowitz, M. Lahav and J. Sagiv, *J. Am. Chem. Soc.*, 1989, **111**, 1436.
187. I. Weissbuch, G. Berkovic, L. Leiserowitz and M. Lahav, *J. Am. Chem. Soc.*, 1990, **112**, 5874.
188. J. Kuther and W. Tremel, *Chem. Commun.*, 1997, 2029.
189. J. Kuther, R. Seshadri, W. Knoll and W. Tremel, *J. Mater. Chem.*, 1998, **8**, 641.
190. A. Bejarano-Villafuerte, M. van der Meijden, M. Lingenfelder, K. Wurst, R. M. Kellogg and D. B. Amabilino, *Chem. – Eur. J.*, 2012, **18**, 15984.
191. A. Bousquet, H. Awada, R. C. Hiorns, C. Dagron-Lartigau and L. Billon, *Prog. Polym. Sci.*, 2014, **39**, 1847.
192. M. R. Topka and P. H. Dinolfo, *ACS Appl. Mater. Interfaces*, 2015, 7, 8053.
193. H. Bellanger, T. Darmanin, E. Taffin de Givenchy and F. Guittard, *Chem. Rev.*, 2014, **114**, 2694.
194. Z. Chu and S. Seeger, *Chem. Soc. Rev.*, 2014, **43**, 2784.
195. D. Rana and T. Matsuura, *Chem. Rev.*, 2010, **110**, 2448.

CHAPTER 2

Tools for Surface Characterisation of Surface-based Supramolecular Systems

2.1 Overview of Analysing Surfaces and Interfaces

If we are able today to perform supramolecular chemistry at interfaces in a rational way, it is almost certainly because of the huge advancement in the analytical tools that are available for probing these regions of space and the theoretical approaches to model them that are increasingly precise. In many cases, it is now not overly challenging to obtain information on the packing, bonding and orientation of molecules at an interface using these techniques, although single techniques alone are very rarely capable of providing a molecular level insight of supramolecular structure. The use of calculations to model experimental observations are required frequently. Many of the guiding principles of molecular chemistry at surfaces were determined by what we might consider rudimentary techniques today, but perhaps it is useful to recall that the initial observations by Blodgett and Langmuir, the pioneer of surface assembly, were relatively simple bulk measurements. Now, studies are performed with force, electrons and photons to give precise information of biological, physical and chemical performance at the single molecule level.[1]

So that the analyses of the supramolecular systems that are detailed in the coming chapters can be appreciated, it is important to have at least qualitative insight into the measurements used to characterise them. When carrying out research into molecules adsorbed under surfaces, it is useful to understand the requirements, limitations and information that can be gleaned from each particular characterisation technique. For example,

Monographs in Supramolecular Chemistry No. 19
Supramolecular Chemistry at Surfaces
By David B. Amabilino

Published by the Royal Society of Chemistry, www.rsc.org

characterisation of solid–liquid interfaces presents some interesting challenges that can be overcome with a number of techniques[2] that are not present for vacuum–solid interfaces. Each has its own margins in terms of resolution, the characteristics of the substrate and the damage that they will inflict upon the surface when they are performed.

2.2 Calorimetry

The thermodynamic characteristics of adsorption give fundamental information in a system comprised of adsorbent and adsorbate. In a supramolecular sense, the true energetic character of binding of adsorbent to adsorpt is revealed, averaged over the whole sample. The molar heat of adsorption is the value that is obtained from calorimetry experiments. This value includes adsorpt–adsorpt and adsorpt–adsorbate interactions. If a solvent is included in the system, evidently all the interactions of this component of the mixture with adsorpt and adsorbent also play a role.

Mixing calorimetry was initially used to study the adsorption of alkanes to graphite, where it was seen that the heat of wetting increased with increasing alkyl chain length.[3] Furthermore, it was found that the heat of adsorption decreases as temperature is increased and that long-chain alcohols have a higher heat of adsorption than the corresponding alkanes.[4] Recently, flow calorimetry was used with heptane solutions of long-chain alkanes and their alcohol, bromide, carboxylic acid and thiol derivatives to study the effects of interactions between all the components of the system including the adsorbent graphite.[5] While it was found that calculated lattice energies for layers of alcohol, thiol and alkane on the substrate ranked their enthalpy of adsorption in the correct order, with the alcohol most strongly adsorbed, the fatty acid and bromide had lower enthalpies of adsorption than expected. This fact was explained by the influence of solution-phase interactions with the adsorpt, and demonstrates the importance of taking into consideration all interactions between the components of the system in order to understand the adsorption processes.

Calorimetry provides a unique insight into the enthalpic origins of the adsorption process, and it is expected that further advancements in cell designs will lead to measurements being carried out on extremely small samples. Isothermal adsorption microcalorimetry has been used to great effect for the study of adsorption of surfactants to polymers[6] as well as glass and graphite.[7] The value obtained does not provide detail concerning the orientation of the molecules on the surface, and other techniques will always be necessary in order to study this feature. It is equally true that inferred and qualitative kinetic and thermodynamic hypotheses concerning adsorption often lack this quantitative value. Indeed, it has been used to establish structure–property relationships in industrially important coatings.[8] There is the opportunity to be able to study phase transformations on surfaces at a macroscopic scale using calorimetric techniques.

2.3 Optical Microscopy

2.3.1 Polarising Optical Microscopy (POM)

Direct observation of monolayers using traditional light microscopy is not possible because the features are below the diffraction (Abbe) limit of visible light. The observation of larger structures in multilayers and thin films may be possible, such as the formation of cracks and so on. The techniques are also useful as a preliminary screen of the formation of adsorbed structures on surfaces by evaporation from cast samples.

To view optically the ordering of the azimuthal tilt angle (how the adsorpts' long axis is angled with respect to the normal to the plane of the interface) in Langmuir–Blodgett layers, a supramolecular passage of order to liquid crystals has been exploited.[9] The birefringence of nematic liquid crystals can be observed in a straightforward manner with a polarising optical microscope, and the directionality of the domains in the mesophases is very often determined by the nature of the surface on which they are applied: the order at the interface of surface and liquid crystal is carried up to the bulk fluid. In the case of Langmuir–Blodgett monolayers of 1-monopalmitoyl-*rac*-glycerol, star-shaped domains (that had previously been observed by atomic force microscopy and were approximately 700 microns in diameter) were seen in the polarising optical microscope with the brightness of the domains being dependent on the relative orientation of the liquid crystal domain and polarising filter orientation. In this case, the smallest tilt domain size was approximately 5 microns. Clearly, a limitation is that the domain sizes in the monolayer have to be sufficiently big to be resolved using the optical microscope.

2.3.2 Fluorescence Microscopy

Fluorescence microscopy is of widespread use in the analysis of biological systems, where light is used to excite chromophores embedded in a disperse way and the emitted light is recorded. Especially confocal microscopy—where specific planes of the sample can be selected precisely by inserting a pinhole to eliminate the light coming from planes above and below the confocal plane—has been used to study biological and synthetic luminescent samples. More recently, super-resolution microscopy has come to the fore, where the same chromophore emission characteristics are looked at in more detail using the non-linear response of the units to incident light. Related techniques, involving transfer of energy resolved in time, are being employed widely for the tracking of fluorophores in biological tissue.

Confocal microscopy has been employed with fluorescence correlation spectroscopy[10,11] for studying phase separation in bilayers of biological relevance.[12] The method requires the employment of two fluorescent probes and spatial and temporal mapping of their distribution in the lipid. The three-dimensional reconstruction of the images from different planes allowed proof of the exact superposition of like-phase domains in monolayers on different sides of the membrane, something unachievable until that point.

In situ fluorescence microscopy was identified relatively early on as a powerful technique for the observation of monolayers at the air–water interface.[13] It can be used in conjunction with other methods and simulation to determine the organisation of molecules in the Langmuir layers.[14] Fundamental mechanical properties of the layers can also be determined.[15]

The main limitation of fluorescence microscopy is the interference in imaging of proximal molecules: imaging monolayers of fluorophores with very precise resolution using optical microscopy is, for the moment at least, not possible. However, when surfaces are patterned, fluorescence imaging can be tremendously useful for determining adsorption to selected regions of a substrate. As an example, patterned self-assembled monolayers with methyl-terminated thiolates (hydrophobic regions) 10, 30, 60, and 90 microns wide separated by 120 microns of protein-resistant regions were coated with a protein that adsorbed only to the methyl-terminated regions (fibronectin).[16] Cells were adsorbed to the surface, and after fixing and staining with fluorescein-labelled phalloidiovine were visualised by both inverted microscopy and fluorescence microscopy. This approach can be applied to the selective adsorption of a wide range of fluorescent systems.

2.3.3 Brewster Angle Microscopy (BAM)

The observation of monolayers on a liquid is readily studied using Brewster angle microscopy (BAM), which can even determine the phase character of different areas in layers, in real space and time.[17,18] The technique works by aiming a light source (usually p-polarised) at the interface and observing the reflection at the Brewster angle (approximately 56°) where the in-plane polarisation (p-polarised) component is zero for water. Therefore, the pure liquid gives no signal because there is no reflection at the Brewster angle. A microscope records the light reflected from the water–air interface when film affects the polarisation of the light and is recorded, in an analogous way to transmission polarised light microscopy. The change in polarisation could be caused by the change of density of the liquid or molecular order. Since the initial designs, compact models have become available[19,20] and now the technique is common for the observation of layers on water's interface with air.[21]

The technique is especially useful for the characterisation of Langmuir layers prior to their transfer onto substrates, as a real idea of film morphology can be gleaned by the *in situ* measurement. For example, the phase diagrams of molecular clips and tweezers (used for molecular recognition of guest molecules) were determined and their morphology at various surface pressures was observed, where spherical and acicular aggregates were observed in the liquid condensed phase.[22] The technique is also capable of discerning multilayers formed in the Langmuir films. The liquid crystalline compounds of the family 4′-*n*-alkyl(1,1′-biphenyl)-4-carbonitrile show multilayers with the formation of spiral-shaped defects in the drop-shaped domains that they form, although the derivation of the thickness is not trivial.[23] The technique does come into its own in the observation of metastable phases that are generated either in spreading or at intermediate pressure intervals during compression of Langmuir monolayers. The amphiphile *N*-hexadecanoyl L-alanine methyl ester and its racemate show an extremely rich display of aggregate morphology when spread under slightly different conditions at the air–water interface.[24] To take one variable, the droplet size of the solution in which the surfactant is applied influences greatly the nature of the first-formed aggregates (Figure 2.1). Smaller drops tend to form isolated branched aggregates, while larger drops, with more material present, tend to form much larger acicular objects that are also branched regularly. Three distinct asymmetric dendritic growth patterns were observed overall. The imaging also showed that the parent *N*-hexadecanoyl L-alanine changes from platelet formation at 298 Kelvin to a dendritic growth mode at 303 Kelvin. The racemates show completely different behaviour. Hydrogen bond formation was implicated in the formation of all these aggregates. Thus, the BAM reveals a level of detail at the macroscopic level *in situ* that not only shows the self-assembly during the film growth process, but is also useful in determining the degree of structural change that can take place during the formation of Langmuir–Blodgett layers. The two-dimensional organisation of streptavidin, which goes through phases that have bright and dark contrast depending on their density,[25] shows how quantitative analysis of BAM contrast can lead to new insight into the behaviour of materials *in situ* at the air–monolayer–water interface.

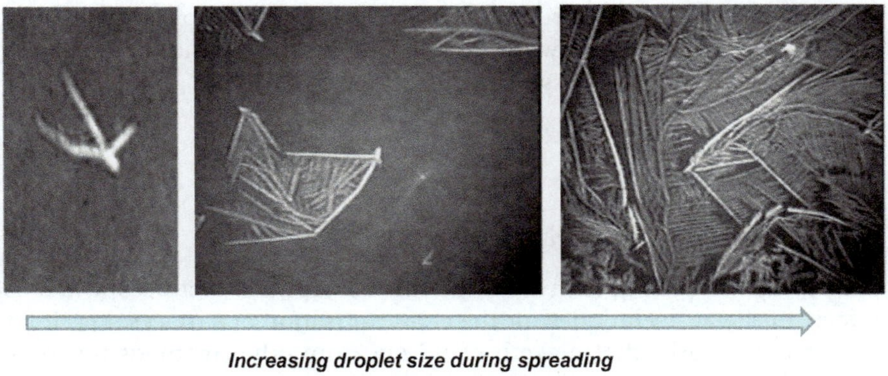

Figure 2.1 Brewster angle microscopy images of *N*-hexadecanoyl L-alanine methyl ester at the air–water interface showing the aggregates formed from increasing droplet size upon spreading.
Reprinted with permission from ref. 24. Copyright 2005 American Chemical Society.

2.3.4 Scanning Near Field Optical Microscopy (SNOM/NSOM)

The resolution limit of light in the far field can be overcome by observing objects from a very short distance, much smaller than the wavelength of light, where the evanescent wave can be detected. Evanescent waves are non-radiative waves formed at interfaces, and are responsible for total attenuated reflection infrared spectroscopy. In the case of the scanning near field optical microscope (SNOM/NSOM), all the contrast tricks of operation for optical microscopy can be used for observing objects at a surface using a tip that is located precisely very close to the sample. In addition, near-field spectroscopy involving either Raman or fluorescence can be employed, although in general these methods require very specific preparative procedures. In addition, the difficulty of measuring samples and avoiding artefacts has meant that the method has had limited general application. There are noteworthy studies that do demonstrate the potential of this group of measurements.

The technique has been used in conjunction with atomic force microscopy (see later) for the observation of lignin's supramolecular structure.[26] The transmission-mode SNOM gave images that showed both hollow and layered supramolecular globular structures in a sample of the lignin model compound. Langmuir–Blodgett layers can be characterised using the technique, as it shows (relatively large scale, sub-micron) changes in morphology upon molecular switching in these layers[27] (see Chapter 8). In the area of chemisorbed self-assembled monolayers, SNOM showed that a layer of a fluorescent Zn(II) Schiff-base complex chemisorbed on a silicon (100) surface showed luminescence similar to that shown by the isolated chromophore in solution.[28] Cast thin films of the same compound did not exhibit any emission. The results lead to the conclusion that the uniform and ordered structure in the SAM give a unique photoluminescent behaviour in a condensed state, not accessible by solution processing. They also show the potential of the method for fluorescent systems that although even single molecules can be observed with these techniques,[29] have not been fully exploited in the area of self-assembled monolayers to date.

2.4 Spectroscopy

2.4.1 Absorption Spectroscopy

It is perhaps surprising to find the humble absorption spectroscopy in a list of state-of-the-art techniques for the characterisation of surfaces. Yet, it has some real strengths that make it stand out from other methods for monitoring film growth. It is readily available, cheap, non-destructive, fast and quantitative. Obviously, the technique is only applicable to substrates that are transparent to the absorption bands of the adsorbate that are being probed to monitor the thickness of the layers, such as glass, indium tin oxide, and the like.

It is especially appropriate for monitoring the growth of films using the layer-by-layer technique. For example, when a pyridyl ligand was used in the growth of multilayers through coordination to either ionic palladium or colloidal palladium, the growth at each step of the process could be monitored by the intensity of the absorption band.[30] The inter-chromophore interactions could also be determined because the J-type aggregates (where molecules are stepped supramolecularly) or H-type aggregates (where the molecules are stacked directly on top of one another) have different absorption maxima. Furthermore, should different coordination environments be present, giving rise to metal-to-ligand charge transfer bands at different positions in the spectrum, the relative abundance of each could be evaluated.[31]

The reflection of light from Langmuir layers at variable angles has also been employed to study the relative orientations of chromophores with respect to the interface.[32] The technique has been used effectively to study the orientation of amphiphilic dyes, and reveals the change in molecular orientation as a function of the surface pressure.

2.4.2 Reflection-Absorption Infrared Spectroscopy (RAIRS)

Infrared spectroscopy of molecules on surfaces gives information on the nature of both the non-covalent bonds as well as molecular conformation of the adsorbates, and can be defined as one of the key non-destructive techniques for certain types of monolayer. Reflection-absorption infrared spectroscopy (RAIRS)—or infrared reflection-absorption spectroscopy (IRRAS)—works by shining a laser beam off a surface and recording the infrared absorption of the reflected beam, where the angle of incidence and reflection are identical. In general, light with polarisation either parallel (p-polarised) or perpendicular (s-polarised) to the plane of incidence is used. The intensity of the vibrations that are observed depends on their orientation with respect to the surface and the polarisation of the light beam. The functional group specificity and richness of IR spectroscopy in general makes the RAIRS/IRRAS method of great utility for the study of supramolecular effects in monolayers, and has the advantage of being a non-destructive method.

At the air–water interface, the technique can be used *in situ* for the characterisation of monolayers,[33] more often than not looking at the order of aliphatic chains.[34] It is of enormous interest in the study of many supramolecular systems, from determining lipid–protein interactions[35] to the binding of metal ions by cyclic hosts[36] in this interface. It can also reveal structural changes that take place during compression of the Langmuir layers, and processes such as monolayer collapse.[37]

On metal surfaces, remembering that the RAIRS spectra only detect vibrational modes with a component of the transition dipole moment oriented vertically to the interface, information on bonding and orientation can be obtained. For example, the UHV study of the amino acid L-alanine on copper (110) showed rich spectra leading to the assignment of two chemisorbed phases.[38] At low coverage, a thermodynamically stable phase characterised by the symmetric ν_s(OCO, 1411 cm^{-1}) and the scissor δ_s(OCO, 771 cm^{-1}) vibrational modes indicated that the carboxylate oxygen atoms were equidistant from the surface. Also, for this phase, the NH_2 wagging band at 1145 cm^{-1} was intense, implying that this group lies parallel to the surface. These observations indicated that all three coordinating atoms were in contact with the copper. At higher coverage, the spectrum is very different, with new bands, most notably that corresponding to the asymmetric stretch of the carboxylate group, ν_s(OCO, 1630 cm^{-1}), indicating that when there is more material on the surface, this moiety tilts up from the surface with one of the two oxygen atoms coordinated to the copper. This initial study was borne out by subsequent in-depth studies using a battery of equipment.[39]

The technique can also be used to study the effects of electrochemistry on chemisorbed monolayers, as for 2-(11-mercaptoundecyl)hydroquinone on gold.[40] For this sample, strong signals between 1700 and 1100 cm^{-1} corresponding to the terminal quinone moiety were recorded using p-polarisation measurements, while s-polarisation did not give any signals for potentials lower than +1.2 V. Above +1.4 V, monolayer decomposition arose, the original bands decreased in p-polarisation but the s-polarisation spectra revealed them. Stretches of alkylsulfonic acid were observed, showing that the adsorbate had been oxidised. Therefore, the technique is sensitive not only to orientation, but to compositional change in dynamic monolayers.

2.4.3 Raman Spectroscopy and Surface Enhanced Raman Spectroscopy (SERS)

Raman spectroscopy is also very powerful for the characterisation of molecules adsorbed on certain substrates, metals in particular. Especially, surface enhanced Raman spectroscopy (SERS) is extremely sensitive, but can only happen at the surface of a metal (the photon couples to the charge density oscillations of the metal, the plasmon) where the dipoles of the adsorbate interact in an appropriate manner with the surface wave.[41] The sensitivity of the technique is sufficient to make it applicable for use in sensor devices, particularly for environmental and biological applications.[42] However, the technique can also be very informative for the structure of self-assembled monolayers on metals as well as their supramolecular chemistry.

Early on, the SERS spectra of adsorbed disulfide compounds with either phenyl or benzyl substituents on silver showed that the molecular fragments lie either in the plane for the phenyl derivative or perpendicular to the surface for the benzyl compound, and importantly confirmed the rupture of the sulfur–sulfur bond during chemisorption to give the sulfides.[43] That particular work was seen as relevant to the use of these compounds as lubricants where the chemistry of the disulfide compounds was not known when in contact with metals. Studies of the diphenyldisulfide on gold also showed that the compound is cleaved upon adsorption, although the diselenide and naphthalene disulfide do not cleave upon adsorption.[44] Therefore, non-trivial effects concerning reactivity at metal surfaces can be probed with great effect by SERS.

SERS has proved particularly useful in the study of monolayers formed on roughened platinum by cyclodextrins bearing pyridyl groups, as well as their mixed monolayers with 1-nonanethiol.[45] The SERS spectrum of the monolayer of the pyridyl-substituted cyclodextrin gave strong peaks at 1035, 1209, 1550, and 1606 cm^{-1} corresponding to the adsorbed pyridine linkers. No significant peaks from the cyclodextrin macrocycle were observed, meaning that the host was oriented away from the surface. When the alkyl thiol was introduced, the resulting SERS spectrum showed both the presence of the thiol and the pyridyl groups, demonstrating that the adsorbate remained on the surface. Somewhat in contrast, the same class of host adsorbed on gold gives a SERS spectrum with carbohydrate peaks in the range 2100 to 2300 cm^{-1} showing that the macrocycle is in close proximity to the surface. It also provides the opportunity to use this signal to report any re-orientation of the cyclodextrin on the adsorbate. This effect was indeed observed upon annealing of the sample. Furthermore, the spectroscopy showed the complex formation between monolayers of cyclodextrin and fullerenes.[46]

2.4.4 Sum Frequency Generation Vibrational Spectroscopy (SFGS)

SFGS is a non-linear optical-based method that can record the vibrational spectra of molecules at interfaces between two bulk phases because of the inherent polarity.[47–49] Analysis of the spectra gives the average tilt angle of the molecules with respect to the normal to the interface plane, the polar orientation and even molecular conformation.[50] It involves combining visible and infrared light in the same space and time, resulting in a coherent beam with vibrational optical peaks. The chiral variant of SFGS can measure the arrangement of enantiomeric compounds at interfaces.[51,52] It has been employed particularly for observing the assembly of peptide-derived materials at interfaces,[53,54] although there is no reason why synthetic supramolecular systems should not be probed more frequently using the same method.

The technique has been used quite extensively to study self-assembled monolayers assembled on solid surfaces as well as Langmuir–Blodgett layers, where the aliphatic chains generally employed are used to determine conformation and orientation of chains, and can even detect adsorption of counter-ions.[55] The spectra provide more specific and complementary information to RAIRS: for example, all-*trans* long-alkyl chains observed present methylene signals that are either IR or Raman active, but not both because of the local centre of symmetry. So, SFGS is very sensitive to the groups at the ends of the chains and to conformational defects.[56]

By way of an example applied to a supramolecular system, the SFGS technique has been used to observe the formation of polar multilayers formed by complementary hydrogen bonds on a silica substrate.[57] The non-covalent network formed from derivatives of melamine and barbituric acid gave spectra in which the methyl signal was particularly strong, indicating the polar order with this moiety pointing "up" from the layer (a random up–down orientation would not give rise to a signal), and broadening of the signal from the symmetrical resonance of the methylene groups showed a very low density of *gauche* defects in the chains. This spectroscopic signature provides a detail that is difficult to derive from any other technique alone. In addition, the development of ultrafast nonlinear coherent vibrational sum-frequency spectroscopy techniques gives important information concerning the thermal conductance of molecules at interfaces, and while technically challenging, will give further and unique insight into fast dynamic processes at interfaces.[58]

2.5 Scanning Probe Microscopy

2.5.1 Atomic Force Microscopy (AFM): Introduction

From the determination of a nanostructure's morphology to the measurement of surface charges and even conducting pathways in materials, this family of probe techniques is tremendously useful for the characterisation of the organisation of molecules, monolayers and thin films. In its most simple variant, a fine tip made of a usually quite hard material (the radius is typically a few tens of nanometres, but can be smaller) is passed across a surface through a piezoelectric positioning device that raster scans (from the Latin *rastrum* for rake) across the surface (Figure 2.2). The deformation of the

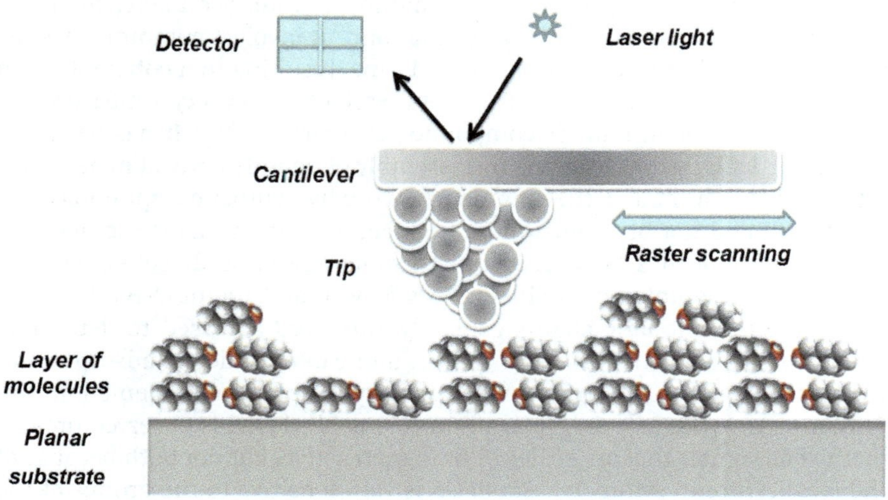

Figure 2.2 A cartoon of an atomic force microscope tip on a cantilever on a layer of molecules on a surface, and the light that is used for recording the image.

cantilever to which this tip is attached is monitored using reflected laser light that is targeted on the back-side of the cantilever with respect to the tip. For precise information to be obtained from any AFM experiment, the surface under study should not be too rough and it is better if the substrate is atomically flat. No molecular level information will be available if the substrate's roughness is bigger than the molecular dimensions. Some of the many modes of AFM are described in turn in the paragraphs that follow. While the AFM techniques are relatively soft, their use can damage surfaces. Measurement of wide areas before and after close-up inspection usually reveal any mechanical (or other) damage to the surface. Indeed, as we shall see, the technique can be used to remove material from a surface in order to judge its depth.

2.5.2 Topographic AFM Analysis

The "bread and butter" AFM analysis is where the microscope provides a topographical map of the surface through contact of the tip with the sample. The microscope measures the height of the protrusions from the surface very accurately (the z direction, perpendicular to the plane of the surface), while the lateral resolution (x and y) is limited by the size of the tip and "convolution" typically arises. This effect is caused by the similar magnitudes of object and tip, and arises because the points around the apex of the tip touch any object before said apex actually reaches it, giving a false impression of width and breadth. There are two main modes that are typically performed: contact mode and intermittent contact mode.

In contact mode, the tip is dragged across the surface by the piezoelectric device and any deviation is measured by deflection of the laser beam on the back-side of the cantilever to the tip. This mode affords precise information on the topography provided the sample is hard enough to resist the mechanical force of the tip dragging across it: often for supramolecular systems on surfaces, this variant is not ideal, as it can often remove material from the surface while scanning. Apart from damaging the surface, the material that is picked up by the tip generates noise, impeding satisfactory imaging. For this reason, other modes are applied more usually to the relatively soft molecular systems.

The intermittent contact mode (often referred to as "tapping") involves the tip of the microscope "touching" the surface as it is lifted up and down perpendicularly to the plane. Practically, this measurement is made by vibrating the tip at a frequency of several kilohertz and measuring the changes in the position of the tip as it is raster scanned across the surface. This method causes far less damage to the sample than the contact mode.

AFM measurements on a surface can be performed in either vacuum, air or liquid. There are no restrictions concerning the nature of the sample, provided that it does not react chemically with the tip. This kind of reaction is a very rare occurrence when unwanted, although it can be sought and used in certain instances to probe molecular properties.

The depth of self-assembled monolayers is impossible to judge using an AFM because only the surface of the monolayer is observed, unless some kind of defect is introduced. An AFM tip can be used, if dragged with enough force, to remove molecules from a surface. Effectively, the tip is used to scratch the surface. Therefore, it can be used as a kind of lithography. The use of AFM-derived techniques as patterning tools will be covered in Chapter 9.

2.5.3 Force Measurements Using an AFM

The AFM does not have to be used as an imaging tool alone, it can be used as a mechanical tool to test the strength of interactions between molecules through friction in the plane or the resistance to elongation when a molecule is attached to the tip of the microscope. In the first case, the friction of a tip of an AFM and the surface can be quantified, and the mechanisms of friction, lubrication and wear—the tribological properties—determined. The results are comparable to bulk measurements of these characteristics but the AFM provides a nanometre-scale view of the effects. This has been done for LB films and chemisorbed structures, showing that the latter are more robust, while the LB films are obviously more prone to wear.[59] Other devices have been used to study the tribological properties of self-assembled monolayers, such as the pin-on-disk tribometer.[60] On the other hand, the force curve of the AFM tip—the force exerted on the cantilever during approach and retraction from the surface—provides information on elasticity, hardness and surface charge.[61] In general, the AFM comes into its own when measuring samples with very different properties in close proximity (on the scale of nanometres) where the phase signal of the AFM image can be used qualitatively to map chemical functionality.

An area where great advances have been made in the understanding of host–guest interactions is in the measurement of force spectroscopy curves for the determination of binding constants between hosts (or guests) in a monolayer and guests (or hosts) attached to an AFM tip (Figure 2.3). In these experiments, the tip is brought into contact with the surface so that host and guest can bind, and the tip is retracted until the tip breaks free from the surface that it was adhered to through the supramolecular interaction, so that the rupture or pull-off force is measured. This force, usually of the order of picoNewtons, gives a measure of the interaction between host and guest directly. Multiple repetitions of this force spectroscopy are necessary to provide statistical weight to the stochastic measurements.

In this way, a variety of host moieties (ferrocenyl, phenyl, adamantyl, and so on) attached to an AFM tip through a thiolate–gold chemisorption were bound to a surface with chemisorbed cyclodextrin derivatives.[62] The forces required in this case were in the range 39 to 102 pN, and the quanta were loading (approach) rate independent showing that fast complexation/decomplexation events occurred. The values for the rupture force had the same trend as the free binding energy for complexation determined in

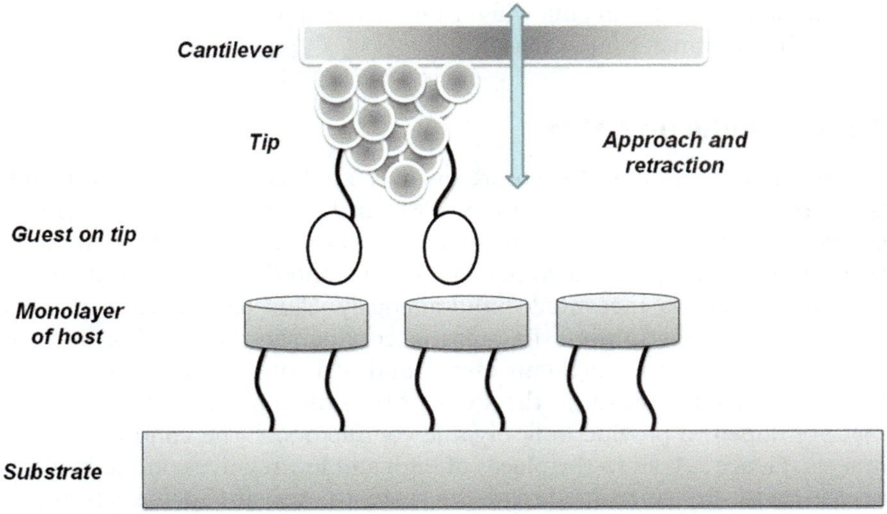

Figure 2.3 A cartoon showing how host–guest interactions can be determined using an atomic force microscope tip with the guest attached to the tip and a layer of host molecules on a surface.

solution on model compounds. This kind of measurement has been extended to a wide range of supramolecular systems, including polymers, and it gives a unique view of the binding between complementary moieties that can be measured in different media.[63,64]

2.5.4 Kelvin Probe Force Microscopy

Kelvin probe force microscopy is one of the non-contact AFM modes where surface potential is imaged.[65,66] In the non-contact modes, the changes in frequency of tip vibration during interaction with the surface can be measured or the amplitude of the tip can be measured at a constant frequency. Electronic characterisation of the surface of organic thin films by Kelvin probe force microscopy is a non-destructive method that maps the local electric potential difference between the tip and the sample. Therefore, the technique (and related non-contact modes) is excellent at providing contrast between donor and acceptor materials in a thin film.

The technique is useful for supramolecular materials where very small differences in height might be apparent although large differences in potential are present.[67] This could be as simple as the small difference in length or packing of alkyl chains chemisorbed to a metal surface,[68] or the electron donor or acceptor properties of an organic material could be imaged, giving great insight into the nature of the aggregates from the point of view of their electronic characteristics. Blending of donor and acceptor materials is of great interest for solar cells, and Kelvin probe force microscopy provides a map of the distribution of the donors and acceptors not only in their ground state, but when irradiated can provide maps of the areas where charge transfer takes place.[69,70]

2.5.5 Conducting AFM

When the tip that is used is covered with a conducting material and when the surface is conducting, then the AFM can be used to generate maps of current passing through the material.[71,72] The tip is scanned across the surface with a bias potential applied between it and the material under study and the current is measured simultaneously. This technique is of huge importance in determining the areas of conductivity in samples, and can provide insight into the homogeneity and structure of materials. It has proven particularly useful in the study of thin films of supramolecular gels that are doped to provide bulk organic conductors.[73] The conducting AFM images (Figure 2.4) show the places of high and low relative conductivity in a xerogel, and the microscope can be used to measure potential–current sweeps at any point. The resolution of the technique is down to molecular width, as appreciated in the close-up image in Figure 2.4 where individual supramolecular wires can be seen. The technique can also be used to measure the changes taking place in a conducting organic material of this type under an applied potential in real time.[74]

2.5.6 Scanning Tunnelling Microscopy (STM): Introduction

STM is the imaging technique *par excellence* for supramolecular systems on surfaces. Despite its limitations—of which more later—it is the only technique able to image individual parts of molecules, in other words to give sub-molecular resolution, at every point on a surface. This resolution reaches the atomic level frequently. Individual defects, domain boundaries and impurities are revealed readily. In either "topographic" or

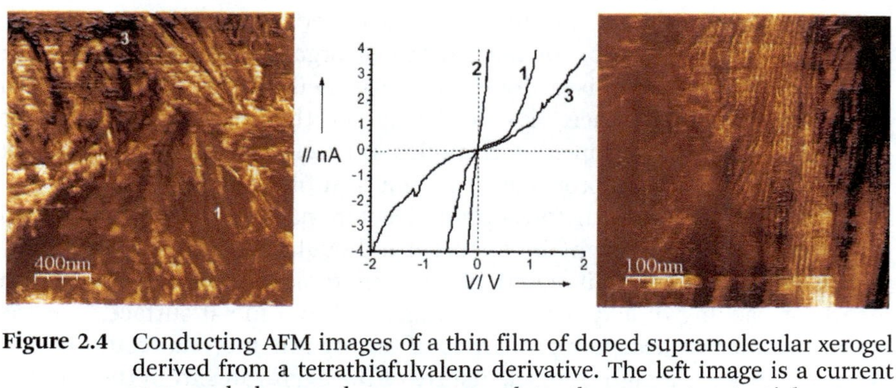

Figure 2.4 Conducting AFM images of a thin film of doped supramolecular xerogel derived from a tetrathiafulvalene derivative. The left image is a current map and the numbers correspond to the current–potential sweeps, which give an indication of the conductivity in the different areas. The image on the right shows how individual supramolecular fibres can be resolved; the lines that can be observed are one molecule wide.

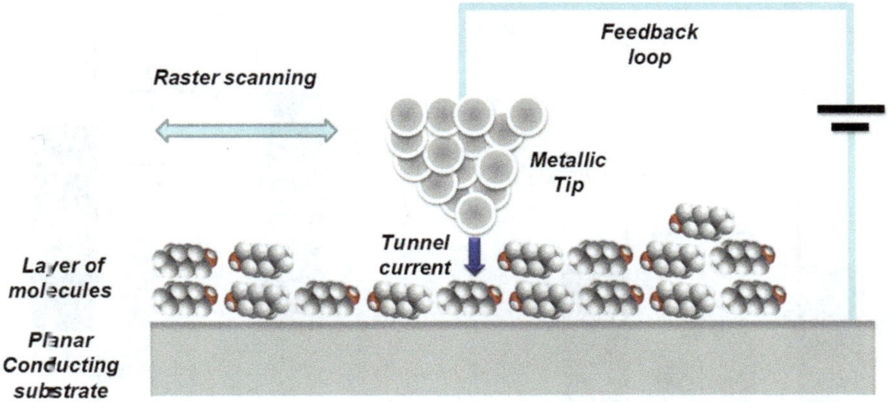

Figure 2.5 A cartoon showing the operation of a scanning tunnelling microscope, where bias is applied between an electrically conducting tip and a conducting surface and the current varies as a function of the material between the two.

electrochemical mode (see later), the microscope employs a metallic tip separated a few picometres from a conducting surface to observe molecules that are adhered to it (Figure 2.5). Herein lies one of the limitations of the technique: a conducting surface is necessary to perform the experiments. In principle, the tip does not actually physically touch the surface. The images that are obtained are a result of the varying current arising from the tunnelling electrons across the tip/adsorbate/adsorbent area, therefore no extra detector is required as in AFM. The tip is raster scanned across the surface with a piezoelectric device and tunnelling current is measured. STM provides a bright contrast—corresponding to a relatively high tunnelling current—from moieties that are able to favour electron tunnelling and/or are closer to the tip of the microscope.

The STM technique is extremely sensitive to the nature of the chemical groups that are in-between conducting tip and surface. It also depends on how these chemical groups interact with the surface, because the tunnelling current depends on the separation between the two conductors (tip and bulk metal), how the orbitals of the molecules are located with respect to the two, and the bias potential used. This situation means that the same chemical functionality at different absorption sites can give different tunnelling currents and therefore different contrast in the two positions. Often, the molecular orbitals of relatively large molecules can be seen individually, and by varying the bias voltage, the HOMO or LUMO can be imaged.[75–77] The way this works is shown schematically in Figure 2.6. As a negative potential is applied to the tip, so its Fermi level rises and when coinciding with the energy level of the LUMO electrons will be able to tunnel through the molecule much more efficiently than when the energy levels are not aligned. When a positive bias is applied to the tip, the Fermi level of the tip is lowered, and when it coincides with the HOMO so tunnelling efficiency through

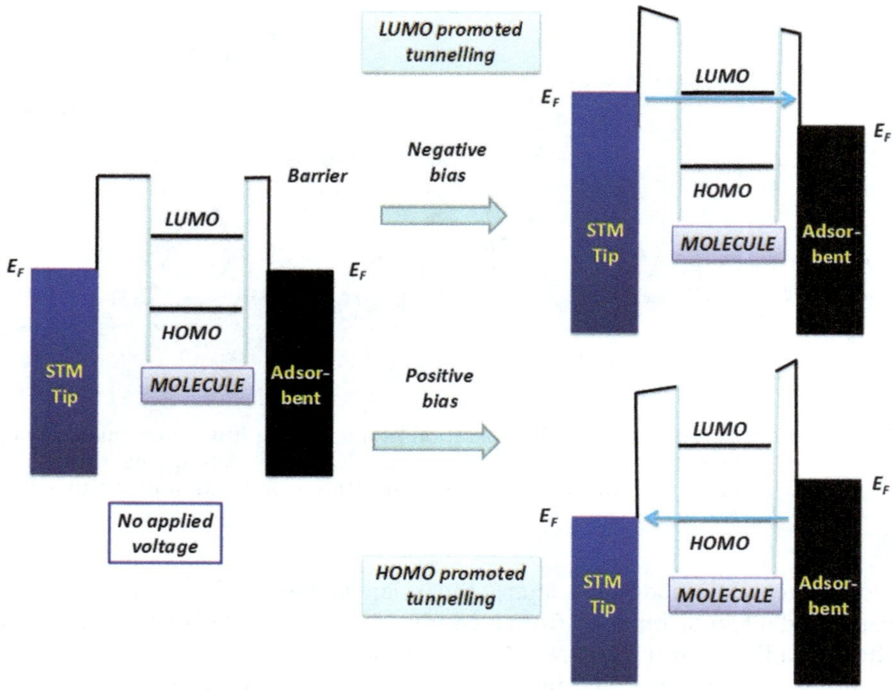

Figure 2.6 A representation of the elastic tunnelling routes in molecules adsorbed on a conducting surface with a metallic tip whose bias with respect to the surface is changed. In this hypothetical example, when no bias is applied, the Fermi levels of tip and surface (E_F, that here have the same energy for simplicity) lie between the HOMO and LUMO energy levels (that are presented as a well in the barrier between tip and adsorbent here). When a negative bias is applied to the tip, the Fermi level rises, and electrons tunnel through the LUMO. Conversely, when a positive bias is applied, electrons tunnel from surface to tip through the HOMO.

this orbital becomes more effective. Therefore, it is clear that one can be observing orbitals selectively rather than the physical shape of the molecule in STM.[78]

The principle limitation of STM is that the adsorbent has to be a bulk conductor. This requirement means that the material to which the molecules under study are adsorbed must be good conductors of electricity. The adsorbent must also be atomically flat, so that the tip of the microscope does not hit the substrate (this would be seen as streaky noise in the image). Typical substrates include, but are not restricted to, silver, gold, graphite, copper, nickel, or molybdenum disulfide. Obviously, some of these must be performed either in an inert atmosphere or under UHV conditions to avoid degradation of the surface.

The STM technique can be carried out under ultrahigh vacuum (UHV) conditions, on "dry" samples under room conditions in air, as well as well as in solution. In UHV, the technique is perhaps at its best, as the system is completely "clean" and no influence of solvent or adhered vapour or gas to the surface can affect the measurements. Additionally, under these conditions, the temperature of the adsorbent can be varied over a very wide range. Adsorbents that are sensitive to the atmosphere (such as metals) can be studied with no influence of oxidation or water adsorption. When measurements are performed in air, adsorption of water vapour should be considered and the adsorbent must be relatively inert (as well as being conducting). These problems can be overcome by performing the STM in a liquid that covers the adsorbent, with the adsorpt in solution. The liquid should either be poorly conducting (in the case of standard measurements) or in an electrolyte (in the case of electrochemical STM, see later).

2.5.7 Purely "Topographical" STM

When the STM tip is brought into electrical feedback by lowering it very close to the conducting surface, two modes can be used to register images: constant height or constant current mode. In the former, the tip is held at a constant height over the sample and raster scanning is performed (in the *x,y* plane) with measurement of the change in current intensity giving the contrast in the resulting image. In constant current mode, the tip is again raster scanned but the piezoelectric controller moves the metallic tip in the *z* direction (perpendicular to the plane of the surface) so as to try and maintain a constant current in the circuit.

It should be noted that an identical chemical functionality in different positions on a conducting surface can appear with different contrast because this technique is sensitive to the electronic coupling between the surface and the molecule at an extremely local level.[79] Therefore the term "topological" is used in quotation marks because—unlike AFM—true topography is not observed: there is a huge influence of the electrical characteristics of the space between tip and adsorbent on the signal being measured. As a result, the *z* distances given in STM images are very approximate, and can even be misleading.

The very high resolution of STM allows measurement of the positions of single atoms. This feature is extremely useful for obtaining the reference axes of crystalline adsorbents. The orientation of the adsorbent can be measured and the relative orientation of any adsorbed molecules can therefore be determined. In the case of physisorbed layers, the tip can even be used to measure the surface atoms after measuring images of complete monolayers on top of the adsorbent by simply lowering it and "sweeping away" the layer.

In UHV, STM can be used to measure the characteristics of single molecules with submolecular resolution[80,81] as well as supramolecular systems where individual components can be discerned. Also, very fine layers of insulating material can be used on the surface of the adsorbent, allowing remarkable resolution in the STM images of the compounds, as for the case of pentacene deposited on a two monolayer thick sodium chloride adsorbent on copper (111).[82,83] Frequently, imaging is not enough (as the positions of the molecules are not immediately obvious), and then theoretical modelling of both adsorption and images comes into its own (see Section 2.11 of this chapter) and the marriage of theory and experiment gives a precise picture of the molecular geometry at the surface.[84] The advantage of UHV STM experiments is the huge range of temperatures that can be explored, allowing locking of the supramolecular structures of even very weakly adsorbed molecules.

STM can also be performed under ambient conditions. In air, the uncontrolled nature of the atmosphere in contact with the sample and other factors often leads to noisy images, however, under a liquid or solution, STM can provide images of remarkable quality. The main limitation of the liquid is that it should be as insulating as possible, and also should not evaporate during the timescale of the experiment (while single images take seconds or minutes, sessions recording different domains can last several hours). The self-assembly of molecules at the solution–solid interface is perhaps the most interesting environment from a supramolecular perspective,[85] and here the layers that are formed are revealed by imaging that shows how molecules pack and implies the non-covalent interactions that can take place.[86] The system is under equilibrium, and therefore molecules can adsorb and desorb on the timescale of the experiment, and processes such as Ostwald ripening can be observed often.[87] The technique is sensitive to the orientation of the molecules, in that molecules lying flat on the surface appear with similar contrast to any surrounding aliphatic chains, as for a tetrathiafulvalene derivative substituted with alkyl groups[88] (Figure 2.7). On the other hand, if non-covalent interactions between molecules are strong compared with the molecule–surface interaction, the rigid parts of the molecule can be oriented quasi-perpendicularly to the interface, as seen for a tetrathiafulvalene with hydrogen bonds between its lateral chains[89] (Figure 2.7).

2.5.8 Electrochemical STM

Electrochemical STM is performed in an electrolyte solution using the tip and the sample of the classical STM as the working electrodes that are controlled against an independent reference electrode (the difference between the potentials of these is the tunnelling bias voltage). The circuit is completed with a counter electrode. Thus, external voltage and tunnel voltage can be controlled independently.[90] Although the technique is not used widely for the study of supramolecular systems, the ability to image molecules at interfaces under these conditions provides a unique submolecular resolution view of adsorbed systems at different electrochemical potentials.[91] For example, the organisation of 2,2′-bipyridine on a gold (111) surface in a sodium perchlorate solution (0.1 M) shows a remarkable transition from a disordered to an ordered state, that is aligned with the gold surface axes.[92] The method has also demonstrated the mesoscopic

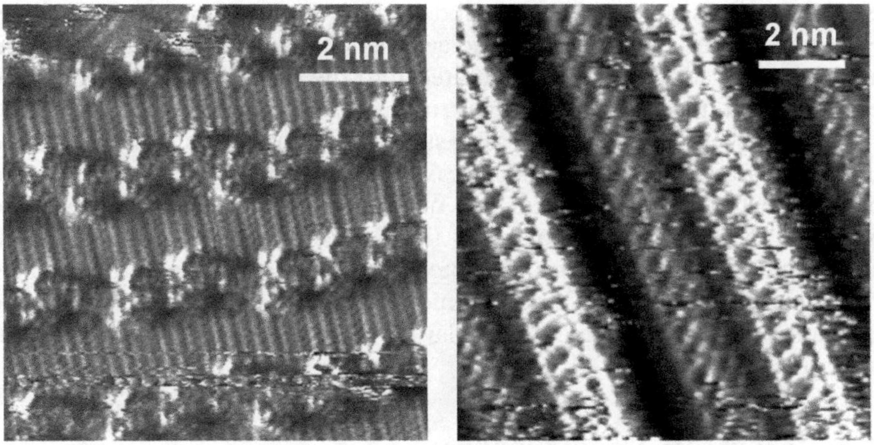

Figure 2.7 STM images (where the bright areas correspond to a high tunnelling current, the large bright areas being tetrathiafulvalene moieties) of molecules lying down on a graphite surface (left), where the alkyl chains can be clearly seen and are interdigitated, and an adsorbed tetrathiafulvalene that stands on edge on graphite because of strong hydrogen bonds between the molecules (right). The aromatic portions appear much brighter than the alkyl chains (that are abutted because of the orientation of the molecule) in this case.

organisation of a self-assembled organometallic cyclophane 3.1 nm long and 1.2 nm high at a gold (111) interface with 0.1 M $HClO_4$.[93] The organisation of the ring was face down on the gold electrode, a big contrast to the side-on organisation seen when the material was physisorbed onto graphite.

2.5.9 Scanning Electrochemical Microscopy (SECM)

SECM is an electrochemical technique where the "tip" is an "ultra-microelectrode" that is scanned at an interface,[94,95] where either current or potential are measured while scanning the tip across the area. As many electrochemical techniques, it is performed in a fluid medium, and can operate either at solid–liquid or liquid–liquid interfaces. The technique is extremely powerful to probe electron transfer processes across monolayers, membranes or particles.[96] The technique provides a map of Faradaic current flow under electrochemical conditions, as opposed to tunnelling current although this can be involved if a complete monolayer isolates the electrolyte solution where the redox process takes place from the bulk electrode surface. Even ionic liquids can be used as the fluid, which is interesting for the study of these materials at interfaces.[97] Because one electrode is used as the tip, the lateral resolution of the experiment is limited compared with other scanning probe techniques, to several hundred nanometres.

One of the most remarkable uses of the technique was to measure the electron transfer rates through alkanethiols chemisorbed on gold electrodes.[98] The experiments included the quantification rates of direct electron transfer through the monolayer, as well as electron transfer mediated by redox active units incorporated into the monolayer. The technique can be used to measure changes in monolayer nature, as shown for the binding of metal ions to self-assembled monolayers of DNA, where the incorporation of metal ions increases electron transport.[99] An advantage of this technique over others for measuring decay of current passage with the nature of the monolayer is that steady state conditions are used, so no charging current or ohmic potential drop over the layer takes place. On a more applied note, the technique can also be used to probe the efficiency of self-assembled monolayers to protect copper metal from corrosion.[100,101]

2.6 Electron Spectroscopy

2.6.1 X-Ray Photoelectron Spectroscopy

This vacuum-based technique determines the chemical bonding taking place at interfaces, and is of great utility in establishing the nature of interactions between substrate and adsorbate as well as the composition of layers. X-Ray photoelectron spectroscopy (XPS, also known as electron spectroscopy for chemical analysis, ESCA) affords spectra that give precise elemental information, in addition to indications regarding the valence of each element. It is also very sensitive to the presence of impurities in the film, although on the downside, radiation damage for molecular systems means that it is a destructive technique. While qualitative assertions can be made from the spectra, in the cases when the valencies of elements change slightly the peak positions, then accurate fitting of the spectra is required to make a quantitative analysis of the data. The technique is especially powerful when covalent modifications are made at a surface. XPS is routinely used for the characterisation of self-assembled monolayers on gold,[102] where the thiolate bound to gold and physisorbed thiol or disulfide can be distinguished.[103] Similarly, conceptually, the interaction of hydroxamic acids on metal oxides was probed.[104] The peak corresponding to the nitrogen 1s orbital showed that on titanium and aluminium oxide, the monolayers were mainly physisorbed, because the peak coincided with that of the pure amphiphile, albeit with a shoulder. This shoulder corresponded to a chemisorbed monodeprotonated state that is clearly the main specie in the monolayers on copper and silver oxides.

Angle-resolved XPS can be used to determine the angles of molecules with respect to the surface plane. Varying the angle of detection gives information at different depths in the layers. In this way, for example, the orientation of molecules in self-assembled monolayers formed by ω-(thiophene-3-ylalkyl) phosphonic acid on different metal oxide surfaces was determined, and the proximity of the polar head group to the inorganic surface was proven.[105]

2.6.2 Auger Electron Spectroscopy

Auger electron spectroscopy (AES) provides an elemental composition of a surface. The principle of AES is that a high energy electron generates a hole in a low lying orbital in an atom (it is not the electron in the HOMO that is removed). That hole is filled in an intra-atom process where a higher energy electron occupies the space, liberating kinetic energy that is transferred to a second electron—the Auger electron—that is emitted (thereby generating a doubly charged atom). The K, L and M nomenclature often used to describe the spectra correspond to 1, 2 and 3 in the chemist's nomenclature for electron shells. The technique probes the first two or three atomic layers of a surface. It is an extremely sensitive technique that can afford quantitative information concerning surface composition.

The technique is particularly useful in the case of changes of surface functionality. Figure 2.8 shows three AES plots corresponding to sequential

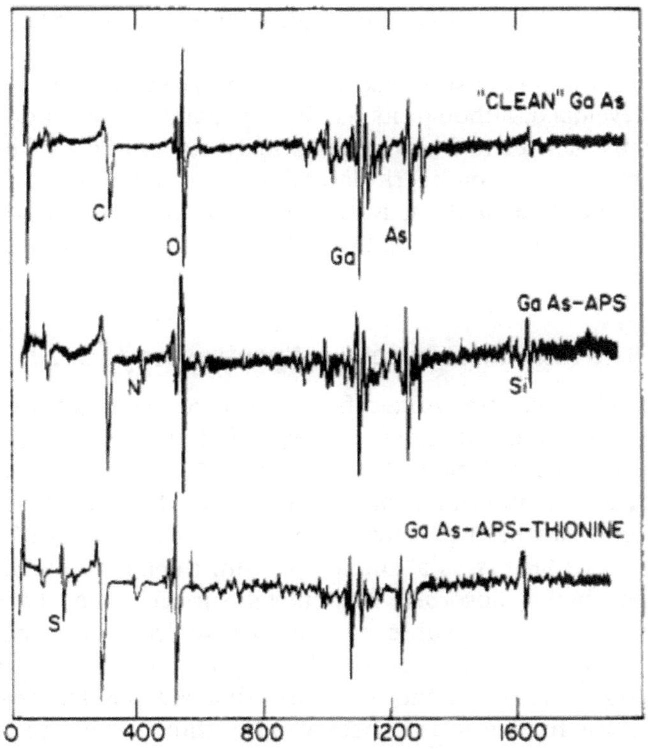

Figure 2.8 Auger spectra of a clean GaAs substrate, and films formed on this surface by reaction with an organic siloxane with an amine terminal group (middle trace) that is subsequently reacted with an isothiocyanate derivative of thionine dye.
Reprinted with permission from ref. 106. Copyright 1978 American Chemical Society.

functionalisation of a hydroxyl-terminated gallium arsenide surface.[106] The top plot shows the "clean" surface, from which it is abundantly clear that the surface is oxidised, as the oxygen peak is seen in addition to those of gallium and arsenic. Furthermore, a clear peak is seen from carbon in some form of impurity at the surface. After this "clean" surface had been treated with 3-aminopropyl-triethoxysilane (APS in the figure), the chemical change at the surface was clearly witnessed by the appearance of signals of the Auger electrons corresponding to nitrogen and silicon (middle trace in Figure 2.8). In addition, there is a clear increase in the intensity of the carbon signal. Finally, reaction of the amine group in that layer to give an isothiocyanate followed by reaction with a dye derivative leads to a film with a spectrum in which sulfur is clearly present. Therefore, the AES technique is a sensitive probe for samples, as in this case, where many surfaces are prepared in parallel. In common with other electron-based ionisation techniques, the damage to the sample will be considerable, and, unlike XPS, no information concerning the valence of the elements is obtained. Perhaps for this reason it is one of the lesser used techniques for supramolecular systems at surfaces.

Apart from the purely spectroscopic variant, scanning Auger microscopy has been developed, although its use in the characterisation of supramolecular systems has yet to be exploited widely. An example of its use is in the characterisation of porphyrin layers on silver, where it was shown that the electron beam actually activates the molecules towards chemical reaction.[107] This kind of activation by the beam could be a limitation in many systems.

2.6.3 Ultraviolet Photoelectron Spectroscopy (UPS)

When photoelectrons are emitted from molecules, some of the kinetic energy can be absorbed, giving rise to spectra that provide the energies of the orbitals in the HOMO (valence) region. The UPS technique is powerful in that it provides an indication of charge transfer between adsorbates and substrates and can be used directly to study the evolution of absorption bands and these factors as a function of film thickness. Thus, for hydroxyphenyl porphyrins adsorbed onto metal surfaces, the charge transfer interaction between the substrate and the molecule can be evaluated directly.[108] For a monolayer of porphyrins on gold (111) the spectra indicated a weak charge transfer to the molecules that are physisorbed, although this feature was hidden in the spectra of the thin film where interactions between porphyrins dominate the packing. On silver (110), a change in the spectra was observed upon annealing that, correlated with XPS data, indicated a chemical change in the adsorbate through the formation of covalent bonds.

The technique is very relevant for the investigation of molecular systems that are potentially useful for solar cells, where the interfaces between materials play a huge role in the charge transport efficiencies. The modification of indium tin oxide electrode cells with electron donors by chemisorption was optimised so as to have a high density of the electroactive molecular component, and UPS played an important role in the understanding of the improved efficiencies of charge transfer observed in these systems.[109] In this case, it showed that modified cation formation and selectivity efficiency were at the root of the improvement. In particular, UPS is useful because it determines HOMO levels in layers, whereas the ionisation potential values from cyclic voltammetry or theory are not in the same range. Indeed, the correlation between oxidation potential from electrochemistry and UPS data has been studied deeply and is a fundamentally important correlation. It has been shown that the oxidation of chemisorbed monolayers on electrodes can be predicted from the UPS-derived energy of the HOMO.[110]

2.7 Diffraction and Related Techniques

2.7.1 Electron Diffraction

Electron diffraction was actually used to study the formation of monolayers on iron, copper and aluminium as early as 1947.[111] The diffraction data showed how a long alkyl chain acid or amine adsorbed to the surfaces of polished metals, by the presence of diffuse diffraction rings giving evidence of the thickness of the films and the slight tilt that the molecules have with respect to the plane perpendicular to the surface.

The diffraction patterns obtained in a transmission electron microscope (TEM) can indicate clearly structural order. It is necessary to transfer the adsorbate material onto the TEM grid that is used for the measurement because any surface would absorb the electrons that traverse the sample in the microscope. An exemplary case is the study of Langmuir–Blodgett layers whereby the films are floated onto the TEM grid using the Walkenhorst–Zingsheim method.[112] The morphology of multilayer systems can also be studied using TEM, for example where the supramolecular structure of chlorosomes was studied[113] using the freeze–fracture technique that is usually associated with biological samples.

The crystalline order of surfaces can be probed using low energy electron diffraction (LEED), which provides precise structural information concerning the order in surfaces in general (it is a good probe for chemical restructuring of metal surfaces[114]) and especially the long-range crystalline packing of molecules on crystalline substrates. As an example of the wide use of this technique for UHV-assembled structures, free base naphthalocyanine on gold (111) gives a LEED pattern from which, through simulation, the unit cell dimensions of the two-dimensional crystal can be determined.[115] Different domains on the surface can be observed in the LEED pattern, because the molecule can pack with any of the three gold axes as the nearest to its own packing. The LEED data in this case, and in many others, was backed up by STM data, and through a combination of these, the supramolecular arrangement can be determined. In the present example, this was through van der Waals interactions between the edges of the naphthalene units of the adsorbates.

LEED can also be used to provide precise two-dimensional crystallographic type and orientation in two-dimensionally organised structures exhibiting different phases. Homochiral tartaric acid on copper (110) undergoes a clear phase transformation with increasing coverage that is revealed blatantly in the LEED pattern, and the alignment of the two phases can be determined and compared with STM data to give a clear view of the supramolecular arrangement on the surface.[116] The beam of the LEED even causes rearrangement of metastable structures *in situ*, which is the case for (*R*,*R*)-tartaric acid on copper (110) in its low coverage phase.[117] The technique is highly sensitive to the generation of homochiral layers, as shown in the use of (*R*,*R*)-tartaric acid to induce chiral crystalline domains in succinic acid monolayers on the same substrate.[118] It should be mentioned, with special relevance to determining locations of materials in mixed monolayers, that the diffraction peaks will only reveal the ordered material and no quantification of any amorphous material on the surface, for which STM comes into its own. LEED is unique in that it provides precise crystallographic information of the two-dimensional crystals with respect to the crystallographic axis/axes of the substrate.

2.7.2 X-Ray Diffraction (XRD)

The determination of structure by diffraction of X-rays is perhaps the gold standard for materials. The technique requires long-range structural order (several microns at least) and the sensitivity depends on the thickness of the sample. With these considerations in mind, it becomes obvious that the use of standard X-ray diffraction for the measurement of monolayers is extremely challenging, especially taking into account the possibility of radiation damage of the sample by the beam. However, as the number of layers increases in the sample, so diffraction with a standard source becomes viable. While standard X-ray diffraction techniques are not used frequently at present, with the advent of synchrotron light sources, they can give useful information.

The control of phase separation of compounds adsorbed at a surface can be witnessed by X-ray diffraction.[119] Albeit at 10 K, a study of the mixing of a range of *n*-alkanes on graphite revealed that while monolayers of compounds with the same space group mixed over the whole range of compositions—the case for *n*-heptane and *n*-nonane—when the compounds belonged to different space groups, they did not. This situation prevailed for mixtures of *n*-nonane and *n*-decane. The diffraction patterns of the layers give a "saw-tooth" line shape in the Bragg peaks, characteristic of two-dimensional materials.[120] The individual compounds have their packing on graphite depicted in Figure 2.9, where the *n*-nonane is commensurate with the graphite substrate, *n*-decane only being uniaxially commensurate. The diffraction patterns of the mixtures on graphite (Figure 2.9a) show that the

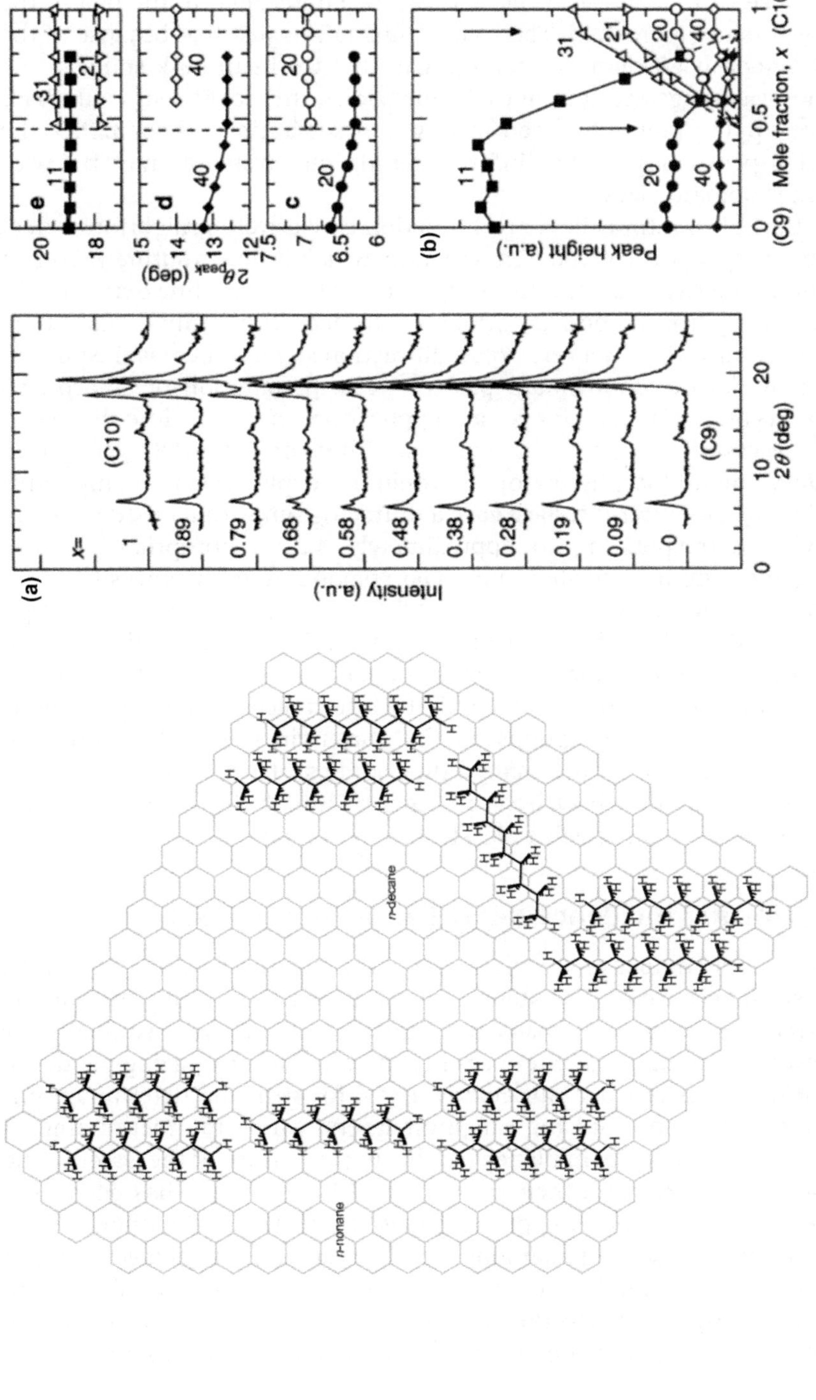

Figure 2.9 The structures of *n*-nonane and *n*-decane on graphite and (a) X-ray diffraction patterns of mixtures of the compounds on this surface. (b) Shows the peak heights as a function of composition.
Parts (a) and (b) reprinted from ref. 119. Copyright 2002, with permission from Elsevier.

diffraction peaks for either compound are at a constant intensity for compositions above 55%, then follow a linear decrease and disappear at 10%. It is clear that there are no peaks other than those of the pure components that are phase-separated on the surface, as can be appreciated in the plot of the peak heights in Figure 2.9b. The lack of mixing was justified because of the lack of a good coefficient of isomorphism that works in bulk solids,[121] although it was suggested that in two dimensions, this coefficient could have different tipping points between mixing and separation. As for LEED, the diffraction peaks only provide information on the crystalline material with significant domain sizes.

The study of thin films by X-ray diffraction is relatively straightforward, at least obtaining a powder type diffractogram, which is done routinely. To give two contrasting cases, the thermal evaporation of azoquinoline derivatives—that form hydrogen-bonded dimers—afforded thin films that were amorphous, as shown by a very broad diffraction signal, whereas the powder of the material did show sharp peaks.[122] The preparation of amorphous materials is also of interest for certain applications, and therefore this result is not necessarily negative. In contrast, films of sandwich-type mixed (phthalocyaninato)(porphyrinato) europium triple-decker compounds formed at the air–water interface gave a diffractogram with a broad signal in the region corresponding to approximately 3.5 Å, appropriate for π–π stacking and a number of sharp peaks corresponding to distances from 1.1 to 2.3 nm, depending on the derivative.[123] Therefore, the layers have a high degree of order. The diffraction data accompanied by absorption and IR spectroscopy suggested supramolecular dimers held together by π–π interaction, and these dimers self-assemble by intermolecular hydrogen bonds between phenol hydroxyl groups on the porphyrin rings. These are contrasting cases, but there are many other studies of thin films by general diffraction methods because of their availability and relatively straightforward—qualitatively at least—interpretation.

2.7.3 Grazing Incidence X-Ray Diffraction (GIXRD)

Synchrotron techniques are revolutionising the characterisation of all kinds of materials, because of the intensity of the light beam hitting the sample and effective sensitivity. Perhaps this statement holds true especially for monolayers and thin films, interfaces in general,[124] where grazing incidence X-ray diffraction is proving particularly useful for structure determination.

The method is powerful for Langmuir layers, where the measurements show the electron density in planes parallel to the surface of the water in the monolayers, and therefore their structure.[125] The technique has been used to show the region of binding of manganese(II) to fatty acid monolayers on water, and prove that the metal ions form a layer with a thickness of about 5 Å indicating tight binding to the polar head of the amphiphile.[126] The thickness corresponds to the diameter of hydrated manganese ions, and the well-defined position indicates a specific interaction. The technique has also been used to determine the structure of multi-chain amphiphilic monodendrons at the water–air interface.[127] A high degree of structural order in the compressed monolayers was indicated by the multiple, intense diffraction peaks. The structures derived from the GIXRD experiments indicate a centred rectangular lattice with molecular axes tilted 27° toward neighbours for a two-chain dendron and an oblique lattice in which the tilted molecular axes are in low-symmetry directions for a three-chain dendron. They are very different to the bulk mesophases formed by the materials, and show the particular environment that the air–water interface provides for organising molecular materials.

Both Langmuir layers and the Langmuir–Blodgett films (after transfer to a hydrophobitised silicon wafer) of amphiphilic hexabenzocoronene derivatives show well-defined supramolecular stacks of the molecule.[128] The GIXRD measurements indicated that in the monolayer on water, the molecule is probably rotated by 30° to the surface normal and 35° to the stacking direction. The diffraction measurements on the Langmuir–Blodgett 40-layer films transferred onto both quartz and silicon (previously treated with a hydrophobic siloxane) give intense peaks, but with maxima in different places: 19.9 Å and 23.3 Å, respectively. On silicon, a single peak was observed, corresponding to a single phase, macroscopically aligned along the dipping direction, as shown by performing diffraction at different orientations with respect to the surface. The molecules were stacked in an eclipsed (H-aggregate type) manner, resulting in columns. The quartz-deposited layer actually showed two peaks, corresponding to polymorphs, possibly the one observed on the water surface and the more stable one that in fact became the only one after annealing, indicating it to be a thermodynamic product. Thus, GIXRD can facilitate the assessment of structural changes during the processing of films on and from water.

2.7.4 Normal Incidence X-Ray Standing Wave (NIXSW)

When an incident wave normal to the surface of a solid is used to generate a reflected standing wave from a single crystal of the material, the wavelength assumes the periodicity of the bulk. The surface changes the reflected wave in a way that depends on the structure and composition when the photon energy passes through the Bragg condition. Inelastic scattering gives emitted photoelectrons or Auger electrons that give the position and elemental signature of the adsorbate material. Used on synchrotron X-ray sources, this technique can afford an atomic-level picture of the surface and adatoms because the intensities of the emitted electrons depend on whether they are at nodes or antinodes.[129] The value of these measurements is demonstrated in the very precise determination of the sulfur atom position in chemisorbed monolayers on a gold (111) surface, where it was demonstrated that alkyl thiols actually adsorbed to a gold adatom sitting at one of the hollow sites on the surface.[130] The sample was prepared by adsorbing hexanethiol from an ethanolic solution, followed by cleaning and transfer to the UHV chamber at a synchrotron, and ditertiary butyldisufide was introduced in the gas phase. These results on mixed layers and different alkyl chain compounds show the strength of the technique, which gives a good reason to question the classical view of adsorption in this kind of system and highlights the need for thorough characterisation to fully understand adsorption models. A subsequent investigation of butylthiolate on the same surface shows that a gold adatom is also involved in the structure, although here, two thiolates coordinate to it from opposite sides and are also attached to unreconstructed surface atoms.[131] The NIXSW measurements are quite unique in the information they provide.

For monolayers of 3,4,9,10-perylene tetracarboxylic dianhydride on gold,[132] silver[133] and copper[134] (all (111) crystal surfaces), the position of the adsorbate was determined to be approximately 3.3, 2.9 and 2.6 Å, respectively, from the metal surface plane. A charge transfer between the molecule and the metal was observed from UPS experiments, increasing from gold to silver and copper, and in concert with the NIXSW measurements, correlate the bonding strength with the distance from the surface in this adsorbed system.[135]

2.8 Ellipsometry

This non-destructive optical technique provides information concerning the structure of homogeneous films where polarisation changes in the light are measured. While precise interpretation of data can require painstaking analysis, qualitative measurements of adsorption can be achieved and are useful in a practical sense. The principal parameter that is obtained is the film thickness.

An aspect of ellipsometry that complicates the interpretation of the results is that the fitting of the data requires knowledge of the refractive index of the materials present. In principle, the measurement can be used to obtain values of both film thickness and refractive index, but for thin films, the technique is not sensitive enough. Often, the refractive index of organic films is approximated to the value of the bulk materials (about $n = 1.45$) that is convenient for films grown on silicon oxide on silicon ($n = 1.462$)[136] where a single value can be used in the film thickness calculations.[137]

2.9 Near-Edge X-Ray Absorption Fine Structure (NEXAFS) Spectroscopy

The orientation and alignment of thin films of molecules on surfaces can be probed using NEXAFS spectroscopy,[138] a synchrotron-based technique. As with most synchrotron experiments, the molecular samples are frequently cooled to avoid damage from the intense radiation. Core electrons are excited into unoccupied molecular orbitals, and the efficiency with which this is done depends on the relative orientation of the X-ray beam and the transition dipole moment (referred to as linear dichroism in adsorption of the X-rays). If a film is oriented, the intensity of the photoemission will vary for the absorption resonances. It has been used to study a wide variety of thin film structures including Langmuir–Blodgett films,[139] π-functional molecules on semiconductors,[140] and it can even reveal the differences in methods used to adsorb dyes onto cotton.[141]

A nice example is that shown in Figure 2.10 of an oligophenyl dinitrile that organises into well-defined supramolecular structures on the (111) face of silver.[142] The intense peak corresponding to the aromatic π orbitals of the carbon atoms in the aromatic groups at low angles with respect to the plane of the surface indicates a coplanar arrangement. The two signals from the nitrile group result from one of the π orbitals being in resonance with the phenyl ring and the other being perpendicular to it. Nonetheless, these signals and their variance in intensity with the angle indicate that they too are coplanar to the surface. This information was not available from any other of the assayed techniques, notably STM where the nitrile groups were not imaged.

NEXAFS can also be used in an imaging mode, even with curved samples, leading to measurements of samples at different orientations with respect to the X-ray beam.[143] Therefore, the approach is suitable for high-throughput approaches to thin films.

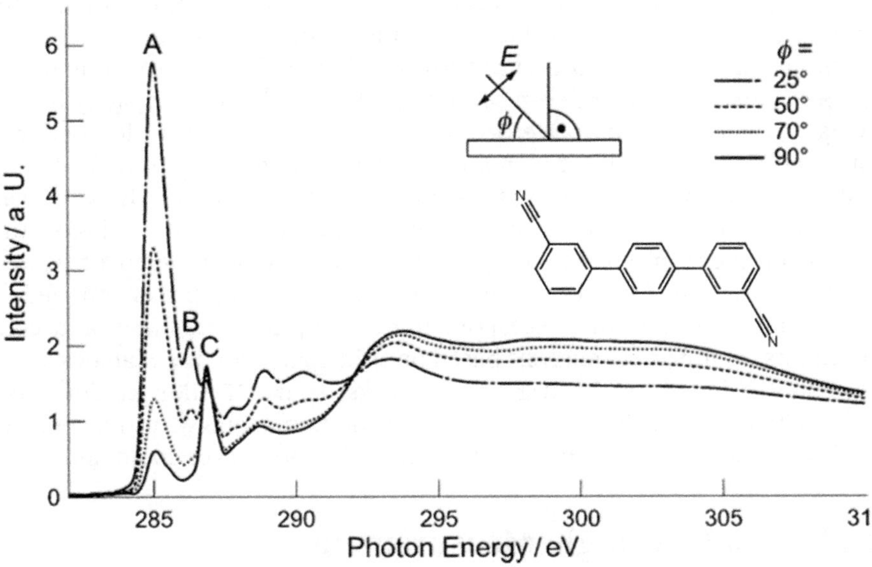

Figure 2.10 NEXAFS spectroscopy on the C1s edge of the molecule shown on a silver (111) surface. Peak A (at 285 eV) is the lowest π* resonance associated with the phenyl ring, B (286.3 eV) and C (286.9 eV) correspond to the mutually perpendicular π orbitals of the nitrile moiety. *E* shows the electric field of the polarised X-ray beam and ϕ the angle between the surface plane and the beam.
Reproduced with permission from ref. 142. J. Wiley and Sons, Copyright © 2010 WILEY-VCH Verlag GmbH & Co. KGaA, Weinheim.

2.10 Time-of-flight Secondary Ion Mass Spectrometry

This destructive technique provides chemical information on the composition of surfaces of all kinds, but especially where organic fragments are attached.[144] The method employs a pulsed ion beam to break off chemical fragments, which are extracted with an electric field and analysed using a mass detector. The calculation of the molecular weight is based on the time the ions take to pass down the flight path from source to detector. The ion beam could be based on ions such as caesium or gallium, which cause very significant damage, or so-called cluster sources based on SF_5 or SF_6, and more recently fullerene and gold clusters. The cluster systems not only reduce damage, but also provide greater sensitivity with high yield and resolution.[145]

The technique has been used to probe the assembly of metal complexes of relatively large terpyridyl ligands (based on perylene diimides) on indium tin oxide surfaces.[146] At each step of the method, which involved complexation of iron to a surface-anchored terpyridyl ligand followed by binding of a second ligand to this newly generated surface, the outermost chemical fragments could be observed. The technique does not give any indication of the number of layers on the surface though (for which absorption spectroscopy was used). TOF-SIMs does, however, allow a monitoring of surface functionalisation and an optimisation of reaction conditions by careful comparison of monolayers prior to and after complex formation.

The technique can also be used for imaging surfaces, because maps of samples can be provided, even of biological tissue.[147] This capacity is particularly interesting for patterned monolayers as well: the dot patterns formed from a chemisorbed carbohydrate using a microarray printer where a reactive monolayer on gold was coupled with a disaccharide (glycan) was mapped using TOF-SIMs.[148] This example is particularly interesting in the context of chemical analysis of adsorbates that might undergo unwanted chemical reactions upon deposition. The spectra showed peaks assigned to the immobilised glycan, the reactive underlayer, gold substrate, and contaminants. Multivariate analysis based on principal component analysis was necessary to determine the origin of the peaks, as in TOF-SIMs, parent peaks are rarely observed. The mapping showed the homogeneity of the spotting method, which is important for any activity the monolayer could show.

2.11 Contact Angle Measurements

The angle that a drop of liquid makes with a surface (Figure 2.11) is a remarkably sensitive and reliable way to judge certain characteristics of the supramolecular arrangement in films. Precisely because it is so sensitive, the interpretation of the result must be considered carefully. Control experiments and repeated measurements are necessary, and can in fact be used to test the reproducibility and variation upon preparative conditions for a certain adsorbate–adsorbent system. The polarity, roughness, and order in

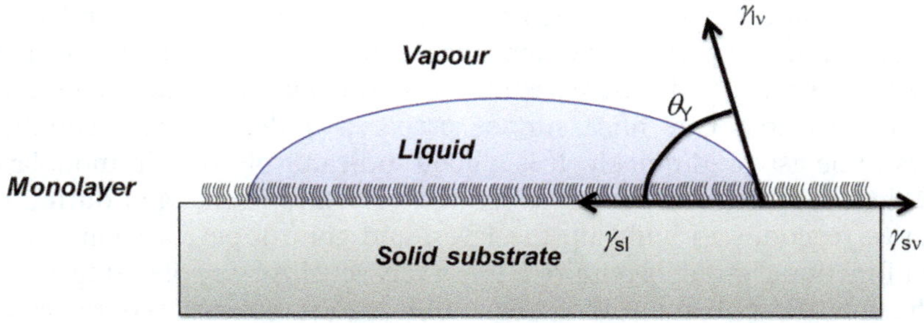

Figure 2.11 A cartoon indicating the parameters defining the contact angle (θ_Y) formed by a drop of liquid on a surface.

dry monolayers can all influence the contact angle. The same adsorbate adsorbed to different surface of the same material with the exception of their roughness will give contrasting results. The difference in contact angles may be a result purely of the non-identical roughness of the initial surfaces, but could also arise from different packing in the monolayer.

A very low contact angle would mean practically perfect wetting and a flat layer of liquid on the surface. This measurement was derived from the Young equation:

$$\gamma_{lv} \cos \theta_Y = \gamma_{sv} - \gamma_{sl}$$

These parameters are defined in Figure 2.11, in which γ_{sl} is the solid–liquid surface tension, γ_{lv} the liquid–vapour surface tension, and γ_{sv} the solid–vapour value. The Young contact angle is θ_Y, for a perfect system with an atomically flat surface. Habitually, surfaces are inhomogeneous and a range of contact angles are observed; a mean is usually given for the static drop contact angle.

Large contact angles (the maximum theoretical value is 180°) mean that the surface is not wetted at all by the liquid, and the liquid requires a significant surface tension for large values. Examples of systems with high contact angles would be water on wax or (new!) non-stick frying pans, or water on a duck's back (where hydrophobicity and roughness play an important role). The liquid used most is water. A drop of this liquid, highly pure, will give wildly different contact angles for polar or apolar surfaces. In general, more polar surfaces give quite low contact angles because the surface is wetted well by the water. On the other hand, very apolar surfaces will result in the formation of what to the naked eye could appear a perfect drop, with a contact angle far in excess of 100°. The experiments are usually performed on chemisorbed layers because the drop can affect physisorbed layers, as demonstrated for self-assembled monolayers of octadecanethiol and octadecylamine in gallium arsenide.[149] The lower contact angle for the physisorbed octadecylamine surface was ascribed to partial dissolution in the water.

The great advantage of the method over others is that it is fast and simple. The technique is particularly appropriate for chemisorbed layers in two situations especially: (i) where a control over the quality of the monolayers of an identical adsorbate on a surface prepared under different conditions needs to be assessed quickly. It is a good indicator of order in monolayers. (ii) When the chemical functionality of the surface is changed during consecutive reactions to build up the layer, and control over a change in the outer functional group is required. It is a very sensitive way of seeing changes in the outermost chemical moieties. A thorough investigation of the contact angles of chemisorbed layers on silicon oxide on a silicon wafer shows how fluoro-termination substituents provide high contact angles (approximately 110°) with water; amine and nitrile are relatively polar, and oligoethylene oxide had the lowest angle (approximately 36°) in the series.[150] The native surface had a contact angle of less than ten degrees for comparison. The work also studied a variety of organic solvents, where in general much lower angles are observed because of their lower surface tension in grand measure, although important differences were seen for all the liquids. The calculation of polar and disperse components of the surface energy from the data provides a reference for understanding the wetting behaviour of liquids on surfaces in general.

2.12 Electrochemical Methods

Self-assembled monolayers on metals have proved extremely useful for the use of electrochemistry as an analytical tool.[151,152] But electrochemistry is also a useful characterisation method for the monolayers on electrodes (or surfaces that can behave as such). The order within self-assembled monolayers can be judged by the capacitance of the surface (its ability to store charge): a well-formed tightly packed layer's capacitance should be independent of the solution composition.[153] For a well-packed layer where the thickness can be increased incrementally—such as that of alkanethiols—the dielectric constant can be calculated from the slope of the reciprocal capacitance *versus* thickness plot.[154] This effect is shown graphically in Figure 2.12, where the cyclic voltammetric (CV) plots of current *versus* potential are shown for a number of monolayers on gold.[155] Two electroactive ions and electrolyte solutions are shown to prove the generality of the approach, and either ferricyanide ($Fe(CN)_6{}^{3-}$) or hexaaquo iron(III) ($Fe(H_2O)_6{}^{3+}$) (which have opposite charge, note) in KCl or perchloric acid, respectively. As can be seen in the graph, for bare gold, an expected CV response is seen for the iron(III) reduction. Dramatically, these redox waves are shifted to more negative potential (1-butylthiol) to the point where any current is reduced dramatically. Particularly, with increasing chain length, there is a decrease in the cathodic currents for both couples. The observations are consistent with a diffusion of charges to more lightly covered surface areas where electron tunnelling is more likely than for most of the interface.[156] Therefore, the technique is almost unique

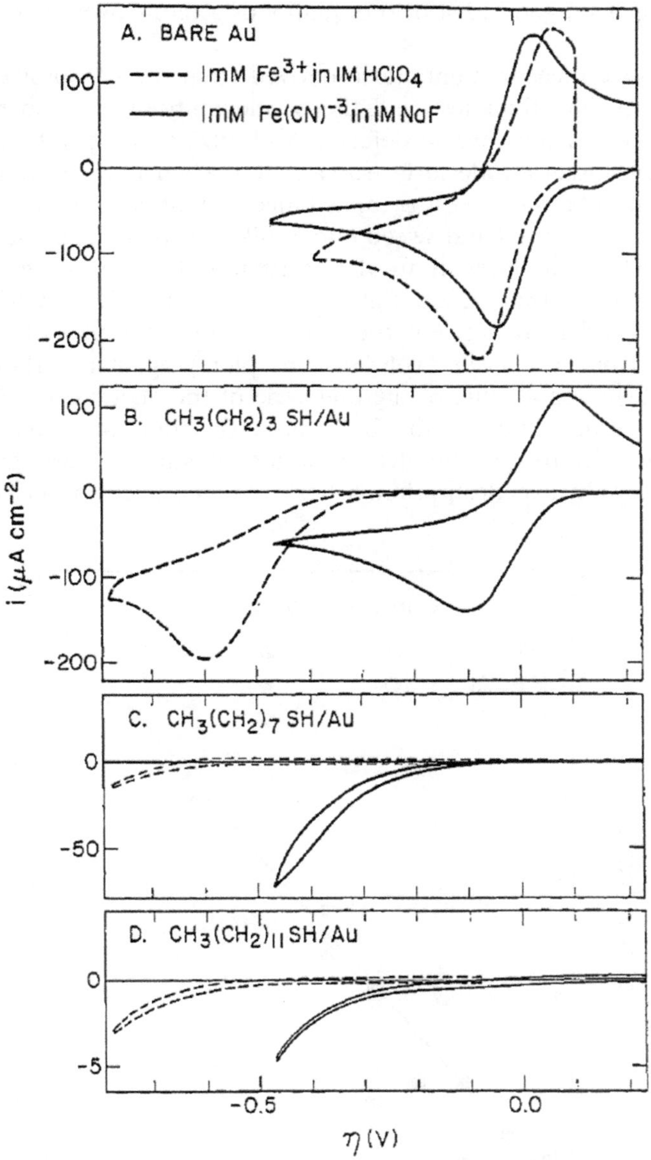

Figure 2.12 Plots from cyclic voltammetric sweeps (100 mV s^{-1}) showing current plotted against over-potential (the difference between the applied working electrode potential and the formal reduction potential) for (A) naked gold, (B) a monolayer of 1-butylthiol on gold, (C) 1-octylthiol on gold, and (D) 1-dodecylthiol on gold. The two lines in each plot correspond to 1 mM $Fe(CN)_6^{3-}$ in KCl (1 M, solid line) and 1 mM $Fe(H_2O)_6^{3+}$ in $HClO_4$ (1 M, dashed line).
Reprinted with permission from ref. 155. Copyright 1987 American Chemical Society.

in that it gives a good idea of the quality of a monolayer over a relatively large scale.

Capacitance measurements (also derived from the CV) provide an indication of the film thickness and the permeability of the film to ions (although it is less sensitive to defects). Multiplying sweep rate, differential capacitance, and electrode surface area gives the capacitive current. Taking a long-chain thiol monolayer, the capacitance current is virtually independent of electrolyte, potential and sweep rate, unlike a classic metal and more like an electrical double layer in an ideal capacitor. The layers are strongly resistant to ion penetration, especially at long chain lengths, where packing is denser and order greater. For these alkanethiols, the plot of capacitance (obtained from the cyclic voltammetry plots) against carbon number (Figure 2.13) gives a value of the dielectric of the alkanes of 2.6, which is similar to that obtained for bulk poly(ethylene). An alternative—but destructive—electrochemical determination of surface coverage of thiols on gold is reductive desorption, where the one-electron scission of the

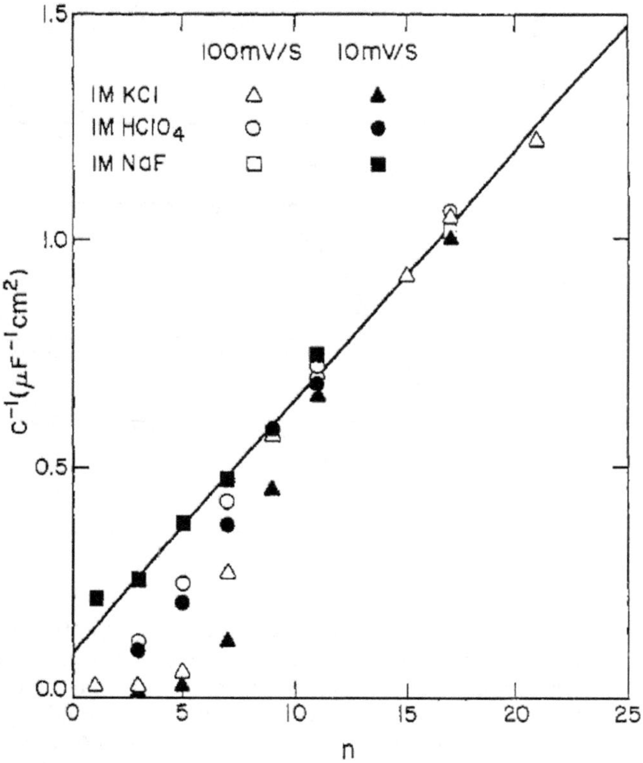

Figure 2.13 A plot of reciprocal capacitance (+0.3 V *vs.* Ag/AgCl, saturated KCl, sweep rates 100 and 10 mV s^{-1}) against carbon number for 1-alkyl thiols on gold. Electrolytes are indicated.
Reprinted with permission from ref. 155. Copyright 1987 American Chemical Society.

gold–thiolate bond in alkaline conditions leads to a wave that can be quantified.[157,158]

On the other hand, electrochemistry has been central in the development of supramolecular chemistry[159] in many molecular recognition systems,[160] and particularly responsive systems.[161] In these cases, either a shift in the redox potential indicates binding (or not) of a guest by a host (either of which can be located at the interface and be electrochemically active) or the electrochemistry is used to effect a change in molecular configuration through rearrangement of the different moieties and/or the supramolecular organisation.

2.13 Quartz Crystal Microbalance, Surface Plasmon Resonance and Other Sensors

When molecules adsorb to a surface, a number of characteristics of the interface can be modified that can be probed in order to give a response for detection. The methods are advantageous in that they do not require labelling of the analyte to give a response. While the majority of these techniques have not found widespread use for specific use in synthetic supramolecular systems—the majority are used in the fast-moving area of biomolecule detection[162,163]—two stand out: the quartz crystal microbalance (QCM) and surface plasmon resonance (SPR) have been developed greatly, exploiting self-assembled monolayers as sensors.[164] Both techniques require chemisorption of a recognition element with known activity from solution onto the modified surface, in principle, as no chemical signal or image is produced. Both techniques are effective, for example, in the detection of condensation of organic vapours when thin films of calixarenes are used as the active layer on the surface of the devices.[165]

A quartz crystal is used as a resonator in QCM, and any addition or removal of material to it will give a change in the resonance. The devices can measure adsorption of small molecule monolayers up to whole cells, and can be used in vacuum or solution.[166] The adsorption of 4-*tert*-butylcalix[4]-arenetetrathiolate onto a gold-coated QCM provided a sensor with great selectivity for alkyl benzenes.[167] The reason for this selectivity was because of the chemisorption orientation of the host ring, with the phenyl rings quasi-perpendicular to the plane of the gold surface (as shown by IR and ellipsometric analysis) and the rigidity of this state, compared with the conformational mobility that a calixarene experiences in solution. The necessity for this kind of specific adsorption was proven by carrying out control experiments with chemisorbed decanethiol and spray-coated 4-*tert*-butyl-calix[4]arene, which shows poor selectivity compared to the chemisorbed calix. The extreme mass sensitivity of this kind of sensor means that even chiral analysis with suitable chiral coatings for the modified quartz can be performed.[168] Non-specific adsorption can be problematic in these devices, and it has been shown that pre-treatment of aluminium oxide with a chemisorbed fluorocarbon adsorpt supressed adsorption of interfering agents in a subsequently coated layer of a supramolecular affinity system and improved the performance of the device.[169]

Surface plasmon resonance sensors detect changes in light reflected from the interface between a metal and a dielectric.[170,171] The charge density oscillation that gives rise to this resonance is an electromagnetic wave whose field is distributed mainly towards the dielectric. Therefore, any chemical change near the dielectric changes the optical wave, and in the multiple configurations, it is this change that is monitored. As in the QCM, selective binders of molecules are generally adsorbed to the gold and the analyte is passed in solution over the surface. The technique has led to commercial instruments that routinely determine binding based on affinity of the analyte for the host adhered to the gold in the monolayer. Above all, biotechnology has exploited the devices to the full, with high throughput screening possible[172] for the identification of host–guest interactions that can subsequently be detailed in solution. In terms of the utility of the technique in supramolecular self-assembled monolayers, there is perhaps broader untapped potential here, but the technique has proved valuable for monitoring the coordination of metal ions to self-assembled monolayers and the subsequent attachment of a functional ligand to the fresh metallo-organic surface.[173] The real time measurement in flow, with the possibility for intermediate washing steps, makes the technique very appropriate for assessing the assembling of complex structures near the surface of metals with a surface plasmon.

2.14 Molecular Modelling

The simulation of the structure of adsorption geometries, surface organisation, and even imaging appearance is often vital in the interpretation of the analytical tools; here, a brief overview to highlight its importance is given, but the increasing power and reliability of computation is providing vital insight into experiments and its utility cannot be overemphasised.

For large supramolecular systems, molecular dynamics simulations are extremely useful for obtaining sub-molecular information concerning the driving forces behind superstructure formation in systems where it is otherwise very complex.[174] The case of bistable [2]rotaxanes on gold (111) is a particularly nice example.[175] The calculations required careful parameterisation of the interactions between the adsorbate and the gold as well as checking of the predicted energies of binding of the components of the system. The work shows the packing of the molecules on the metal after dynamics and optimisation predicts the relative stability of the states of the compound on the surface, and provides the tilt of the molecule with respect to the normal to the metal surface, giving the thickness of the layer. Also, the surface tension calculated with the best organisation in the SAM for each of the two states agreed remarkably well with experimental results obtained on films.

For physisorbed supramolecular systems on graphite, the combination of molecular mechanics and molecular dynamics has been used to explore the preference for chiral molecules to form a particular handedness of hydrogen-bonded rosette (Figure 2.14).[176] Because the adsorbate is chiral, there is an energetic preference for one handedness of the rosette, which the theoretical work estimated to be approximately 81 kcal mol^{-1}. The difference in energy comes purely from the van der Waals interactions between adsorbates and adsorbates and surface, the hydrogen bonding energy was identical in the two cases. The origin of the favoured twist is the orientation of the methyl substituents on the stereogenic centre in the molecules. This ability to pick out the contributions of distinct non-covalent interactions facilitates the explanation of selectivity in assembly and also the design of new systems. Molecular dynamics can also be used for modelling the growth of physisorbed monolayers and films, a nice case being that of sexithiophene on C_{60} fullerene.[177] Adsorption of the rod-like molecule on the (001) surface of crystalline fullerene led to the formation of tilted monolayers where the long axis of the adsorbate was quasi-perpendicular to the surface. The theory showed that the nucleation and growth process on top of these monolayers passed through defective islands with chiral edges and very complex metastable arrangements, that upon heating ion the molecular dynamics up to 600 K, produced a drastic rearrangement to give multilayers with the long axis of the adsorbate parallel to the interface. This observation drives experiment, because it was suggested that if a liquid crystalline state of materials could be accessed, it could lead to large structural reorganisation on surfaces.

Force-field methods are not suitable for the determination of electronic properties, and the relatively recent development of density functional theory (DFT) has made far more ready the *ab initio* study of smaller molecules at surfaces. The studies of cobalt(II) octaethylporphyrin adsorbed as a

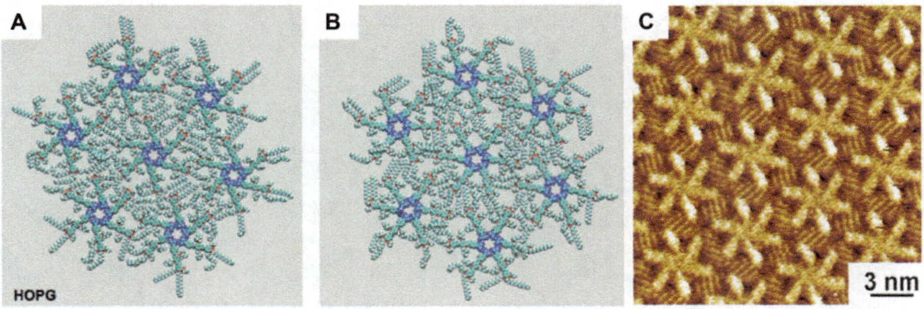

Figure 2.14 The low energy structures of monolayers of a chiral rosette-forming compound on graphite (shown as a grey background for clarity) in clockwise (A) and anticlockwise (B) arrangements. The experimental STM image at the graphite-1-octanol interface is shown in C for comparison.
Reproduced from ref. 176 with permission from the Royal Society of Chemistry.

monolayer on Au (111) and HOPG (0001) reveal the reasons for the packing on substrates and allow calculation—at different applied potentials—of the STM images that agree with those observed.[178] A periodic boundary condition—that supposes a continuum of structure in each crystallographic direction for a commensurate structure—shows that the molecules lie flat on both substrates and are located at the three-fold and six-fold centres on gold (111) and graphite (0001), respectively. A tighter binding was predicted for adsorption to gold over graphite. The work function of the conducting surfaces was observed to be lowered by the adsorbed molecule using photoelectron spectroscopy, a feature of the system that was predicted by the DFT studies.

In a more strongly bonded—but more dynamic—system, DFT proved central in the interpretation of chiral switching in monolayers of bisuccinate on Cu (110).[179] The monolayer of the di-deprotonated succinic acid on the metal is chemisorbed so that it forms chiral domains, and in these layers, molecules were observed by STM to be switching between different orientations, as can be seen in the experimental images in Figure 2.15. STM suggested that pairs of molecules were involved. DFT theory showed that it is not possible to accommodate two molecules of the original chirality of the domain in a different orientation other than the starting one. Therefore, upon reorganisation, the molecules must switch chirality. Two possible outcomes with a double chirality switch were with the two molecules adsorbed in an adjacent fashion (closed pair flip, Figure 2.15a) or separated by an atomic distance (open pair flip, Figure 2.15c) within the vacancy. Another

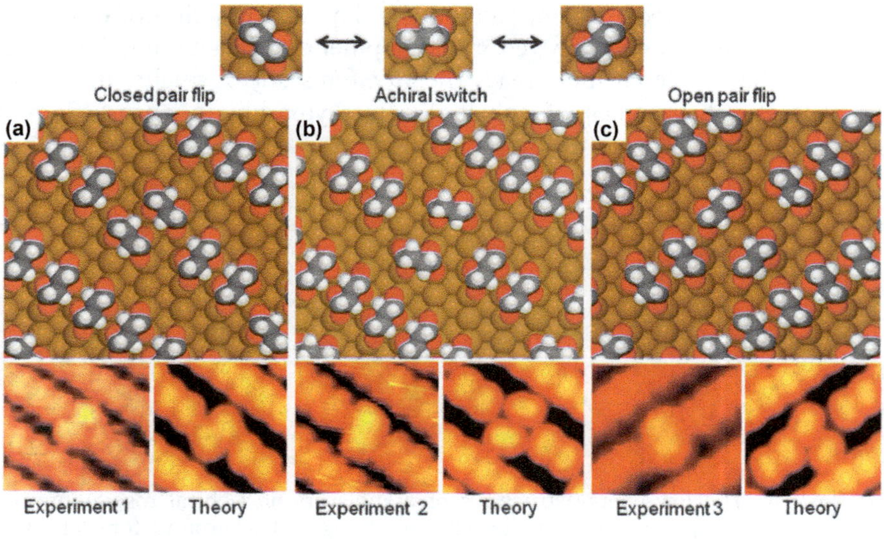

Figure 2.15 Models showing the different kinds of double molecule chiral flips in monolayers of bisuccinate on Cu (110), and the calculated and experimental STM images for each situation.
Reproduced from ref. 179 with permission from the Royal Society of Chemistry.

possible adsorption geometry is an achiral one (Figure 2.15b), the molecules being adsorbed with the oxygen atoms in a rectangular footprint on the surface. The differences in energy of adsorption were extremely small, and only slightly higher than the bulk monolayer energy, explaining the observation of all of them experimentally. It is difficult to imagine determining this explanation on a purely experimental basis on the physical monolayer.

2.15 Techniques Compared and Combined for Studies of Supramolecular Systems

The combination of techniques for interface characterisation can paint a relatively accurate picture of supramolecular organisation and bonding in this challenging environment. It is really only since the end of the 1980s that quite a wide variety of techniques has become available to look at microstructure under certain conditions.[180] Dubois and Nuzzo summarised very nicely the different techniques that have been combined to give an accurate picture of the structure of chemisorbed alkyl thiolates on metals,[181] as reinterpreted here and shown in graphical form in Figure 2.16. In more detail, Table 2.1 shows techniques and their characteristics.

The combination of experimental and theoretical tools is a hallmark of the in-depth study of supramolecular systems on surfaces. To highlight this point, and before going on to the individual cases of surface assemblies in detail, let us consider a few examples. The functionalisation of gold with chemisorbed self-assembled monolayers of 11-mercaptoundecanoic acid was recently probed in real time using PM-IRRAS, cyclic voltammetry, AFM and a quartz crystal microbalance in order to get a full picture of the development of the monolayers with the aim of having the densest layer that gave the lowest possible electrochemical signal.[182] The real-time measurements allowed the optimum immersion time for the thiol to be determined. Thus, the analytical tools become a preparative aid just as traditional solution characterisation techniques have done.

STM measurements of physisorbed layers at the solution–solid interface can give very high resolution images, but details of the orientation of molecules are often of very difficult derivation. Therefore, the study of a long-chain derivative of isophthalic acid and its deuterated analogue using both

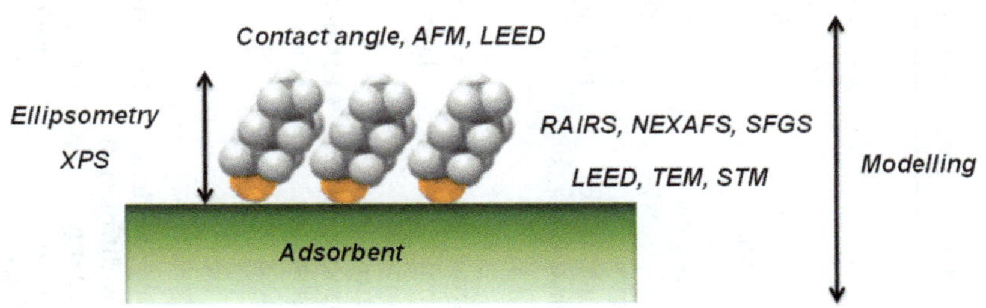

Figure 2.16 The areas of interfaces that are probed with diverse techniques.

Table 2.1 Characterisation techniques, applicability and information obtained.

Technique	Information obtained	Requirements	Ultra-high vacuum	Solution/air	Resolution
Atomic force microscopy: topographic	Morphology of outermost surface of a material	Low roughness surface	Yes	Both	Typically a few nanometres
Brewster angle microscopy	Morphology of domains, real time identification of phase changes	Mainly used at air–water interface, order or density difference to liquid	No	Air–liquid interface	Several microns
Calorimetry	Heat of adsorption	Plenty of sample	No	Solution	Macroscopic
Contact angle measurements	Polarity and/or roughness of the surface	Any dry surface	No	Drop of liquid on dry surface in air	Macroscopic, polarity of outermost molecular unit
Electro-chemical scanning tunnelling microscopy	Electronic topography of layer with surface	Low roughness conducting surface	No	Solution	Sub-nanometre
Electrochemistry	Permeability of monolayer	Conducting surface	No	Solution	Macroscopic
Fluorescence microscopy	Homogeneity and location of fluorophores	Fluorescent adsorpt	No	Both	Microns for bulk samples, several nanometres for isolated chromophores
Grazing incidence X-ray diffraction	Structure	Crystallographic order	Yes	Solution–liquid interface	Microns
Low energy electron diffraction	Crystallographic description of monolayer packing	Long range order (several tens of nanometres)	Yes	No	Unit cell dimensions (size of area sampled a few microns)

Molecular modelling	Precise information on intermolecular interactions, dynamics, electronic distribution and STM appearance	Computing facilities	Yes	Yes	Atomic
Normal incidence X-ray standing wave	Position and chemical nature of adsorbate	Ordered substrate	Yes	No	Microns
Reflection absorption infrared spectroscopy	Chemical functional groups and their orientation	Vibrationally active groups	Yes	Yes	Several microns
Scanning near field optical microscope	Position and morphology of chromophores	Low roughness surface, vibrationally active or fluorescent groups	Yes	Both	Hundreds of nanometres
Scanning tunnelling microscopy (simple topography)	Electronic topography of layer with surface	Low roughness conducting surface	Yes	Both	Sub-nanometre
Sum frequency generation microscopy	Molecular order	Vibrationally active groups	Yes	Both	Several microns
Surface enhanced Raman spectroscopy	Chemical functional groups and their orientation	Rough metallic surface and proximity of dipoles to it	Yes	Yes	Several microns
Time-of-flight secondary ion mass spectrometry	Composition of surface	Patterned surfaces are preferred for internal reference	Yes	No	Several microns
X-ray photoelectron spectroscopy	Chemical bonding	Universal	Yes	No	Microns

STM and sum frequency generation spectroscopy is of particular interest.[183] The SFGS results indicated a tilting of the central part of the molecule with respect to the plane of the interface, an apparently minor effect but one not detected by molecular modelling. The subtlety of the effect shows the interest of SFGS for this study, and the necessity of introducing solvent in the molecular modelling of this kind of system where it is believed that fleeting interactions can affect molecular orientation. In a quite different setting, a study of an ionic liquid salt at the silver (111)–vacuum interface using STM was able to locate cations and anions in an ordered form, the one next to the other, and XPS and modelling show the distribution of charge in the system.[184] It was shown that the main electrostatic interactions take place between the ions of the adsorbate, and the interactions with the substrate are mainly van der Waals type in character, as indicated by the mobility of the compound on the surface seen in the STM images. Each technique helped paint part of the picture of the interface.

A quite beautiful study involving high resolution force spectroscopy, Kelvin probe microscopy, scanning tunnelling microscopy and DFT-based simulations was able to unravel with precision the non-covalent bonding and electronic states of a rod-like molecule with relatively minor differences between its two ends.[185] The molecule (4-(4-(2,3,4,5,6-pentafluorophenylethynyl)-2,3,5,6-tetrafluoro-phenylethynyl)phenylethynylbenzene) was deposited in vacuum onto a gold (110) surface where it formed supramolecular tapes up to 80 nm long. Kelvin probe microscopy was able to detect differences in potential evident in the maps it generated that implied hydrogen bonding between fluorine and hydrogen atoms at opposite ends of neighbouring molecules, and the theory fully supported this idea.

References

1. S. A. Claridge, J. J. Schwartz and P. S. Weiss, *ACS Nano*, 2011, **5**, 693.
2. F. Zaera, *Chem. Rev.*, 2012, **112**, 2920.
3. D. H. Everett and G. H. Findenegg, *Nature*, 1969, **223**, 52.
4. J. H. Clint, J. S. Clunie, J. F. Goodman and J. R. Tate, *Nature*, 1969, **223**, 51.
5. R. A. Barnard and A. J. Matzger, *Langmuir*, 2014, **30**, 7388.
6. G. J. Fox, D. M. Bloor, J. F. Holzwarth and E. Wyn-Jones, *Langmuir*, 1998, **14**, 1026.
7. Z. Király and G. H. Findenegg, *Langmuir*, 2000, **16**, 8842.
8. C. Kan, M. Keefe, K. Olesen and P. Saucier, *J. Coat. Technol. Res.*, 2011, **8**, 1.
9. J. Fang, U. Gehlert, R. Shashidar and C. M. Knobler, *Langmuir*, 1999, **15**, 297.
10. S. T. Hess, S. H. Huang, A. A. Heikal and W. W. Webb, *Biochemistry*, 2002, **41**, 697.
11. I. V. Perevoshchikova and Y. N. Kotova, *Biochemistry*, 2011, **76**, 497.

12. J. Korlach, P. Schwille, W. W. Webb and G. W. Feigenson, *Proc. Natl. Acad. Sci. U. S. A.*, 1999, **96**, 8461.
13. C. M. Knobler, *Science*, 1990, **249**, 870.
14. R. C. Sarasij, P. Srivastava and M. Rao, *Phys. Rev. E: Stat., Nonlinear, Soft Matter Phys.*, 2012, **85**, 041920.
15. B. L. Stottrup, A. H. Nguyen and E. Tuzel, *Biochim. Biophys. Acta*, 2010, **1798**, 1289.
16. M. Mrksich, L. E. Dike, J. Tien, D. E. Ingber and G. M. Whitesides, *Exp. Cell Res.*, 1997, **235**, 305.
17. D. Hoenig and D. Moebius, *J. Phys. Chem.*, 1991, **95**, 4590.
18. S. Henon and J. Meunier, *Rev. Sci. Instrum.*, 1991, **62**, 936.
19. M. A. Cohen Stuart, R. A. J. Wegh, J. M. Kroon and E. J. R. Sudholter, *Langmuir*, 1996, **12**, 2863.
20. G. Marshall, M. Dennin and C. M. Knobler, *Rev. Sci. Instrum.*, 1998, **69**, 3699.
21. D. Vollhardt, *Curr. Opin. Colloid Interface Sci.*, 2014, **19**, 183.
22. H. Rehage, B. Achenbach, F.-G. Klärner, M. Kamieth and J. Panitzky, *Tenside, Surfactants, Deterg.*, 2004, **41**, 121.
23. M. N. G. de Mul and J. A. Mann, Jr., *Langmuir*, 1998, **14**, 2455.
24. F. Hoffmann, K. J. Stine and H. Huhnerfuss, *J. Phys. Chem. B*, 2005, **109**, 240.
25. W. Frey, W. R. Schief, Jr. and V. Vogel, *Langmuir*, 1996, **12**, 1312.
26. M. Micic, K. A. Radotic, M. Jeremic, D. Djikanovic and S. B. Kammer, *Colloids Surf., B*, 2004, **34**, 33.
27. H. Tachibana, Y. Yamanaka and M. Matsumoto, *J. Phys. Chem. B*, 2001, **105**, 10282.
28. S. Di Bella, G. Consiglio, G. La Spina, C. Oliva and A. Cricenti, *J. Chem. Phys.*, 2008, **129**, 114704.
29. E. Betzig and R. J. Chichester, *Science*, 1993, **262**, 1422.
30. M. Altman, A. D. Shukla, T. Zubkov, G. Evmenenko, P. Dutta and M. E. van der Boom, *J. Am. Chem. Soc.*, 2006, **128**, 7374.
31. S. Richter, C. H.-H. Traulsen, T. Heinrich, J. Poppenberg, C. Leppich, M. Holzweber, W. E. S. Unger and C. A. Schalley, *J. Phys. Chem. C*, 2013, **117**, 18980.
32. R. C. Roldan-Carmona, C. Rubia-Paya, M. Perez-Morales, M. T. Martin-Romero, J. J. Giner-Casares and L. Camacho, *Phys. Chem. Chem. Phys.*, 2014, **16**, 4012.
33. R. A. Dluhy, *J. Phys. Chem.*, 1986, **90**, 1373.
34. R. Mendelsohn, J. W. Brauner and A. Gericke, *Annu. Rev. Phys. Chem.*, 1995, **46**, 305.
35. R. Mendelsohn, G. Mao and C. R. Flach, *Biochim. Biophys. Acta*, 2010, **1798**, 788.
36. E. Waidely, C. Pumilia, A. Malagon, E. F. Vargas, S. Li and R. M. Leblanc, *Langmuir*, 2015, **31**, 1368.
37. T. E. Goto and L. Caseli, *Langmuir*, 2013, **29**, 9063.
38. J. Williams, S. Haq and R. Raval, *Surf. Sci.*, 1996, **368**, 303.

39. S. M. Barlow, S. Louafi, D. Le Roux, J. Williams, C. Muryn, S. Haq and R. Raval, *Surf. Sci.*, 2005, **590**, 243.
40. S. Ye, A. Yashiro, Y. Sato and K. Uosaki, *J. Chem. Soc., Faraday Trans.*, 1996, **92**, 3813.
41. S. Schlücker, *Angew. Chem., Int. Ed.*, 2014, **53**, 4756.
42. T. Vo-Dinh, *Sens. Actuators. B*, 1995, **29**, 183.
43. C. J. Sandroff and D. R. Herschbach, *J. Phys. Chem.*, 1982, **86**, 3277.
44. K. Bandyopadhyay, K. Vijayamohanan, M. Venkataramanan and T. Pradeep, *Langmuir*, 1999, **15**, 5314.
45. C. T. Mallon, R. J. Forster, A. McNally, E. Campagnoli, Z. Pikramenou and T. E. Keyes, *Langmuir*, 2007, **23**, 6997.
46. A. McNally, R. J. Forster and T. E. Keyes, *Phys. Chem. Chem. Phys.*, 2009, **11**, 848.
47. E. B. Eisenthal, *Chem. Rev.*, 1996, **96**, 1343.
48. S. Ye, Y. J. Tong, A. M. Ge, L. Qiao and P. B. Davies, *Chem. Rec.*, 2014, **14**, 791.
49. G. L. Richmond, *Annu. Rev. Phys. Chem.*, 2001, **52**, 357.
50. A. G. Lambert, P. B. Davies and D. J. Neivandt, *Appl. Spectrosc. Rev.*, 2005, **40**, 103.
51. T. Petralli-Mallow, T. M. Wong, J. D. Byers, H. I. Yee and J. M. Hicks, *J. Phys. Chem.*, 1993, **97**, 1383.
52. L. Fu, J. Liu and E. C. Y. Yan, *J. Am. Chem. Soc.*, 2011, **133**, 8094.
53. Z. Wang, L. Fu and E. C. Y. Yan, *Langmuir*, 2013, **29**, 4077.
54. E. C. Y. Yan, L. Fu, Z. Wang and W. Liu, *Chem. Rev.*, 2014, **114**, 8471.
55. C. D. Bain, *J. Chem. Soc., Faraday Trans.*, 1995, **91**, 1281.
56. R. N. Ward, D. C. Duffy, P. B. Davies and C. D. Bain, *J. Phys. Chem.*, 1994, **98**, 8536.
57. M. S. Johal, Y. W. Cao, X. D. Chai, L. B. Smilowitz, J. M. Robinson, T. J. Li, D. McBranch and D. Q. Li, *Chem. Mater.*, 1999, **11**, 1962.
58. J. A. Carter, Z. Wang and D. D. Dlott, *Acc. Chem. Res.*, 2009, **42**, 1343.
59. B. Bhushan, A. V. Kulkarni, V. N. Koinkar, M. Boehm, L. Odoni, C. Martelet and M. Belinm, *Langmuir*, 1995, **11**, 3189.
60. J. Liu, J. Li, B. Yu, B. Ma, Y. Zhu, X. Song, X. Cao, W. Yang and F. Zhou, *Langmuir*, 2011, **27**, 11324.
61. H.-J. Butt, B. Cappella and M. Kappla, *Surf. Sci. Rep.*, 2005, **59**, 1.
62. T. Auletta, M. R. de Jong, A. Mulder, F. C. J. M. van Veggel, J. Huskens, D. N. Reinhoudt, S. Zou, S. Zapotoczny, H. Schönherr, G. J. Vancso and L. Kuipers, *J. Am. Chem. Soc.*, 2004, **126**, 1577.
63. P. Samori, *Chem. Soc. Rev.*, 2005, **34**, 551.
64. T. Ito, I. Grabowska and S. Ibrahim, *TrAC, Trends Anal. Chem.*, 2010, **29**, 225.
65. W. Melitz, J. Shen, A. C. Kummel and S. Lee, *Surf. Sci. Rep.*, 2011, **66**, 1.
66. V. M. Palma and P. Samori, *Adv. Mater.*, 2006, **18**, 145.
67. C. Musumeci, A. Liscio, V. Palermo and P. Samori, *Mater. Today*, 2014, **17**, 504.

68. A. Liscio, E. Orgiu, J. M. Mativetsky, V. Palermo and P. Samori, *Adv. Mater.*, 2010, **22**, 5018.
69. V. Palermo, M. B. J. Otte, A. Liscio, E. Schwartz, P. A. J. de Witte, M. A. Castriciano, M. M. Wienk, F. Nolde, G. De Luca, J. J. L. M. Cornelissen, R. A. J. Janssen, K. Müllen, A. E. Rowan, R. J. M. Nolte and P. Samorì, *J. Am. Chem. Soc.*, 2008, **130**, 14605.
70. S. Watanabe, Y. Fukuchi, M. Fukasawa, T. Sassa, A. Kimoto, Y. Tajima, M. Uchiyama, T. Yamashita, M. Matsumoto and T. Aoyama, *ACS Appl. Mater. Interfaces*, 2014, **6**, 1481.
71. R. Berger, H. J. Butt, M. B. Retschke and S. A. L. Weber, *Macromol. Rapid Commun.*, 2009, **30**, 1167.
72. T. W. Kelley, E. L. Granstrom and C. D. Frisbie, *Adv. Mater.*, 1999, **11**, 261.
73. J. Puigmartí-Luis, V. Laukhin, A. Pérez del Pino, J. Vidal-Gancedo, C. Rovira, E. Laukhina and D. B. Amabilino, *Angew. Chem., Int. Ed.*, 2007, **46**, 238.
74. C. Munuera, J. Puigmartí-Luis, M. Paradinas, L. Garzón, D. B. Amabilino and C. Ocal, *Small*, 2009, **5**, 214.
75. S. Richter and Y. Manassen, *J. Phys. Chem.*, 1994, **98**, 2941.
76. R. Strohmaier, J. Petersen, B. Gompf and W. Eisenmenger, *Surf. Sci.*, 1998, **418**, 91.
77. C. J. Villagomez, T. Zambelli, S. Gauthier, A. Gourdon, S. Stojkovic and C. Joachim, *Surf. Sci.*, 2009, **603**, 1526.
78. S. N. Magonov and M.-H. Whangbo, *Adv. Mater.*, 1994, **6**, 355.
79. L. C. Giancarlo and G. W. Flynn, *Acc. Chem. Res.*, 2000, **33**, 491.
80. F. Moresco and A. Gourdon, *Proc. Natl. Acad. Sci. U. S. A.*, 2005, **102**, 8809.
81. S. Chiang, *Chem. Rev.*, 1997, **97**, 1083.
82. J. Repp, G. Meyer, S. M. Stojkovic, A. Gourdon and C. Joachim, *Phys. Rev. Lett.*, 2005, **94**, 026803.
83. L. Gross, N. Moll, F. Mohn, A. Curioni, G. Meyer, F. Hanke and M. Persson, *Phys. Rev. Lett.*, 2011, **107**, 086101.
84. P. Donovan, A. Robin, M. S. Dyer, M. Persson and R. Raval, *Chem. – Eur. J.*, 2010, **16**, 11641.
85. S. De Feyter and F. C. De Schryver, *Chem. Soc. Rev.*, 2003, **32**, 2003.
86. S. De Feyter and F. C. De Schryver, *J. Phys. Chem. B*, 2005, **109**, 4290.
87. A. Stabel, R. Heinz, F. C. De Schryver and J. P. Rabe, *J. Phys. Chem.*, 1995, **99**, 505.
88. M. M. S. Abdel-Mottaleb, E. Gomar-Nadal, M. Surin, H. Uji-I, W. Mamdouh, J. Veciana, V. Lemaur, C. Rovira, J. Cornil, R. Lazzaroni, D. B. Amabilino, S. De Feyter and F. C. De Schryver, *J. Mater. Chem.*, 2005, **15**, 4601.
89. J. Puigmartí-Luis, A. Minoia, H. Uji-i, C. Rovira, J. Cornil, S. De Feyter, R. Lazzaroni and D. B. Amabilino, *J. Am. Chem. Soc.*, 2006, **128**, 12602.
90. A. A. Gewirth and B. K. Niece, *Chem. Rev.*, 1997, **97**, 1129.
91. D. Wang and L.-J. Wan, *J. Phys. Chem. C*, 2007, **111**, 16109.

92. F. Cunha and N. J. Tao, *Phys. Rev. Lett.*, 1995, **75**, 2376.
93. J.-R. Gong, L.-J. Wan, Q.-H. Yuan, C.-L. Bai, H. Jude and P. J. Stang, *Proc. Natl. Acad. Sci. U. S. A.*, 2005, **102**, 971.
94. *Scanning Electrochemical Microscopy*, ed. A. Bard and M. Mirkin, Marcel Dekker, New York, 2001.
95. S. Amemiya, A. J. Bard, F.-R. F. Fan, M. V. Mirkin and P. R. Unwin, *Annu. Rev. Anal. Chem.*, 2008, **1**, 95.
96. R. W. Murray, *Chem. Rev.*, 2008, **108**, 2688.
97. D. A. Walsh, K. R. J. Lovelock and P. Licence, *Chem. Soc. Rev.*, 2010, **39**, 4185.
98. B. Liu, A. J. Bard, M. V. Mirkin and S. E. Creager, *J. Am. Chem. Soc.*, 2004, **126**, 1485.
99. B. Lin, A. J. Bard, C.-Z. Li and H.-B. Kraatz, *J. Phys. Chem. B*, 2005, **109**, 5193.
100. C. Li, L. Li, C. Wang, Y. Zhu and W. Zhang, *Corros. Sci.*, 2014, **80**, 511.
101. C. Li, L. Li and C. Wang, *Electrochim. Acta*, 2014, **115**, 531.
102. M. Zharnikov and M. Grunze, *J. Phys.: Condens. Matter*, 2001, **13**, 11333.
103. D. Castner, K. Hinds and D. W. Grainger, *Langmuir*, 1996, **12**, 5083.
104. J. P. Folkers, C. B. Gorman, P. E. Laibinis, S. Buchholz, G. M. Whitesides and R. G. Nuzzo, *Langmuir*, 1995, **11**, 813.
105. B. Adolphi, E. Jahne, G. Busch and X. Cai, *Anal. Bioanal. Chem.*, 2004, **379**, 646.
106. I. Haller, *J. Am. Chem. Soc.*, 1978, **100**, 8050.
107. F. Vollnhals, P. Wintrich, M.-M. Walz, H.-P. Steinrück and H. Marbach, *Langmuir*, 2013, **29**, 12290.
108. L. Smykalla, P. Shukrynau, C. Mende, H. Lang, M. Knupfer and M. Hietschold, *Chem. Phys.*, 2015, **450–451**, 39.
109. C. K. Song, K. A. Luck, N. Zhou, L. Zeng, H. M. Heitzer, E. F. Manley, S. Goldman, L. X. Chen, M. A. Ratner, M. J. Bedzyk, R. P. H. Chang, M. C. Hersam and T. J. Marks, *J. Am. Chem. Soc.*, 2014, **136**, 17762.
110. Y. Yokota, Y. Mino, Y. Kanai, T. Utsunomiya, A. Imanishi, M. A. Wolak, R. Schlaf and K.-i. Fukui, *J. Phys. Chem. C*, 2014, **118**, 10936.
111. L. O. Brockway and J. Karle, *J. Colloid Sci.*, 1947, **2**, 277.
112. H. P. Zingsheim, *Scanning Electron Microscopy*, ed. O. M. Johari, 1977, vol. 1, p. 357.
113. L. A. Staehelin, J. R. Golecki and G. Drews, *Biochim. Biophys. Acta*, 1979, **589**, 30.
114. K. Heinz, *Surf. Sci.*, 1994, **299**, 433.
115. T. A. Pham, F. Song and M. Stöhr, *Phys. Chem. Chem. Phys.*, 2014, **16**, 8881.
116. M. Ortega Lorenzo, V. Humblot, P. Murray, C. J. Baddeley, S. Haq and R. Raval, *J. Catal.*, 2002, **205**, 123.
117. M. Ortega Lorenzo, S. Haq, T. Bertrams, P. Murray, R. Raval and C. J. Baddeley, *J. Phys. Chem. B*, 1999, **103**, 10661.
118. M. Parschau, S. Romer and K.-H. Ernst, *J. Am. Chem. Soc.*, 2004, **126**, 15398.

119. A. Inaba, S. M. Clarke, T. Arnold and R. K. Thomas, *Chem. Phys. Lett.*, 2002, **352**, 57.
120. J. K. Kjems, L. Passell, H. Taub, J. G. Dash and A. D. Novaco, *Phys. Rev. B: Solid State*, 1976, **13**, 1446.
121. A. I. Kitaigorodsky, *Mixed Crystals*, Springer-Verlag, Berlin, 1984.
122. N. A. El-Ghamaz, E. M. El-Menyawy, M. A. Diab, A. A. El-Bindary, A. Z. El-Sonbati and S. G. Nozhac, *Solid State Sci.*, 2014, **30**, 44.
123. J. Gao, G. Luc, J. kan, Y. Chen and M. Bouvet, *Sens. Actuators, B*, 2012, **166–167**, 500.
124. L. Cristofolini, *Curr. Opin. Colloid Interface Sci.*, 2014, **19**, 228.
125. J. Als-Nielsen, D. Jacquemain, K. Kjaer, F. Leveiller, M. Lahav and L. Leiserowitz, *Phys. Rep.*, 1994, **246**, 251.
126. J. Daillant, L. Bosio, J. J. Benattar and C. Blot, *Langmuir*, 1991, **7**, 611.
127. W.-J. Pao, F. Zhang, P. A. Heiney, C. Mitchell, W.-D. Cho and V. Percec, *Phys. Rev. E: Stat., Nonlinear, Soft Matter Phys.*, 2003, **67**, 021601.
128. B. W. Laursen, K. Nørgaard, N. Reitzel, J. B. Simonsen, C. B. Nielsen, J. Als-Nielsen, T. Bjørnholm, T. I. Sølling, M. M. Nielsen, O. Bunk, K. Kjaer, N. Tchebotareva, M. D. Watson, K. Müllen and J. Piris, *Langmuir*, 2004, **20**, 4139.
129. R. G. Jones, A. S. Y. Chan, M. G. Roper, M. P. Skegg, I. G. Shuttleworth, C. J. Fisher, G. J. Jackson, J. J. Lee, D. P. Woodruff, N. K. Singh and B. C. C. Cowie, *J. Phys.: Condens. Matter*, 2002, **14**, 4059.
130. M. Yu, N. Bovet, C. J. Satterley, S. Bengio, K. R. J. Lovelock, P. K. Milligan, R. G. Jones, D. P. Woodruff and V. Dhanak, *Phys. Rev. Lett.*, 2006, **97**, 166102.
131. A. Chaudhuri, D. C. Jackson, T. J. Lerotholi, R. G. Jones, T.-L. Lee, B. Detlefs and D. P. Woodruff, *Phys. Chem. Chem. Phys.*, 2010, **12**, 3229.
132. S. K. M. Henze, O. Bauer, T.-L. Lee, M. Sokolowski and F. S. Tautz, *Surf. Sci.*, 2007, **601**, 1566.
133. A. Hauschild, K. Karki, B. C. C. Cowie, M. Rohlfing, F. S. Tautz and M. Sokolowski, *Phys. Rev. Lett.*, 2005, **94**, 036106.
134. A. Gerlach, S. Sellner, F. Schreiber, N. Koch and J. Zegenhagen, *Phys. Rev. B: Condens. Matter Mater. Phys.*, 2007, **75**, 045401.
135. S. Duhm, A. Gerlach, I. Salzmann, B. Bröker, R. L. Johnson, F. Schreiber and N. Koch, *Org. Electron.*, 2008, **9**, 111.
136. R. J. Archer, *J. Electrochem. Soc.*, 1957, **104**, 619.
137. I. Haller, *J. Am. Chem. Soc.*, 1978, **100**, 8050.
138. J. Stohr, *NEXAFS Spectroscopy*, Springer Series in Surface Science, 2003, vol. 25.
139. D. A. Outka, J. Stöhr, J. P. Rabe and J. D. Swalen, *J. Chem. Phys.*, 1988, **88**, 4076.
140. C. Castellarin-Cudia, P. Vilmercati, R. Larciprete, C. Cepek, G. Zampieri, L. Sangaletti, S. Pagliara, A. Verdini, A. Cossaro, L. Floreano, A. Morgante, L. Petaccia, S. Lizzit, C. Battocchio, G. Polzonetti and A. Goldoni, *Surf. Sci.*, 2006, **600**, 4013.

141. G. P. Hastie, J. Johnstone and K. J. Roberts, *J. Mater. Sci. Lett.*, 1998, **17**, 1223.
142. M. Marschall, J. Reichert, K. Seufert, W. Auwarter, F. Klappenberger, A. Weber-Bargioni, S. Klyatskaya, G. Zoppellaro, A. Nefedov, T. Strunskus, C. Woll, M. Ruben and J. V. Barth, *ChemPhysChem*, 2010, **11**, 1446.
143. J. E. Baio, C. Jaye, D. A. Fischer and T. Weidner, *ACS Comb. Sci.*, 2014, **16**, 449.
144. A. Benninghoven, *Angew. Chem., Int. Ed. Engl.*, 1994, **33**, 1023.
145. R. Michel and D. G. Castner, *Surf. Interface Anal.*, 2006, **38**, 1386.
146. V. Spampinato, S. Vitale, S. Quici, A. Torrisi, V. Stepanenko, F. Würthner and A. Licciardello, *Surf. Interface Anal.*, 2012, **45**, 206.
147. D. Touboul, F. Halgan, A. Brunelle, R. Kersting, E. Tallarek, B. Hagenhoff and O. Laprévote, *Anal. Chem.*, 2004, **76**, 1550.
148. K. M. Bolles, F. Cheng, J. Burk-Rafel, M. Dubey and D. M. Ratner, *Materials*, 2010, **3**, 3948.
149. N. Belman, K. Jin, Y. Golan, J. N. Israelachvili and N. S. Pesika, *Langmuir*, 2012, **28**, 14609.
150. D. Janssen, R. De Palma, S. Verlaak, P. Heremans and W. Dehaen, *Thin Solid Films*, 2006, **515**, 1433.
151. J. J. Gooding, F. Mearns, W. Yang and J. Liu, *Electroanalysis*, 2003, **15**, 81.
152. D. Mandler and S. Kraus-Ophir, *J. Solid State Electrochem.*, 2011, **15**, 1535.
153. C. E. D. Chidsey and D. N. Loiacono, *Langmuir*, 1990, **6**, 682.
154. V. L. Lanza and D. B. Herrman, *J. Polym. Sci.*, 1958, **28**, 622.
155. M. D. Porter, T. B. Bright, D. L. Allara and C. E. D. Chidsey, *J. Am. Chem. Soc.*, 1987, **109**, 3559.
156. C. Amatore, J. M. Saveant and D. Tessier, *J. Electroanal. Chem.*, 1983, **147**, 39.
157. M. M. Walczak, D. D. Popenoe, R. S. Deinhammer, B. D. Lamp, C. Chung and M. D. Porter, *Langmuir*, 1991, 7, 2687.
158. D. E. Weisshaar, B. D. Lamp and M. D. Porter, *J. Am. Chem. Soc.*, 1992, **114**, 5860.
159. A. E. Kaifer and M. Gomez-Kaifer, *Supramolecular Electrochemistry*, Wiley-VCH Verlag, 1999.
160. O. N. Oliveira, R. M. Iost, J. R. Siqueira, Jr., F. N. Crespilho and L. Caseli, *ACS Appl. Mater. Interfaces*, 2014, **6**, 14745.
161. A. N. Shipway, E. Katz and I. Willner, *Struct. Bonding*, 2001, **99**, 237.
162. X. Fan, I. M. White, S. I. Shopova, H. Zhu, J. D. Suter and Y. Sun, *Anal. Chim. Acta*, 2008, **620**, 8.
163. M. N. Velasco-Garcia, *Semin. Cell Dev. Biol.*, 2009, **20**, 27.
164. S. Flink, F. C. J. M. van Veggel and D. N. Reinhoudt, *Adv. Mater.*, 2000, **12**, 1315.
165. A. V. Nabok, A. K. Hassan and A. K. Ray, *J. Mater. Chem.*, 2000, **10**, 189.
166. K. A. Marx, *Biomacromolecules*, 2003, **4**, 1099.

167. M. T. Cygan, G. E. Collins, T. D. Dunbar, D. L. Allara, C. G. Gibbs and C. D. Gutsche, *Anal. Chem.*, 1999, **71**, 142.
168. H. S. Guo, *J. Nanosci. Nanotechnol.*, 2014, **14**, 1884.
169. M. Brutschy, D. Lubczyk, K. Müllen and S. R. Waldvogel, *Anal. Chem.*, 2013, **85**, 10526.
170. J. Homola, S. S. Yee and G. Gauglitz, *Sens. Actuators, B*, 1999, **54**, 3.
171. J. Homola, *Chem. Rev.*, 2008, **108**, 462.
172. J. A. Maynard, N. C. Lindquist, J. N. Sutherland, A. Lesuffleur, A. E. Warrington, M. Rodriguez and S.-H. Oh, *Biotechnol. J.*, 2009, **4**, 1542.
173. O. Shekhah, N. Roques, V. Mugnaini, C. Munuera, C. Ocal, J. Veciana and C. Wöll, *Langmuir*, 2008, **24**, 6640.
174. M. Linares, A. Minoia, P. Brocorens, D. Beljonne and R. Lazzaroni, *Chem. Soc. Rev.*, 2009, **38**, 806.
175. S. S. Jang, Y. H. Jang, Y.-H. Kim, W. A. Goddard III, A. H. Flood, B. W. Laursen, H.-R. Tseng, J. F. Stoddart, J. O. Jeppesen, J. W. Choi, D. W. Steuerman, E. Delonno and J. R. Heath, *J. Am. Chem. Soc.*, 2005, **127**, 1563.
176. A. Minoia, Z. Guo, H. Xu, S. J. George, A. P. H. J. Schenning, S. De Feyter and R. Lazzaroni, *Chem. Commun.*, 2011, **47**, 10924.
177. G. D'Avino, L. Muccioli and C. Zannoni, *Adv. Funct. Mater.*, 2014, **25**, 1985.
178. B. Chilukuri, U. Mazura and K. W. Hipps, *Phys. Chem. Chem. Phys.*, 2014, **16**, 14096.
179. N. Liu, G. R. Darling and R. Raval, *Chem. Commun.*, 2011, **47**, 11324.
180. M. Möhwald, *Thin Solid Films*, 1988, **159**, 1.
181. L. H. Dubois and R. G. Nuzzo, *Annu. Rev. Phys. Chem.*, 1992, **43**, 437.
182. R. Marques de Oliveira, J. Ferreira, M. J. L. Santos, R. M. Faria and O. N. Oliveira Jr., *ChemPhysChem*, 2011, **12**, 1736.
183. A. A. McClelland, S. Ahn, A. J. Matzger and Z. Chen, *Langmuir*, 2009, **25**, 12847.
184. F. Buchner, K. Forster-Tonigold, B. Uhl, D. Alwast, N. Wagner, H. Farkhondeh, A. Groß and R. J. Behm, *ACS Nano*, 2013, 7, 7773.
185. S. Kawai, A. Sadeghi, X. Feng, P. Lifen, R. Pawlak, T. Glatzel, A. Willand, A. Orita, J. Otera, S. Goedecker and E. Meyer, *ACS Nano*, 2013, 7, 9098.

CHAPTER 3

Supramolecular Systems on the Surface of Water

3.1 Whence, Why and How: Supramolecular Layers on Water

The formation of layers of molecules on the surface of water was observed qualitatively in all likelihood millennia ago in bogs, fens, swamps, mires and marshes where decomposition of organic matter could result in oil formation on the tranquil water, even though it cannot have been understood until much later. As mentioned in the general introduction, it was Agnes Pockels[1] who first articulated thoughts concerning the nature of films and the component molecules at the water–air interface. The advances in determining the true monomolecular thickness of the films by Henri Devaux[2] and Irving Langmuir[3] laid a foundation for the exploitation of this medium for assembling molecules.

In many ways, it is the most versatile way of studying molecules on surfaces, for several reasons: the interface between water and air is completely flat when still; there is no restriction on molecular weight of the compound; multicomponent systems are possible within a layer; the layers can be passed onto substrates using a variety of transfer modes, forming multilayers if desired; the formation of a layer is almost certain, provided the compound is immiscible in water. Naturally, the principal restriction is that the compound should not react with water and, in principle, flat substrates are used. The quality of the films that are formed depends on the particular compound being used, but can be extremely high when specific non-covalent interactions lead to self-assembly, rather than the forced compression that happens in the Langmuir trough.[4]

Monographs in Supramolecular Chemistry No. 19
Supramolecular Chemistry at Surfaces
By David B. Amabilino

Published by the Royal Society of Chemistry, www.rsc.org

The procedure for making Langmuir monolayers is simple, in principle, and it can be used as a teaching tool in an estimation of Avogadro's number.[5] An organic solution of the target monolayer material is dropped very gently onto the uncontaminated surface of purified water. Dust should be scrupulously avoided, and vibrations minimised. Once the solvent evaporates, the adsorbates are left at the surface either as a kind of two-dimensional gas, or with islands of aggregated molecules, depending upon the affinity of the molecules for one another relative to their affinity with the surface. Compression of the surface can lead to single molecule thick layers that can have the characteristics of a solid. During compression (see Section 1.4), the monolayer experiences a series of phase changes in many ways analogous to those of bulk matter, from gas, to liquid to solid. The situation concerning the state of the material in the monolayer is not always clear-cut though, and analogous systems to bulk disordered solids, such as glasses, may form. The phase diagrams can be determined though, exploiting the battery of analytical methods in addition to the isotherm and molecular modelling. They provide a detailed view of how compression and temperature can affect the structures of simple amphiphiles at the air–water interface and how the phases interconvert,[6] and they have also been applied to some degree in more intricate supramolecular systems. Specifically, the compressibility of the monolayers—obtained from the isotherms—gives the type of phase being formed, for example liquid expanded or solid.[7]

The classic design of compounds that will assemble on a water surface has a hydrophobic section attached to a polar group, generally acid, base, oligoether or alcohol, charged "-ates" or "-iums" (organic anions or cations). The high affinity of the polar regions of the adsorbate for the water forces the orientation of the molecule. The general names for this kind of molecule are, of course, "surfactant" (molecules that lower the tension at any interface between like or different states of matter) or "lipid" (naturally occurring hydrophobic or amphiphilic molecules) and the design of the basic shape and polarity of the molecules in question can affect the packing that is generated,[8] aside from any specific non-covalent interaction. The sub-family of polar molecules termed "hydrotropes" also falls under this classification.[9] In general, the great advantage of assembling amphiphilic molecules is precisely that anisotropic polar order is highly probable with this type of design.[10] No method based on self-assembly provides such a high assurance of well-oriented supramolecular organisation or a virtual guarantee that only one layer will be formed. However, there are also a number of multicomponent scenarios that one could envisage for the self-assembly of supramolecular systems on water, some of which are shown in Figure 3.1.

The boon of the assembly on the surface of water that no restriction on molecular weight of the adsorpt exists has meant that the aggregation and interactions of big biological and synthetic molecules and even macromolecules can be studied there. The processing of polymers *via* the Langmuir–Blodgett technique[11] provides unique characteristics compared

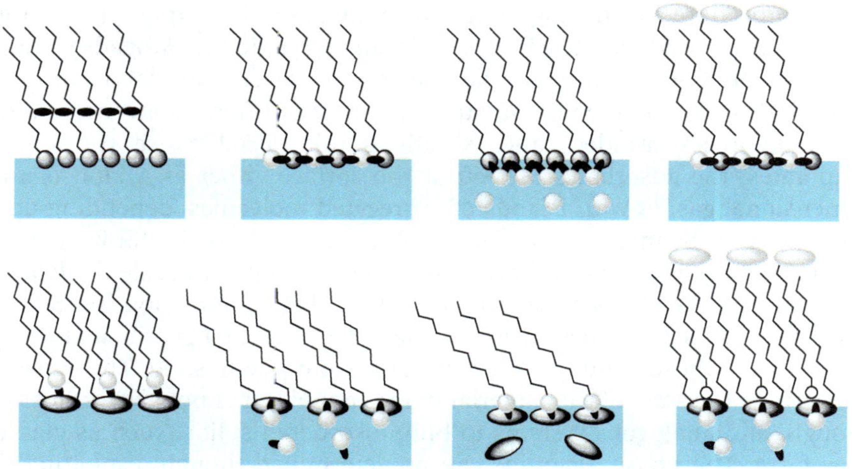

Figure 3.1 Cartoons of possible supramolecular constructs for the air–water interface. The black ellipses represent specific non-covalent interactions, larger light and dark spheres polar head groups. When objects are shown in the bulk water (the subphase), they are in equilibrium with the monolayer. The lighter ellipses above the layer represent relatively bulky apolar groups.

with other film-forming routes (such as spin coating) because the pressure at which the monolayers are transferred can influence the packing and orientation. In general, polymers have very different morphology to their component monomers in Langmuir monolayers,[12] largely because of the restrictions that a polymer backbone puts on organisation of any particular unit. Other deposition routes not involving the use of chemisorption are at the mercy of the inherent interactions between the adsorbent and the adsorpt in the medium of choice. The water subphase used in the preparation of the films can be manipulated to influence the monolayer, through binding or chemically controlled oxidation and reduction, as we shall see later on.

The most common way to study layer formation on water is with the Langmuir trough with an incorporated surface pressure sensor: it is a most powerful tool for determining the development of layers as the mobile barrier compresses them, as presented in Chapter 1. However, the information it provides about molecular-level assembly can be limited. The Brewster angle microscope (BAM) is a tremendously important piece of equipment for seeing organisation of molecules on the water's surface over long scales (microns) both when the films are compressed as well as prior to complete compression, as seen in Chapter 2. More sophisticatedly, the use of synchrotron radiation provides detailed information concerning order and molecular orientation at the interface. These techniques probe the supramolecular order of the films at the interface of water and air as they are prepared, and the first part of this chapter describes these situations.

Transfer of these layers onto solid substrates results in films that can be studied using the general battery of techniques open to surfaces. The interface is now an artificial one, in the sense that the monolayer is forced to be in contact with the solid. However, the surface of the monolayer that is exposed to air can be studied using methods that are not viable on water's surface. These transferred layers will be the subject of the second half of this chapter. Finally, an overview of more contemporary work at liquid interfaces will be presented, to show that while the methods of monolayer formation are a century old, there are still great opportunities for exploiting assembly at the surfaces of liquids for studying supramolecular phenomena in new supramolecular systems.

3.2 Non-covalent Interactions at Play in the Formation of Monolayers at the Water–Air Interface

The dominant force in the—autonomous or barrier controlled—assembly of molecules on water's interface with air is the hydrophobic effect.[13,14] This effect is particularly acute at the interface because a full quota of each hydrogen bond for the interface water molecules is not fulfilled, hence the high surface tension in the absence of surfactants and the dramatic decrease when they are added. The highest order is achieved when the organic molecules have polar groups that interact with the water subphase and apolar groups that can pack well together—effectively forming a two-dimensional crystal—with the aid of van der Waals interactions when oriented way from the interface, as if they were a separate phase covalently linked to the head groups.

While the overriding strength of the hydrophobic effect is evident, the packing of the organic section is determined by its shape and any specific non-covalent interactions that can take place, such as hydrogen bonding or π stacking interactions. The weaker interactions between alkyl chains—for simple lipids—is tremendously selective and powerful over the monolayers. It has been shown by theory—and confirmed by comparison with experimental data—that the angle formed between alkyl chains and the interface plane takes on discrete values that are determined by the van der Waals interactions between the methylene groups, and discontinuous values of the tilt are obtained depending on the exact structure of the surfactant.[15] Aliphatic alcohols, for instance, align virtually perpendicular to the interface and have hexagonal packing according to theory and experiment.[16] Bent-core molecules show a singular arrangement at the surface of water in Langmuir films, with the aliphatic chain bearing the polar head group virtually perpendicular to the interface and the aromatic cores in a face-to-face H-type aggregate, leaving the "top" alkyl chains highly tilted with respect to the monolayer–air interface.[17] A quite dramatic effect is seen in the behaviour of hyperbranched surfactants in Langmuir layers, where they show interfacial properties similar to those of fluorocarbons because of their relatively high density.[18]

Chemically and constitutionally similar molecules can present very different hierarchical superstructures both within domains of monolayers and in the morphology of these domains. The two phospholipids drawn in Figure 3.2 are a good example, dipalmitoylphosphatidylcholine contains a quaternary ammonium head group while dipalmitoyl-phosphatidylethanolamine has a primary ammonium group and the two show very different arrangement and domain structure because of the dipoles of the molecules.[19] The quaternary ammonium compound shows domains comprised of curved arms in triskelion structures in the images of the monolayers (under equilibrium conditions) from a Brewster angle microscope (BAM) (Figure 3.2). The curving of the arm-like

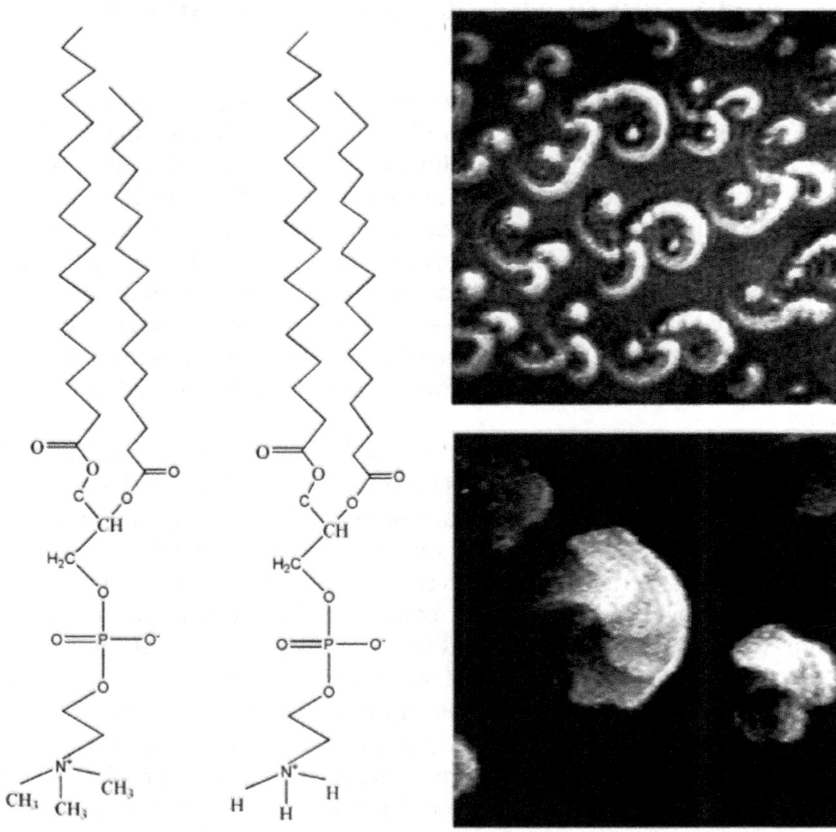

Figure 3.2 The chemical structures of two very similar phospholipids and BAM images of their monolayers on water. The top image corresponds to the trimethyl-ammonium surfactant and the bottom image to the ammonium one. Reprinted with permission from ref. 19. Copyright 2007 American Chemical Society.

domains arises from the chirality of the head group, that is transferred into the mesoscopic shape.[20,21] On the other hand, the analogue displays rounder domains that have internal curved defects where differences in orientation between areas are apparent. Molecular modelling was used in an effort to explain these important differences in mesoscopic morphology from the interaction taking place between the surfactants in the monolayers. The quaternary ammonium surfactant was calculated to have a bigger dipole than the primary derivative, because of the former's more extended conformation, as well as repulsive dipole interaction. Additionally, it was calculated (using quantum chemical methods) that on average, it also has practically double the in-plane dipole moment of the primary ammonium compound in the monolayers. It was suggested that the non-covalent interactions within and between molecules can determine the mesoscopic shape of domains in monolayers, and that this correlation between microstructure and mesostructure could take place in conceptually related biological systems in general and specifically membranes.

Molecular geometry also plays an important role when molecules assemble at the air–water interface, as shown beautifully for unsaturated fatty acids, where van der Waals interactions determine the packing in the monolayer. The effect of having either *cis* or *trans* isomers was known to have an effect on packing from studies of the isotherms of a variety of long-chain acids and alcohols.[22] In this early paper, it was also suggested that the shape influences whether long-range hydrogen bonding can be formed or not. The *cis* double bond hinders molecular rotation and the molecules pack in a bent shape, making packing difficult and causing a low melting temperature.[23] High supersaturation is attained even at moderate compression speed in the Langmuir trough. The morphology of crystallites formed under these far from equilibrium conditions is very different between molecules and from the *trans* isomers, leading to dendritic growth, showing how the molecular shape has an influence on the supramolecular arrangement and even condensate morphology.

Complementary hydrogen bonding pairs that can form supramolecular tapes in the bulk solid state have also been proven to exist at the air–water interface.[24] It was shown that an amphiphile with an adenine moiety as the polar head group formed a complex with thymine in the aqueous subphase,[25] and it was subsequently shown that a wide range of hydrogen bond donors and acceptors familiar from other environments can also form complexes at the air–water interface.[26] The complementary derivatives and barbituric acid that have been tested and proven in many two-dimensional environments are also effective at the interface of air and water.[27–29] Therefore, despite the fact that water is the subphase and the area where these hydrogen bonding interactions take place, it is not an impediment to their formation, in most cases because of the hydrophobic effect with its roots in the stability of the hydrogen bonded network in water.

The effect of the proportion of hydrophobic and hydrophobic sections of a polymer have been varied in size systematically with the aim of determining the influence of this parameter on monolayer formation and structure on water,[30] although the study actually revealed the significance of the ratio of hydrophobic block to the end group. The polymer was poly(1,1-diethylsilacyclobutane)-*b*-poly(methacrylic acid) (Figure 3.3) in which the methacrylic acid section is, in principle, ionised in the water subphase. X-Ray reflectivity measurements of the monolayer on water showed beautifully how the layer structure varies with the applied surface pressure and the relative length of the hydrophilic and hydrophobic blocks. This data showed that for virtually any surface pressure above a threshold, the polymers with an excess of one particular block provided a more structured film (at *m*–*n* ratios of 29 : 22, 29 : 37 at one extreme and 33 : 93 at the other). The surface pressure–areas isotherms showed that these layers were relatively well behaved, but an important observation is that the polymer with an *m*–*n* ratio of 34 : 50 actually presented the largest area per molecule. The result showed that when an excess of one block is present, it either curls up away from the surface (hydrophobic) or enters the aqueous phase (in the case of the hydrophilic block) forming a brush-like superstructure. In all cases, the methacrylic acid block forms a carpet-like layer parallel to the interface, but when the ratio of this hydrophilic section increases, it enters the water near the interface while keeping the carpet-like layer intact, shielding, in a way, the hydrophobic block from the water. It would be interesting to be able to study this kind of phenomenon in small molecules, but the tendency of the hydrophobic sections in small molecules often precludes generalisations. In this polymer's case, the poly(1,1-diethylsilacyclobutane) section is well above its glass transition temperature and effectively acts like cooked spaghetti, oozing flat over the interface in contact with the air.

The addition of counter-ions in the liquid subphase to reinforce the structuring and increase the robustness of monolayers is a useful approach that was recognised early on in the preparation and transfer of layers, as shall be discussed later in the discussion of the transfer of the films. Indeed, it was shown that the films of stearic acid are extremely sensitive to the presence of either aluminium or copper in the aqueous subphase,[31–33] and amine-based surfactants display a similar response to organic acids.[34] This is not the only case that demonstrates that monolayers are a unique area for supramolecular chemistry on water.

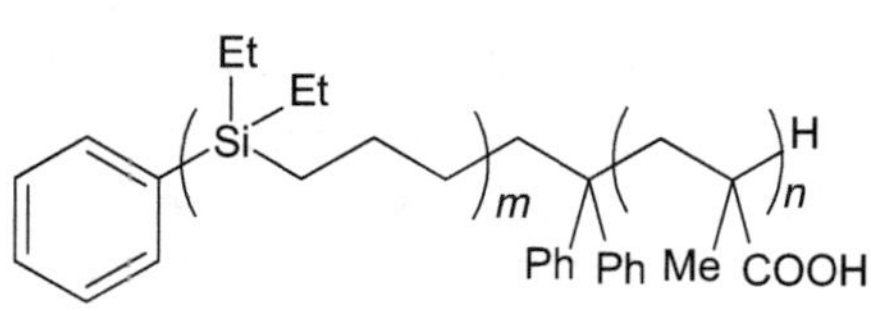

Figure 3.3 Structure of poly(1,1-diethylsilacyclobutane)-*b*-poly(methacrylic acid).

3.3 Discrete Self-assembled Structures and Monolayer Superstructure at the Water–Air Interface

The deposition of molecules onto a liquid surface is usually performed by dissolving a tiny amount of the adsorpt in a volatile liquid, immiscible with the substrate, that evaporates leaving a two-dimensional gas, in principle. What the molecules actually do on the liquid surface is not actually known in many cases. From a supramolecular perspective, there is no reason why one could not design—by astute choice of specific interactions or control of polarity or packing characteristics—compounds that would come together in a certain way without the need to saturate the surface or apply a surface pressure. In this section, cases will be presented where the molecules self-assemble at water's interface with air giving discrete nanostructures.

Langmuir observed the formation of "oil lenses" in expanded monolayers of tetradecane on water.[35] Phospholipids show an amazing variety of organic forms when viewed and water's surface using fluorescence microscopy in the presence of tiny proportions of fluorescent label.[36] Many of these shapes are the result of the chirality of the head group in these naturally occurring compounds.[37] A beautiful and important example is the domains formed on water's interface with air by dipalmitoylphosphatidylcholine with small amounts of cholesterol introduced.[38] The pure lipid shows blobby, somewhat Paisley pattern-like, domains with feature sizes of approximately 40 microns in length when compressed so that disordered and ordered domains are present (Figure 3.4). The pressure applied to the surface will affect both the order and tilt angle of the surfactants in the layer.[39] When 0.2 mole percent of cholesterol is added, spirals with tighter bends are observed, with diameters of approximately 20 microns. Up to 0.4 mole percent, now these shapes are longer, lines with tight spirals branching off them are observed by both epifluorescence and AFM. The width of the domains was decreased from 5 to approximately 1 micron. The origin of these increasingly intricate shapes—which implies an increased interface area and therefore intuitively higher energy—is a result of reduced line tension. The line tension is the energy required to create a wetting line; that is, the interface between the phases (where three phases are in contact[40]). It is important in general in determining the lateral organisation in all kinds of lipid membranes.[41] In the present case, this reduction in energy is aided by domains that are between 10 and 100 nm in diameter, not visible in the images in Figure 3.4 but observed by AFM of transferred monolayers at the interfaces between the ordered and disordered pure lipid domains. The fraction of the area occupied by these domains scaled linearly with the concentration of cholesterol, implying a ratio of the lipid to cholesterol of 6 : 1. These specific assemblies, presumably comprising a cholesterol molecule surrounded by six lipids, are very significant in the understanding of treatment of respiratory distress

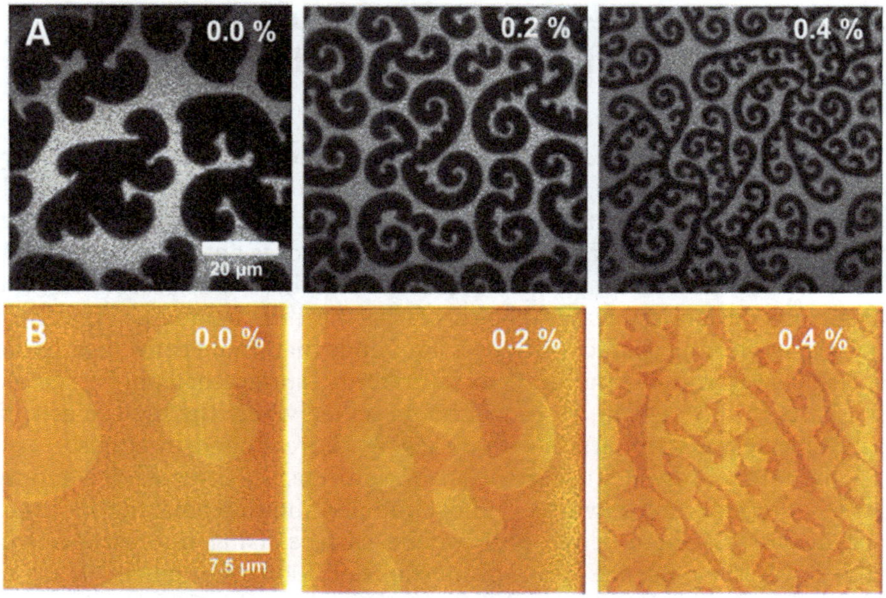

Figure 3.4 Epifluorescence images (top row) of dipalmitoylphosphatidylcholine monolayers on water with the indicated mole percentages of cholesterol present (the lighter areas are the less ordered phase). The images in B are AFM of transferred monolayers on mica (here the brightest areas are brighter).
Reproduced with permission from the American Association for the Advancement of Science, ref. 38.

syndrome, since the lipid used comprises about 60% of lung surfactant,[42] and cholesterol is implied as being important in medication.

However, the direct observation of very small (nanometre size) supramolecules is extremely challenging *in situ*, because the optical microscopies are unable to observe them and unless reasonably long-range order is induced upon compression, scattering techniques are also left wanting. Therefore, transfer of the films is unavoidable practically. Here, it shall be assumed that what is transferred is a reasonable reflection of what is adsorbed at the surface, which in certain cases has proven to be the case. There is theoretical support for the general principles of the self-assembly of molecules at the water–air interface, taking into account both thermodynamic and packing considerations.[43] Although applying pressure and measuring the surface area–pressure isotherm intuitively might not show, in this calculation it was shown that the formation of surface aggregates prior to compression can give rise to distinct features in the curve. A continuous phase transition can take place from small aggregate to full compressed monolayer and the comparison of this theoretical approach with experiment can in principle be used to determine aggregation numbers, assuming equilibrium states.

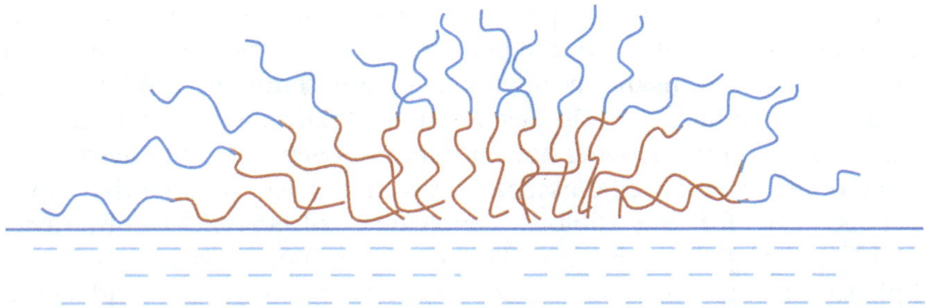

Figure 3.5 A cartoon representation of a surface micelle formed by a (macro)-molecule with a block-type chemical structure on the surface of a liquid.

The separation of phases using block-type molecules or polymers can give rise to the formation of "surface micelles" (Figure 3.5).[44] The (macro)-molecules should have a structure where two very chemically dissimilar units are linked by a covalent bond. Just as in bulk block copolymers[45,46]—one containing poly(styrene) as one block and poly(vinyl pyridine) as the other would be a suitable example—the separate sections tend to form separate microscopic phases, though complete phase separation is not possible because of the covalent link between the two sections. In the case of the block systems deposited on the surface of a liquid, provided this system is at sufficient dilution, the molecules will tend to self-assemble into separated structures of the type shown as a cartoon in Figure 3.5. Just as in bulk block copolymers, the nature of the aggregate will depend on the relative lengths of the blocks and their propensity to mix. The fact that this kind of structure has only been observed relatively recently is surely a result of the difficulty in characterising these systems.

The layers of the block copolymer poly(styrene)-*b*-poly(4-vinylpyridine) (where *b* indicates block) with the pyridine units were fully quaternised with decyl iodide (leading to the christening as block "ionomer") and with a monomer ratio of 240 to 260 show the formation of surface micelles on water.[47] The Langmuir film balance measurements at the air–water interface reveal surface pressure/area isotherms with a sharp transition at high pressures. TEM measurements on the monolayers transferred from the microscope's carbon-coated copper onto grids revealed regular patterns of circular objects which would correspond to approximately 120 individual block copolymer molecules. At relatively low surface pressures, the inter-micelle distance indicates that the decyl-4-pyridinium block extends in a radial fashion from a polystyrene core comprising coiled hydrophobic polymer chains. *In situ* synchrotron X-ray reflectivity measurements of the diblock copolymer monolayers at the air–water interface showed that there is no significant hydration of the layer and confirmed the hexagonal packing observed in transferred samples.[48] The virtually identical structure of films *in situ* and transferred to surfaces shows how robust these kinds of systems are and supports other measurements on transferred films where *in situ* measurements have not been possible. Even at the lowest surface pressure detectable in the set-up used, the density of surface micelles was estimated at one hundred thousand million per square centimetre. When a high pressure was applied, the micelles submerged partially into the water and became less ordered, showing how pressure can actually act in detriment to surface order when specific assembly can take place. A somewhat similar effect is seen for the diblock copolymer poly(styrene)-*b*-poly(ethylene oxide),[49] although in this case the layer does not apparently enter the subphase, but the hydrophilic block may undergo a pressure-induced dehydration process. The spontaneous formation of surface micelles virtually in the absence of surface pressure also shows the generality of the assembly of block copolymers at water's interface with air.[50]

The presence of a relatively highly hydrophilic block is not a prerequisite for this phenomenon.[51] As an example, the diblock copolymers poly(styrene)-*b*-poly(*n*-butyl-methacrylate) and poly(styrene)-*b*-poly(*t*-butyl-methacrylate) when deposited on water self-assemble affording surface micelles with starfish, rod and planar morphologies, as determined by TEM and AFM.[52] While these shapes are also observed for the ionic analogues, in these purely hydrophobic and neutral polymers, the boundaries between the different forms appear at a lower poly(styrene) block proportion than the ionic systems in the phase diagram. Studies of the monolayer formation at different temperatures revealed indications of reversible phase transitions with entropy changes much lower than those of the ionic counterparts. The phase change in these non-ionic materials was believed to arise from a reorientation of the side chains from an orientation parallel to the interface to one that was perpendicular. The work proved the generality of this kind of interface aggregation, even for simple non-ionic polymers.

In situ data from static and dynamic and static evanescent wave light scattering studies of poly(styrene)-*b*-poly(methylmethacrylate) (with molecular weights of 880 and 290 kDa, respectively) at the air–water interface showed that disk-shaped aggregates were formed upon spreading a solution of the polymer on water.[53] It was proposed that these objects comprised approximately 240 polymer molecules and had a "furry disk" morphology, comprising a dense core of twisted and self-entangled poly(styrene) segments on top of a layer of more extended poly(methylmethacrylate) block in contact with the water. The authors associated the uniform size of the aggregates to a collective association driven by large energy and entropy change in competition, and implied that the stretching of the methacrylate chains was at least partially responsible for this situation. It was proven that at low coverage, these self-assembled disks diffuse freely across the width and breadth of the plane of the interface. The disks are also capable of a further level of hierarchical aggregation prior to full monolayer formation where they form large two-dimensional bunches. They are immobile at higher coverages. It was proven that the relation between blocks determines whether the surface micelles form: if the blocks contained the same number of monomer units, no surface micelles were observed.

The partial deuteration of poly(2-(dimethylamino)ethyl methacrylate)-*b*-poly(methyl methacrylate) (70 mol% of the former and mean molecular weight 10 000 Daltons) facilitated a neutron reflectivity measurement study of the macromolecule at the interface of water and air thereby allowing determination of the changing phases of this material in this environment and indicated the presence of surface micelles.[54] At relatively low concentration, the polymer is oriented perpendicular to the interface plane and the amine-containing block (that is more hydrophilic than the methyl methacrylate block) penetrates the water subphase, but at higher relative concentration, micelle-like aggregates are formed at the surface with the hydrophobic block at the core. The adsorption isotherm revealed a surface phase transition from a monolayer 2 nm thick to a structurally distinct layer 4 nm wide at the interface that coincides with a concentration where in the bulk solution there is an onset of aggregation to form micelles. In other systems, it has been established that this balance between formation of dispersed micelles in the bulk solvent and assembly at interfaces can be controlled through the hydrophobicity of the more apolar block.[55] It has also been shown that π-functional polymers can be incorporated into block copolymers that form specific nanostructures at the air–water interface, the case being poly(isoprene)-*b*-poly(styrene)-*b*-poly(3-hexylthiophene), whose thiophene segment displayed significant photoluminescence.[56]

Molecules incorporating fluorocarbon and hydrocarbon fragments have a great tendency to form clusters both in solution and at the water–air interface because of the incompatibility, or reciprocal phobicity, of the two sections.[57] Long-chain acids with hydrocarbon segments between ten and twenty-two carbon atoms between the polar group and a fluorocarbon chain with between six and nine units gave monodisperse flat clusters a few tens of nanometres in diameter (Figure 3.6).[58] The experiments were formed at pH 7 using cadmium(II) in the subphase, and the incorporation of cadmium into the monolayer was proven. Clusters as small as seventeen nanometres in diameter were observed by AFM after transfer of the objects onto solid substrates.[59] Steric hindrance arising from the bigger van der Waals radius of fluorine than half the inter-carbon atom bond length makes the chain take up a semi-rigid 15/7 helical conformation,[60] whereas the hydrocarbon chains have a linear zig-zag form.[61] This situation means that the cross section of the methylene chain segment is approximately 20 Å^2, about two thirds the value of the fluorocarbon segment. The transfer of the films of compounds with different hydrocarbon chain lengths onto flat glass using the "scooping up" method (see Section 3.7) at different compression values revealed amazingly uniform clusters for all compounds whose sizes increase with the length of the chains (Figure 3.6). The diameters are approximately 17 nm for $CF_3(CF_2)_7(CH_2)_{10}COOH$, 33 nm for $CF_3(CF_2)_7(CH_2)_{16}COOH$ and 70 nm for $CF_3(CF_2)_7(CH_2)_{22}COOH$, but although the sizes of the clusters are uniform within each object, no crystalline order could be detected. The smallest clusters are composed of

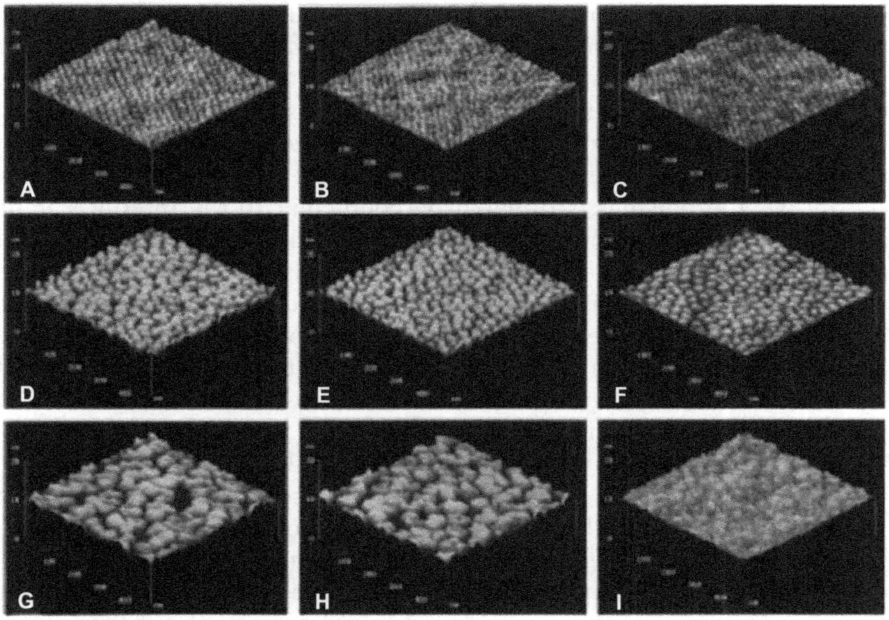

Figure 3.6 AFM images of Langmuir films of $CF_3(CF_2)_7(CH_2)_{10}COOH$ (A–C), $CF_3(CF_2)_7(CH_2)_{16}COOH$ (D–F) and $CF_3(CF_2)_7(CH_2)_{22}COOH$ (G–I) transferred at different and increasing surface pressures from left to right. The images are 500 nm × 500 nm.
Reprinted with permission from ref. 58. Copyright 1998 American Chemical Society.

approximately 700 molecules. Before surface pressure was detected using the balance on the Langmuir trough, clear clusters are seen with domain sizes on the order of millimetres, as they are just after pressure is detected, with little change in their morphology, suggesting that these layers are already present after spreading. At higher compression, the molecules with longer hydrocarbon chains seem to rearrange on a large scale, but with sixteen and ten methylene groups, the clusters maintain their packing, although a deformation from a circular to a hexagonal shape is observed. The type of cation in the subphase has a very significant effect on the shape of the clusters: when monovalent potassium was introduced, poorly-defined clusters with no regular arrangement were observed by AFM after transfer, but lanthanum(III) did result in the formation of extremely well-defined hexagonally-packed clusters. For molecules with shorter fluorocarbon segments and longer hydrocarbons, doughnut-shaped clusters were observed with nanometre-range holes (Figure 3.7) that did not disappear even at quite large surface pressures. The mechanism of formation of these circular clusters is awaited, although it was suggested they could be formed by smaller bunches. It was proposed that the mechanism of formation of all these clusters was through a rapid nucleation-growth route that is limited by the mass transport of the molecules on the water, and the

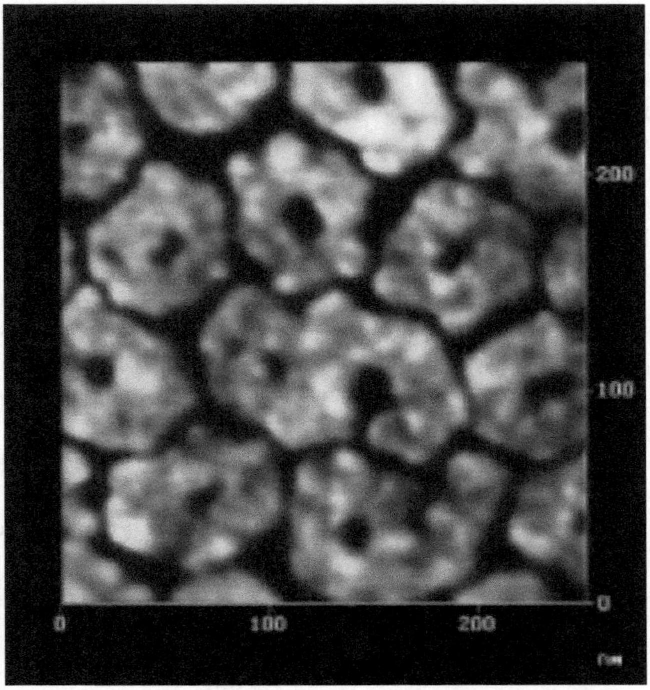

Figure 3.7 AFM images of a transferred Langmuir film of $CF_3(CF_2)_3(CH_2)_{22}COOH$. The image is 250 nm×250 nm.
Reprinted with permission from ref. 58. Copyright 1998 American Chemical Society.

homogeneity is possible because of the speed of solvent evaporation. To prove the generality of the formation of clusters on fast evaporation of solvent, the authors used the same technique and observed small clusters of stearic acid and dipalmitoylphosphatidic acid, although these fused over time, unlike the fluorinated surfactants.

It turns out that the acid group is not necessary: non-polar molecules with fluoro- and hydrocarbon blocks form similar nanodisks.[62] The diblock fluoroalkane–alkane "nonpolar amphiphile" molecule $C_8F_{17}C_{16}H_{33}$ when spread from chloroform solution onto pure water forms a layer that could be reversibly compressed and expanded. When compressed, a relatively stable layer is formed with a limiting area that corresponds approximately to the cross section of the fluorocarbon block (30 Å^2). AFM images of films transferred onto a silicon substrate at different surface pressures—corresponding to a fully compacted layer, and two prior stages—clearly showed round objects of uniform size that were independent of the surface pressure, implying that they existed before the two-dimensional liquid state had been reached. X-Ray reflectivity measurements on the transferred compact layer were fitted to a model where the alkane segment was facing the hydrophilic surface and the fluorocarbon chain was on top (a single component fitting of the electron density did not provide a good comparison). The segregation is believed to be caused by the density mismatch of the segments. The length of the hydrocarbon block is the dominant part in determining the structure of the nanoscale objects.[63] Direct evidence for the formation of organised networks of the micelles at the water–air interface was achieved[64] using grazing incidence small-angle X-ray scattering (GISAXS) at a synchrotron light source, that proved that they measured approximately 30 nm for the compound $C_8F_{17}C_{16}H_{33}$. These results were in full concurrence with the size of the hemimicelles observed after transfer from the interface onto solid substrates.[65] Subsequent *in situ* GISAXS measurements also indicated the strong interaction of the first layer with the water subphase as a driving force for the nanostructure formation.[66]

A partially-fluorinated alkane attached to a naphthyl ring through a urethane group (Figure 3.8) has a great tendency to self-assemble at water's surface.[67] The molecule has an unusual geometry in that the middle hydrocarbon segment has a smaller cross sectional diameter than the fluorocarbon segment, but this in turn is smaller than that of the naphthyl group. Transfer of the Langmuir layers at increasing surface pressure results first in the formation of isolated spirals (with external diameters between

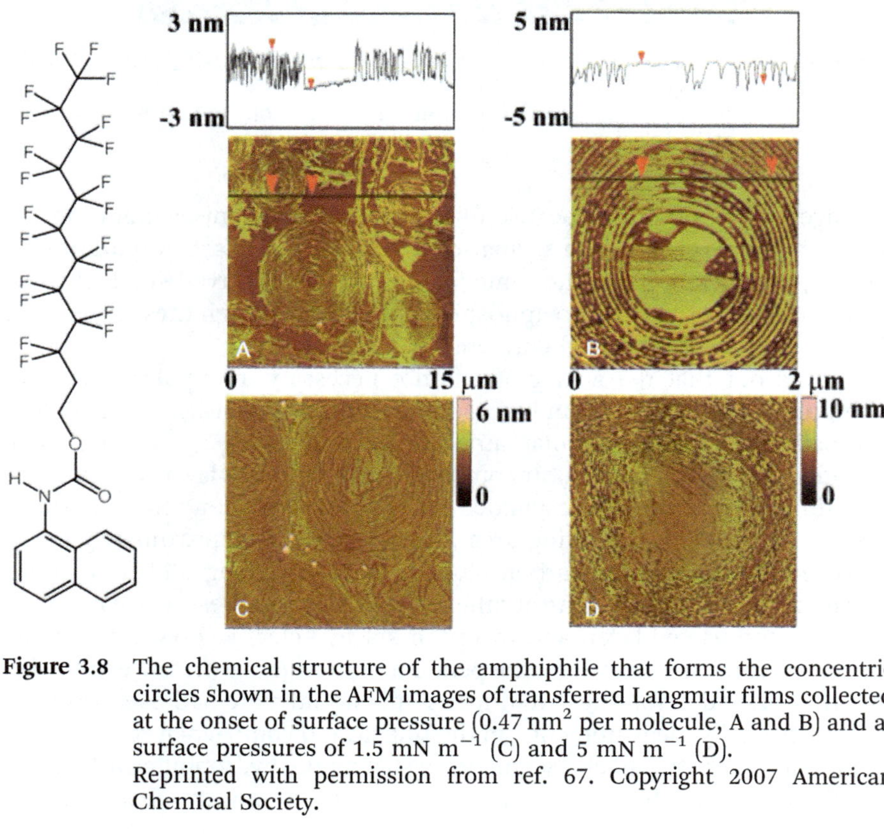

Figure 3.8 The chemical structure of the amphiphile that forms the concentric circles shown in the AFM images of transferred Langmuir films collected at the onset of surface pressure (0.47 nm^2 per molecule, A and B) and at surface pressures of 1.5 mN m^{-1} (C) and 5 mN m^{-1} (D).
Reprinted with permission from ref. 67. Copyright 2007 American Chemical Society.

one and three microns) and fibres of similar width and height of approximately 2.1 nm (coinciding with the approximate length of the molecule), followed by dense layers where the spirals are still evident (Figure 3.8). IR spectroscopy of the LB film of the compound shows that the fluorocarbon chain is poorly ordered when compared with structures where this segment is believed to be regularly packed. The urethane groups are clearly forming hydrogen bonds in the system, and it is reasonable to assume that these combine to form a one-dimensional hydrogen-bonded chain parallel to the interface. Interestingly, the films also showed circular dichroic effects, suggesting a spontaneous symmetry breaking during film formation by the intrinsically achiral molecule. Work with an analogous compound with a phenyl head group shows that the nature of this part of the molecule is very important to the assembly route. The results of the work show clearly that the combination of bulky head group and hydrogen bonding interaction lead to totally different superstructures to the simpler systems described earlier.

The structure of monolayers formed from a variety of quaternary ammonium-based amphiphiles at the air–water interface has revealed the importance of both non-covalent bonding between chains and the subphase composition.[68] The π–A isotherm curves and SFG spectroscopy of these compounds showed that the dialkyl dimethylammonium compounds are not so densely packed as other derivatives; the presence of many *gauche* defects in the conformations of the alkyl chains causes this situation. The conformational order and the density of the layers was increased when halide ions were introduced into the water subphase, even for compounds with ester and amide moieties in the chains of the surfactant, where hydrogen bonds help structure the in-plane organisation. In the less dense phases, the electrostatic repulsion between the ammonium head groups appears to be dominant in preventing ordering of the alkyl chains, whose flexibility allows space to be filled in the layer in the less dense packing. The quaternary ammonium surfactants are perceived as biodegradable compounds that are used widely commercially, and this study was aimed at understanding their structure and relating it with their action.

Gemini amphiphiles are interesting surfactants for the formation of a series of Langmuir layers with functional counter-ions, such as porphyrins,[69] polyoxometallates,[70] or even DNA.[71] They are attractive because the space they occupy on the surface can be varied through chemical design, thereby influencing the space that the counter-ion has available at the interface and therefore its organisation. As a quite stunning example—in terms of its demonstration of the influence of molecular constitution on supramolecular structure and nanoscale object morphology—the complex formed between a selenacarbocyanine dye and different imidazolium surfactants serves to demonstrate and inspire.[72] The imidazolium-containing surfactants—all bearing a heptadecyl chain at the 2-position of the macrocycle—were either of the gemini type and linked by alkyl chains of even carbon number between two and ten or a single imidazolium ring, as shown in Figure 3.9.

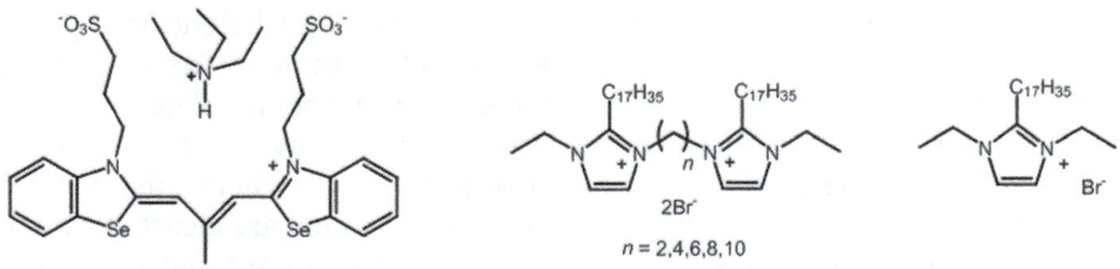

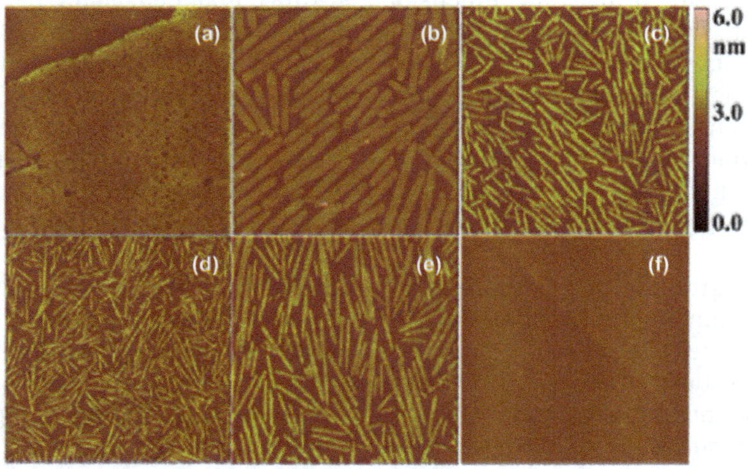

Figure 3.9 AFM images of a transferred Langmuir film of the selenacarbocyanine dye on monolayers of the surfactants indicated, where a–e correspond to adsorption to the gemini surfactant monolayers with $n = 2$, 4, 6 and 8, respectively, and f to the single imidazolium moiety-containing surfactant. All images' sides are 10 microns.
Reprinted with permission from ref. 72. Copyright 2006 American Chemical Society.

These surfactants were spread in chloroform onto a water subphase containing the selenacarbocyanine dye, whose structure is shown alongside the surfactants in the same figure, at a concentration where spontaneous aggregation did not take place. The dye spontaneously adhered to the monolayers formed by the surfactants, probably through electrostatic interactions with the surfactant and π–π interactions between the dyes themselves. Topographic AFM measurements of transferred monolayers showed the dramatic effect that the spacer in the gemini surfactants had on the morphology of the aggregates, all of which were different to the single imidazolium-containing surfactant molecule. All the images in Figure 3.9 correspond to layers transferred onto freshly-cleaved mica at 20 °C at a fixed load. The gemini surfactant with a two carbon atom spacer does not lead to a well-defined morphology, a layer with holes is transferred, possibly because of a mismatch between the space between the cationic head groups and the packing demands of the dye. On the other hand, all the other gemini surfactants give rod-like aggregates of various widths and lengths. It is interesting that the lateral growth is most favoured by the gemini with a four carbon atom spacing, that might correspond well with the space between polar groups in the dye, and therefore favouring a commensurate structure. When the transfer was performed at 30 °C, the fibres formed were thinner and longer, perhaps indicating that mass transfer and/or reorganisation in the layer aids the growth in a favoured direction determined by the π-stacking of the dye. The gemini surfactants have a quite unique effect, because the single chain surfactant produces a uniform layer. The modulation of the size of organic objects in this way is potentially interesting for the optical properties that they might possess, and the work demonstrates well that by modulating both surfactant constitution and temperature of the subphase, a vast array of nanometre-scale objects is accessible from a single dye component. The work fits into a broader area of amphiphiles for the preparation of nanostructured molecular materials, one of increasing interest and promise.[73]

3.4 Molecular Recognition at the Interface of Water and Air

The interactions between water-soluble hosts and amphiphilic guests (or *vice versa*) can be probed with great precision using the Langmuir trough and the nature of the surface pressure isotherms that the layers present.[74] An early example of this potential use was demonstrated for the binding of a synthetic biotin derivative to streptavidin dissolved in water.[75] The protein streptavidin comprises four identical subunits, each one capable of binding biotin derivatives very favourably (the binding constant, *K*, is approximately 10^{15} M^{-1}). The surface–pressure isotherms of the biotin-derived lipids are very different before and after introduction of streptavidin into the aqueous subphase in the Langmuir trough. The calculated surface area per molecule is far larger after binding, clearly indicating an interaction between host and guest. Control experiments (where the streptavidin was saturated by binding to non-amphiphilic biotin) ruled out non-specific binding as the source of the difference in these isotherms. Fluorescently-labelled streptavidin was also used as a probe for the binding of the protein to the surface. The intensity of the luminescence depended greatly upon the polarisation of the light relative to the orientation of the ordered domains (Figure 3.10). This anisotropy indicates clearly that the biotin lipid-streptavidin domains are highly oriented in the two-dimensional self-assembled system. The domains were approximately fifty microns in width and breadth. In fact, these domains were formed prior to compression of the layer, when the biotin was in the two-dimensional gas-like state, showing that their formation is a true self-assembly phenomenon rather than a result of preorganisation of the lipid on the water by compression. The domains were destroyed when a force of approximately 20 m Nm^{-1} was applied to the compressed layer, a feature seen in the surface pressure isotherms. Subsequently, transfer of these films

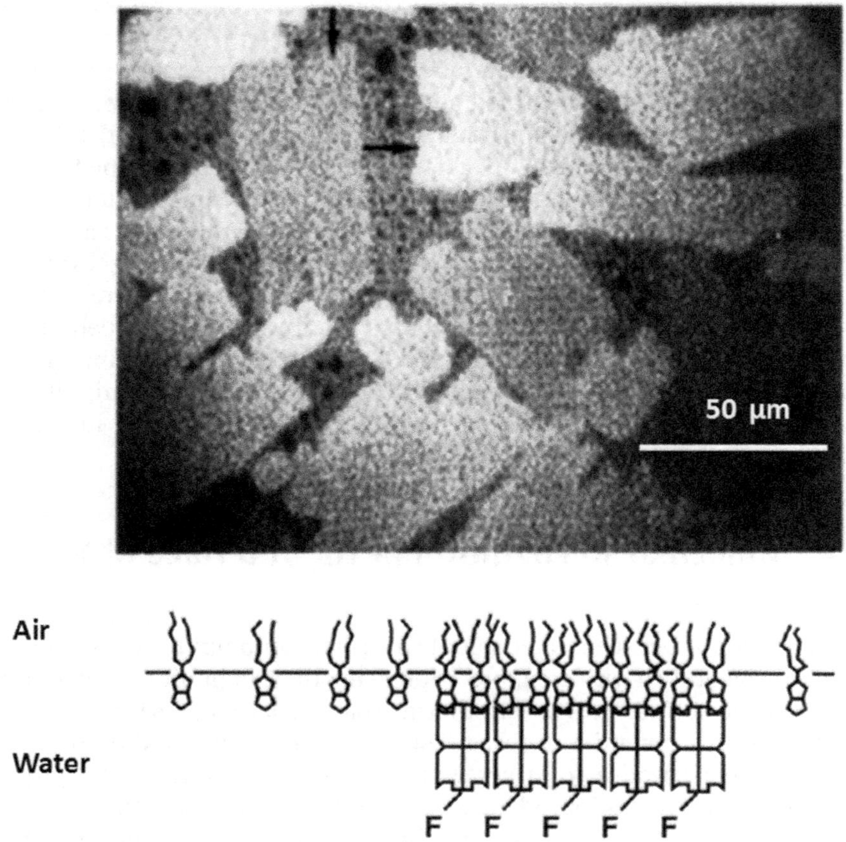

Figure 3.10 A fluorescence microscope image (top) of a monolayer at the air–aqueous solution interface (shown schematically below) of fluorescently-labelled streptavidin (the rectangular object comprising four blocks where F is a fluorescent probe) bound to a lipid derived from biotin (the fused pentagon in the cartoon). The black arrows in the micrograph indicate ordered domains that are oriented perpendicularly to one another.
Reprinted with permission from ref. 75. Copyright 1989 American Chemical Society.

onto TEM grids and their observation by microscopy and electron diffraction proved the presence of crystallographically-ordered arrays of streptavidin attached to the biotin monolayer in the domains.[76]

An electrostatic interaction between streptavidin proteins that adsorb at the interface leads to a significant effect of the ionic strength of the subphase on the streptavidin two-dimensional crystallisation.[77] A molecular modelling study implied that interactions between amine-containing side-chains of neighbouring proteins were responsible for this effect, a hypothesis confirmed by creating a mutant protein without these groups, the morphology of the two-dimensional crystals formed by these mutant proteins was rectangular unlike the wild-type protein that formed butterfly-shaped morphology under identical conditions.

Mixed Langmuir layers of the N-confused porphyrin (5,10,15,20-tetraphenyl-2-aza-21-carbaporphyrin) and methyl octadecanoate upon compression with subphases containing different halide salts of sodium showed a selective binding for one: iodide.[78] The surface pressure–area isotherm gave a more condensed film than the pure organic materials, and absorption spectroscopy indicated formation of both J-aggregates in the presence of this halide, specifically. H-aggregates were observed for the other halides, therefore the iodide appears to favour a different supramolecular arrangement of the chromophores within the monolayer. X-Ray photoelectron spectroscopy of the transferred film also showed the incorporation of iodide into the layer, but not other halides. The single protonation of the porphyrin at neutral pH was proposed as the driving force for this recognition, and the phenomenon is a kind of "aggregation-mediated" anion recognition, and *vice versa*.

By transferring concepts from crystal engineering, the combination of an anionic surfactant with a cation capable of forming a two-dimensional network with it has proven to be a host network for neutral lipophilic guests.[79] GIXRD showed that the two-component system of guanidinium ions and organosulfonate surfactant forms a two-dimensional network at low pressures (A in Figure 3.11) that when squeezed becomes a bilayer system with the cation submerged in the water but still in contact with the sulfonate group (B in Figure 3.11). The system is most robust when there is synergy between the two-dimensional hydrogen-bonded network and host–guest packing parallel to it, as shown by inclusion of an alkyl biphenyl derivative that forms a specific three-component system (C in Figure 3.11), as shown from the surface–pressure isotherm as well as GIXRD, which proved the order in the layer. The interactions between biphenyl units are important to the type of structure, because a straight chain alkane does not penetrate fully into the supramolecular host network (A in Figure 3.11). A much shorter chain on the neutral guest in the system does not lead to intercalation, indicating that the van der Waals interactions between the chains also play an important part in sustaining the three-component superstructure.

The separation of certain chemical functionalities within monolayers through like–like interactions can lead to phase separation, shown beautifully for aliphatic amino acid derivatives in an (*S*)-1,2-dipalmitoyl-glycero-3-phosphoethanolamine mixed layer (2 : 1 ratio of amino acid to phospholipid, Figure 3.12).[80] While *in situ* grazing incidence X-ray diffraction (GIXD, at a synchrotron facility) revealed that α-stearyl glutamic acid formed domains containing one or other enantiomer (enantiomorphous spontaneously resolved domains) as well as a racemic phase with the phospholipid crystallised separately, the N^{ε}-stearoyllysine showed a conglomerate (that is, separate domains of one or the other enantiomer). The conclusions had to be reached carefully by comparing the GIXD patterns of each of the components separately and comparing them with the mixed layer, which was prepared by spreading a mixed solution of the components in chloroform.

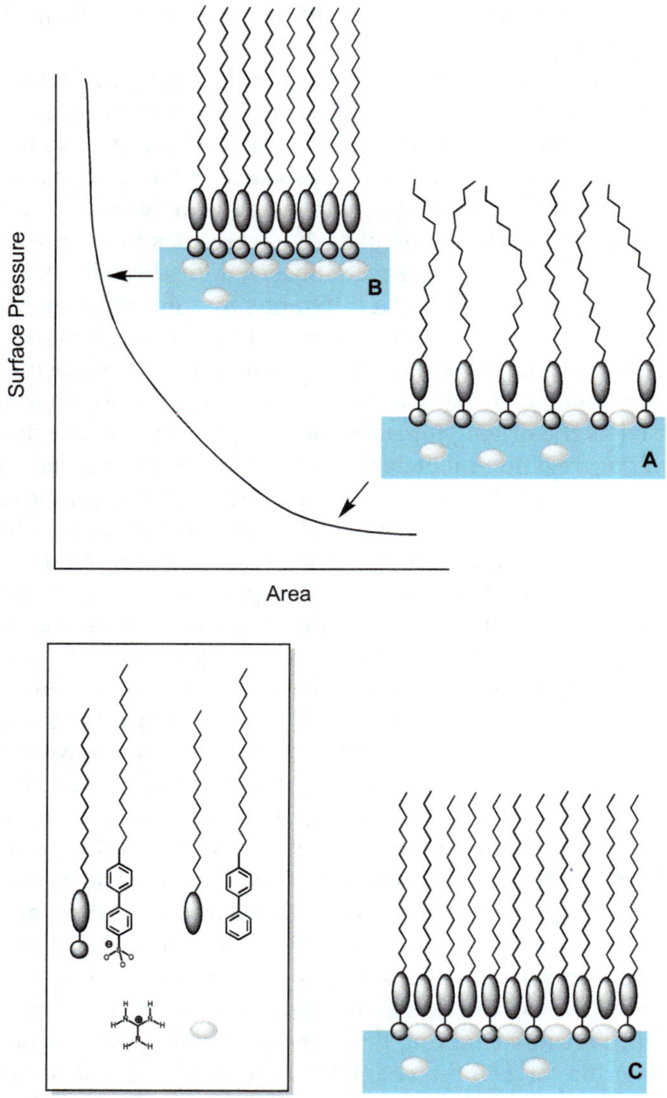

Figure 3.11 Representations of the monolayers of a bicomponent mixture of an organosulfonate anionic surfactant (see key in the box) and guanidinium cation at low (A) and high (B) relative surface pressures, and the three-component system that in addition incorporates a neutral alkyl biphenyl.

The lysine derivative had already been shown to form a conglomerate on its own at the air–water interface,[81] although in the mixed layer while spontaneous resolution is again observed, the packing is slightly modified. Nonetheless, the robustness of the like-enantiomer interaction in the head group of the amino acid derived head group—between the zwitterionic ammonium–carboxylate interaction as well as a hydrogen-bonded chain

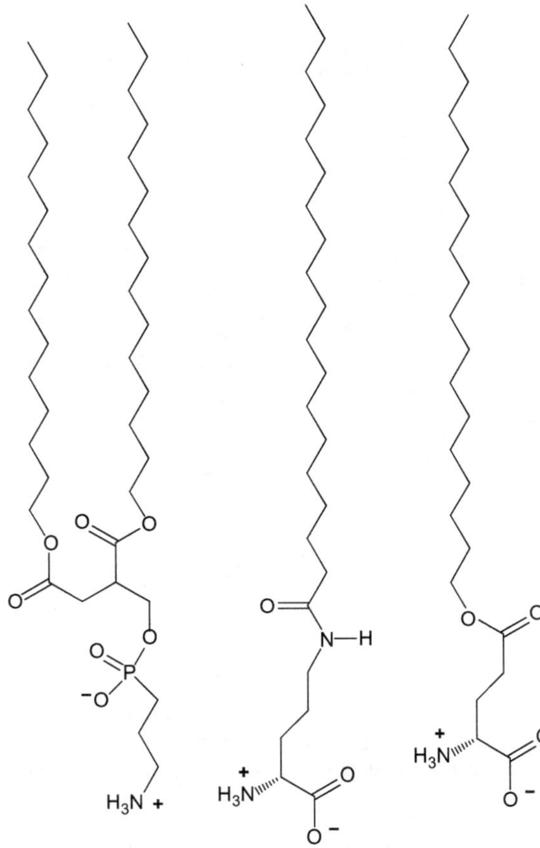

Figure 3.12 From left to right: (*S*)-1,2-dipalmitoyl-glycero-3-phosphoethanolamine, N^{ε}-stearoyl lysine and α-stearyl glutamic acid.

between the amide moieties—means that the derivative maintains the non-covalent bonding scenario largely with a difference in the tilt angle of the alkyl chains with respect to the interface's plane. The non-racemic (7 : 3 ratio of enantiomers) glutamic acid derivative also formed separate domains, enantiomorphic ones and a racemic area, both in the pure amphiphile as well as in the phospholipid mixture as revealed by the GIXD patterns. The results go some way to explaining the formation of homochiral strands of covalently-bonded molecules from related amphiphiles.[82]

The phase separation of racemic compounds into domains of left- or right-handed in these kinds of systems can be determined by the non-covalent interactions taking place in the layer, where hydrogen bonds can favour separation (Figure 3.13). In Langmuir layers, only the translation and a glide plane normal to the surface can give rise to racemic compounds, whereas the inversion centre, the glide with its plane parallel to the surface and the related two-fold screw symmetry (the most common symmetry elements in 3D) are not possible. However, the alkyl chain packing tends to be of herringbone

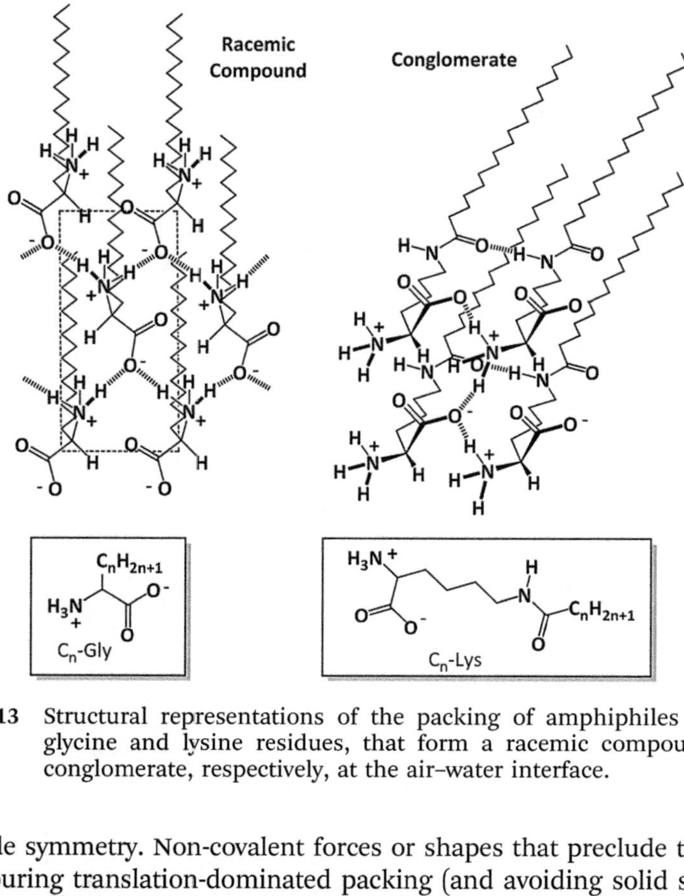

Figure 3.13 Structural representations of the packing of amphiphiles based on glycine and lysine residues, that form a racemic compound and a conglomerate, respectively, at the air–water interface.

with glide symmetry. Non-covalent forces or shapes that preclude this element, favouring translation-dominated packing (and avoiding solid solutions) are required.[83] Synchrotron GIXRD showed that the enantiopure amphiphiles C_n-Gly ($n = 12$ or 16) have crystalline domains with an oblique unit cell (chiral in two dimensions) wherein molecules are related by translation only. The corresponding racemates pack in racemic domains with symmetry elements of herringbone packing (Figure 3.13). Contrastingly, both enantiomerically pure and racemic C_n-Lys ($n = 17$ or 21) have homochiral domains with oblique unit cells, characteristic of a conglomerate. The linking of molecules through hydrogen bonds between the amide units incorporated in the amphiphilic chains gives translation symmetry. The chiral selection in assembly has been taken into more complex systems in monolayers,[84] whereby chiral amphiphiles with acidic or basic head groups where spontaneous separation of enantiomers can also be designed.[85]

It has been shown that in the binding of adenine, mono- and tri-phosphate have dramatically higher binding constants to guanidinium-based amphiphiles at the air–water interface than they do in the bulk or even in curved membranes.[86–88] Diaminotriazine-derived amphiphiles also bind nuclear bases with exceptional binding constants at the air–water interface, comparable with those in chloroform.[89] The stabilisation of hydrogen bonding interactions at the air–water interface is somewhat surprising in some ways, in that water might be considered a competitor for these interactions. Theoretical studies using quantum chemical analysis modelled the interface as two areas with contrasting dielectric constant (that of the lipid and that of the water) and showed that non-covalent bonding interactions are enhanced at the interface when compared with bulk water.[90] But even when the binding site is exposed to water, binding can be enhanced. The models used were guanidinium binding to phosphate and diaminotriazine binding to thymine. The reason for the large binding constants was assigned to modification of the interfacial potential by the reaction field arising from the hydrophobic region. It was proposed that this enhancement is a universal effect. An analysis based on the Poisson–Boltzmann equation reinforced this idea.[91] Indeed, the formation of supramolecular hydrogen-bonded tapes at the interface has been shown many times.[92,93] The hydrogen bonding at the interface can be disrupted when a polar co-solvent, such as dimethylsulfoxide, is present.[94] The stripe-like aggregates seen for the 2-amino-4,6-dioctadecylamino-1,3,5-triazine amphiphile in the presence of cyanuric acid are not evident in BAM images when this solvent is present, although dioxane's presence in the subphase apparently reinforces the non-covalent interactions between the complementary components.

The binding selectivity of a host in a monolayer on water's surface can be modulated in a mechanical fashion.[95] A host molecule comprising a hydrophilic triazacyclononane "head" bearing cholesterol "legs" (Figure 3.14) shows a moderate preference for uridine over thymidine (differing only by a

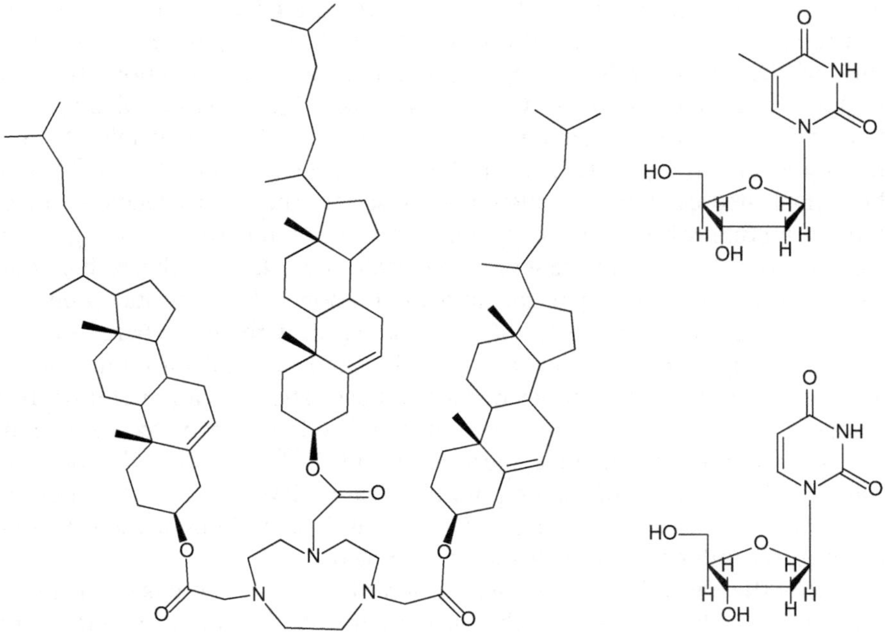

Figure 3.14 A molecular host (left) that recognises uridine (bottom right) at the expense of thymidine (top right) in Langmuir layers when lithium chloride is present in the subphase.

methyl substituent) binding when compressed at the interface of water and a solution of the guests. However, when lithium chloride was introduced into the subphase, a quite dramatic increase in binding constant for uridine was observed, while that for thymidine remained essentially unchanged. Therefore, selectivity was increased tremendously. The selectivity also changed as a function of surface pressure applied to the layer, being greatest just before the layer collapsed. The function of the lithium is to bind to the triazacyclononane "head" thus rigidifying the binding site, and evidently making the binding pocket more suitable for uridine binding. Thus, supramolecular rigidification of the host by lithium ion binding coupled with mechanical compression optimises molecular recognition in the system. The possible generality of this kind of host system was shown in the enantioselective recognition of amino acids, where compression of monolayers containing the sodium complex of a cholesterol-legged cyclen showed an inversion in selectivity between L and D valine or L and D phenylalanine upon compression.[96]

The organic organisation at the surface of water can be transferred to inorganic materials during their nucleation at the interface between water, monolayer and air. Calcium carbonate—because of its importance in the natural world, where it forms part of shells, bones and teeth—has been the subject of great attention regarding nucleation and growth, above all because metastable states are accessible and it is believed that understanding their formation will lead to a greater appreciation of the formation of crystals in general. In the introductory chapter, it was mentioned that a film of 5-hexadecyloxyisophthalic acid on a supersaturated calcium bicarbonate can template the formation of the metastable form aragonite.[97] The structure of the amphiphiles changes the nucleation phenomenon; the use of the octa-acids rccc-4,6,10,12,16,18,22,24-octakis-*O*-(carboxymethyl)-2,8,14,20-tetra(*n*-undecyl) resorc[4]arene and 5,11,17,23,29,35,41,47-octakis(1,1,3,3-tetramethylbutyl)-49,50,51,52,53,54,55,56-octa(carboxymethoxy) calix[8]arene both nucleated calcium carbonate crystallisation: the former leads to aragonite formation while the latter templates calcite nucleation and growth, although they do so at different areas per molecule.[98] It was believed here that the determination of polymorph and orientations of the crystals depends on the average charge density of the monolayer, with high charge density favouring aragonite or vaterite, and that the shape and calcium ion coordination nature of the amphiphiles was less critical. The nucleation of inorganic compounds at the air–monolayer–water interface is not restricted to calcium salts; other examples include barium fluoride[99,100] or sulfate[101] and iron salts.[102] The structure of the amphiphile clearly plays a role in determining the charge density (of carboxylate groups per unit area) though, and therefore has a direct influence on nucleation and growth.

During the crystallisation of these inorganic materials, it has been shown that both the head group of the surfactant and the mobility within the monolayer are important for effective nucleation.[103] A series of substituted carboxylic acids, derived from octadecanoic acid, with substituents at either the α (OH, Br, CH_3, and COOH) or β (OH) position were used to form monolayers at the air–water interface and were employed for the crystallisation of calcium carbonate. The stronger binders of calcium ions, the surfactants with a carboxylate or hydroxyl group at the α position, had a dominant electrostatic interaction with the metal ion and unassisted crystal growth was observed. In contrast, the Langmuir layers of the surfactants bearing methyl or bromine groups at the α position showed preferred growth, with oriented growth leading to truncated rhombohedra. It is believed that the relatively weak interaction between head group and metal ion and the flexibility of the molecules in the film at the interface are responsible for this effect, in line with other observations suggesting that a degree of adaptability of the nucleation-assisting ligand is required for efficient crystallisation and morphology control.[104] The reason for this effect appears to be the participation of the co-ion in the nucleation and growth (the bicarbonate ion can induce the formation of a cation-mediated network of hydrogen bonds[105]), so that strong binding negates this effect largely.

Langmuir monolayers can also be used for the templated growth of organic crystals. Chiral monolayers of surfactants derived from α amino acids template the formation of glycine which is oriented according to the stereochemistry of the chiral compound.[106] The *R* configuration α-amino acid surfactant promoted growth of one enantiotopic face of the achiral glycine that can be observed optically in the morphology of the crystals.[107] The growing face was used for the incorporation of glutamic acid from the solution, and from the racemate only the *S* enantiomer was incorporated. When the opposite enantiomer of the surfactant was used in the Langmuir layer, the *R* enantiomer of the glutamic acid was included preferentially in the glycine crystal (Figure 3.15). Consequently, the aqueous solution supporting the Langmuir layer was enriched in the opposite enantiomer to that included in the glycine, providing a route for amplification of chirality through the growth of an achiral compound.

The structural changes of amphiphile aggregates at surfaces—very difficult to observe directly—can actually be monitored by the transfer of their structure to templated organic crystal growth and the detection of the orientation of the crystals by second harmonic generation.[108] Crystallisation of 4-hydroxybenzoic acid in water gives crystals at the bottom of the dish, while small amounts of 4-methoxybenzoic acid lead to crystallisation at the air–water interface. The surfactants 4-(hexadecyloxy)benzoic acid, 4-(hexadecylamino)benzoic acid and 4-(hexadecanoyloxy)benzoic acid all induce the crystallisation of 4-hydroxybenzoic acid at the water–air interface. It was proposed that the formation of a hydrogen bonded dimer between the aliphatic acids at the interface and parallel to it promoted the growth of the crystals there. This hypothesis was borne out using the nonlinear optical effect that witnessed a reorientation of the templating amphiphiles—that are perpendicular to the interface on water—when 4-hydroxybenzoic acid was present in the aqueous subphase.

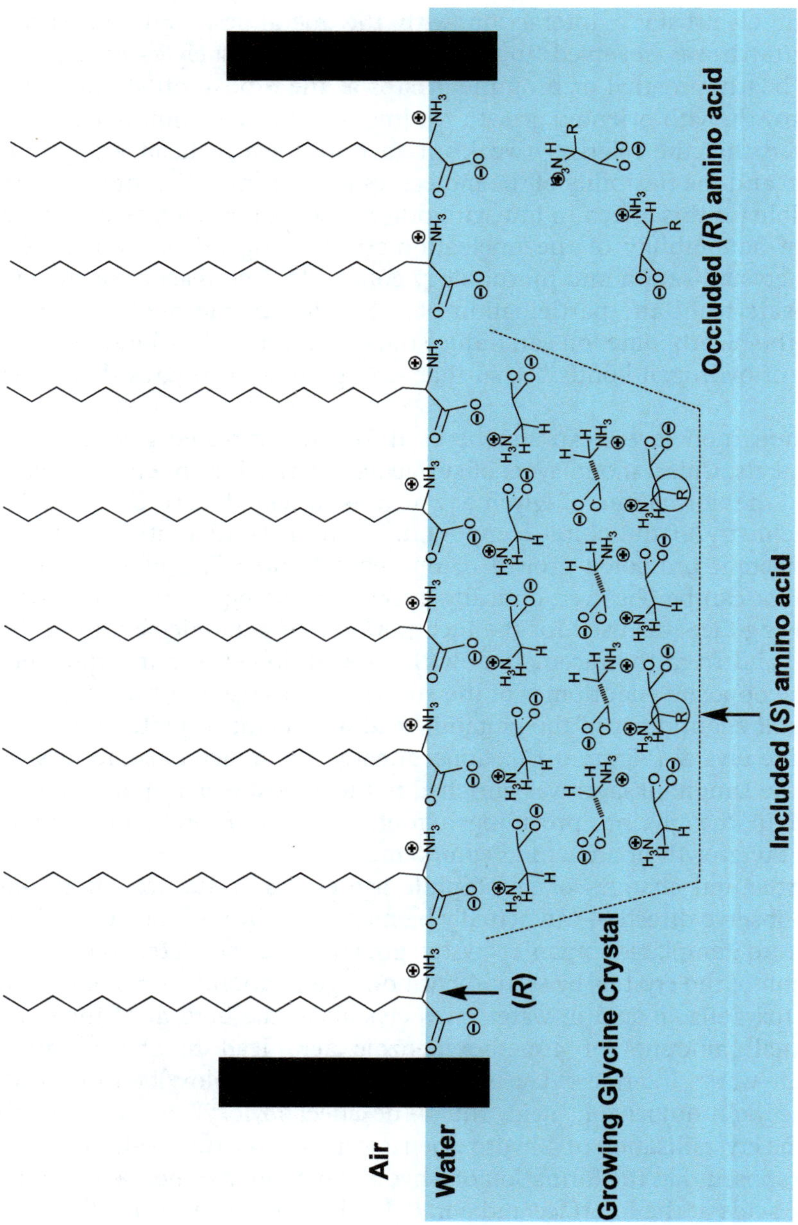

Figure 3.15 The templating of glycine crystal growth by a Langmuir layer formed by a chiral surfactant derived from an amino acid where the growing glycine crystal selectively includes one enantiomer of an amino acid from the aqueous subphase, leaving the opposite enantiomer in solution.

3.5 Supramolecular Manifestations of Molecular Conformational Change in Langmuir Layers

Small aggregates of molecules can be formed at sub-monolayer coverage on water with conformations determined by their thermodynamically most stable sate assuming equilibrium. Compression of these layers with the barrier in the Langmuir trough can change the orientation and conformation of the adsorbate. Relatively high lateral pressure could even straighten conformational bends or force macrocycles to close their preferred open conformation, or tilt the rings from a flat to perpendicular arrangement with respect to the interface plane. Therefore the supramolecular function of the components of the monolayers can be hugely influenced by squeezing them in the plane, reducing the area per molecule.

The compression of layers on water under air can lead to conformational changes within the amphiphile molecules as well as to supramolecular interactions between them.[109] The compression of layers of a derivative of the carbohydrate methyl xylopyranoside bearing three 4-(pyrene-1-yl)benzoyl groups in a mixture with methyl stearate or methyl 2,3,4-tri-*O*-stearoyl-β-D-xylopyranoside demonstrate the supramolecular influence on conformation. The latter is a rigid matrix, and induced the xylopyranose ring to adopt its metastable 1C chair conformation, while the softer matrix included the molecule in its thermodynamically more stable C1 conformation. The fluorescence characteristics of the two conformations were very different, and therefore this kind of medium-sensitive optical response within the monolayer as a result of the supramolecular surroundings might prove useful for exploration of the nature of different phases of monolayers of other surfactants.

The opening and closing of spiropyran-based macrocycles incorporating oligoethyleneoxy chains show differing responses to metal ions at the air–water interface.[110] Above pH 3, the spiropyran unit is "open" (Figure 3.16) and there is a charge located on the molecule. As a result, it occupies less surface area on the water presumably because of the more open geometry and the charges on the cycles cause a more upright orientation at the interface. Despite the lack of alkyl chains on the molecules, they do form

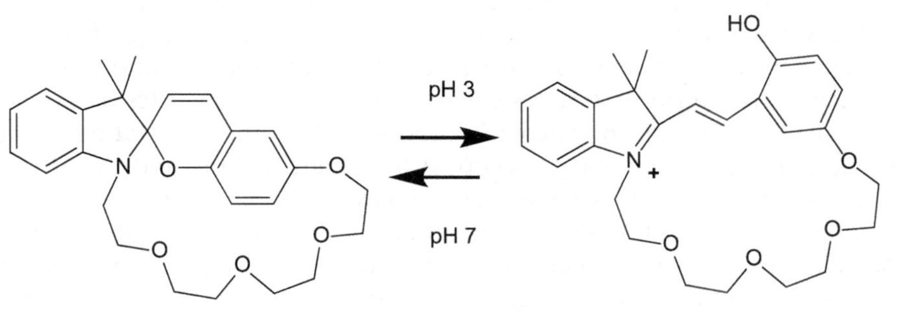

Figure 3.16 A macrocyclic spiropyran derivative that binds sodium ions selectively at the air–water interface in its closed form at neutral pH.

monolayers although the isotherms are not very well defined with or without cations. However, clear differences are seen in the surface area–pressure isotherms when lithium, sodium, potassium or copper(II) ions are used in the subphase. In the closed form at neutral pH, a clear binding preference for potassium was observed, with the onset of pressure detection starting at 170 $Å^2$ per molecule, while the lithium and sodium had an onset at approximately 120, copper(II) at 100, and the pure host at approximately 70 $Å^2$ per molecule. It was proposed that a sandwich-type complex of the sodium by the macrocycle might be taking place. In the open form of the molecule at pH 3, the onset of detectable surface pressure was at 45 $Å^2$ per molecule. Exposure of this form to the metal ions in the subphase caused an onset of surface pressure at approximately 90 $Å^2$ per molecule for all cases, indicating a non-specific interaction possibly because of the presence of the cationic form of the molecule at the interface. The work shows how response to substrate ions depends greatly upon the conformation and charge of the host at the interface.

Large macrocycles substituted with steroid derivatives (cholic acid) produce chiral cavity arrays at the air–water interface that can also be deformed (closed) upon applying pressure with the mobile barrier in the trough.[111] The host comprised a 1,6,20,25-tetraaza[6.1.6.1]paracyclophane core linked to four steroid groups by an L-lysine spacer, which is flexible and pH responsive. When the host was spread and compressed, as the layer formed, fluorescence intensity was observed for derivatives of both cholic acid, which has three hydroxyl groups giving it a hydrophilic and hydrophobic face, and cholanic acid, which is purely hydrophobic. The cholic acid derivative showed two steps in its isotherm, the first presumably with the steroid hydrophilic surface in contact with the water, and the second with these arms forced upwards. In this latter step, the fluorescence intensity of the incorporated dye increases dramatically. The fluorescent dye was seen to reversibly enter and exit the cavity of the cholic acid derivative at the interface upon passing between these two adsorbed forms. The phenomenon was pH dependent; at high values, the formation of free amine (instead of ammonium) is believed to render the macrocycle less dynamic. The pressure-induced fluorescence change was repeatable over several cycles, suggesting a possible use as a kind of actuator, showing piezoluminescence.[112]

Large molecules with multiple interaction sites can undergo significant changes in their orientation and non-covalent bonds upon compression of the layers they form. Rotaxanes with multiple recognition sites are a good example. The [2]rotaxane shown in Figure 3.17—with two "stations", one tetrathiafulvalene one that is occupied by the cyclophane and the other a dioxynaphthalene residue—forms a liquid expanded phase under moderate compression.[113] In this state, the naphthalene ring that is not at the centre of the cyclophane component may enter alongside π stacking interactions with the 4,4′-bipyridinium moieties, so that an overall tilted orientation is adopted. Under compression, the molecules are forced into adopting a perpendicular arrangement with respect to the air–water interface.

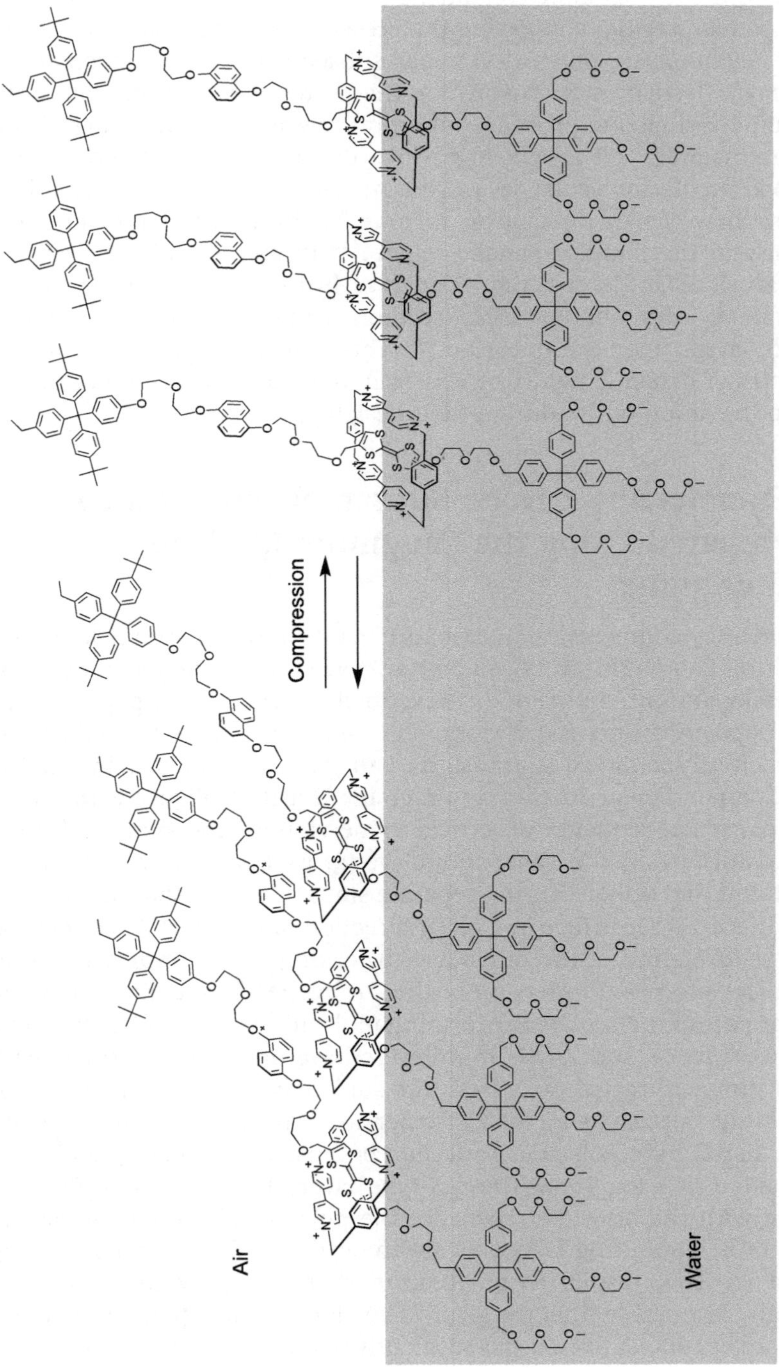

Figure 3.17 A [2]rotaxane that passes from a liquid expanded region in its Langmuir layer where the naphthalene ring that is not at the centre of the cyclophane component may enter alongside π stacking interactions, which under compression, converts into a solid phase in which the components are quasi-perpendicular to the interface.

The isotherms of the layers showed that compared with the free thread component, the tetracationic cyclophane favours the formation of a defined liquid expanded region, supporting the presence of favourable interactions between neighbouring molecules. It has subsequently been demonstrated using X-ray reflectivity at a synchrotron that water of hydration around the tetracationic cyclophane plays an important role in the structuring of this kind of monolayer.[114] These changes in hydration have been verified and understood at the molecular level using molecular dynamics calculations that show how the cyclophane is surrounded by water at the interface, particularly in the liquid expanded region.[115] In amphiphilic [2]rotaxanes incorporating pillar[5]arene, the balance of hydrophilic and hydrophobic regions along with similar cross sections leads to very well behaved Langmuir layers upon compression.[116] This contrast highlights how the possibilities of dynamic nature or very well defined layer formation can be built into the amphiphiles during their design.

3.6 Transferring Layers: Supramolecular Systems Prepared Using the Langmuir–Blodgett Technique

Once formed by compression of cast surfactants at the air–water interface, the resulting Langmuir layers can be transferred using the technique first developed by Katharine Blodgett[117] described in detail in her paper "Films Built by Depositing Successive Monomolecular Layers on a Solid Surface" where calcium stearate was transferred to glass and metals by vertical dipping (perpendicular to the air–water interface).[118] Figure 3.18 shows how the particular sequence of vertical immersion or emersion can lead to different films using the same monolayer(s) as materials for the films provided that they adhere during separate steps (it can be that monolayers do not adhere to the surface in a given direction of transfer). Each method requires re-formation of the monolayer by casting and compression after each transfer process. Immersion of the surface leads to contact between the apolar part of the monolayer-forming molecule and the solid. It can be appreciated that X- and Z-type multilayers have opposite polarities, with the polar groups directed away from the surface because of immersion or away from the surface as a result of emersion. The Y-type multilayers, on the other hand, have polarities cancelled because of alternating emersion or immersion at each step. These Y-type layers have a clear difference compared with the other two, because the polar head groups are always adjacent in neighbouring layers, as are the hydrophobic sections, leading to a bilayer. This results in a situation that is clearly more thermodynamically favourable than the X- or Z-type layers where polar and apolar sections are in contact but are not covalently bonded. In all three cases, it is possible to incorporate different molecules or even mixtures of them at each step.

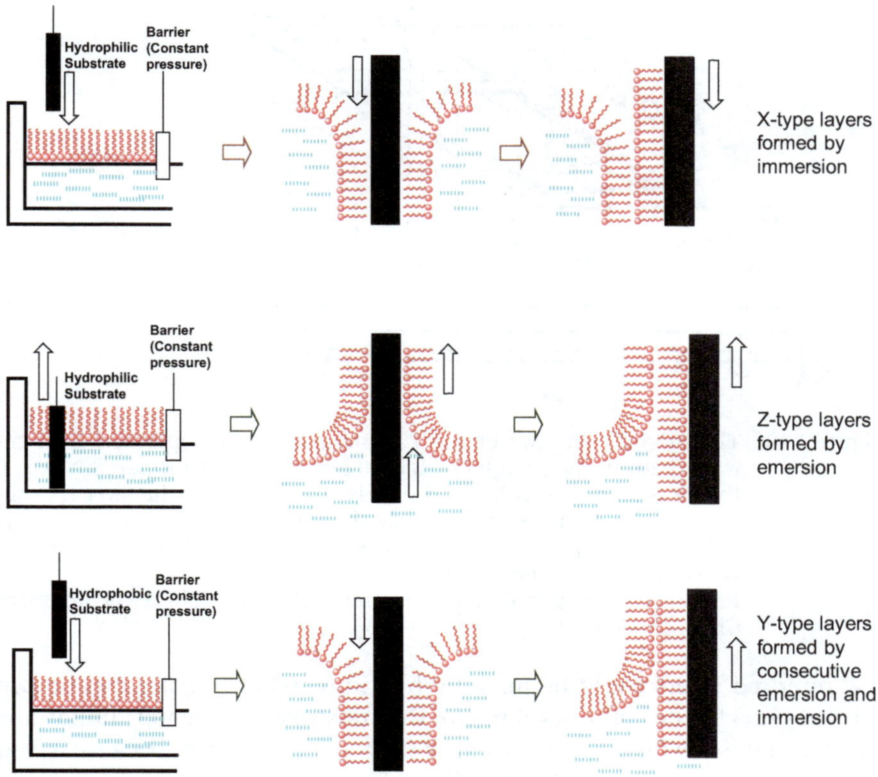

Figure 3.18 The three different generic types of Langmuir–Blodgett layers and the dipping sequences that lead to them in the Langmuir trough.

The apparatus used for the first time is shown in the drawing from the original paper in Figure 3.19. It can be observed that a handle was used by Blodgett to wind up and down the substrate to which the Langmuir layer was transferred. The remainder of the set-up is that typical of a Langmuir trough.

The seminal paper by Blodgett and Langmuir in 1937 demonstrated the transfer of no less than 3001 layers of barium-copper stearate onto glass.[119] The paper describing these Y-type films laid down many of the important principles, possibilities and difficulties in preparing multilayers using the transfer technique that bears their names. Deposition of the first layer, the monolayer upon which all subsequent layers are built, onto surfaces of metal or glass supports involves different considerations to the bulk of the film, because it must adhere well to the support. In the case of glass, stearate adsorbed better at alkaline pH, although adhesion at neutral pH could be improved by treating the glass with paraffin wax or stearate salts, effectively making the surface hydrophobic and favouring adhesion. The subsequent transfers involved monolayers adhering to similar sheets. The process is very sensitive to the counter-ion to the stearate anion; thousands of layers were only possible with barium, and not calcium, magnesium or zinc, for which a

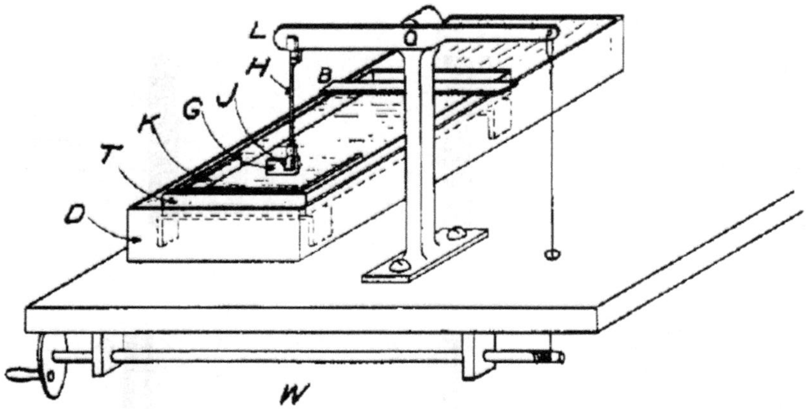

Figure 3.19 The original drawing of the Langmuir–Blodgett apparatus: D is the water bath in the trough (T), B the barrier coated with paraffin wax, G the glass slide that is lowered or raised, operated by the lever L or the windlass W connected to the slide through a rod (H) with a clamp (J). Detachable strips of glass (K) stopped the organic solution reaching the waxed surfaces of the device.
Reprinted with permission from ref. 118. Copyright 1935 American Chemical Society.

few hundred layers could be made. While immersion transfer can be performed quickly, the withdrawal rate of the substrate is extremely important, because if it is too fast, water pulls back the transferring layer because it wets the film very well when the hydrophilic side of the layer faces away from the substrate and upon emersion can flow back to the surface taking the layer with it. The presence of approximately 10% copper salts relative to barium in the subphase was found to be essential to prevent cracking and "fogging" of multilayers, that is the presence of unidentified specks giving the film a cloudy appearance: no other salt was found to be so effective. Because the films prepared at pH 7 contain both stearate and steric acid, the neutral organic can be washed out with benzene leaving a so-called "skeleton film" that is also birefringent and has a much lower refractive index than the original film. The thicknesses of the films were determined using optical measurements of the interference of light reflected from the top and bottom surfaces of the multilayer films, and for the barium stearate gave a bilayer spacing of 24.4 Å. All the films were found to be uniaxial crystals, whose optic axis was perpendicular to the substrate's surface plane.

An early example of the formation of functional Y-type layers was that of the formation of multilayers of chlorophyll on glass or chromium.[120] Up to 600 layers were formed, each with a calculated thickness of approximately 14 Å (assuming a refractive index of 1.64). Higher pressures on the film resulted in films that were assumed to contain water, because the double layer was up to 60 Å thick. The surface of the film with emersion as the last step was hydrophobic, with a contact angle with a water drop of 80°. While chlorophyll is highly fluorescent in dilute solution, in the mono- or multi-layers the fluorescence is quenched because of the proximity of the chromophores. The monolayers of chlorophyll can be transferred onto multilayers of barium stearate giving a contact angle of 90°, significantly more than the pure chromophore multilayers and showing how the substrate can influence the multilayer structure.

The ability to change composition and polarity in the LB films led to their wide use to study supramolecular materials for optoelectronics.[121] In materials for non-linear optics and electronics especially, the alignment of molecules that the LB method provides is useful,[122] as it rules out the tendency of many molecular systems to organise in centrosymmetric ways.[123] This realisation drove the development of more sophisticated troughs for the preparation of non-centrosymmetric layers.[124] The use of alternating layers in the Y type films is a logical and successful strategy for the preparation of thin films that show non-linear optical activity: a two-component system comprising ω-tricosenoic acid and a highly polarisable merocyanine dye[125] or LB films comprising a subphthalocyanine and behenic acid are good examples.[126] The alternating chromophore–polyelectrolyte sequence (reminiscent of layer-by-layer assembly) is similarly successful for the preparation of noncentrosymmetric LB films.[127]

It is clear that the transfer of the Langmuir layers onto solid substrates gives films that can be probed with a wide range of surface science methods, but one should consider that the new interface is somewhat artificial, in that the monolayer is forced into contact with the solid. Evidently, if there is a repulsive force between the side of the monolayer that is transferred and the solid's surface, the layer will either transfer poorly or will not be stable. Therefore the choice of support material is important. An example of this is the adhesion of chemically-modified tetraether lipids caldarchaeol-PO_4 on silicon oxide surfaces that had been modified with chemisorbed layers of different types.[128] The monolayer films on hydrophobic substrates were more resistant to washing with organic solvents than the substrates terminated with amine groups in the chemisorbed SAMs. While there can be sensitivity to a substrate, this dependence is not always present and in certain cases the formation of complexes in the Langmuir trough can lead to reliably transferred films, and the solid substrate precision probing can be carried out.

The complex formed by stearic acid monolayers with branched poly(ethyleneimine) introduced in the subphase produced a uniform layer.[129] The domains were liquid crystalline, and were observed by fluorescence microscopy in the first instance. However, transfer of the complex in its liquid crystalline phase onto mica allowed a more precise study by AFM.[130] The films showed mean domain sizes and shapes that varied with the rate of monolayer compression, being more polydisperse and smaller at high velocities. The AFM clearly showed the mechanical difference between liquid-like and liquid crystal-like domains through applied force spectroscopy with phase-sensitive detection. Nanometre-scale morphological detail within the stearic acid domains was discerned using the probe technique that was beyond the optical microscopic techniques of the time.

Thus the edges of the domains appeared more ordered (particularly for fast compression) and very small points were observed within the domains (that were also observed between the ordered domains in the liquid-like regions). Importantly, AFM was also able show that the presence of the fluorescent dye did affect the structure and morphology of the aggregates. The dye-containing mixture had fibre-like morphology while without the dye, a fat distorted star-like domain shape was seen. Therefore, when transfer of the monolayer is reliable, the AFM can reveal important morphological detail not perceived optically in the Langmuir monolayer and show the effect of small quantities of additive on film structure and morphology.

The structure of chromophores determines their supramolecular organisation in Langmuir and LB films and therefore their properties, such as fluorescence. The family of hemicyanines shown in Figure 3.20 were used to prepare LB films and their fluorescence was measured in order to understand the effect of structure, orientation and assembly on this property.[131] The compounds have quite different constitutions, one of which has a quite classical "head–tail" type, where the polar head group might be expected to be close to the water subphase with the alkyl chain upright. In fact, this compound and the "tail–head" analogue both show similar isotherm characteristics as monolayers, and not dissimilar absorption profiles, although the LB film of the latter shows a morphology with flower-like structures when studied by AFM.[132] The behaviour of the compound with two tails (tail–head–tail configuration) is quite different in its isotherm characteristics as well in its transfer properties to solids. It can be transferred onto both hydrophilic and hydrophobic surfaces at relatively high surface pressure compared with the other compounds, and forms very smooth hydrophobic films with different thicknesses: between 5 and 6 nm on the hydrophobic surface and between 2.5 and 3.5 nm on the hydrophilic one. This observation suggested a more extended structure on the hydrophobic surface that actually gives much higher fluorescence intensity than either the same compound deposited on the hydrophilic surface or indeed the other compounds. In a separate piece of work concerning the

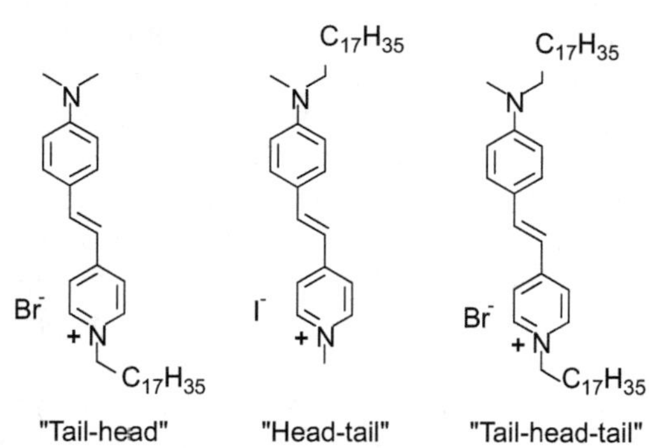

Figure 3.20 The chemical structures of a family of hemicyanines that show the importance of structure–property dependence in LB films.

fluorescence of LB films transferred from a subphase at different pH, it was shown by absorption spectroscopy that a proportion of the transferred films comprised non-fluorescent H-aggregates.[133] Also, time-resolved measurements of the fluorescent species showed three decay processes, two of them associated with isomerisation of a stilbene unit in the chromophore. Thus, photochemistry can reveal multiple supramolecular situations within the LB films. The unique combination of molecular structure and supramolecular arrangement determined by the substrate, as well as the use of subphase pH control, demonstrate the importance of the material used to transfer the layers.

The transfer of layers can produce multiple environments, as has been seen in the previous example, but annealing can yield more homogeneous films. When an LB film was made from zinc(II) phthalocyanine by casting its solution in *N*-methyl pyrrolidone (NMP) and chloroform onto water and transferring the sample by vertical dipping onto quartz, glass or ITO, absorption spectroscopy revealed the presence of H and J aggregates as well as monomer.[134] This film prepared near room temperature was annealed at 65 °C resulting in a layer that showed the presence of H-aggregate (characteristic of the α-phase of the solid) and monomer only in the absorption

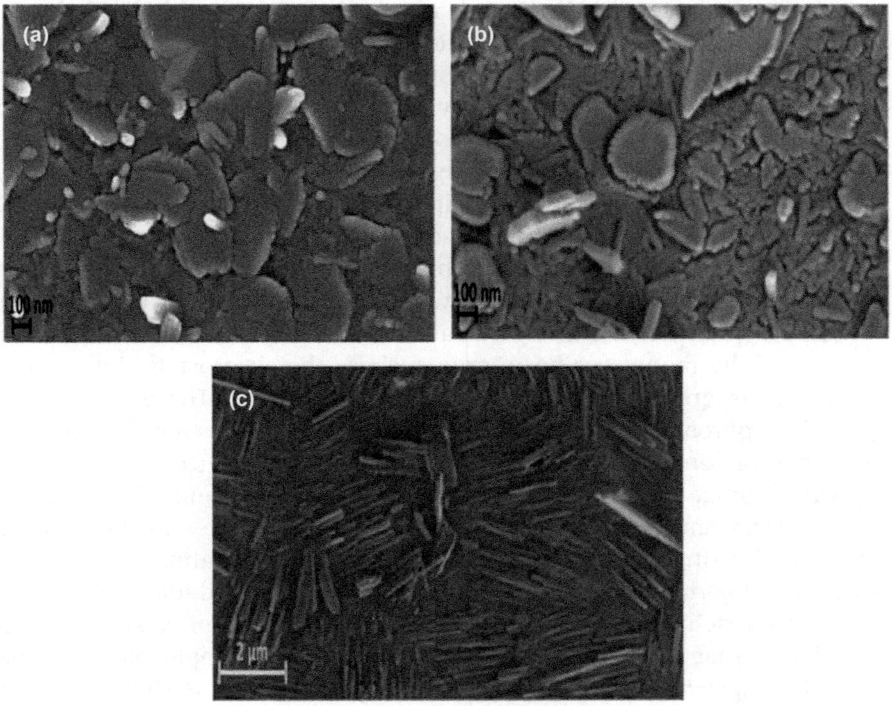

Figure 3.21 FESEM images of LB films made of 40 layers of zinc(II)phthalocyanine (a) as deposited and annealed at (b) 65 °C and (c) 290 °C.
Reproduced from ref. 134 with permission from the Royal Society of Chemistry.

spectra (Figure 3.21). In addition, field emission scanning electron microscopy (FESEM) of the films showed a clear change in their morphology, from predominantly plates to aggregates with a more acicular sub-structure. Further heating to 290 °C resulted in a structural phase transition (to the more stable β polymorph[135]) driven by π–π stacking of the aromatic macrocycle. There is an accompanying change in the morphology of the films to a network of needles that stack locally parallel to one another. This structural and morphological phase change results in improved conductivity in the film. In another kind of π-functional material, tetrapyridylporphyrins have been shown to have different inter-chromophore orientations in LB films made of the compounds or their cadmium(II) complexes that were formed at the air–water interface before transfer.[136] While the orientation of the molecules was similar in the LB films of the pure compound and its cadmium complex, the coordination of the metal ion by the pyridyl nitrogen atoms forces a different packing of the chromophores in the layers, showing an alternative to the control of chromophore packing in this kind of film.

The family of complexes and interlocked molecules derived from cyclo bis(paraquat-*para*-phenylene) and π electron rich aromatic guests have been explored extensively because of their relevance for the preparation of devices.[137] The parent cyclophane is not amphiphilic, but can be organised at the surface of water-forming monolayers using the phospholipid dimyristoylphosphatidic acid as a counter-ion and "anchor".[138] The monolayers could be prepared either by direct co-spreading, or by the prior formation of a monolayer of the phospholipid followed by injection of the cyclophane into the subphase and its adsorption. An optimum ratio of the two components was that expected for a perfect charge balance between the tetracationic cyclophane and the mono-anionic lipid, as revealed by surface potential determination. The kinetics of the adsorption of the cyclophane (at different subphase concentrations) from the bulk aqueous phase to the Langmuir layer of lipid is diffusion limited, and were explained appealing to the Nernst diffusion layer model, assuming that as soon as the cyclophanes are within a distance of the monolayer head groups where the electrostatic interaction energy is far greater than kT, they are bound there. However, the co-spreading approach was favoured because of the use of less material and the avoidance of long adsorption times. These films were transferred using vertical dipping, and laid the foundation for the formation of multilayers of pseudorotaxanes (comprised of a thread-like molecule with no stopper groups and a ring) at the air–water interface.[139] The combination of cyclo bis(paraquat-*para*-phenylene), the same phospholipid used for this host and a π electron rich aromatic amphiphile (Figure 3.22) when co-spread form monolayers that can also be transferred and form multiple bilayers. The absorption spectra of sixteen bilayers of the pseudorotaxane showed a clear charge transfer band with a maximum at approximately 540 nm, while the cyclophane multilayers with no π electron rich guest show an absorption maximum at approximately 400 nm with no hint of a charge transfer band. The use of the phospholipid is essential; in its absence, co-spreading of

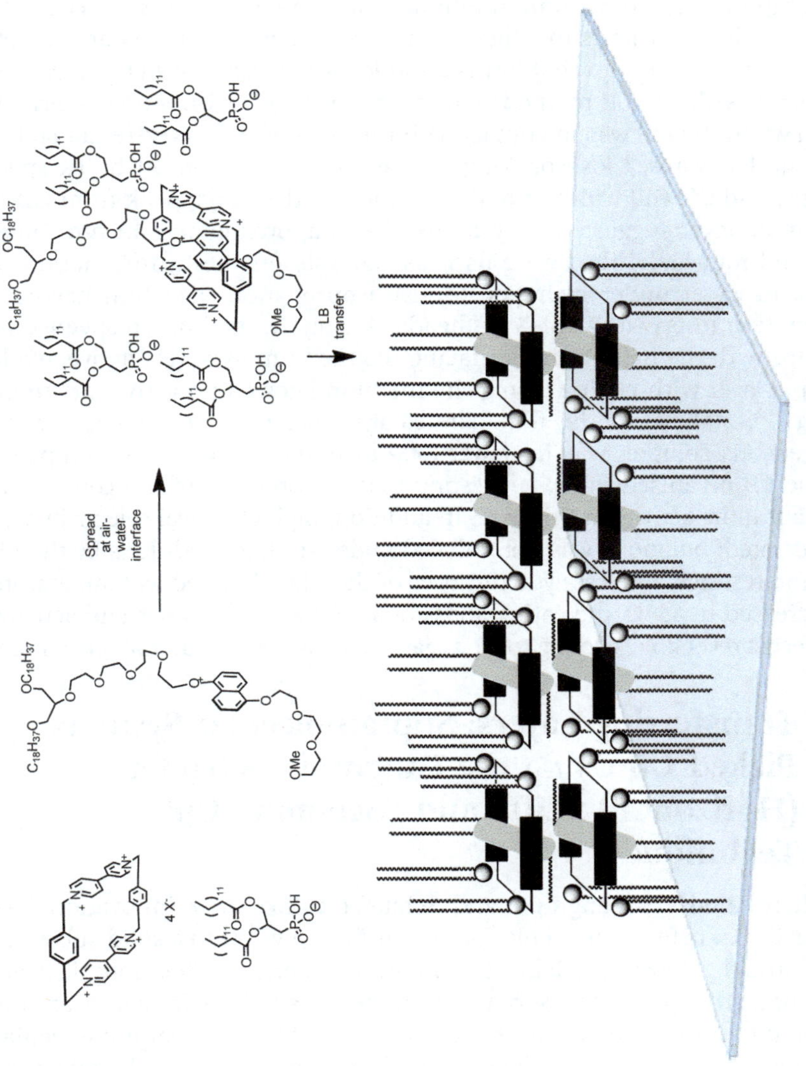

Figure 3.22 The components of a self-assembling pseudorotaxane that forms at the air–water interface, and can be transferred onto hydrophobised quartz.

cyclophane and thread results in decomplexation as a result of the favourable hydration of the tetracation. A similar approach using the phospholipid has been employed for the formation of multilayers of [2]catenanes, where the periodic layered structure was proven using small angle X-ray scattering as well as absorption spectroscopy.[140]

The order in LB films has proven to be influential in the catalytic action of amphiphilic (2,2′-bipyridine)rhodium(hexadiene) complexes.[141] The 2,2′-bipyridine ligand when substituted with two heptadecyl chains at the 4,4′ positions gives a complex that forms a stable monolayer on water that can be transferred with the LB method onto hydrophilic or hydrophobic glass, although when the tail was in contact with the glass, the films were not stable except under water. Close packing in the layers was inferred by IR spectroscopy, and the full width at half maximum of the Bragg peak from GIXD of the monolayer suggested a crystallite size of approximately 4.5 nm. One-, three- and four-layer films were used as catalysts for the hydrogenation of acetone in water under hydrogen to give isopropanol, each film having a different terminus (with "tails" or "heads" facing the medium, respectively). The films with the metal complex facing the medium were very active, while the triple layer with chains facing the medium showed no activity, demonstrating the stability of the film. The catalytic activity of the complexes in homogeneous solution was far lower (500 turnovers in two days) compared with the LB organised films, suggesting that the ordering of the complex is important in its activity in this case. In addition, high chemical selectivity was demonstrated: butanone was not reduced under similar conditions by the LB films. Interestingly, the catalytic turnover of the films lowered as temperature was increased to 45 °C, indicating a loss of supramolecular order and activity. The effect is reversible, cooling to 30 °C gave an active layer material once more.

3.7 Transferring Layers: Supramolecular Systems Picked Up Using the Langmuir–Schaefer (Horizontal) "Lift" and "Scooping Up" Technique

The lift method—as Langmuir and Schaefer called their innovation[142]—involves the transfer of molecules on the surface of water to a solid substrate that is brought down parallel to the air–water interface. Therefore, it is actually not a prerequisite to apply a pressure to the surface with a barrier. It is perhaps the most suited transfer method for "fishing" the supramolecular structures that are formed at the air–liquid interface. In their original study, Langmuir and Schaefer developed this method partly to ensure that the hydrophilic side of enzyme monolayers would face away from a solid substrate and therefore maintain its catalytic activity. Specifically, they studied the clotting of milk by pepsin. Therefore, also, a reason for developing the lift method was that relatively high surface pressures could reduce the activity of the enzymes; therefore, the vertical transfer methods are not suitable

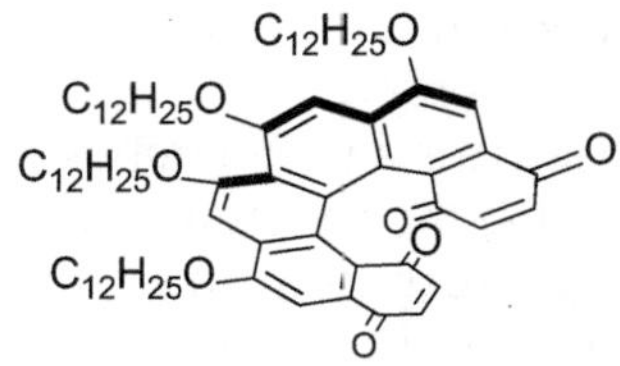

Figure 3.23 A helicene derivative that forms columnar aggregates that align in monolayers on water and can be transferred onto hydrophobic glass by horizontal dipping.

because the proteins are presumably deformed during the processes. The lift method allows transfer of fragile monolayers while maintaining their structure.

The technique is useful for the study of both sub-monolayers and for the formation of multilayers. A particularly interesting case of the formation of multilayers of fibrils is that of a helicene derivative (Figure 3.23) that self-assembles into columns in solution and on the surface of water when deposited from chloroform solution.[143] The films were compressed and allowed to stand for thirty minutes before being transferred onto hydrophobitised surfaces (glass, quartz or silicon for different studies) by horizontal dipping up to sixty times. Vertical dipping was also employed, as it has the advantage of being faster and easier, though it is far more sensitive to changes in temperature and humidity. AFM showed that the layers comprised columns several hundred nanometres long and approximately five nanometres wide. Both non-racemic and racemic samples were used, with the order being greater in the former case. In addition, the second harmonic generation by the films of the non-racemic sample were approximately thirty times that of the racemic case, showing the clear influence that chirality has on non-linear optical effects.

Metal–organic frameworks comprising cobalt(II)tetra-(phenylcarboxylato)-porphyrin coordinated through the carboxylate moieties to copper(II) ions made on water have been used as the two-dimensional base for the growth of a film by horizontal dipping.[144] Two of the metal ions coordinate four of the anionic groups and two pyridyl ligands in a paddlewheel-type arrangement, while the pyridyl ligands also bind to the central metal ion of the porphyrin leading to a square two-dimensional sheet. The material was made by casting a solution of the porphyrin and pyridine in a mixture of chloroform and methanol onto the water containing copper(II), in an approach reminiscent of one used for zirconium octadecylphosphonate films prepared by the Langmuir–Blodgett transfer method.[145] In the present case, they were transferred by horizontal dipping onto silicon (100) or quartz. Rinsing and repetitions of this procedure gave the films that were characterised by absorption spectroscopy in the first instance, but also by synchrotron X-ray diffraction, which showed a remarkable degree of structural order, probably because of the steric and non-covalent demands of packing the sheets—inter-sheet π–π interactions are formed. It was claimed that this approach to metal–organic frameworks was unique in its capacity to provide both out-of-plane and in-plane crystalline order.

The combination of porphyrins and fullerenes bearing oppositely charged side chains was achieved using the Langmuir–Schaefer method onto indium tin oxide (ITO), and a photocurrent was observed upon illumination (Figure 3.24).[146] The formation of the film was novel, in that the fullerene was spread at the air–water interface (from a mixture of chloroform and DMSO) and then the porphyrin was injected into the subphase and complexed with the Langmuir layer of the cation. Interestingly, both BAM of the Langmuir layer and AFM of the transferred film indicate that the complex is far more homogeneous than the fullerene on its own, because—presumably as a result of its overall shape and lack of directional interactions—it does not form a homogeneous layer. The efficiency of the photocurrent generated by the bilayer increased with increasing surface pressure being applied during the horizontal lift-off. The higher pressure causes a larger angle between the average plane of the substrate and the porphyrin, leading to a more favoured photoinduced electron transfer process. The work shows nicely how the combination of non-covalent interactions and application of surface pressure can optimise thin film performance, in a way not so readily done with other layer-forming techniques.

The complex formed between a zinc(II)phthalocyanine bearing four oligo(ethylene oxide) chains and an imidazole-derived C_{60} was employed successfully for the formation of LB films that were used as the photoactive material in photoelectrochemical cells.[147] The complex (Figure 3.25) forms Langmuir layers that show a very smooth isotherm, indicative of a liquid expanded phase. It was proposed that the oligo(ethylene oxide) chains gradually change their conformation from being in plane with the air–water interface to one where they are submerged in the subphase. The phthalocyanine component also shows this kind of behaviour, while the fullerene shows a sharp transition more characteristic of a solid, demonstrating that the liquid-like phase arises from the oxygen-containing chains. The complex between the fullerene and the phthalocyanine was transferred by the LB technique onto a fluorine-doped tin oxide surface whose surface had been made hydrophobic with a self-assembled chemisorbed monolayer. Absorption spectroscopy proved the presence of both components in the multilayer films. Vibrational spectroscopy using PM-IRRAS of LB films transferred at different surface pressures showed that the plane of the zinc(II)phthalocyanine was tilted with respect to the interface, and the oligo(ethylene oxide) chains were quasi-perpendicular at higher pressures. Transient absorption spectroscopy in the multilayers proved electron transfer from the electron-donating macrocycle to the fullerene, with a fast forward electron transfer and much slower recombination rate. The LB films of the complex on the siloxane-coated fluorine-doped tin oxide electrode functioned either as a photoanode or photocathode (depending on which mediator was employed and the applied potential). Greater efficiency was observed as a photoanode. The pure phthalocyanine could could not be

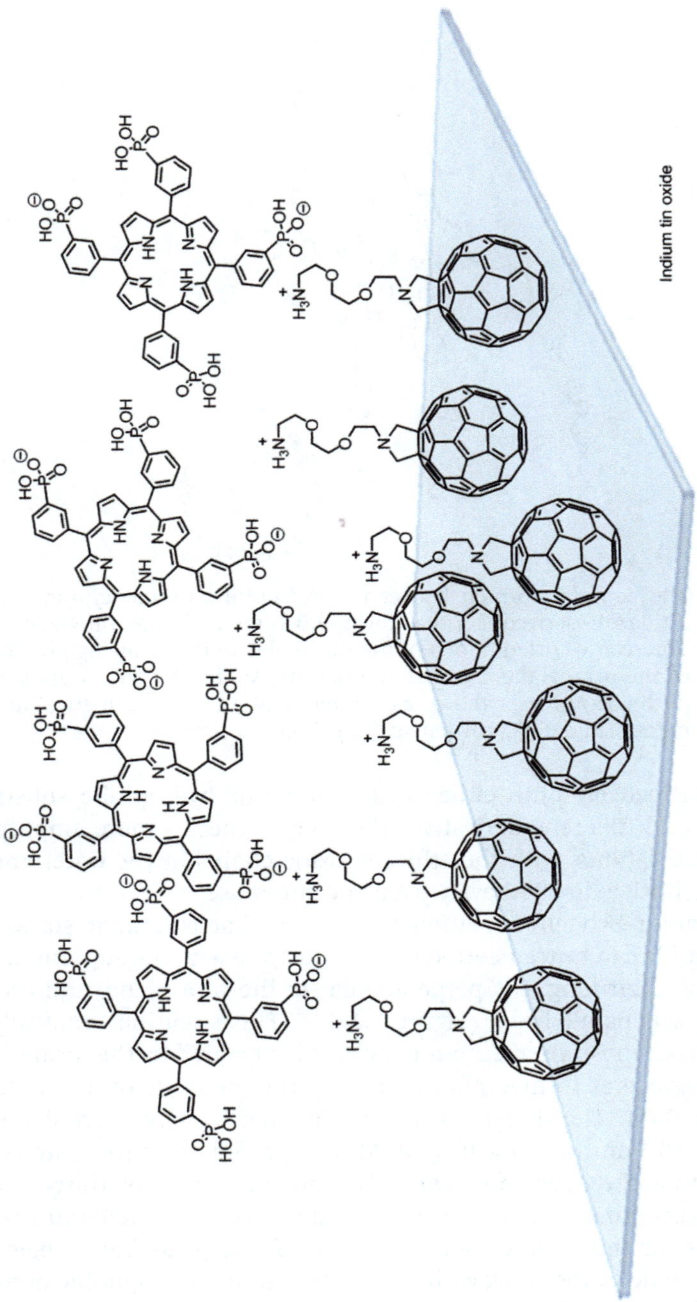

Figure 3.24 The film prepared by the Langmuir–Schaefer method onto indium tin oxide from a complex of fullerene and porphyrin in a Langmuir layer.

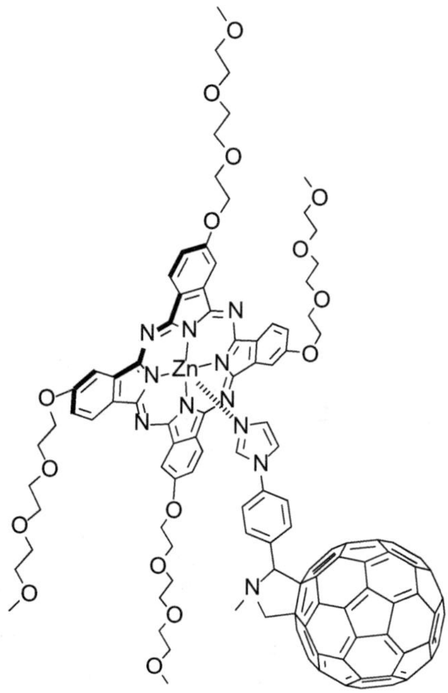

Figure 3.25 The complex formed between a hydrophilic phthalocyanine and an imidazole-derived C_{60} that forms LB films acting as photoelectrodes. The relative orientations of the molecules in this drawing give an idea of the arrangement in the multilayers. While the substitution of the phthalocyanine is shown as symmetrical, in fact a statistical arrangement of the oligo(ethylene oxide) chains is present.

transferred effectively onto either hydrophobic or hydrophilic substrates. Thus the bound fullerene derivative effectively induces a more amphiphilic character to the films, with the ethylene oxide chains in the water and the fullerene derivative directed away from the subphase.

The Langmuir–Schaeffer method has shown that columnar stacks of a perylenediimide can be transferred onto a substrate with the supramolecular stacks having their long axis perpendicular to the film plane, and the molecular long axis parallel to it (Figure 3.26).[148] This situation—radically different to other deposition methods assayed—is favoured by the formation of "rafts" of aggregates formed after spreading the molecule on the water, as observed by BAM. The thickness of the film that was transferred was approximately 10.5 nm (as shown by AFM and profilometer measurements), suggesting that the aggregates comprised in the region of thirty stacked molecules (assuming a 3.5 Å π–π stacking distance). It is believed that the driving force for this arrangement is contact of the polar imide regions of the first molecule in the stack with the water and the hydrophobic nature of the many alkyl chains that avoid contact with the polar subphase. The film is not perfect; holes are present and some thicker regions are evident from AFM imaging. The spontaneous formation of columns perpendicular to the

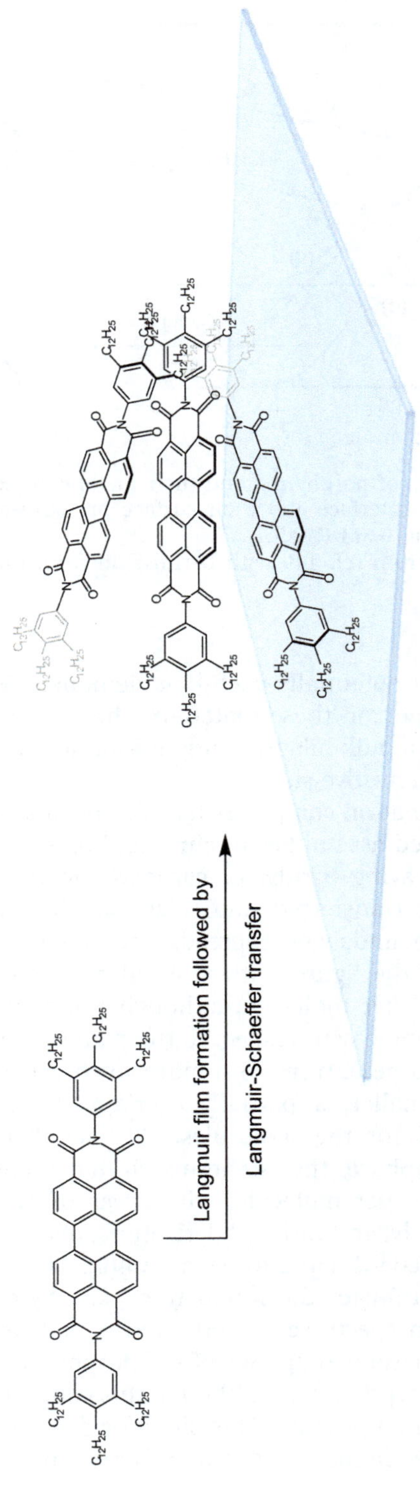

Figure 3.26 A perylenediimide derivative that forms aggregates on water that can be transferred by the Langmuir–Schaeffer method onto surfaces so that the columnar aggregates stand up from the substrate. AFM and profilometer measurements suggested these stacks comprised approximately 30 molecules, not the three that are shown to represent the system here.

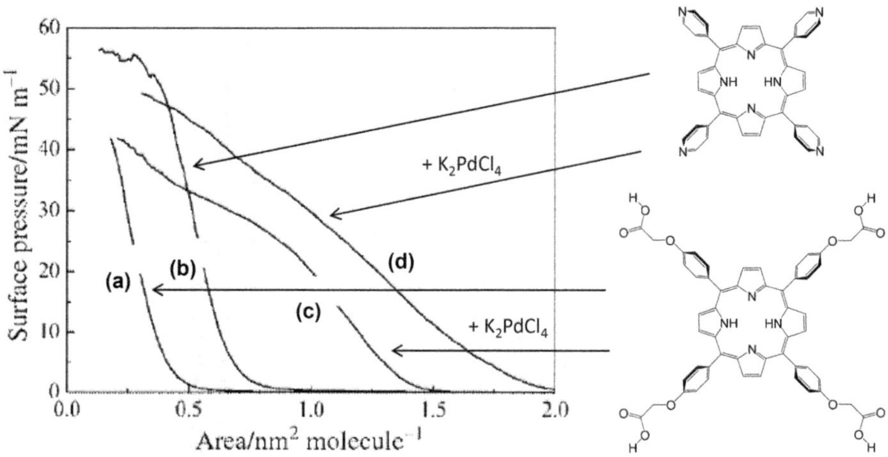

Figure 3.27 The structures of porphyrins and their metal complexes assembled at the air–water interface and their surface pressure–area isotherms on pure water and 0.3M K_2PdCl_4.
Reproduced from ref. 149 with permission from the Royal Society of Chemistry.

interface is unusual and potentially useful, as demonstrated in the photo-induced electrical transport in these materials when deposited into device interfaces, as shown for a bulk bilayer configuration for sixty fullerene with these perylenediimide derivative stacks.

The transfer of coordination complexes has also been shown to be useful for the transfer of ordered assemblies of chromophores onto surfaces. Two porphyrin derivatives, having pyridyl or carboxyl functions (Figure 3.27), have been studied and transferred onto glass as their palladium complexes.[149] The two compounds when spread at the air–water interface give the isotherms shown in the figure, with molecular cross-sections of about 0.5 nanometres squared per molecule, although the acid area is smaller indicating a higher tilt angle with respect to the plane of the interface. The area per molecule would be 0.9 nm^2 for a purely vertical compound. Since the actual area was smaller, a partially overlapped bilayer-type superstructure was proposed for the molecules.[150] When 0.3M K_2PdCl_4 was introduced into the subphase, the isotherms shifted dramatically, corresponding to higher areas per molecule, with areas of 1.75 and 1.35 nm^2 per molecule for the pyridyl and carboxyl derivatives, respectively, indicating the formation of arrays coordinated to palladium(II). The horizontal lift-off technique was used to transfer the monolayers, one by one, onto hydrophobic glass. Absorption spectroscopy indicated a "side-to-side electronic coupling" of the coordination complexes of the porphyrins as seen in a red shift in the principle absorption peak, although this coupling is weaker than in the layers of the pure porphyrins. This situation led the authors to propose the formation of metallomacrocycles in the layers that were grown up to eight layers thick. The characteristics of the porphyrins prompted the authors to suggest potential applications of these thin layer materials.

An alternative transfer to the lift-off technique method is the "horizontal scooping up" method or modified inverted Langmuir–Schaefer (MILS), where a substrate is held submersed in the subphase and parallel to the interface with air while compression is performed, and is then lifted out of the liquid through the layer in the horizontal position.[151] The technique is related to the inverted Langmuir–Schaefer configuration—where a horizontal substrate is pushed up through the monolayer from beneath—that was developed to study monolayer structure in fluorescence mode during transfer.[152] This method has been employed to study the collapse of stearic acid monolayers in the presence of a variety of metal ions.[153] After the transfer process to silicon just after collapse, X-ray reflectivity gave electron density profiles indicating three layers, in which the one nearest the interface comprised two molecules on the same "side" of the cation. Long after collapse, the same measurement showed that this structure relaxes to a more symmetrical one where the alkyl tails are directed in opposite directions away from the metal centre. A "folding and sliding mechanism" was conjectured to explain these observations that are caused by a rearrangement in the coordination of the metal ion.

3.8 Whither Langmuir Layers?

The development of Langmuir–Blodgett layers, long before the dawning of self-assembled monolayers, meant that they were recognised and exploited as tools for the study of single molecule thick layers of compounds for many reasons, but the possibilities today are as wide as they ever were because of the understanding of their formation and the tools for their manipulation and characterisation.[154] The early study on the preparation of "perforated monolayers", where mercurated calixarenes—with between four and six aryl units in the ring—were assembled at the air–water interface, provided solution-accessible cavities at the centre of macrocycles when they were in the plane of the monolayer.[155,156] The monolayers were stabilised by complexation of the metal with malonic acid, which caused an increase in the surface viscosity because of bonding between the rings. The work shows two trends that continue to be topical: the generation of porous materials and the post-assembly linking of molecules in monolayers.

The current trend towards the preparation of covalent bidimensional networks (partially inspired by the remarkable findings in graphene[157,158] and other "two-dimensional" materials[159]) is gaining pace, and some notable achievements have been made recently using water as the surface on which the networks are made. Coordination chemistry can be used for the preparation of free-standing sheets of two-dimensional polymeric complexes. The hexaterpyridyl ligand[160] shown in Figure 3.28 forms a uniform monolayer at the air–water interface, and studies on complexation of simpler compounds with iron(II) implied that it could be used for the formation of a

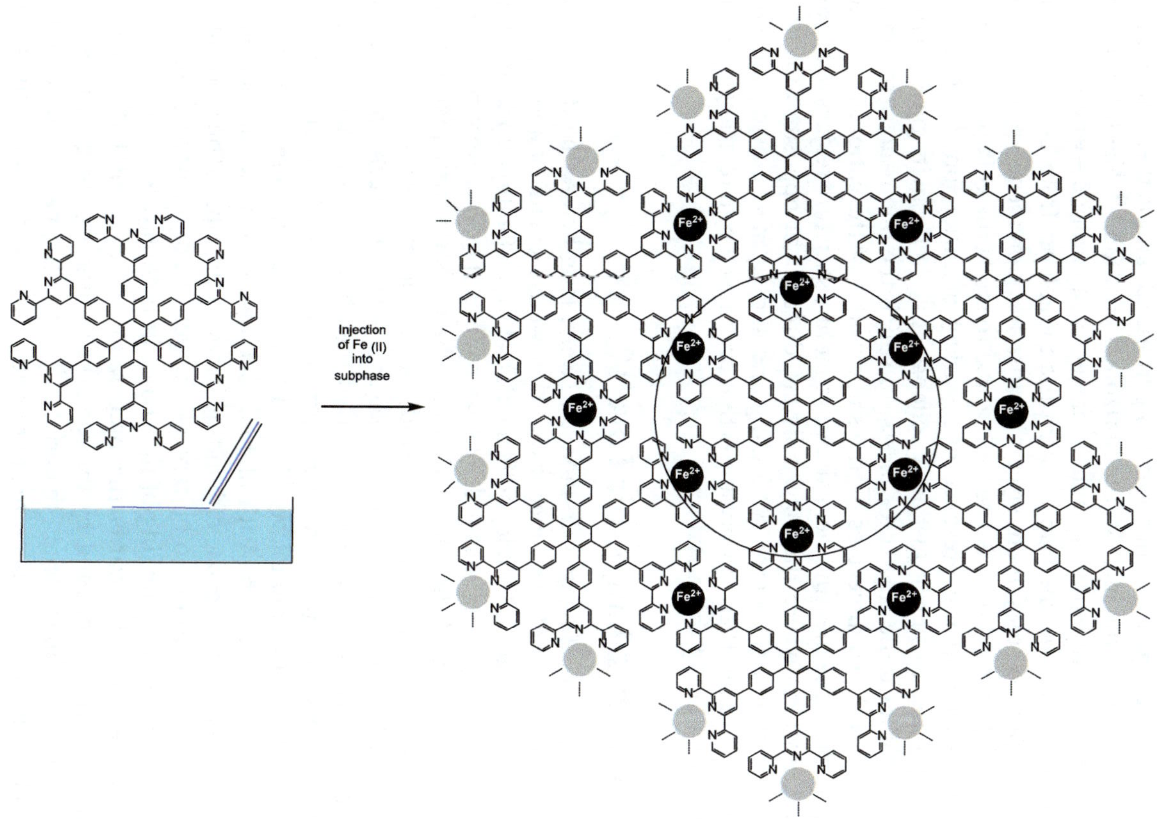

Figure 3.28 The formation of free-standing nanosheets at the air–water interface by coordination of a hexaterpyridine to iron(II) ions.

two-dimensional highly reticulated layer.[161] The number of variables available for the introduction of the metal ion into the layer of organic ligands is daunting; surface pressure, temperature, the metal salt used and its concentration in the water subphase, the pH of the subphase, and how to deposit the monolayer and introduce the metal salt! Once formed, a host of solid supports and transfer methods are available. As the authors put it, because of the "manifold of parameters, the impact of which is only partially understood, the decisions on how to proceed were made with a portion of intuition"! The subsequent procedure involved formation of the monolayer followed by injection of an iron(II) salt into the subphase. Concerning the surface pressure, an intermediate level was applied to allow lateral room to the surface molecules to be able to rotate and slide to form the complexes but at the same time to avoid drastic contraction, which can result in poor film structure.[162] It was recognised that the local concentration in the water fluctuates violently in the initial stages using this route, but it is envisaged as more controlled than adding the salt before spreading. The formation of the layer was followed by *in situ* absorption spectroscopy where the metal-to-ligand charge transfer band appeared after coordination polymer formation. The lack of a signal in fluorescence measurements also showed the absence of the free ligand. Transfer of the polymerised films onto quartz or hydrolysed silicon by vertical transfer and onto holey carbon TEM grids by horizontal transfer allowed microscopic investigation of the material. AFM revealed pieces of sheet 1.4 nm high corresponding well with the expected thickness taking into account counter-ions to the metal ion and possible water of hydration. Folds three times that height were evident, which can be envisaged as local pleats as in a piece of paper or cloth. The pieces were hundreds of microns in width. Large-scale sheets were also observed in low magnification TEM, although the film is extremely sensitive to the electron beam and therefore detailed structural data could not be gleaned from the technique. The large sheets can support themselves over the twenty micron squares in the TEM grid, and thus—although precise structural information could not be attained—it was concluded that the reversible binding of iron(II) to the terpyridyl ligand leads to monolayer thick sheets that have mechanical stabilities that the components do not. The careful study of adsorption spectra did indicate that practically all available coordination sites were occupied in the film.

Injection of benzenehexathiol in ethyl acetate onto the surface of water containing nickel(II) acetate results in the formation of nanosheets of the hexagonal coordination polymeric network nickel bis(dithiolene) complex (Figure 3.29) in which three transition metal ions are coordinated around each benzenehexathiolate hub.[163] The networks were actually formed without application of any surface pressure, leading to a two-dimensional growth of islands, something that would not be possible in the bulk because of the flexibility of the coordination polymer leading to closed three-dimensional structures. Measurement of the surface pressure under compression was used to judge the amount of sheet-like material formed for various

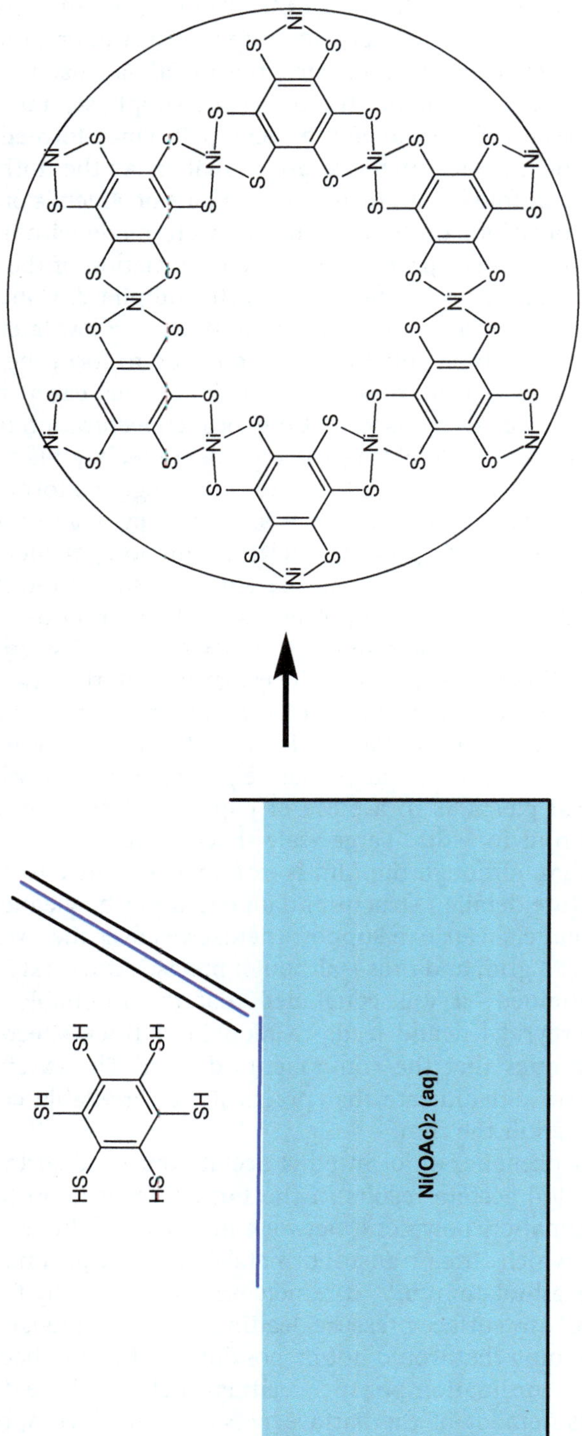

Figure 3.29 A cartoon showing the deposition of benzenehexathiol onto the surface of an aqueous bath containing nickel(II)acetate that leads to the formation of the two-dimensional hexagonal network shown on the right.

experiments with different amounts of benzenehexathiol introduced onto the water surface, with a gradual increase in the onset of surface pressure being observed with increasing coverage of the component, indicating the generation of the covalent network.[164] At significant pressures the films collapsed, presumably because of folding of the sheets. The films of these sheets at the water surface have been transferred using the Langmuir–Schaefer vertical lift-off technique onto HOPG and the layers were shown to be one molecule thick using AFM, and STM showed the hexagonal network clearly. The sheets of the network were up to tens of microns in width and breadth. Transfer of the network from water onto ITO was proven using absorption spectroscopy where a clear charge transfer band was observed that increased in intensity with consecutive transfers, indicating that this potentially interesting semiconducting material had been achieved. On the other hand, when the formation of the complex was performed at a liquid–liquid interface, micrometre thick films were formed. Both the monolayers and the multilayers could be reduced using an organic reagent, demonstrating the interest of the films in molecular electronics.

The potential in the use of Langmuir layers for a contemporary drive towards robust two-dimensional systems has also been demonstrated for polymerisation of a tris-anthracene amphiphile.[165] The trigonal star-shaped monomer molecule (Figure 3.30)—that can be viewed as a derivative of triptycene as the core—has a diethylene glycol chain that orients and anchors it at the air–water interface, and at 25 °C forms a dense monolayer upon compression. But at 1 °C, a phase corresponding to a phase occupying a far greater area is observed, corresponding to an open supramolecular framework (which collapses to the densely-packed phase under surface pressure). Photopolymerisation of the open phase gave a robust film which could be transferred by the Langmuir–Schaefer method to a TEM grid, where optical microscopy showed the film to be robust and birefringent—or by the Langmuir–Blodgett method to a silicon wafer where AFM was performed. The films have a thickness of one molecule (approximately 1.2 nm) and were prepared over several square centimetres, and have a total pore area of

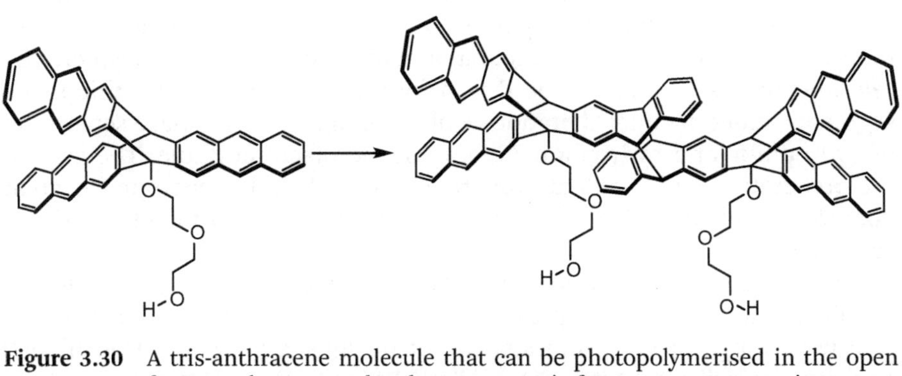

Figure 3.30 A tris-anthracene molecule that can be photopolymerised in the open framework supramolecular structure it forms on water to give a two-dimensional porous sheet (only one connection is shown here).

approximately 30% open area. They are strong enough to support their own weight in an open atmosphere, and stable to hydrolysis. STM proved that the sheets have a hexagonal cell network.

The Langmuir–Blodgett assembly method has a very important place in the development of molecular electronics,[166–168] both in the past as well as for future developments, as we shall see later in this book. A particular case is the electron acceptor 7,8-dimethyl-10-dodecyl isoalloxazine (a flavin derivative), which forms domains in its LB films with electron transport properties that depend on their shape.[169] The use of a Langmuir trough is not necessary for materials that display robust self-assembly on the surface of water when deposited from organic solvents. Slow evaporation on the surface of water of a toluene solution of a block polymer comprising regioregular poly(3-hexylthiophene) and poly(ethylene glycol) leads to the formation of a free-standing film composed of nanowires of the π-functional block in the hydrophilic section.[170] The pure polythiophene derivative has a great tendency to crystallise, with a high melting point, and accessing long range ordered structure under near equilibrium conditions is very challenging; when films are cast, only very short fibres are formed as a result of the dominant kinetic effects. However, the amphiphilic block copolymer forms domains with remarkably large regions of mutually-aligned hydrophilic and hydrophobic sections when formed by evaporation of a solution in toluene on water. The films were sensitive to the concentration of the block copolymer in the organic solution, with the materials containing larger domains formed from the more dilute solutions. These domains—of up to ten square microns in area—had aligned areas of polythiophene sections that were approximately 7.4 nm wide according to TEM images (Figure 3.31): that width was in close accord with the calculated width of a fully extended polymer chain resulting from a polymer with the molecular weight used in the synthesis of the block copolymer. A noteworthy feature of the material is that between 82 and 91% by weight was the polythiophene block, a fact that ensures that the conduction is not "diluted"; in fact, the material's electrical transport in FETs was witnessed to be orders of magnitude higher when compared with disordered films of like thickness (approximately 4 nm determined by AFM, that corresponds to multilayers in the region of three molecule thick films) of the same material or pure conducting polymer. The self-assembly is believed to be driven by the π-stacking of the thiophene segments, aided by interdigitation of alkyl chains and the phase separation of hydrophilic and hydrophobic blocks in the macromolecule. The ease of the preparation and excellent properties of the films demonstrate the great potential of self-assembly at water surfaces for the preparation of functional supramolecular films.

The combination of variation in relative sizes of the blocks in a diblock copolymer and the surface pressure in the Langmuir trough allows access to a wide range of sizes and spacing of polymer aggregates. The diblock system comprising poly(styrene)–poly(ethylene oxide) with varying size of the two moieties was transferred at different surface pressures to a surface by

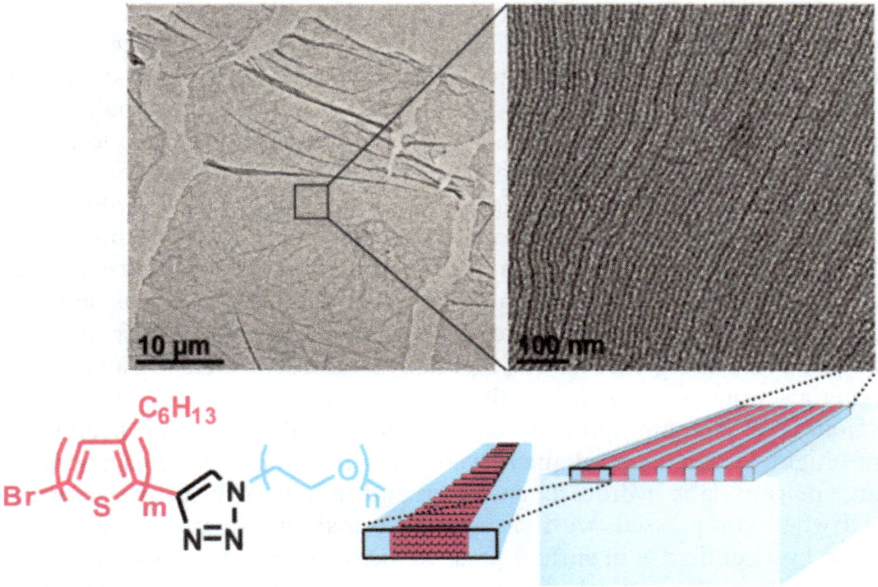

Figure 3.31 TEM images (top) of the block copolymer (drawn below) that self-assembles at a water surface when a toluene solution of the macromolecule is evaporated slowly.
Reprinted with permission from ref. 170. Copyright 2014 American Chemical Society.

the Langmuir–Blodgett technique.[171] Using these variables, the surface density of the poly(styrene) aggregates in the material can be tailored. The poly(ethylene oxide) block has to be sufficiently big in comparison to the more hydrophilic block so that the latter are separate from one another. For a given polymer composition, the deposition pressure has a linear relationship with the spacing of the poly(styrene) aggregates, and the dominance of the surface phase diagram allows a precise selection of the dimensions of the hydrophobic areas. This non-lithographic method for patterning is surely attractive for the preparation of materials and for the study of supramolecular chemistry in these organic systems.

Interestingly, when the reverse situation is taken, with a very low proportion of hydrophilic polymer, then the assembly is apparently dominated by the evaporation of the solvent used to deposit the material on the water surface, with different types of aggregates being seen whose propensity was largely independent of surface pressure and compression speed.[172] Though much is known and sizes of block copolymer aggregates can be controlled, there is much to be learned still concerning the effects of the amount of hydrophilic and hydrophobic sections and their relationships with the types of aggregates formed by polymers and smaller molecules, and the importance of kinetic factors including the evaporation of solvent. In addition, the combination of polymers in the Langmuir trough and their study after transfer raises the chance to study phase diagrams of complex polymer systems in the two-dimensional space of the air–water interface, as done for the triblock copolymer polystyrene-*block*-poly(ethylene oxide)-*block*-polystyrene combined with the diblock copolymer polystyrene-*block*-poly(methyl methacrylate), where circular micelles were observed.[173] Compression of the layers resulted in splitting of the large well-packed aggregates into smaller and more uniform agglomerations, demonstrating that the triblock copolymer increased homogeneity of the structural morphology in this system.

The effect of flow when Langmuir monolayers are formed and transferred can be important; supramolecular polymers can be aligned just as polymers are.[174] The effect of flow in aggregation phenomena in general is an emerging area that is of great interest, especially for the controlled organisation of materials when solutions, liquids or suspensions are moved in one direction.[175–178] An amphiphilic hexabenzocoronene derivative has been made into highly ordered and aligned layers upon transfer.[179] Monolayers of the large polyaromatic hydrocarbon with an anthraquinone head group formed well when compressed, with clear sharp transitions in the isotherm, followed by a gentle rise in surface pressure before collapse. Repeated transfer of monolayers by vertical dipping onto a quartz substrate, made hydrophobic with a chemisorbed layer, afforded multilayers that were highly birefringent; the polarised optical micrographs showed large differences with the relative rotation of crossed polarisers with respect to the film. Diffraction studies confirmed this effect. An irreversible phase transition to a single hexagonally-packed arrangement of columns—still macroscopically aligned with the original dipping direction—was achieved by warming the sample. The rearrangement to co-facially stacked disks corresponded to a packing with the planes perpendicular to the column axes. Therefore, the original LB transfer gave a metastable phase on the substrate, a not uncommon phenomenon but one particularly relevant to the LB method compared with other film-forming techniques because the films do not end up on a surface on which they are made.

When Langmuir layers are compressed, it has been shown that two opposing circular flows exist, like a "vortex", on either side of the trough.[180] The direction of these flows depends on which end barrier is moved. The chirality is generated because of the broken symmetry of each circular motion at the interface, with the subphase on one side of the circle and the air on the other. These flows are such that Langmuir–Blodgett deposition of the adsorpt on a solid small substrate located either side of the trough will result in a material deposited which has been moving in a determined direction, clockwise or anticlockwise (Figure 3.32). It was shown using an achiral porphyrin derivative at the air–HCl(aq) interface (where the adsorpt is protonated at the pyrrole positions) that deposition in different areas of the compressed trough led to superstructures with opposing optical activity in circular dichroism spectroscopic studies. The results show how local effects can affect the organisation of material in the Langmuir–Blodgett technique, and could be responsible for the formation of domains in certain systems.

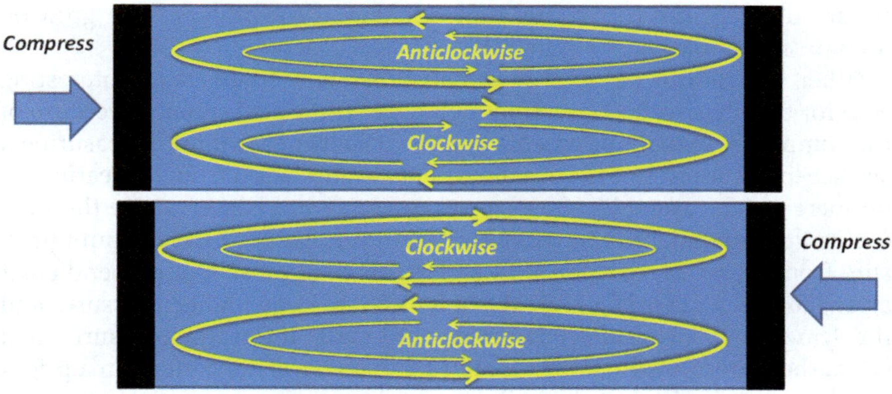

Figure 3.32 A simplified cartoon of the proposed vortex flow that leads to local chiral structures in the compression of Langmuir layers.

The emergence of chirality induced from a chiral anionic surfactant to a prochiral amphiphile has been observed in a relatively simple system that demonstrates a surprising and intriguing variety of complicated morphologies at the air–water interface.[181] The oppositely charged phospholipid dimyristoyl phosphatidate and 2-methyl-1-octadecylquinolinium, when combined in equimolar proportion and spread on water, form a homogeneous mixture. In the pressure–area isotherm of this mixture, an "overshoot" is observed, where as the layer is compressed, the surface pressure goes through a maximum and then reduces to a plateau value. This region in the isotherms coincided with the observation of small aggregates in the BAM images of the layer. The behaviour was interpreted as a nucleation and growth process with different kinetic characteristics that become comparable when the pressure drops upon the initial nucleation and loss of pressure. Depending on the sequence of applied surface pressure steps that the mixed Langmuir monolayer is exposed to, different domain shapes—mostly with curved arms or internal features—are observed by BAM. The kinetics play an important role in the appearance of these metastable structures, and depend on the rate of compression and time held at a given position on the isotherm for the structure to evolve. Circular dichroism spectra of all the structures indicated a single chiral arrangement arising from the supramolecular arrangement of the quinolinium head groups induced by the chiral counter-ion. The arrangement was modelled with theoretical simulations (molecular mechanics) that showed the inclination of the quinolinium moiety to form chiral arrangements. The structures were interpreted as a result of the balance between electrostatic interactions of the quinoline and phosphatidate head groups, the stacking of the aromatic cations, and the van der Waals forces acting between the quasi-parallel alkyl chains directed away from the interface. The assembly is sensitive to the head group in the cationic component; when a hemicyanine was employed, highly ordered curved domains were observed where the tilt of this unit is believed to influence the growing domain morphology.[182]

Other liquids can be used as the adsorbents. Mercury is an interesting case for the physisorbed layers, though little studied presumably because of the complexity of the experiments and the toxicity of the liquid. Measuring a surface–pressure isotherm in analogy to the water-based trough, stearic acid on mercury, for example, shows a clearly stable monolayer where the molecules lie co-parallel to the surface of the liquid metal.[183] At medium pressure from a barrier the molecules apparently reorganise to a perpendicular arrangement sequentially, but the layer collapses at higher pressure, and the lying down orientation is greatly favoured. Interestingly, lauric acid (12 carbon atom chain rather than 18 in stearic acid) prefers an upright arrangement, indicating that for longer chains, the van der Waals interactions with the surface dominate supramolecular organisation. This tendency has subsequently been proven for *n*-alkanes on mercury.[184] Adsorption of surfactants at the mercury–cyclohexane solution interface can also take place, and in this case, orientations perpendicular to the interface were observed.[185] When the head group was an ionic trimethylammonium derivative, multilayer formation was favoured,[186] presumably because of the great tendency of these materials to form lamellae. A synchrotron study of thiol layers on mercury showed that although a dense film is formed, there is no long range order in the plane of the interface, presumably because the motion of the disordered liquid subphase dominates the ordering effect of the van der Waals interactions between the alkyl chains.[187]

A remarkable bi-component trilayer system prepared in one step in a Langmuir trough has been described[188] that may point to novel designs for functional multicomponent multilayers synthesised in a single step. At low surface pressures, the equimolar mixture of *N*-10-dodecyl acridine and stearate forms a monolayer where pairs of the surfactants alternate their orientation in the plane, as indicated in structure A in Figure 3.33. When surface pressure is increased, a collapse point is reached, followed by a clear plateau: this plateau was proved to correspond to trilayer structure B in Figure 3.33. The first and third layers of the film both have their polar head groups directed toward the water phase, while the middle layer is pointed away from it. This situation is favoured because of the tendency of the acridine head groups to aggregate as well as the mismatch in alkyl chain length between the two components, which leads to an ordered interdigitation between the alkyl chains of the middle layer and that in contact with the water. This ordering of the alkyl chains was witnessed using reflection IR spectroscopy, where the signals corresponding to methylene groups increased dramatically in intensity upon formation of the trilayer. The thickness of the trilayer was shown to be approximately 7.6 nm, according to X-ray reflectivity measurements at a synchrotron, three times that of the monolayer (2.5 nm). The surface potential just above the layer was determined using a surface-potential probe equipped with a vibrating plate

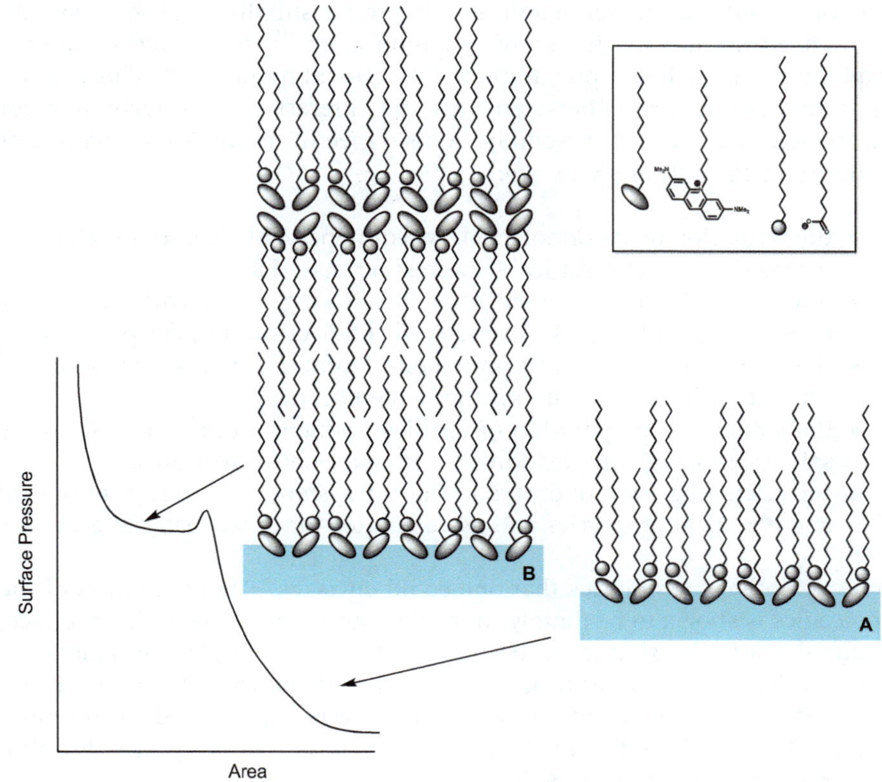

Figure 3.33 A representation of the surface pressure–area isotherm at 21 °C of a layer formed by an *N*-10-dodecyl acridine and stearate equimolar mixture on water that is a monolayer at low surface pressure (A) but collapses in a specific way to give a trilayer (B) as surface pressure is increased.

simultaneously with the surface pressure isotherm, which showed that as the layer was compressed, the dipole moment perpendicular to the interface decreased dramatically, starting to lower before the collapse point and continuing after it, consistent with the proposed pressure-induced phase change at the interface. The ordered structure of the trilayer was also evident in the BAM images of the film in the collapse pressure, where areas of monolayer and trilayer were present. Bright hexagonal patterns were observed with internal structure, with two-fold symmetry at a certain polarisation. Designing systems such as this that might undergo controlled multilayer formation under pressure is an intriguing prospect.

All these fascinating cases of undoubted contemporary relevance show that, far from being a mature technique with little new science or potential, the unique environment of the air–water interface affords practically limitless scope for discovery and the transfer techniques offer huge space for inventive construction of supramolecular materials.[189] While it has been demonstrated that the temperature of the water subphase can have significant effects on the morphology of Langmuir layers,[190] this is perhaps a little exploited variable in the preparation of functional monolayers. This feature, in combination with others, ensures the interest of this technique for supramolecular surface assembly. A few specific examples of particular promise in the author's view are:

- the formation of supramolecular two-dimensional systems, of relevance for porous and electronically interesting materials;
- orientational control of adsorbates allowing superstructure–activity relationships, of interest for the study of electron transfer processes;
- the preparation of multicomponent layers for the study of phase separation in natural and synthetic membranes;
- the single step preparation of multicomponent multilayers, to explore self-assembly of complex films for transfer onto substrates;
- the oriented growth of ordered thin film systems, of interest for optical and electrical properties and as substrates for other material growth,

amongst other possibilities that one could envisage. The orientation of the molecules is shown to be mainly controlled out of the plane of the interface, though clearly the simultaneous self-assembly of multiple components at the interface and coming from bulk solution is an area that begs further exploration. The toolkit of supramolecular systems, inspired by solution-state self-assembly and solid-state structures, is sure to bring new thin film materials that present unique behaviour.

References

1. A. Pockels, *Nature*, 1891, **43**, 437.
2. H. Devaux, *Annu. Rep. Smithsonian Institute*, 1913, 261.
3. I. Langmuir, *J. Am. Chem. Soc.*, 1917, **39**, 1848.
4. A. Ulman, *An Introduction to Ultrathin Organic Films: From Langmuir–Blodgett to Self-Assembly*, Academic Press, Boston, 1991.
5. L. C. King and E. K. Neilsen, *J. Chem. Educ.*, 1958, **35**, 198.
6. V. Kaganer, H. Möhwald and P. Dutta, *Rev. Mod. Phys.*, 1999, **71**, 779.
7. J. T. Davies and E. K. Rideal, *Interfacial Phenomena*, Academic Press, London, 1961, pp. 217–281.
8. A. Czajka, G. Hazell and J. Eastoe, *Langmuir*, 2015, **31**, 8205.
9. M. H. Hatzopoulos, J. Eastoe, P. J. Dowding, S. E. Rogers, R. Heenan and R. Dyer, *Langmuir*, 2011, **27**, 12346.
10. I. R. Peterson, *J. Phys. D: Appl. Phys.*, 1990, **23**, 379.
11. M. Breton, *J. Macromol. Sci., Rev. Macromol. Chem. Phys.*, 1981, **21**, 61.
12. J. Adams, W. Rettig, R. S. Duran, J. Naciri and R. Shashidhar, *J. Phys. Chem.*, 1993, **97**, 2021.
13. B. Widom, P. Bhimalapuram and K. Koga, *Phys. Chem. Chem. Phys.*, 2003, **5**, 3085.

14. F. Biedermann, W. M. Nau and H. J. Schneider, *Angew. Chem., Int. Ed.*, 2014, **53**, 11158.
15. Y. B. Vysotsky, E. S. Kartashynska, E. A. Belyaeva, D. Vollhardt, V. B. Fainerman and R. Miller, *J. Phys. Chem. C*, 2015, **119**, 5523.
16. E. S. Fomina, Y. B. Vysotsky, E. A. Belyaeva, D. Vollhardt, V. B. Fainerman and R. Miller, *J. Phys. Chem. C*, 2014, **118**, 4122.
17. I. Giner, I. Gascón, J. Vergara, M. C. López, M. B. Ros and F. M. Royo, *Langmuir*, 2009, **25**, 12332.
18. M. Sagisaka, T. Narumi, M. Niwase, S. Narita, A. Ohata, C. James, Yoshizawa, E. Taffin de Givenchy, F. Guittard, S. Alexander and J. Eastoe, *Langmuir*, 2014, **30**, 6057.
19. K. Thirumoorthy, N. Nandi and D. Vollhardt, *Langmuir*, 2007, **23**, 6991.
20. N. Nandi and D. Vollhardt, *J. Phys. Chem. B*, 2002, **106**, 10144.
21. N. Nandi and D. Vollhardt, *Chem. Rev.*, 2003, **103**, 4033.
22. J. Marsden and E. K. Rideal, *J. Chem. Soc.*, 1938, 1163.
23. K.-i. Iimura, Y. Yamauchi, Y. Tsuchiya, T. Kato and M. Suzuki, *Langmuir*, 2001, **17**, 4602.
24. I. Kuzmenko, H. Rapaport, K. Kjaer, J. Als-Nielsen, I. Weissbuch, M. Lahav and L. Leiserowitz, *Chem. Rev.*, 2001, **101**, 1659.
25. H. Kitano and H. Ringsdorf, *Bull. Chem. Soc. Jpn.*, 1985, **58**, 2826.
26. K. Ariga and T. Kunitaki, *Acc. Chem. Res.*, 1997, **31**, 3371.
27. T. M. Bohanon, S. Denziger, R. Fink, W. Paulus, H. Ringsdorf and M. Weck, *Angew. Chem., Int. Ed. Engl.*, 1995, **34**, 58.
28. Y. Ebara, K. Itakakura and Y. Okahata, *Langmuir*, 1996, **12**, 5165.
29. K. Matsuura, Y. Ebara and Y. Okahata, *Langmuir*, 1997, **13**, 814.
30. E. Mouri, Y. Furuya, K. Matsumoto and H. Matsuoka, *Langmuir*, 2004, **20**, 8062.
31. I. Langmuir and V. J. Schaefer, *J. Am. Chem. Soc.*, 1937, **59**, 2400.
32. M. Muramatsu and T. Sasaki, *Bull. Chem. Soc. Jpn.*, 1952, **25**, 21.
33. M. Muramatsu and T. Sasaki, *Bull. Chem. Soc. Jpn.*, 1952, **25**, 25.
34. T. Sasaki and M. Muramatsu, *Bull. Chem. Soc. Jpn.*, 1953, **26**, 96.
35. I. Langmuir, *J. Chem. Phys.*, 1933, **1**, 756.
36. H. Möhwald, *Annu. Rev. Phys. Chem.*, 1990, **41**, 441.
37. N. Nandi and D. Vollhardt, *Chem. Rev.*, 2003, **103**, 4033.
38. K. Kim, S. Q. Choi, Z. A. Zell, T. M. Squires and J. A. Zasadzinski, *Proc. Natl. Acad. Sci. U. S. A.*, 2013, **110**, E3054.
39. W. Lin, A. J. Clark and F. Paesani, *Langmuir*, 2015, **31**, 2147.
40. A. Amirfazli and A. W. Neumann, *Adv. Colloid Interface Sci.*, 2004, **110**, 121.
41. A. J. García-Sáez, S. Chiantia and P. Schwille, *J. Biol. Chem.*, 2007, **282**, 33537.
42. W. Bernhard, J. Mottaghian, A. Gebert, G. A. Rau, H. von Der Hardt and C. F. Poets, *Am. J. Respir. Crit. Care Med.*, 2000, **162**, 1524.
43. J. Israelachvili, *Langmuir*, 1994, **10**, 3774.
44. K. Albrecht, A. Mourran and M. Moeller, *Ordered Polymeric Nanostructures at Surfaces*, ed. G. J. Vancso and G. Reiter, Book Series: Advances in Polymer Science, 2006, vol. 200, p. 57.

45. H. A. Klok and S. Lecommandoux, *Adv. Mater.*, 2001, **13**, 1217.
46. M. J. Fasolka and A. M. Mayes, *Annu. Rev. Mater. Res.*, 2001, **31**, 323.
47. J. Y. Zhu, A. Eisenberg and R. B. Lennox, *J. Am. Chem. Soc.*, 1991, **113**, 5583.
48. Z. Li, M. W. Zhao, J. Quinn, M. H. Rafailovich, J. Sokolov, R. B. Lennox, A. Eisenberg, X. Z. Wu, M. W. Kim, S. K. Sinha and M. Tolan, *Langmuir*, 1995, **11**, 4785.
49. J. K. Cox, K. Yu, A. Eisenberg and R. B. Lennox, *Phys. Chem. Chem. Phys.*, 1999, **1**, 4417.
50. J. K. Cox, K. Yu, B. Constantine, A. Eisenberg and R. B. Lennox, *Langmuir*, 1999, **15**, 7714.
51. S. Li, S. Hanley, I. Khan, S. K. Varshney, A. Eisenberg and R. B. Lennox, *Langmuir*, 1993, **9**, 2243.
52. S. Li, C. J. Clarke, R. B. Lennox and A. Eisenberg, *Colloids Surf., A*, 1998, **133**, 191.
53. B. H. Lin, S. A. Rice and D. A. Weitz, *J. Chem. Phys.*, 1993, **99**, 8308.
54. S. W. An, T. J. Su, R. K. Thomas, F. L. Baines, N. C. Billingham, S. P. Armes and J. Penfold, *J. Phys. Chem. B*, 1998, **102**, 387.
55. O. Théodoly, M. Jacquin, P. Muller and S. Chhun, *Langmuir*, 2009, **25**, 781.
56. L. Zhao, C. Feng, X. Pang, J. Jung, M. C. Stefan, P. Sista, R. Han, N. Fang and Z. Lin, *Soft Matter*, 2013, **9**, 8050.
57. P. Lo Nostro, *Curr. Opin. Colloid Interface Sci.*, 2003, **8**, 223.
58. T. Kato, M. Kameyama, M. Ehara and K.-i. Iimura, *Langmuir*, 1998, **14**, 1786.
59. T. Kato, M. Kameyama and M. Kawano, *Thin Solid Films*, 1996, **273**, 23.
60. C. W. Bunn and E. R. Howell, *Nature*, 1954, **174**, 549.
61. C. W. Bunn, *Trans. Faraday Soc.*, 1939, **35**, 482.
62. M. Maaloum, P. Muller and M. P. Krafft, *Angew. Chem., Int. Ed.*, 2002, **41**, 4331.
63. G.-F. Zhang, M. Maaloum, P. Muller, N. Benoit and M. P. Krafft, *Phys. Chem. Chem. Phys.*, 2004, **6**, 1566.
64. P. Fontaine, M. Goldmann, P. Muller, M.-C. Fauré, O. Konovalov and M. P. Krafft, *J. Am. Chem. Soc.*, 2005, **127**, 512.
65. M. P. Krafft, *Acc. Chem. Res.*, 2012, **45**, 514.
66. P. Fontaine, M.-C. Fauré, L. Bardin, E. J. M. Filipe and M. Goldmann, *Langmuir*, 2014, **30**, 15193.
67. P. Chen, X. Ma, Y. Zhang, K. Hu and M. Liu, *Langmuir*, 2007, **23**, 11100.
68. A. Ge, Q. Peng, H. L. Wu, H. Liu, Y. Tong, T. Nishida, N. Yoshida, K. Suzuki, T. Sakai, M. Osawa and S. Ye, *Langmuir*, 2013, **29**, 14411.
69. X. Zhai and M. Liu, *J. Colloid Interface Sci.*, 2006, **295**, 181.
70. M. Jiang, X. Zhai and M. Liu, *Langmuir*, 2005, **21**, 11128.
71. X. Chen, J. Wang, N. Shen, Y. Luo, L. Li, M. Liu and R. K. Thomas, *Langmuir*, 2002, **18**, 6222.
72. G. Zhang, X. Zhai and M. Liu, *J. Phys. Chem. B*, 2006, **110**, 10455.

73. M. Ramanathan, L. K. Shrestha, T. Mori, Q. Ji, J. P. Hill and K. Ariga, *Phys. Chem. Chem. Phys.*, 2013, **15**, 10580.
74. K. Ariga and J. P. Hill, *Chem. Rec.*, 2011, **11**, 199.
75. R. Blankenburg, P. Meller, H. Ringsdorf and C. Salesse, *Biochemistry*, 1989, **28**, 8214.
76. S. A. Darst, M. Ahlers, P. H. Meller, E. W. Kubalek, R. Blankenburg, H. O. Ribi, H. Ringsdorf and R. D. Kornberg, *Biophys. J.*, 1991, **59**, 387.
77. T. C. Edwards, S. Koppenol, W. Frey, W. R. Schief Jr., V. Vogel, R. E. Stenkamp and P. S. Stayton, *Langmuir*, 1998, **14**, 4683.
78. K. Ariga, T. Kunitake and H. Furuta, *J. Chem. Soc., Perkin Trans. 2*, 1996, 667.
79. D. J. Plaut, S. M. Martin, K. Kjaer, M. J. Weygand, M. Lahav, L. Leiserowitz, I. Weissbuch and M. D. Ward, *J. Am. Chem. Soc.*, 2003, **125**, 15922.
80. I. Weissbuch, I. Rubinstein, M. J. Weygand, K. Kjaer, L. Leiserowitz and M. Lahav, *Helv. Chim. Acta*, 2003, **86**, 3867.
81. I. Weissbuch, M. Berfeld, W. G. Bouwman, K. Kjaer, J. Als-Nielsen, M. Lahav and L. Leiserowitz, *J. Am. Chem. Soc.*, 1997, **119**, 933.
82. I. Rubinstein, G. Bolbach, M. J. Weygand, K. Kjaer, I. Weissbuch and M. Lahav, *Helv. Chim. Acta*, 2003, **86**, 3851.
83. I. Weissbuch, I. Kuzmenko, M. Berfeld, L. Leiserowitz and M. Lahav, *J. Phys. Org. Chem.*, 2000, **13**, 426.
84. H. Rapaport, I. Kuzmenko, M. Berfeld, K. Kjaer, J. Als-Nielsen, R. Popovitz-Biro, I. Weissbuch, M. Lahav and L. Leiserowitz, *J. Phys. Chem. B*, 2000, **104**, 1399.
85. I. Kuzmenko, K. Kjaer, J. Als-Nielsen, M. Lahav and L. Leiserowitz, *J. Am. Chem. Soc.*, 1999, **121**, 2657.
86. D. Y. Sasaki, K. Kurihara and T. Kunitake, *J. Am. Chem. Soc.*, 1991, **113**, 9685.
87. D. Y. Sasaki, M. Yanagi, K. Kurihara and T. Kunitake, *Thin Solid Films*, 1992, **210/211**, 776.
88. D. Y. Sasaki, K. Kurihara and T. Kunitake, *J. Am. Chem. Soc.*, 1992, **114**, 10994.
89. K. Kurihara, K. Ohto, Y. Honda and T. Kunitake, *J. Am. Chem. Soc.*, 1991, **113**, 5077.
90. M. Sakurai, H. Tamagawa, Y. Inoue, K. Ariga and T. Kunitake, *J. Phys. Chem. B*, 1997, **101**, 4810.
91. H. Tamagawa, M. Sakurai, Y. Inoue, K. Ariga and T. Kunitake, *J. Phys. Chem. B*, 1997, **101**, 4817.
92. V. Marchi-Artzner, F. Artzner, O. Karthaus, M. Shimomura, K. Ariga, T. Kunitake and J.-M. Lehn, *Langmuir*, 1998, **14**, 5164.
93. M. Weck, R. Fink and H. Ringsdorf, *Langmuir*, 1997, **13**, 3515.
94. Q. Huo, K. C. Russell and R. M. Leblanc, *Langmuir*, 1998, **14**, 2174.
95. T. Mori, K. Okamoto, H. Endo, J. P. Hill, S. Shinoda, M. Matsukura, H. Tsukube, Y. Suzuki, Y. Kanekiyo and K. Ariga, *J. Am. Chem. Soc.*, 2010, **132**, 12868.

96. T. Michinobu, S. Shinoda, T. Nakanishi, J. P. Hill, K. Fujii, T. N. Player, H. Tsukube and K. Ariga, *Phys. Chem. Chem. Phys.*, 2011, **13**, 4895.
97. A. L. Litvin, S. Valiyaveettil, D. L. Kaplan and S. Mann, *Adv. Mater.*, 1997, **9**, 124.
98. D. Volkmer, M. Fricke, C. Agena and J. Mattay, *J. Mater. Chem.*, 2004, **14**, 2249.
99. J. Kmetko, C. Yu, G. Evmenenko, S. Kewalramani and P. Dutta, *Phys. Rev. Lett.*, 2002, **89**, 186102.
100. S. Kewalramani, J. Kmetko, G. Dommett, G. Evmenenko, H. Mo and P. Dutta, *Thin Solid Films*, 2007, **515**, 5627.
101. B. R. Heywood and S. Mann, *Langmuir*, 1992, **8**, 1492.
102. D. C. F. Wieland, P. Degen, M. Paulus, M. A. Schroer, S. Bieder, C. J. Sahle, J. Möller, S. Leick, Z. Chen, B. Struth, H. Rehage and M. Tolan, *Colloids Surf., B*, 2013, **109**, 74.
103. C. Lendrum and K. M. McGrath, *Cryst. Growth Des.*, 2010, **10**, 4463.
104. D. C. Popescu, M. M. J. Smulders, B. P. Pichon, N. Chebotareva, S. Y. Kwak, O. L. J. van Asselen, R. P. Sijbesma, E. Di Masi and N. A. J. M. Sommerdijk, *J. Am. Chem. Soc.*, 2007, **129**, 14058.
105. C. Lendrum and K. M. McGrath, *J. Colloid Interface Sci.*, 2009, **331**, 206.
106. E. M. Landau, M. Levanon, L. Leiserowitz, M. Lahav and J. Sagiv, *Nature*, 1985, **318**, 353.
107. E. M. Landau, S. G. Wolf, M. Levanon, L. Leiserowitz, M. Lahav and J. Sagiv, *J. Am. Chem. Soc.*, 1989, **111**, 1436.
108. I. Weissbuch, G. Berkovic, L. Leiserowitz and M. Lahav, *J. Am. Chem. Soc.*, 1990, **112**, 5874.
109. K. Ariga, T. Mori, M. Akamatsu and J. P. Hill, *Thin Solid Films*, 2014, **554**, 32.
110. G. Ma, Q. Zhou, X. Zhang, Y. Xu and H. Liu, *New J. Chem.*, 2014, **38**, 552.
111. K. Ariga, T. Nakanishi, Y. Terasaka and J.-i. Kikuchi, *J. Porous Mater.*, 2006, **13**, 427.
112. K. Ariga, Y. Terasaka, D. Sakai, H. Tsuji and J.-i. Kikuchi, *J. Am. Chem. Soc.*, 2000, **122**, 7835.
113. I. C. Lee, C. W. Frank, T. Yamamoto, H.-R. Tseng, A. H. Flood, J. F. Stoddart and J. O. Jeppesen, *Langmuir*, 2004, **20**, 5809.
114. K. Nørgaard, J. O. Jeppesen, B. W. Laursen, J. B. Simonsen, M. J. Weygand, K. Kjaer, J. F. Stoddart and T. Bjørnholm, *J. Phys. Chem. B*, 2005, **109**, 1063.
115. S. Soon Jang, Y. Hee Jang, Y.-H. Kim, W. A. Goddard III, J. Wook Choi, J. R. Heath, B. W. Laursen, A. H. Flood, J. F. Stoddart, K. Nørgaard and T. Bjørnholm, *J. Am. Chem. Soc.*, 2005, **127**, 14804.
116. T. M. N. Trinh, I. Nierengarten, M. Holler, J.-L. Gallani and J.-F. Nierengarten, *Chem. - Eur. J.*, 2015, **21**, 8019.
117. K. B. Blodgett, *J. Am. Chem. Soc.*, 1934, **56**, 495.
118. K. B. Blodgett, *J. Am. Chem. Soc.*, 1935, **57**, 1007.
119. K. B. Blodgett and I. Langmuir, *Phys. Rev.*, 1937, **51**, 964.
120. I. Langmuir and V. J. Schaefer, *J. Am. Chem. Soc.*, 1937, **59**, 2075.

121. D. R. Talham, T. Yamamoto and M. W. Meisel, *J. Phys.: Condens. Matter*, 2008, **20**, 184006.
122. G. J. Ashwell, R. C. Hargreaves, C. E. Baldwin, G. S. Bahra and C. R. Brown, *Nature*, 1992, **357**, 6377.
123. C. P. Brock and J. D. Dunitz, *Chem. Mater.*, 1994, **6**, 1118.
124. M. F. Daniel, J. C. Dolphin, A. J. Grant, K. E. N. Kerr and G. W. Smith, *Thin Solid Films*, 1985, **133**, 235.
125. I. R. Girling, P. V. Kolinsky, N. A. Cade, J. D. Earls and I. R. Peterson, *Opt. Commun.*, 1985, **55**, 289.
126. M. V. Martínez-Díaz, B. del Rey, T. Torres, B. Agricole, C. Mingotaud, N. Cuvillier, G. Rojo and F. Agulló-López, *J. Mater. Chem.*, 1999, **9**, 1521.
127. M. S. Chandra, Y. Ogata, J. Kawamata and T. P. Radhakrishnan, *Langmuir*, 2003, **19**, 10124.
128. S. Vidawati, J. Sitterberg, U. Rothe and U. Bakowsky, *Colloids Surf., B*, 2011, **87**, 209.
129. L. F. Chi, R. R. Johnston and H. Ringsdorf, *Langmuir*, 1991, **7**, 2323.
130. L. F. Chi, M. Anders, H. Fuchs, R. R. Johnston and H. Ringsdorf, *Science*, 1993, **259**, 213.
131. K. Rajesh, K. Rajendra and T. P. Radhakrishnan, *J. Phys. Chem. B*, 2010, **114**, 849.
132. M. S. Chandra, M. G. Krishna, H. Mimata, J. Kawamata, T. Nakamura and T. P. Radhakrishnan, *Adv. Mater.*, 2005, **17**, 1937.
133. E. Abraham, C. Grauby-Heywang, S. Selector and G. Jonusauskasa, *J. Photochem. Photobiol., B*, 2008, **93**, 44.
134. D. Roy, N. Mani Das, N. Shakti and P. S. Gupta, *RSC Adv.*, 2014, **4**, 42514.
135. L. Gaffo, M. R. Corderio, A. R. Freitas, W. C. Moreira, E. M. Girotto and V. Zucolotto, *J. Mater. Sci.*, 2010, **45**, 1366.
136. D. J. Qian, C. Nakamura and J. Miyake, *Langmuir*, 2000, **16**, 9615.
137. P. M. Mendes, A. H. Flood and J. F. Stoddart, *Appl. Phys. A: Mater. Sci. Process.*, 2005, **80**, 1197.
138. R. C. Ahuja, P.-L. Caruso, D. Möbius, G. Widburg, H. Ringsdorf, D. Philp, J. A. Preece and J. F. Stoddart, *Langmuir*, 1993, **9**, 1534.
139. R. C. Ahuja, P.-L. Caruso, D. Möbius, D. Philp, J. A. Preece, H. Ringsdorf, J. F. Stoddart and G. Wildburg, *Thin Solid Films*, 1996, **284**, 671.
140. C. L. Brown, U. Jonas, J. A. Preece, H. Ringsdorf, M. Seitz and J. F. Stoddart, *Langmuir*, 2000, **16**, 1924.
141. K. Töllner, R. P. Biro, M. Lahav and D. Milstein, *Science*, 1997, **278**, 2100.
142. I. Langmuir and V. J. Schaefer, *J. Am. Chem. Soc.*, 1938, **60**, 1351.
143. T. Verbiest, S. Van Elshocht, M. Kauranen, L. Hellemans, J. Snauwaert, C. Nuckolls, T. J. Katz and A. Persoons, *Science*, 1998, **282**, 913.
144. R. Makiura, S. Motoyama, Y. Umemura, H. Yamanaka, O. Sakata and H. Kitagawa, *Nat. Mater.*, 2010, **9**, 565.
145. H. Byrd, J. K. Pike and D. R. Talham, *Chem. Mater.*, 1993, **5**, 709.

146. S. Conoci, D. M. Guldi, S. Nardis, R. Paolesse, K. Kordatos, M. Prato, G. Ricciardi, M. G. H. Vicente, I. Zilbermann and L. Valli, *Chem. – Eur. J.*, 2004, **10**, 6523.
147. I. Obraztsov, K. Noworyta, A. Hart, H. B. Gobeze, C. B. KC, W. Kutner and F. D'Souza, *ACS Appl. Mater. Interfaces*, 2014, **6**, 8688.
148. A. Tolkki, E. Vuorimaa, V. Chukharev, H. Lemmetyinen, P. Ihalainen, J. Peltonen, V. Dehm and F. Würthner, *Langmuir*, 2010, **26**, 6630.
149. D.-J. Qian, C. Nakamura and J. Miyake, *Chem. Commun.*, 2001, 2312.
150. H. Chou, C. T. Chen, K. F. Storck, P. W. Bohn and K. S. Suslick, *J. Phys. Chem.*, 1994, **98**, 383.
151. T. Kato, N. Matsumoto, M. Kawano, N. Suzuki, T. Araki and K. Iriyama, *Thin Solid Films*, 1994, **242**, 223.
152. K. Y. C. Lee, M. M. Lipp, D. Y. Takamoto, E. Ter-Ovanesyan and J. A. Zasadzinski, *Langmuir*, 1998, **14**, 2567.
153. S. Kundu, A. Datta and S. Hazra, *Langmuir*, 2005, **21**, 5894.
154. K. Ariga, Y. Yamauchi, T. Mori and J. P. Hill, *Adv. Mater.*, 2013, **25**, 6477.
155. M. A. Markowitz, R. Bielski and S. L. Regen, *J. Am. Chem. Soc.*, 1988, **110**, 7545.
156. M. A. Markowitz, V. Janout, D. G. Castner and S. L. Regen, *J. Am. Chem. Soc.*, 1989, **111**, 8192.
157. M. J. Allen, V. C. Tung and R. B. Kaner, *Chem. Rev.*, 2010, **110**, 132.
158. S. Guo and S. Dong, *Chem. Soc. Rev.*, 2011, **40**, 2644.
159. P. Miro, M. Audiffred and T. Heine, *Chem. Soc. Rev.*, 2014, **43**, 6537.
160. T. Bauer, A. D. Schlüter and J. Sakamoto, *Synlett*, 2010, 877.
161. T. Bauer, Z. Zheng, A. Renn, R. Enning, A. Stemmer, J. Sakamoto and A. D. Schlüter, *Angew. Chem., Int. Ed.*, 2011, **50**, 7879.
162. J. Sakamoto, J. van Heijst, O. Lukin and A. D. Schlüter, *Angew. Chem., Int. Ed.*, 2009, **48**, 1030.
163. T. Kambe, R. Sakamoto, K. Hoshiko, K. Takada, M. Miyachi, J.-H. Ryu, S. Sasaki, J. Kim, K. Nakazato, M. Takata and H. Nishihara, *J. Am. Chem. Soc.*, 2013, **135**, 2462.
164. K. Hoshiko, T. Kambe, R. Sakamoto, K. Takada and H. Nishihara, *Chem. Lett.*, 2014, **43**, 252.
165. D. J. Murray, D. D. Patterson, P. Payamyar, R. Bhola, W. Song, M. Lackinger, A. D. Schlüter and B. T. King, *J. Am. Chem. Soc.*, 2015, **137**, 3450.
166. T. M. Ginnal, *Ind. Eng. Chem. Prod. Res. Dev.*, 1985, **24**, 188.
167. T. Bjørnholm, T. Hassenkam and N. Reitzel, *J. Mater. Chem.*, 1999, **9**, 1975.
168. K. Nørgaard and T. Bjørnholm, *Chem. Commun.*, 2005, 1812.
169. J. K. Lim, *Thin Solid Films*, 2013, **531**, 499.
170. M. H. M. Cativo, D. K. Kim, R. A. Riggleman, K. G. Yager, S. S. Nonnenmann, H. Chao, D. A. Bonnell, C. T. Black, C. R. Kagan and S.-J. Park, *ACS Nano*, 2014, **8**, 12755.
171. S. M. Baker, K. A. Leach, C. E. Devereaux and D. E. Gragson, *Macromolecules*, 2000, **33**, 5432.

172. C. E. Devereaux and S. M. Baker, *Macromolecules*, 2002, **35**, 1921.
173. X. Wang, G. Wen, C. Huang, Z. Wanga and Y. Shi, *RSC Adv.*, 2014, **4**, 49219.
174. Z.-R. Chen and J. A. Kornfield, *Polymer*, 1998, **39**, 4679.
175. S. Marre and K. F. Jensen, *Chem. Soc. Rev.*, 2010, **39**, 1183.
176. K. Okano and T. Yamashita, *ChemPhysChem*, 2012, **13**, 2263.
177. J. M. Ribo, Z. El-Hachemi and J. Crusats, *R. Lincei, Sci. Fis., Nat.*, 2013, **24**, 197.
178. J. Puigmarti-Luis, *Chem. Soc. Rev.*, 2014, **43**, 2253.
179. B. W. Laursen, K. Nørgaard, N. Reitzel, J. B. Simonsen, C. B. Nielsen, J. Als-Nielsen, T. Bjørnholm, T. I. Sølling, M. M. Nielsen, O. Bunk, K. Kjaer, N. Tchebotareva, M. D. Watson, K. Müllen and J. Piris, *Langmuir*, 2004, **20**, 4139.
180. P. Chen, X. Ma, K. Hu, Y. Rong and M. Liu, *Chem. - Eur. J.*, 2011, **17**, 12108.
181. C. Rubia-Payá, J. J. Giner-Casares, M. T. Martín-Romero, D. Möbius and L. Camacho, *J. Phys. Chem. C*, 2014, **118**, 10844.
182. A. M. González-Delgado, C. Rubia-Payá, C. Roldán-Carmona, J. J. Giner-Casares, M. Pérez-Morales, E. Muñoz, M. T. Martín-Romero, L. Camacho and G. Brezesinski, *J. Phys. Chem. C*, 2010, **114**, 16685.
183. A. H. Ellison, *J. Phys. Chem.*, 1962, **66**, 1867.
184. H. Kraack, B. M. Ocko, P. S. Pershan, E. Sloutskin and M. Deutsch, *J. Chem. Phys.*, 2003, **119**, 10339.
185. D. S. Ambwani, R. A. Jao and T. Fort Jr., *J. Colloid Interface Sci.*, 1973, **42**, 8.
186. M. Matsumoto, H. Yoshimura, V. S. Kulkarni and K. Nagayama, *Colloid Polym. Sci.*, 1990, **268**, 1174.
187. O. M. Magnussen, B. M. Ocko, M. Deutsch, M. J. Regan, P. S. Pershan, D. Abernathy, G. Grübel and J.-F. Legrand, *Nature*, 1996, **384**, 250.
188. C. Rubia-Payá, E. Jimenez-Millán, J. J. Giner-Casares, G. Brezesinski, M. T. Martín-Romero and L. Camacho, *Langmuir*, 2013, **29**, 4796.
189. J. Y. Park and R. C. Advincula, *Soft Matter*, 2011, 7, 9829.
190. H. Tachibana, Y. Yamanaka and M. Abe, *Thin Solid Films*, 2003, **440**, 94.

CHAPTER 4

Physisorbed Layers at Interfaces

4.1 Introduction

The weak non-covalent interactions that lead to attachment and organisation of molecules at interfaces provide the supramolecular scientist with abundant opportunities to study self-assembly and molecular recognition at the (sub)molecular level when using scanning tunnelling microscopy (STM, see Chapter 2).[1] In physisorbed layers, the molecules typically lie with their long axes parallel to the adsorbent, and therefore present their thinnest side to the tunnelling electrons. This orientation is optimum in terms of the technique since it allows precise imaging and also allows the supramolecular scientist to observe the plane where the intermolecular interactions between adsorbates take place. Hydrogen bonding in particular can lead to a variety of supramolecular topologies often in similar ways to solution or bulk crystals.[2] The weakly interacting adsorbent does not usually interfere dramatically with these non-covalent bonds that are located in a plane parallel to the interface. The fact that their structural dimensionality is between that of complexes in solution and bulk crystals also means that their stability is also intermediate: when a complete ordered monolayer is formed, it is often referred to as a two-dimensional crystal, that is stabilised by the surface adhesion, that can be engineered[3] in analogy to three-dimensional crystals, perhaps with greater ease given the restricted dimensionality.

The characterisation of these physisorbed systems is, in general, not as simple as chemisorbed systems: their near equilibrium state means that aggressive analytical methods can disrupt the order in the layers. For example, atomic force microscopy can often disrupt the structures, moving or removing material with the tip, particularly for weakly adsorbed sub-monolayer films. Nonetheless non-contact methods and the use of

Monographs in Supramolecular Chemistry No. 19
Supramolecular Chemistry at Surfaces
By David B. Amabilino

Published by the Royal Society of Chemistry, www.rsc.org

molecules adsorbed to the tip can provide very detailed information. However, scanning tunnelling microscopy (STM) comes into its own as the tip can be passed at a distance large enough so that it interacts relatively weakly with the molecules and it reveals supramolecular structure with stunning detail.[4]

The use of long alkyl chains to adhere molecules to substrates reversibly is a reliable strategy, and the quasi-epitaxial alignment of methylene chains with the graphite surface is particularly trustworthy. Alkanes with no further functional group form two-dimensional crystals on graphite,[5] and non-covalent interactions between the head groups can influence dramatically the packing of the molecules.[6] As an example of the use of long alkyl chains aiding the adsorption of aromatic molecules to surfaces, that of oligoalkyl porphyrins and phthalocyanines is particularly interesting given the high degree of order in the layers.[7] Both a copper phthalocyanine substituted with eight octyloxy chains and a porphyrin with four tetradecyloxy chains form extremely crystalline monolayers with the aromatic cores separated from one another at the 1-phenyloctane–HOPG interface. Varying the bias in the STM imaging allows observation of either HOMO or LUMO areas in the aromatic molecules. The stability of the monolayer is important to perform these measurements reliably and the interdigitation of the alkyl chains in the films presumably aids in this respect. The presence of large aromatic components such as porphyrins[8] does aid in the characterisation of the layers, because in general the cores are identified easily and sub-molecular resolution can give a clear idea concerning structure at the surface.

Whether complete or partial layers are formed, in order to be able to observe supramolecular ordering, the free energy of adsorption must be sufficient to avoid desorption, which could lead either to diffusion across the surface or to the immediate departure of the adsorpt from the interface. Therefore, as in all true supramolecular systems, thermodynamic factors determine whether layers will physisorb specifically at an interface or not. The structure of the layers can also be influenced by the rate at which adsorption and desorption takes place, thus kinetic factors can also be very important, as they are in the formation of bulk crystals. In the following section, these aspects of physisorbed layers will be discussed.

4.2 Thermodynamic and Kinetic Considerations of Physisorption

The growth of supramolecular structures on the surfaces of solids that interact with them weakly is a situation akin to aggregation of molecules in a weakly interacting solvent. Molecules adsorb, but can diffuse around the interface provided there is sufficient thermal energy and low enough coverage. For this reason, the formation of monolayers that are physisorbed to substrates are subject principally to thermodynamic effects, mass transport being relatively fast, although over larger length scales, kinetic effects are important.[9] The use of Ostwald ripening[10,11] to generate large domains is a case where the system lowers its energy by minimising the interfaces between two-dimensional crystal domains, which are thermodynamically not favoured. The phase boundaries are minimised in order to reduce the energy of the system. Alkyl derivatives of anthraquinone and octathiophene adsorbed on graphite from solution display this phenomenon.[12] Domains that have edges tens of nanometres in length appear in the STM images of the systems early after adsorption, but are either desorbed or reorganised so that larger domains grow in size. In these cases, the timescale of the process was less than a minute, but temperature, concentration and the strength of the molecule–surface interaction will all influence the rate of ripening, as shown for coronene on gold (111)—a quite noble metal surface[13] favouring physisorption of flat aromatics—where polymorphs seen in the as-formed film are replaced by another upon thermal annealing.[14]

When molecules adhere to a substrate from solution, the thermodynamic driving force is clearly the greater free energy of adsorption of the adsorpt when compared to the solvent molecules, but the desolvation of the adsorpt must also be effective. This makes the situation quite different to adsorption from the gas phase where the reverse desorption is negligible for large adsorbate molecules.

The enthalpic contribution to binding can be quantified using a Born–Haber cycle analysis.[15] This process involves determining—in this case both experimentally and theoretically—each component in the system alone and in combination with others to establish the enthalpy of each contribution to the overall energy change. The principle is shown in Figure 4.1. Consider first a monolayer adsorbed on a surface and that this system is in a vacuum. Measuring the heat involved in evaporating the monolayer will give the

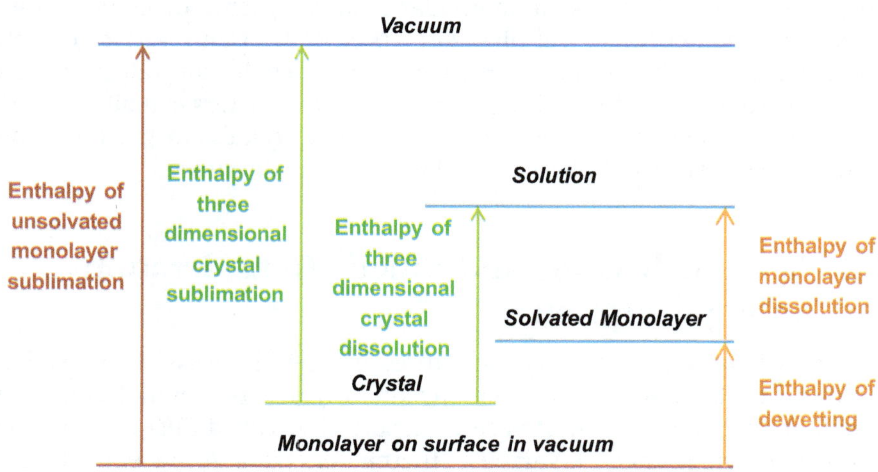

Figure 4.1 A representation of a Born–Haber cycle used for the analysis of the enthalpic contributions to monolayer physisorption, compared with a bulk crystal of the adsorbate material (not to scale).

enthalpy of sublimation of that monolayer. In practice, this measurement can be done by temperature-programmed desorption (employing a quartz crystal microbalance to measure the weight loss at different temperatures) and fitting the data to a Van't Hoff equation. The process is hard for large molecules, with low vapour pressure, because they must be removed to the vacuum.

The situation in solution is far more complex than for a monolayer in a vacuum. In principle, and for thermodynamic arguments to apply, the monolayer is in equilibrium with a solution of the adsorpt. The amount of interface is increased over the monolayer in the vacuum, so that two interfaces can be dissected, the surface–monolayer and monolayer–solvent (Figure 4.2). Therefore the energy of this system—that would ideally comprise a solvated monolayer—relative to the monolayer in the vacuum is raised by the enthalpy of dewetting. This factor actually refers to the difference of the penalty of the solvent dewetting the graphite surface with the gain in energy where the solvent interacts with the adsorbate layer. Until now, this value has to be determined through modelling because of the difficulty in measuring the interaction of solvent with the adsorbate monolayer. The solvated monolayer is now related to a solution of the adsorpt over the monolayer by the enthalpy of monolayer dissolution, and it is this value that can be determined from the Born–Haber cycle to understand the role of the solvent. Whereas microflow calorimetric measurements show one value of the overall enthalpy change, the Born–Haber analysis picks out a detailed picture of the enthalpic contributions.

The entropic contribution to the system was evaluated taking into account translation, rotation, conformation and vibration of solvent and adsorpt as well as the concentration of the latter. In the case of rigid adsorpts, the loss of conformational and vibrational entropy is negligible, upon adsorption translational and rotational entropy are lost though. As far as the solvent is concerned, though, dewetting from the graphite regains conformational, rotational and translational entropy. Since the ordering of solvent on an adsorbed monolayer is believed to be low, the entropy gain upon desorption of the solvent is akin to the difference between the solid and liquid material. Thus, the entropy gain of dewetting favours adsorption of large, rigid molecules.

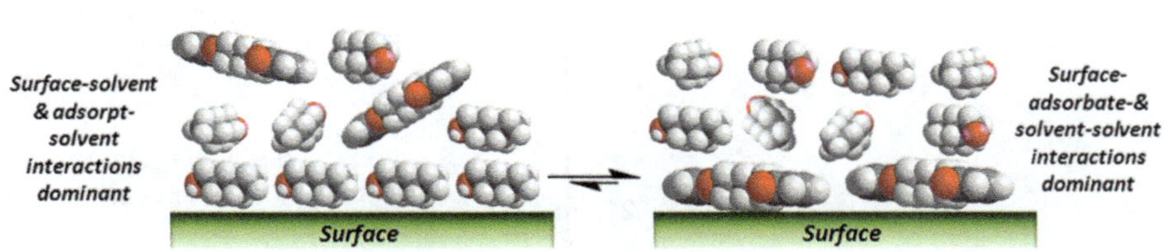

Figure 4.2 The equilibrium between putative adsorbate in solution and a layer of solvent on the surface of the substrate and the situation in which the adsorbate forms a monolayer that is solvated.

Overall, the situation is a balance as shown in Figure 4.2, where equilibrium exists between a stable solvent layer on the substrate and the putative adsorpt solvated in bulk solution and another state where the adsorbate forms the monolayer and the solvent lies above this layer. At present, it is not clear whether this solvent layer is ordered or not, but no evidence up to now has shown an ordered solvent secondary layer and it is believed that the solvent is usually dynamic on the timescale of the STM experiments that are usually used to observe the monolayers.[16]

An analysis of the enthalpic terms has been performed for both terephthalic acid and 4,4′-stilbenedicarboxylic acid on graphite (Figure 4.3).[17] The works show how the solvent can facilitate the desorption process from the adsorbent, leading to exchange and establishing an equilibrium position at the thermodynamic minimum. The energy difference between solution and solvated monolayer with accompanying solvent-covered substrate is quite small, so the solvent lowers the enthalpic desorption barrier dramatically compared with the vacuum. Comparing terephthalic acid and 4,4′-stilbenedicarboxylic acid in nonanoic acid on graphite, the larger molecule clearly has a larger monolayer binding energy, by 50%, as a result of the extended π-system interacting through stacking with the graphite. In addition, the larger molecule has a higher dissolution enthalpy, while the entropy differences between the two systems are negligible, enhancing the stability of the physisorbed layer of the larger molecule.

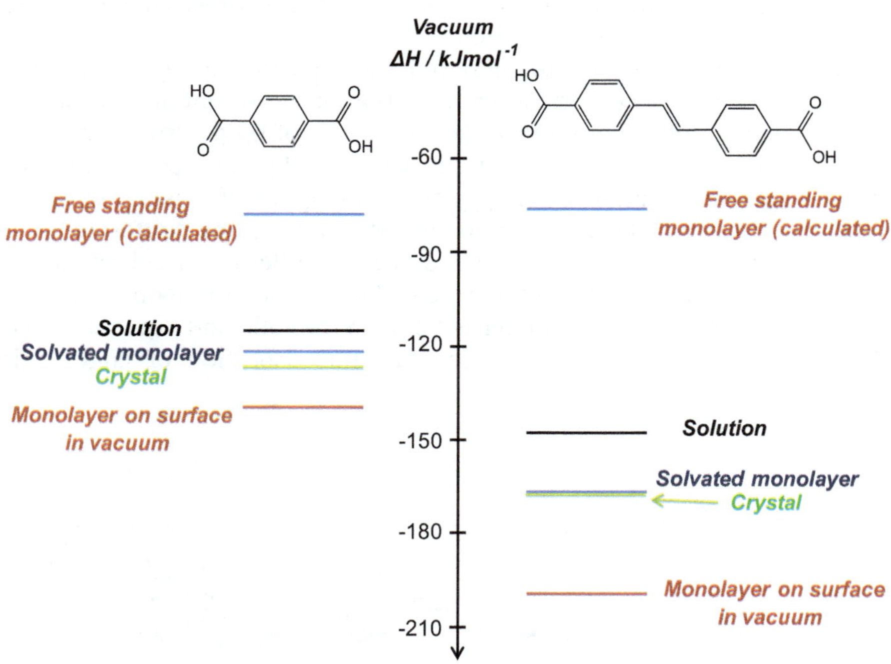

Figure 4.3 Values of the enthalpic terms for physisorbed layers of terephthalic acid and 4,4′-stilbenedicarboxylic acid on graphite in a vacuum and in nonanoic acid.

One should not assume that because molecules physisorb at interfaces that kinetics do not play a role: they can be absolutely dominant.[18] It was first shown—using STM imaging of monolayers heated at different temperatures—that metallooctaethylporphyrins have very slow exchange with a solution of the compounds at 135 °C (Figure 4.4), and that at 25 °C there is essentially no desorption once the monolayer is formed.[19] Cobalt(II) and nickel(II) octaethylporphyrins when adsorbed onto a gold (111) surface at room temperature from a sub-saturated solution in 1-phenyloctane form a monolayer extremely quickly, within ten seconds! The fact that the two molecules were distinguishable by their contrast in the STM (the cobalt porphyrin appears higher and brighter in the images, emphasising the caution that is required when determining height using the technique) made it possible to observe that they form a perfect solid solution when co-adsorbed in varying ratios. This effect is surely caused by the strong adsorbate–adsorbent interaction as well as the fact that their two-dimensional unit cells are virtually identical. This combination of features made it possible to show that no observable change in the positions of the adsorbates took place over a timescale of several hours. Even introducing an excess of one of the components in solution did not lead to any exchange. It was necessary to heat the sample to 135 °C (followed by cooling to trap the state kinetically) before imaging showed the replacement of the nickel porphyrins on a monolayer of the cobalt(II) complexes. Quantification of the rate (6.7×10^{-5} s^{-1} at 35 °C) showed that this non-covalent system actually exchanges more slowly than certain thiolates on gold.[20] Interestingly, the desorption of nickel(II) porphyrins happens at specific sites on the surface (edges and defects), so this monolayer could not be used to reliably model

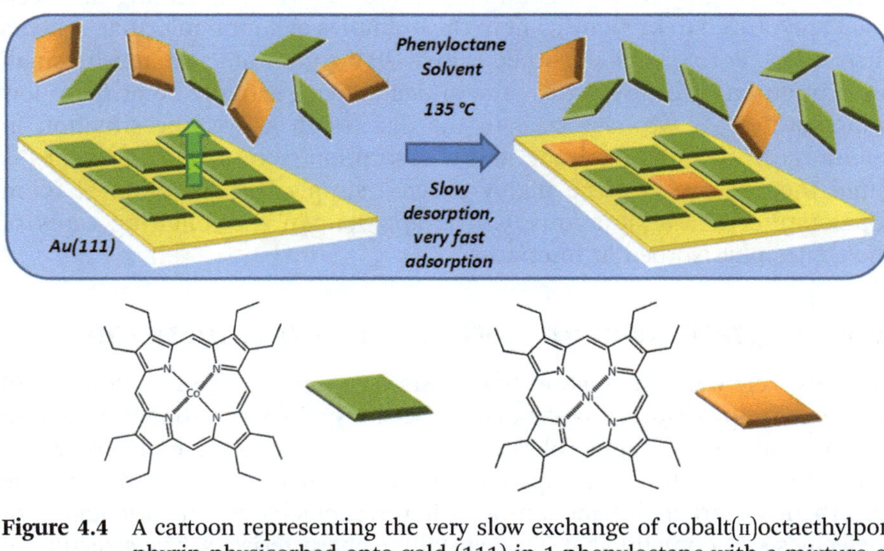

Figure 4.4 A cartoon representing the very slow exchange of cobalt(II)octaethylporphyrin physisorbed onto gold (111) in 1-phenyloctane with a mixture of cobalt(II) and nickel(II) porphyrins in the liquid at 135 °C. The exchange is essentially non-existent at 25 °C.

the exchange. Therefore, the kinetics of monolayer formation can lead to metastable layers, which can mean the formation of monolayers with contrasting structure and properties, as for the case of porous or dense layers formed by star-shaped molecules.[21]

Fast adsorption can distort the thermodynamically-favoured situation in mixtures when the metastable states have a significant lifetime. When a mixture of the two quite dissimilar compounds *n*-tetracontane ($C_{40}H_{82}$) and 4′-octyl-4-biphenylcarbonitrile are deposited from a fluid phase, STM shows that domains of the pure compounds are formed at the graphite–liquid interface.[22] Adsorption of the alkane is favoured over that of the biphenyl-containing compound, but the former generates two-dimensional crystals that nucleate at a far faster rate than the other component: indeed, even at a 600 : 1 solution ratio of the biphenyl derivative to the alkane, the *n*-tetracontane shows high interface coverage as a result of polycrystalline domain formation soon after deposition. The movement of domain boundaries means that over a period of days the—initially small—4′-octyl-4-biphenylcarbonitrile grows in size and populates increasing areas of the surface. It was shown that the kinetics of domain boundary movement is proportional to the roughness of the interface, a rougher interface favouring the growth of the thermodynamically-preferred domain.

Monolayer formation at a low temperature can lead to many domain boundaries and disorder because of fast adsorption and slow in-plane diffusion because of high coverage. At a higher temperature, these defects can be annealed out to favour very structurally regular domains. Nonetheless, certain systems are more prone to kinetic effects—that can be seen in the type of domain boundaries—than others, as shown for long-chain alkyl derivatives on graphite.[23] It shall be pointed out when annealing has been used and proven to be significant in the examples of supramolecular systems that follow. For systems adsorbed under UHV conditions, where very small adsorption energies are acting, the annealing can actually be at quite low temperatures. As we have seen in the discussion above, the situation in solution can be quite different, and if desorption from metastable states is rapid then packing can be highly ordered simply by adsorption at room temperature. These situations will be apparent in the following cases of molecules physisorbed at interfaces.

4.3 Supramolecular Systems in Two Dimensions

The relatively weak interaction that exists between adsorbate molecule and adsorbent in physisorbed layers allows the study of what are often referred to as two-dimensional supramolecular systems, and reveals unique characteristics of the environment compared with three dimensions. Even if weak, the adsorbent–adsorbate interactions are often responsible for the orientation of molecules and domains that have crystalline characteristics. The structure of physisorbed monolayers depends on the surface to which they are gently bound. Therefore, particular symmetry conditions apply to the molecules that are arranged planar to the substrate in physisorbed layers. Since the absolute value of any strain resulting from the mismatch of substrate surface and layer periodicity will increase with coverage, the influence of the substrate on packing of the adsorbate can change and provide different packing features depending on the amount of adsorbed material. Therefore, submonolayer systems will be discussed first, followed by higher coverages.

4.3.1 Sub-monolayer Assemblies

When the coverage of an adsorbate at an interface is low, well below that of a monolayer, the mobility of molecules in the plane of the surface can be high: diffusion is only hindered by the strength of the interaction between the single (or few) molecule(s) and the surface atoms of the substrate. Therefore, observation of submonolayers of physisorbed systems must be performed when diffusion is limited. Solvent is usually absent and often, a low temperature is required. Most information is gleaned from studies performed at a low temperature under UHV conditions, where pure adsorbates can be identified in the absence of any atmospheric contamination and the diffusion can be controlled.

When 1-nitronaphthalene is adsorbed under UHV onto a reconstructed Au (111) surface built up of fcc and hcp stacking uniaxial domains that alternate with the formation of elbows and the surface is cooled to 50 K at a coverage of 0.2 ML, decamer aggregates self-assemble (Figure 4.5).[24,25] The clusters pin at the fcc elbows of the metal's surface. Closer inspection shows that the decamers are chiral, with both left- and right-handed clusters being observed. The chiral decamers can not only be observed, they can be sorted into areas of the surface with the STM tip in a nanometre-scale version of Pasteur's triage of enantiomorphic supramolecular objects.[26] The chirality of the decamers is maintained during the process. This feat is incredible, and emphasises the weak interaction between the adsorbent and the adsorbate clusters. The molecules were deposited close to room temperature—where fast diffusion occurs—and the system was cooled down leading to this thermodynamic situation with narrow distribution of aggregate size. Other, minority, structures exist (chiral tetramers and undecamers are also observed) and, at a lower coverage of 0.03 ML, either isolated molecules or trimers can be observed, again indicating the true equilibrium nature of the system,[27] a truly textbook case of supramolecular chemistry on a surface.

Careful observation of high resolution images aided by calculations confirms that the decamer aggregates are globally chiral, although the two chiral arrangements of the adsorbate are incorporated in an 8 : 2 ratio.[28] Hydrogen bonds between the nitro-group oxygen atoms and the hydrogen atoms attached to the aromatic rings appear to be responsible for the stability of the decamer. Increasing coverage results in the formation of chains of molecules with a head-to-tail homochiral dimer. Increased coverage also implies increased defect density, causing kinks in the chain. At full monolayer coverage, a racemate forms on the fcc domains, and a conglomerate forms in the

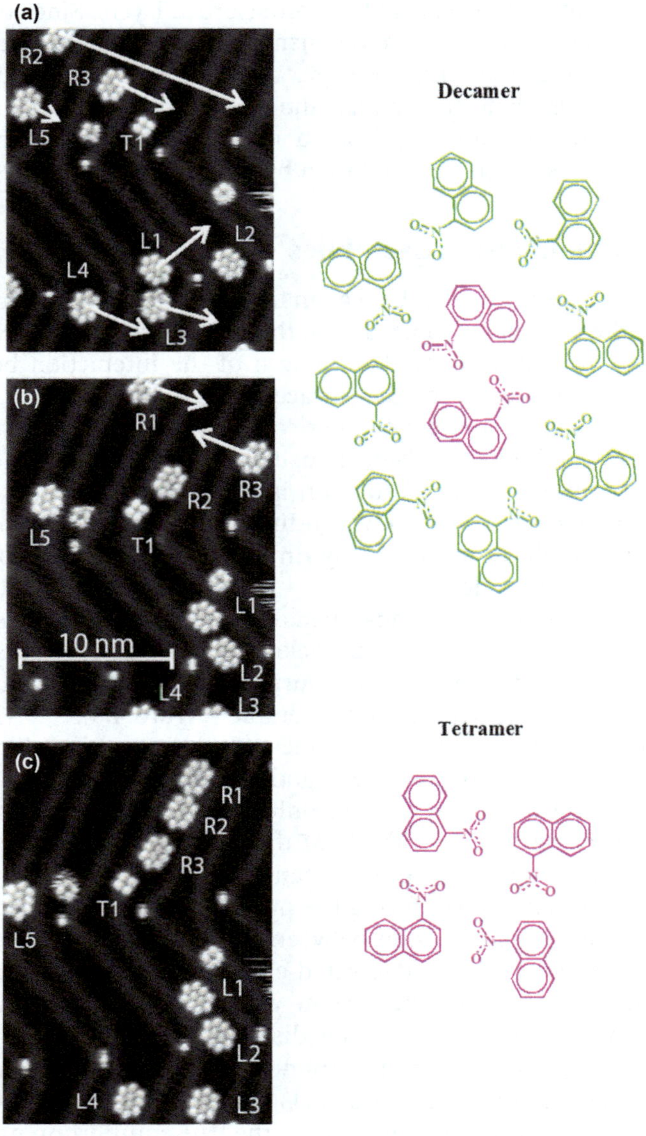

Figure 4.5 Consecutive STM images of the chiral decamers of 1-nitronaphthalene being sorted with the tip of the microscope on the Au (111) surface, and structures of a decamer and tetramer. (a) As prepared, with arrows indicating movement of the clusters to the subsequent image; (b) part way sorted. (c) The right-handed decamers are at the top and the left-handed ones are below. A homochiral tetramer (T1) is also present.

hcp domains of the surface. Overall, electrostatic interactions favour like enantiomer aggregation and stronger adsorption to fcc domains. The racemic compound monolayer in the fcc domains are denser than the double chains: hydrogen bonds link the surface enantiomers.

One of the earlier and more beautiful examples of the self-assembly of supramolecular nano-objects on surfaces is that of a series of porphyrins where cyano groups were introduced at the 4-positions of the phenyl rings attached at the *meso* positions of the tetrapyrrolic core and the molecules were deposited on the herringbone reconstruction of gold (111).[29] STM images of molecules bearing one or two cyanophenyl moieties—the latter as two different isomers—revealed the determining effect that non-covalent interactions have on the supramolecular structures that are formed (Figure 4.6). The porphyrin bearing one cyanophenyl group assembles to give cyclic trimers on the surface, while the compound with two of these groups at the 5 and 10 positions (12 and 3 o'clock) gave a cyclic tetramer because of the steric restrictions of the constitution of the molecule for non-covalent interactions between molecules. The assembly can be viewed as the interaction of the cyano groups coming antiparallel to an energy minimum structure, which was modelled by the authors, as well as a weak hydrogen bond between the cyano group nitrogen atom and protons attached to the phenyl rings. The role of the tertiary butyl groups should not be understated, as they are believed to prevent "side-on" aggregation of the systems, something confirmed in the monolayer of the compound where all the phenyl rings of the porphyrin were substituted with these groups and no significant contact was observed by STM on the gold surface under the same

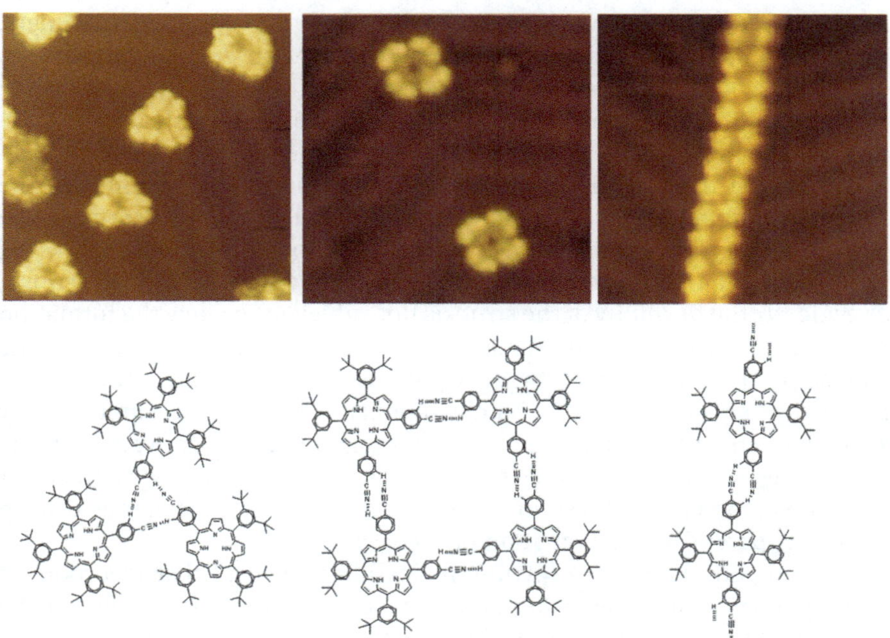

Figure 4.6 STM images on Au (111) at 63 K of the porphyrin self-assembled structures indicated below each image.
Images reproduced with permission of Macmillan Publishers Ltd., *Nature*, Copyright 2001, ref. 29.

conditions as the aggregates. For the compound with the cyanophenyl moieties at the 5 and 15 positions of the porphyrin (12 and 6 o'clock), a linear chain structure was formed, again, because of the assembly restrictions given the topography of the molecule. A study varying the nature of the substituents on the remaining phenyl rings demonstrated the formation of networks and the robustness of the interaction between the cyanophenyl moieties on copper (111).[30] This work showed that the dispersion interactions between alkoxy chains can influence significantly the assembly route, either through essentially one-dimensional structures or directly to two-dimensional networks.

The same principle of constitution-selective assembly based on aryl porphyrins was employed to prepare a variety of elegant structures in which there is an interaction between pyridyl moieties and copper adatoms on the copper (111) surface,[31] something that was later confirmed for the porphyrin bearing four pyridyl groups on the same surface.[32] In this case, the compound bearing one pyridyl substituent and three bulky aromatic units at the other porphyrin *meso* positions led to the formation of isolated linear dimers. Dimer formation on the analogous silver substrate did not give these structures, confirming the specific role of the metal in the assembly. The compound bearing two pyridyl group-terminated struts at the 5 and 10 positions of the porphyrin ring assembles on copper (111) to give a series of beautiful metallomacrocycles, whose propensity was coverage dependent: STM showed that trimers were favoured at low coverages while trimers, tetramers (mainly), pentamers and hexamers were observed at higher coverage. Closed structures are clearly favoured, indicating a thermodynamic situation with all coordinating groups satisfied, and the polydispersity indicates that the outcome of the self-assembly depends upon a delicate balance between the pyridyl–copper–pyridyl interactions, the flexibility of the L-shaped components and the surface's interaction with the adsorbate. The isomer with pyridyl struts at the 5 and 15 positions of the porphyrin ring produced linear structures through the same pyridine–copper interactions, as evidently the flexibility of the struts is not sufficient to allow the formation of cyclic systems. The combination of the 5,10 and 5,15 isomers on the copper (111) surface viewed by STM produced a spectacular array of macrocyclic systems incorporating both components, with the 5,10 units mainly at the corners of the squares and rectangles and the 5,15 derivatives in the connecting sides. An example of these surface-adsorbed supramolecular metallomacrocycles is shown in Figure 4.7. Again, a thermodynamic driving force is apparent in the formation of the cycles.

The compound 4-(*trans*-2-(pyrid-4-yl-vinyl))benzoic acid that incorporates both hydrogen bond donor and acceptor located at opposite "ends" of the molecular structure forms homochiral twin chains on the (111) surfaces of gold and silver (Figure 4.8).[33] The antiparallel chains of molecules are formed by hydrogen bonding between the acid and pyridine moieties and are apparently joined through relatively weak hydrogen bonds between the carbonyl oxygen atom and the aromatic C–H groups. The prochiral

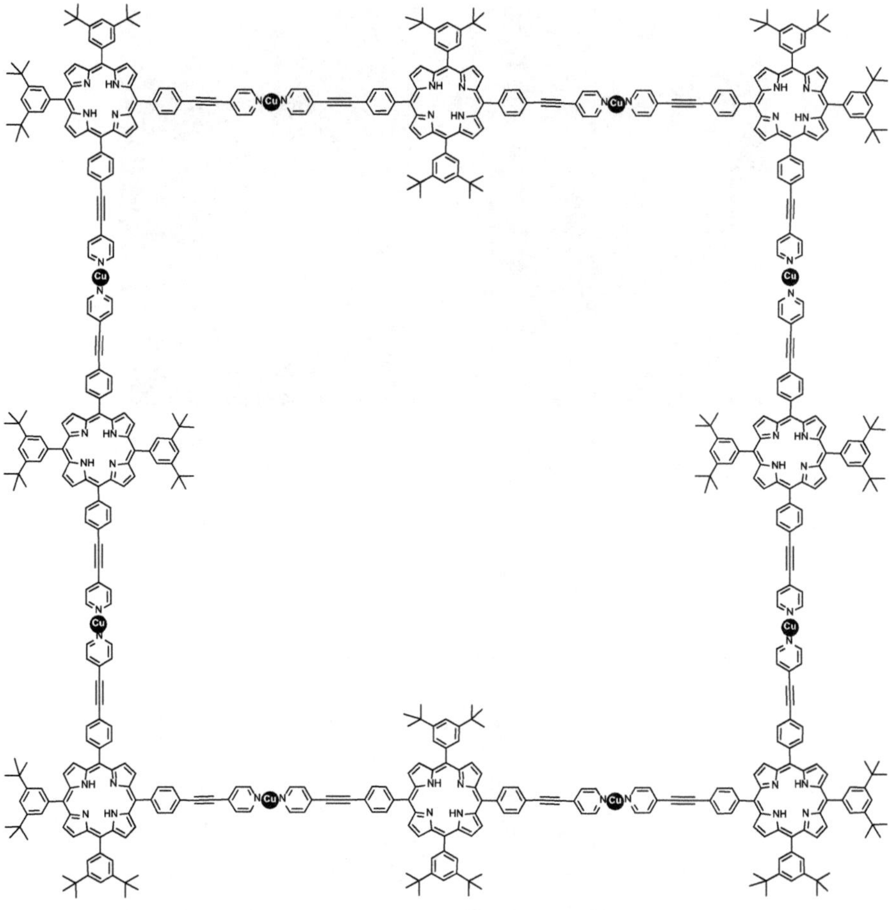

Figure 4.7 The structure of a 4:4 two-component metallomacrocycle formed on Cu (111) with copper adatoms acting as the nodes to which the pyridyl groups coordinate.

molecules are chiral on the surface because of the orientation of the central double bond with respect to the long axis of the molecule. This chirality is transferred to the chains. Incredibly, and despite the large distances between the twin chains (2.5–5 nm) in the domains, homochiral gratings were observed over micrometre scales by STM on the gold (111) and silver (111) surfaces.[34] The formation of metastable triple chains appears to be the source of the transfer of chirality during the growth of the domains. Some chains deviate away from a seed chain and initiate the growth of a new parallel one with the same chirality as the original one. NEXAFS was subsequently used, in conjunction with modelling, to show the slight twisting at submonolayer coverage[35] that might assist in the transfer of chirality. This molecule also resolves spontaneously, though in a quite different way, in the square clusters it forms at submonolayer coverage on copper (100).[36]

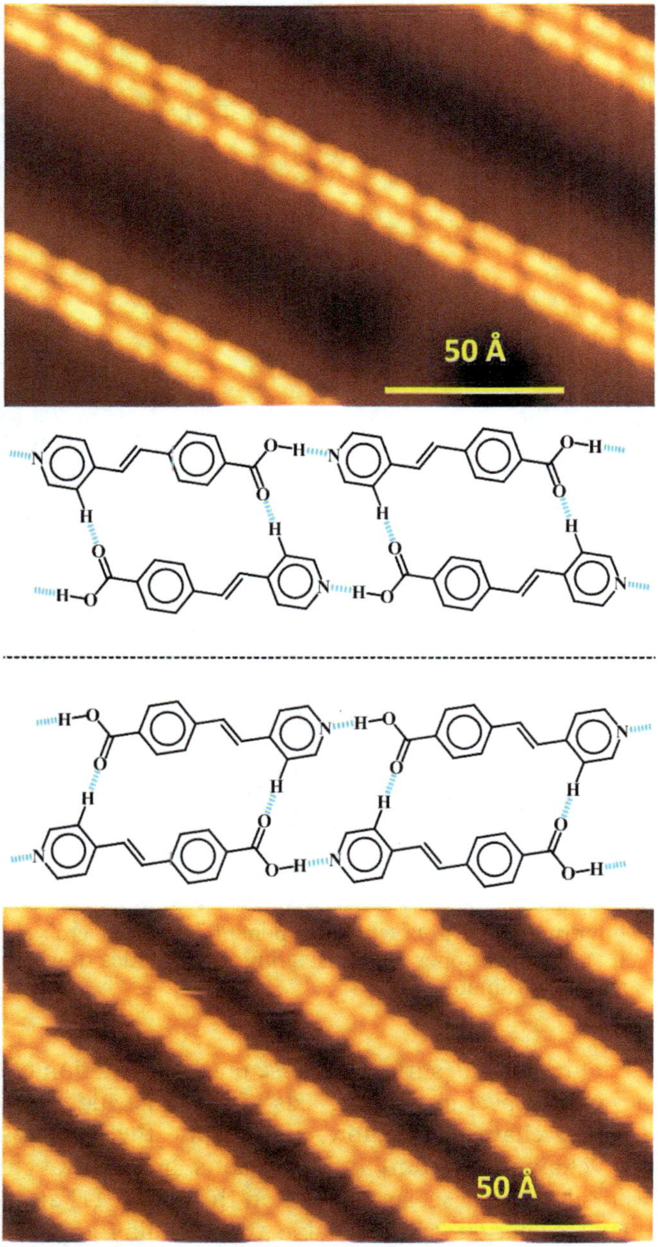

Figure 4.8 A representation of the twin chains of 4-(*trans*-2-(pyrid-4-yl-vinyl))benzoic acid stabilised on the surface of Au (111) and Ag (111) and the observed STM images of the enantiomeric tracks of the compound.

Additionally, and curiously, the molecules form discrete dimers through the carboxylic acid residues on palladium (111),[37] highlighting the very great influence of the adsorbent on supramolecular structure.

The aromatic hydrocarbon rubrene (5,6,11,12-tetraphenyltetracene) displays a truly spectacular example of the formation of clusters with a pentagonal shape as well as spontaneous resolution of conformers at sub-monolayer coverage on a gold (111) surface in UHV at 5 K.[38] This face of gold has a characteristic pattern, because it undergoes a rearrangement of the atomic layer nearest the interface with respect to the next layer into the bulk of the metal, giving it a characteristic reconstruction with alternating areas of hexagonal close-packed and face-centred cubic stacking of the outermost atomic layer. Domain walls, that rotate 120° periodically at elbows, protrude slightly and show a herringbone pattern with a spacing of approximately 6.3 nm.[39,40] The adsorpt, rubrene, has chiral conformers because of the non-planar structure forced by intramolecular contacts between the phenyl ring substituents attached to the acene (Figure 4.9), and these conformational enantiomers can be distinguished in the STM images of the isolated molecules. Three distinct lobes are observed by STM, in clockwise or

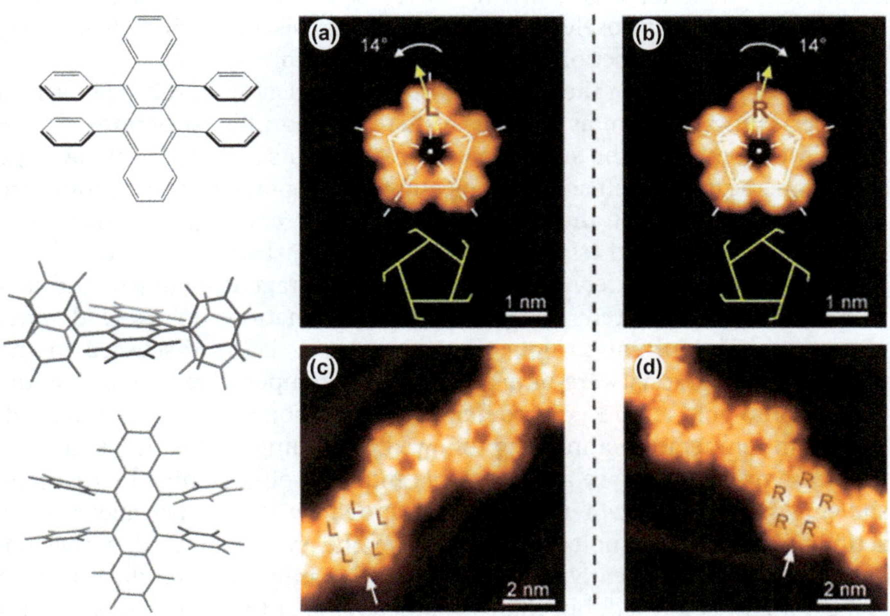

Figure 4.9 The chemical structure of rubrene and representations of the twisted conformation that can exist as a pair of enantiomers. The images (a)–(d) are STM micrographs of isolated supramolecular cyclic pentamers ((a) and (b)) that show left- and right-handedness (denoted L and R here) and the self-assembled chiral chains comprising non-covalently interacting pentamers of only one of these chiralities.
STM images reproduced with the kind permission of Wiley-VCH, Copyright © 2010 WILEY-VCH Verlag GmbH & Co. KGaA, Weinheim, 2005, ref. 38.

anticlockwise arrangement, which define the chirality of each molecule on the surface. These features, which do not interconvert on the timescale of the experiment at the temperature indicated, show the stereochemistry of the higher assemblies. While at very low coverages the molecules adsorb at the step edges exclusively, at a coverage of 0.3 monolayers, strings and rings of cyclic structures are observed (Figure 4.9). The cyclic structures comprise five molecules of identical chirality and have a regular shape. The strings and decamer rings of these pentamers seem to be formed by diffusion of the cyclic pentamers across the surface as one. They unite with one another to give gear-like wheels with nanometre dimensions. The interdigitation of the "teeth" of the neighbouring supermolecules implies a rotation of exactly 36° with respect to each other, resulting in a centre-to-centre distance of approximately 3 nm between their centres. Further supramolecular assembly of the cyclic pentamers leads to the formation of linear higher order structures and occasionally to cyclic decamers (a result of ten unidirectional unions between pentamers) that comprise fifty rubrene molecules with the same chiral conformation at the interface. The structures are incommensurate with the metal's surface, apparently ruling out relatively strong adsorbent–adsorbate interactions as a driving force for the formation of the superstructures. The enantioselective assembly is consistent with self-assembly driven by C–H–π interactions and van der Waals forces.

At higher coverage on the (111) gold surface, and at 50 K, rubrene forms a complex tiling pattern as well as other polymorphic structures.[41] The monolayer comprises the supramolecular pentagons seen at lower coverage and temperature, as well as hexagons and filled heptagons: these motifs are distributed randomly in the tiling pattern. At high coverage, the monolayer comprises close-packed islands of the racemic mixture conformers.[42] The compound shows a variety of polymorphs as coverage is changed, and it is believed that the rubrene has different conformations in these distinct phases. It is important to note that the rubrene layers discussed in the preceding paragraphs were annealed at room temperature: when it is deposited onto gold at 50 K, disordered two-dimensional clusters are formed. The room temperature annealing followed by cooling allows the molecules to diffuse over the surface and assemble into the self-assembled structures, showing how the growth process is governed by the non-covalent interactions between the molecules that drive the system toward a thermodynamic minimum. Clearly, there are close-lying minima given the variety of cyclic structures formed, but this is characteristic of the relatively weak and poorly directional interactions at work in the assembly. When the compound was adsorbed onto different surfaces, the outcome was distinct: dense packed phases with adsorbent-dependent packing were observed. On the more corrugated gold (100) surface—that unlike the (111) crystal face has apparent ridges—at a coverage of approximately 0.2 ML, the supramolecular pentamers are also observed, showing the relative strength of the interaction between molecules compared with the surface-molecule force. That said, two types of islands are observed where the packing of the rubrene molecules is different to that present on the (111) face even at this low coverage. The island phases are homochiral or randomly chiral, and quasi-commensurate and incommensurate with the surface, respectively. On the (111) and (100) faces of silver, only densely-packed islands with heterochiral order of oriented molecules growing from step edges was observed, and this structure is independent of the coverage. On copper crystal faces, the interaction of the rubrene adsorpt with the surface is much stronger than for gold or silver, and the orientation and packing of the molecule is influenced greatly by this stronger adsorption.[43–45]

The dimerisation of the imide group can be used to self-assemble chains of very complex molecules, the so-called "molecular landers", which have a nominally flat aromatic central board and aromatic side groups perpendicular to this region intended to minimise the interaction between surface and board region.[46,47] The molecular lander shown in Figure 4.10 was deposited onto gold (111) under UHV conditions and annealed on the surface for ten minutes at 400 K such that diffusion and self-assembly could take place.[48] The result of measuring the structure at 170 K is hydrogen

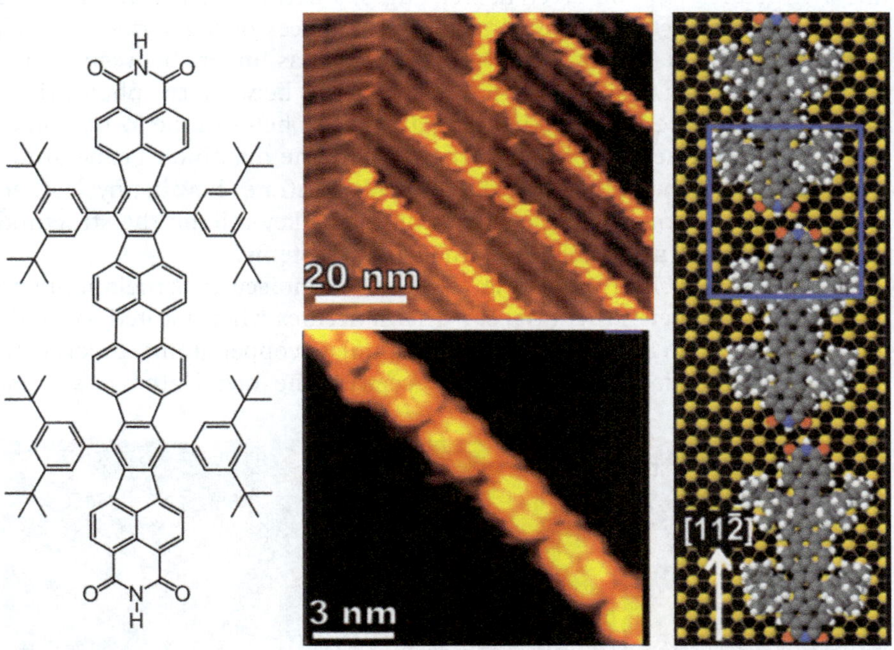

Figure 4.10 A "lander" molecule that self-assembles into hydrogen bonded chains one molecule wide on Au (111). The STM images recorded at 170 K show a larger scale image with the chains running along the reconstructed terraces, and the closer image the detail of a chain where four lobes make up each molecule. The modelling image shows the hydrogen bonds between the imide moieties.
Reproduced from ref. 48 with permission from the Royal Society of Chemistry.

bonded chains of molecules where the diimides unite in a linear arrangement (as shown in the model in Figure 4.10), differently to ones formed by perylene-3,4,9,10-tetra-carboxylic diimide that can give a complex structural outcome.[49,50] For the lander molecule on gold (111), the chains follow the ridges of the herringbone reconstruction of the surface along the fcc areas and the legs prevent side-on interactions at low coverage. Oligomers tens of molecules long are obvious in the STM images.

The supramolecular chains are actually dynamic at 220 K, demonstrating the weak interaction of the molecules with the surface, but chains were still observed. Figure 4.11 actually shows how these supramolecular chains can reorient on the timescale of the experiment. At higher coverages, the chains start to dimerise (see Figure 4.11) and eventually form densely-packed islands with parallel rows of the hydrogen bonded chains.

Subsequently, complementary lander molecules have been co-deposited also on gold (111) in an aim to build more complex chains, but in fact generating a two-dimensional network structure.[51] The same diimide discussed earlier was combined on the gold surface with a lander bearing two diamino-triazine functionalities disposed in a linear topology. Although linear structures were observed at step edges, a network structure as shown in Figure 4.12 is actually formed on the flat terraces of the surface. In this network, the bis-diamino-triazines essentially act as linkers between chains of the diimide. The flexibility of the torsion angles between the phenyl rings in order to optimise hydrogen bonding apparently helps in the formation of this structure, rather than a linear chain where the diamino-triazine group would be coplanar with the surface. The result illustrates how in physisorbed systems the conformational flexibility can play a key role on the structural outcome, even for nominally two-dimensional systems.

A quite remarkable effect with lander kinds of molecules is their ability to restructure a copper (110) surface.[52] The molecules when located at single atom deep steps on the metal actually accumulate copper atoms underneath them—by interaction of the large π face with the metal atoms—so that

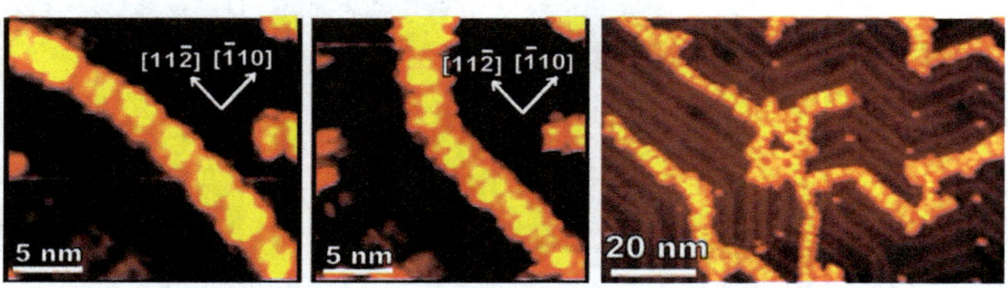

Figure 4.11 Consecutive STM images—10 seconds apart, left and middle—of the "lander" molecule on Au (111) showing the dynamic nature of the chains in the surface at 220 K, and right at higher coverage (0.15 ML), where branching and initial formation of a belt-like superstructure two molecules wide can be seen.
Reproduced from ref. 48 with permission from the Royal Society of Chemistry.

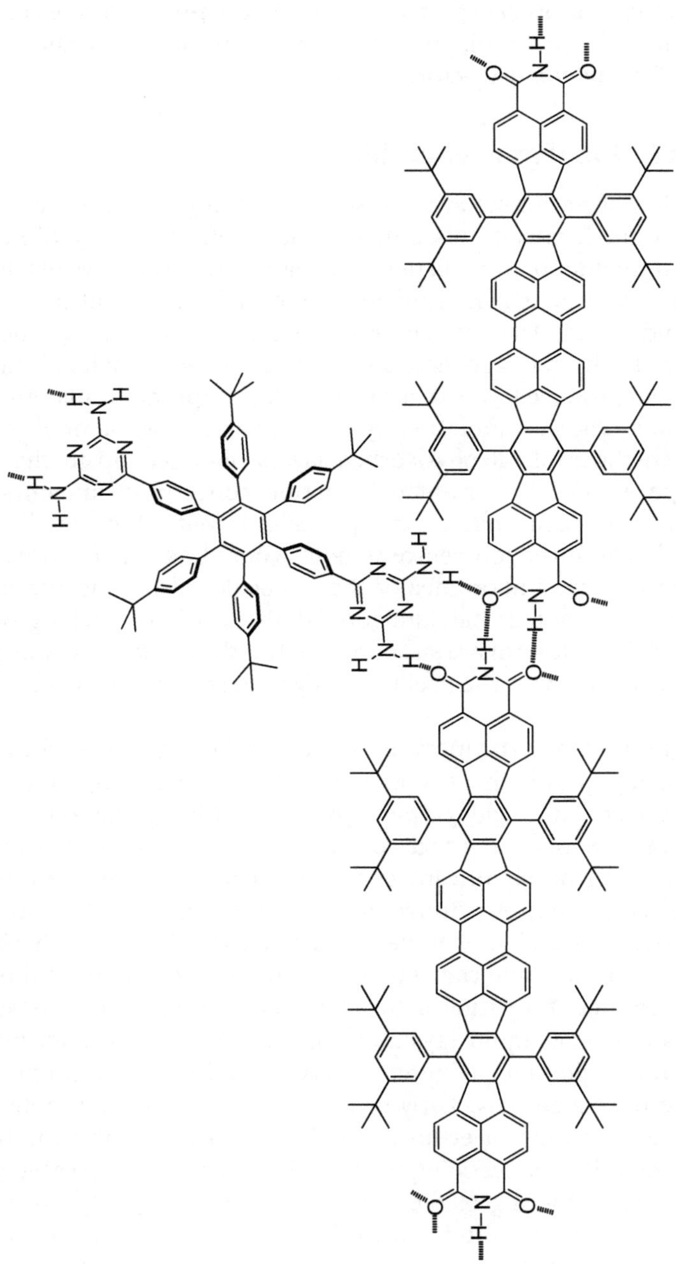

Figure 4.12 A representation of part of the proposed network structure formed by complementary lander molecules on gold (111).

"teeth" two atoms wide and several atoms long are formed. The result shows how weakly adsorbed molecules can actually template the underlying surface when the atoms in it are sufficiently mobile, and is perhaps the inverse situation to the intuitively expected case where organic molecules are driven by the surface. The reversible interaction between the lander and the metal is essential for this nanostructuring taking place.

4.3.2 Full Monolayer Assemblies

The formation of full monolayers of self-assembling molecules on surfaces often leads to the formation of domains that are effectively two-dimensional crystals. And the monolayers display the characteristics of crystals, like long range order, defects, polymorphism, twinning, solid solutions, Ostwald ripening, and so on. The interactions between the primary synthons that might make up discrete aggregates of zero- or one-dimensional nature on the surface can produce layers that follow certain supramolecular and crystal packing tenets. The accumulation by Matzger and co-workers of the structural characteristics of all physisorbed layers obtained up to the time of the paper gave some important information concerning full physisorbed monolayers.[53] Certain plane groups stand out, but especially *p*2 (Figure 4.13). The inversion centre is not favoured as it is in three dimensions, because of the importance of the molecule–adsorbent interface, but nonetheless it was noted that Kitaigorodskii's ideas[54] of packing being favoured by the densest form are supported in two dimensions. As was pointed out, in physisorbed layers the packing is highly responsive to any variance in molecular structure.

The polymorphism in the monolayers can be abundant, as in the case of a simple tetraalkyl porphyrin.[55] STM revealed that this compound is tetramorphic at the octanoic acid–graphite interface, although all four structures share unit cell vectors. This fact means that the different structures can join together with no significant energy penalty, as if they were the two-dimensional equivalent of twinned crystals with non-equivalent structures. The conformational flexibility of the alkyl chains and the ease with which the superstructures as a whole can rotate on the graphite surface is thought to lead to the seamless transition between the polymorphs that is observed.

Full monolayers of liquid crystalline molecules were studied by STM[56] because of the importance of interface structure in determining the properties of the bulk materials. Early work showed a correlation between the nature of the anchoring layer on molybdenum disulfide and the bulk behaviour as nematic or smectic liquid crystal.[57] Another remarkable piece of work showed the ordering of the liquid crystal *trans*-4 pentyl-(4-cyanophenyl)-cyclohexane on an indium tin oxide substrate coated with the photopolymer poly-vinyl-4-methoxy-cinnamate. STM images, though of limited resolution because of the nature of the system, did show the packing of the molecule where dipoles of intercalated molecules have their cyanophenyl groups cancelling one another's dipoles. It was believed that the anisotropic

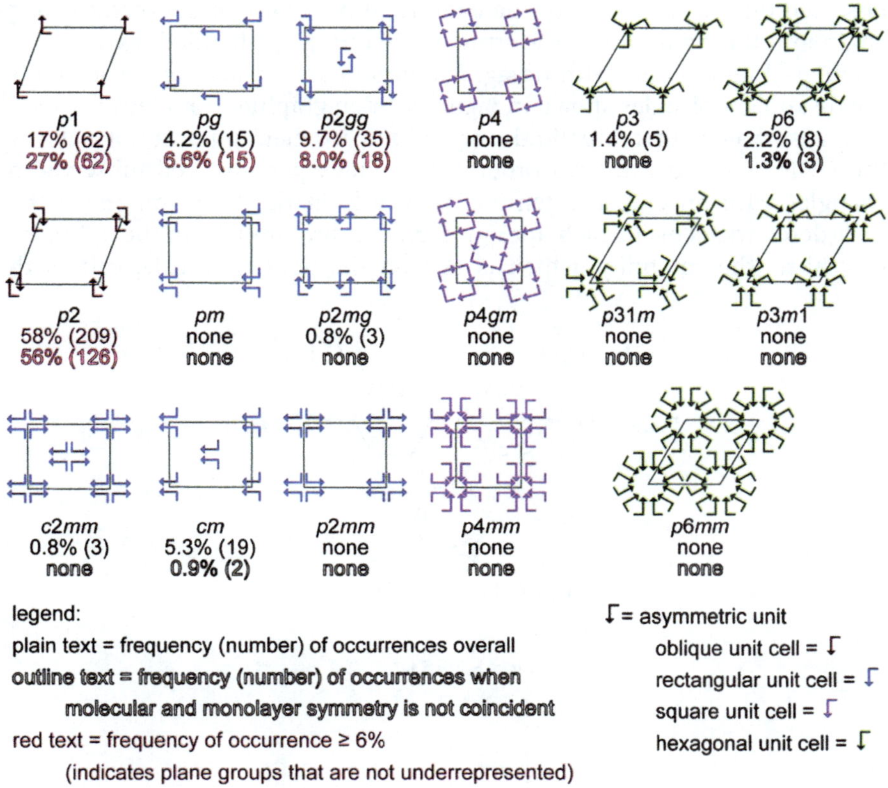

Figure 4.13 Cartoons showing the seventeen plane groups employed to classify monolayer packing. The percentage of each group (followed by the total number) is given followed by the values corresponding to cases where molecules are not at a symmetry element position, in 2007.
Reprinted with permission from ref. 53. Copyright 2007 American Chemical Society.

dispersive surface interaction dominated the packing of the molecule on this surface, which contrasts with the organisation of the same molecule on graphite or molybdenum disulfide. An opportunity still exists to study supramolecular systems on related substrates perhaps using photopolymers as spacing layers that are of so much relevance in organic solar cells.

The study of chiral liquid crystals physisorbed at interfaces of solids with the bulk fluids by STM is also of importance for devices, especially for the ferroelectric mesogens. Comparison revealed a number of fascinating features of the packing of these materials.[58] Perhaps one of the breakthrough results in this area was the observation that racemic mixtures of chiral liquid crystal molecules can form enantiomorphous domains at the interface of the material with graphite.[59] Although the STM images did not reveal the individual chirality of each molecule in this system, the supramolecular stereochemistry was clearly unique in the different domains that were observed on the surface, with the conformational kink in-between the alkyl chain and the rigid part of the structure defining the handedness.

Ordered monolayers with hexagonal lattices are formed by adsorption of the disk-like molecules shown in Figure 4.14 on graphite from tetradecane.[60] The adsorbates (that are liquid crystalline in their pure state) pack on the HOPG surface with two ordered molecules per unit cell (observed as triangles with an ellipse extending outwards in the STM images) with a disordered molecule at each apex (observed as bright disks in the STM). It is very likely that mobility, which is fast on the STM timescale, causes the

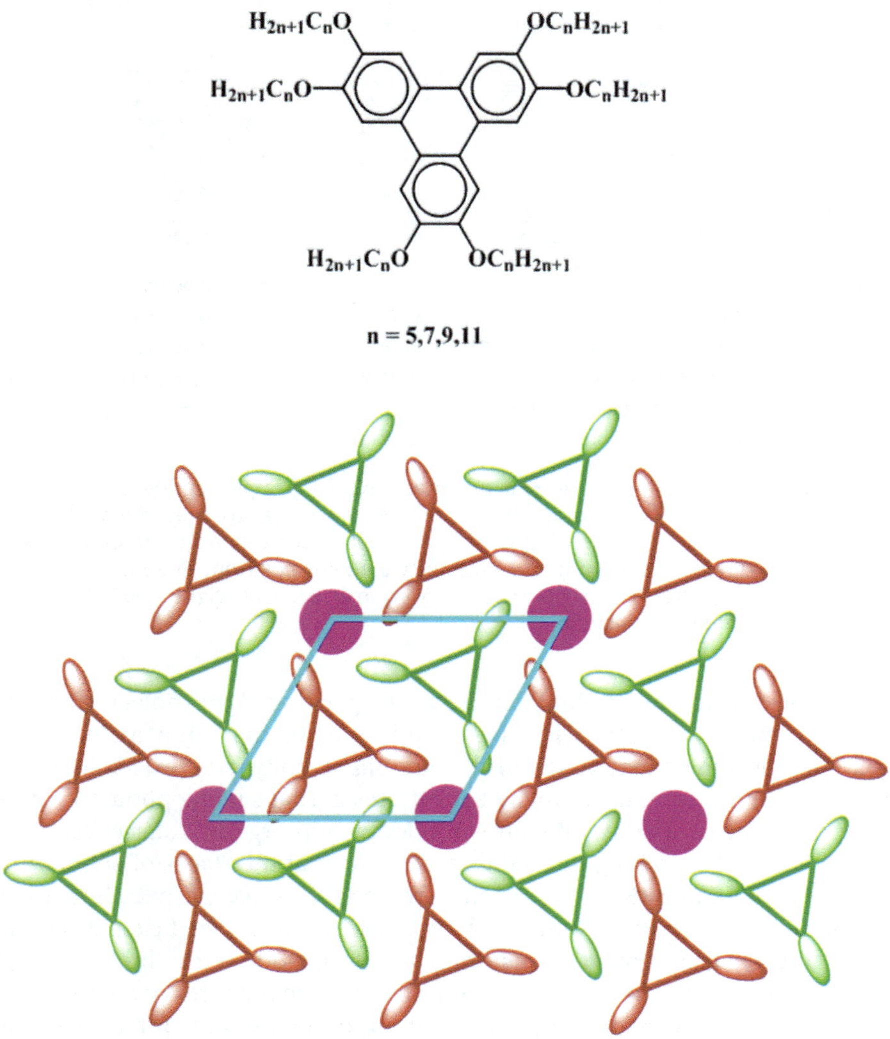

Figure 4.14 Discotic molecules that self-assemble on graphite giving an ordered hexagonal lattice with disordered molecules (represented by purple circles). The unit cell is indicated.

disorder in the molecule surrounded by the six ordered ones, probably because of frustration in packing. Left- and right-handed domains were imaged equally over the surface, the local chirality arising from the packing of the chains that form the local cyclic hexamers that result in a mesh.

A number of studies have been performed on conjugated oligomers and polymers physisorbed at interfaces, because of their importance in the area of optoelectronic materials.[61,62] It is interesting to compare monolayers formed by somewhat similar structures where covalently joined systems, supramolecular rods or supramolecular "stars" are physisorbed onto graphite. The three kinds of derivatives shown in Figure 4.15—that are all chiral thanks to (*S*)-2-methyl butoxy groups at the 2 and 5 positions of the

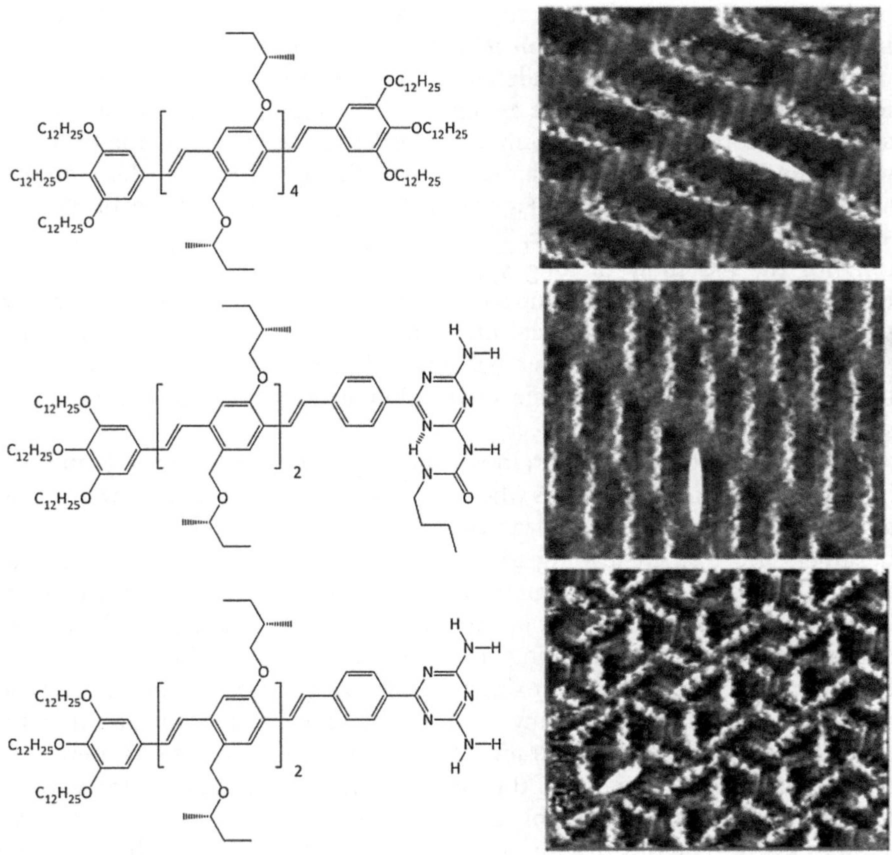

Figure 4.15 Oligo(4-phenylene vinylene) derivatives comprising a purely covalent network in a calamitic-shaped structure (top), a ureido-*s*-triazine compound that dimerises (middle) and a 2,5-diamino-triazine that forms supramolecular chiral hexamer (bottom) at liquid–graphite interfaces. The STM images shown were recorded under 1,2,4-trichlorobenzene for the top two cases and 1-phenyloctane for the bottom case. The oligophenylenevinylene segments are indicated by the white ovals overlaid on the STM images.

phenylenevinylene units—all form self-assembled physisorbed monolayers at liquid–graphite interfaces. The covalent calamitic molecule assembles through interdigitated alkyl chains forming lamellae where the rigid phenylenevinylene core is at a high angle to the alkyl group direction.[63] The aromatic groups apparently lie coplanar to the surface and do not interact with each other, since the chiral side-chains—although not seen in the STM images—must occupy space on the surface. These short chiral chain substituents are rarely imaged well at this kind of interface, because of their small size and degeneracy in terms of physisorption to the graphite. Even so, the chiral side-chain did force a single diastereoisomeric packing structure at the interface. The three dodecyloxy units fill space, but cannot all adsorb to the graphite, so in this case and the subsequent ones, it appears that one of the chains is not fully adhered, but is over the monolayer plane (it is not imaged in any case). Adsorption of a compound bearing a ureido-*s*-triazine moiety—that has a strong tendency to dimerise through the formation of four hydrogen bonds[64]—at one extreme of the oligo(phenylenevinylene) and a single tri-dodecyloxyphenyl unit at the other led to the formation of SAMs where the linear supramolecular dimer was clearly present on the graphite surface.[65] The STM images (Figure 4.15) are similar to those of the C_2 symmetrical covalent structure. Therefore, hydrogen bonding is a valid option for the assembly of long superstructures at surfaces, and the four hydrogen bonds help in forming a C_2 complex that gives the lamellar kind of packing that is favoured on graphite by molecules bearing long alkyl chains. Again, the chirality of the side-chains influences both the packing of the molecules as well as their orientation with respect to the graphite symmetry axes.

A completely different supramolecular topology is seen at a graphite–liquid interface for compounds where 2,5-diamino-triazine groups were used as a terminal ring to the linear oligo(phenylenevinylene) systems: cyclic hexamers are formed leading to assemblages that resemble a six-bladed windmill wheel.[66] These hexamers are a result of the formation of hydrogen bonds between the amino group hydrogen atoms and the pyridyl-type nitrogen atoms in the triazine ring. There is very little space in the middle of the supramolecular ring, approximately seven Ångstroms. Compounds with both one and two phenylenevinylene units bearing chiral groups arrange in this manner. Although the chirality at the stereogenic centres was identical, the supramolecular chirality of the cyclic hexamer was opposite for one with respect to the other, presumably because of a kind of odd–even effect. STM shows clearly that the rosettes form a hexagonal lattice on the graphite surface, with interdigitation of the alkyl chains of neighbouring supramolecular objects (Figure 4.15). The compound with no phenylenevinylene units between the triazine and the trialkoxybenzene forms isolated supramolecular dimers.[67] Therefore, the spacing phenylenevinylene unit in combination with the 2,5-diamino-triazine head group favours superstructures where many hydrogen bonds contribute to the cyclic hexamer formation. It is believed that the key structural features are the ratio of backbone length to the number of carbon atoms in the alkyl chain or the distance between the hydrogen-bonding moieties and the aliphatic region. The chirality expressed in these systems has been explored at the sub-molecular level by molecular modelling showing how the orientation of the stereogenic centres drives the twist in the whole supramolecular structure.[68] The delicate balance of hydrogen bonding, aliphatic chain van der Waals interactions and the subtle influence of the stereogenic centres and the substrate orientation as well as space filling in the monolayer plane all contribute to the exotic packing observed.

One of the most frequently studied moieties that has been exploited for the assembly of supramolecular monolayer structures physisorbed on surfaces is the humble carboxylic acid, which remarkably provides an incredibly rich superstructural collection, particularly for adsorpts with multiple carboxylic groups.[69] A molecule containing a single carboxylate group can in principle self-assemble with the acid forming simple dimers or trimers or a catena (chain) supramolecular polymer (Figure 4.16). When multiple carboxylic acid groups are present in a single molecule, it is clear that these supramolecular synthons[70,71] can be present in different parts of the assemblies that result.

Linear alkyl carboxylic acids have been proven by X-ray and neutron diffraction to pack on graphite (at various coverages) with the carboxylic acid groups forming head-to-head dimers with the alkyl chains in the lamellae interdigitated.[72] The dimerisation of 4-alkoxy benzoic acids leads to lamellar structures in their monolayers at the graphite-1-phenyloctane interface, where the dimer essentially results in the formation of a calamitic supramolecule.[73] The dimers line up, but not in a strictly linear fashion: STM shows that the molecules with even numbered alkyl chains have quite regular kinks in their lamellae (Figure 4.17). The chemical structure of these lamellae is shown above the STM images, and it can be appreciated that

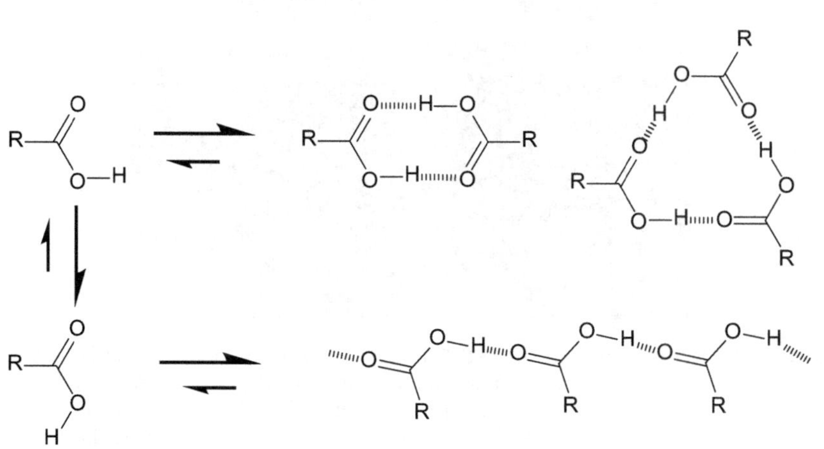

Figure 4.16 The assembly modes of simple carboxylic acids; common supramolecular synthons in surface self-assembly.

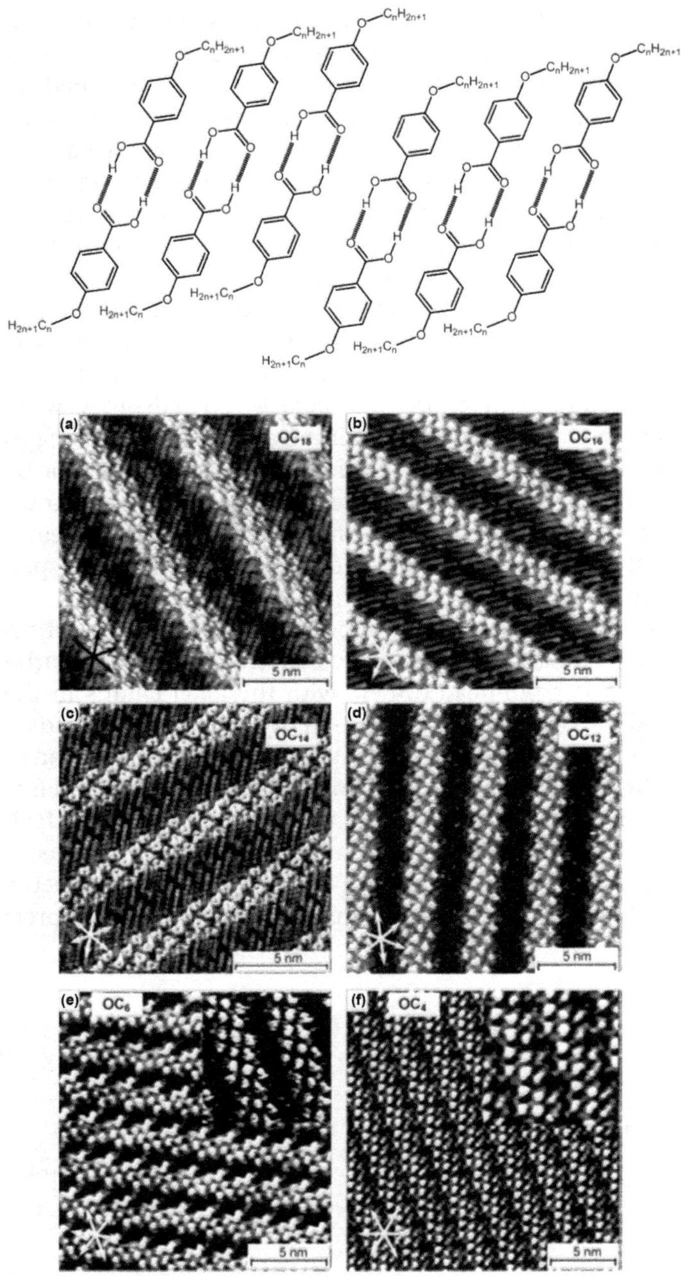

Figure 4.17 Room temperature STM images of alkoxy benzoic acids with even numbers of carbon atoms in their alkyl chains (indicated in each image) at the HOPG–1-phenyloctane interface. The structure at the top shows three of the proposed dimers of the benzoic acids that form a kink every three units, as indicated.

Reproduced with permission from Wiley-VCH, Copyright © 2010 WILEY-VCH Verlag GmbH & Co. KGaA, Weinheim, ref. 73.

there is no directional non-covalent interaction that obviously favours this situation. Rather, it is the van der Waals interactions that cause the lamellae to "jump" every three dimer units. The reason for this situation is that the alkyl chains cannot pack efficiently because of the dimensions of the benzoic head group that lies coplanar to the HOPG surface. The high angle between the alkyl chains and the benzoic acid moiety and the kinks are a structural solution to efficient packing within the lamellae. The structure is not disordered, as kinks in one lamella are copied in the neighbouring ones on either side. It is clear from the STM images that the alkyl chains are abutted (rather than interdigitated) between neighbouring lamellae and that space is filled in the monolayer.

The compounds with an odd number of carbon atoms in the alkyl chains behave quite differently to the even numbered homologues. At low carbon number chains (five and seven for example), the kinks occur every pair of dimer units. This kind of odd–even effect is quite general—with slightly different manifestations—for many monolayer systems where alkyl chains play an important structural role.[74] However, at longer odd alkyl chain lengths, the periodicity in the kinks becomes every three dimer units as for the even carbon atom chains, because the van der Waals interactions—between adsorbates themselves as well as between HOPG and adsorbate—become dominant in determining the packing. The odd–even effects that are observed persist when the acids form a complex with a hydrogen bond acceptor (4,4′-bipyridine) and form monolayers.

The kinked structures are not formed when octanoic acid is used as the solvent for the adsorpts: the solvent co-adsorbs leading to completely straight regular lamellae with partial interdigitation of the alkyl chains (Figure 4.18). In these structures, models propose a heterodimer where the benzoic acid derivative hydrogen bonds to the octanoic acid. It is an interesting surface-based solvate, and parallels efforts in three dimensions to influence packing by the incorporation of small molecules in the structure.

A somewhat similar situation of an irregular structure was observed in a chiral amphiphile derived from lactic acid with a naphthalene spacer to a long alkoxy chain, where kinks were observed at random in the lamellae.[75] In its enantiomerically pure form as either (*R*)- or (*S*)-isomers, the kind of kink lamellae discussed previously were observed except that alkyl chains were interdigitated and therefore did not drive the structural irregularity in the same way. There is preferred frequency but a definite non-periodicity in the kink location along the lamellae (Figure 4.19). Again, in the direction perpendicular to the lamellae, order is conserved, so "one-dimensional crystals" are seen. It is noteworthy that although there is non-periodicity, the methylene groups of the octadecyl chains are extremely well resolved indicating their pseudo-epitaxy with the HOPG. The chirality of the sample has a determining effect on the supramolecular structures that are seen on the surface: when the racemate forms a monolayer, STM reveals that only very short range order—of the order of ten nanometres—is present in a very limited number of regions in the monolayer at the interface. The alkyl chains

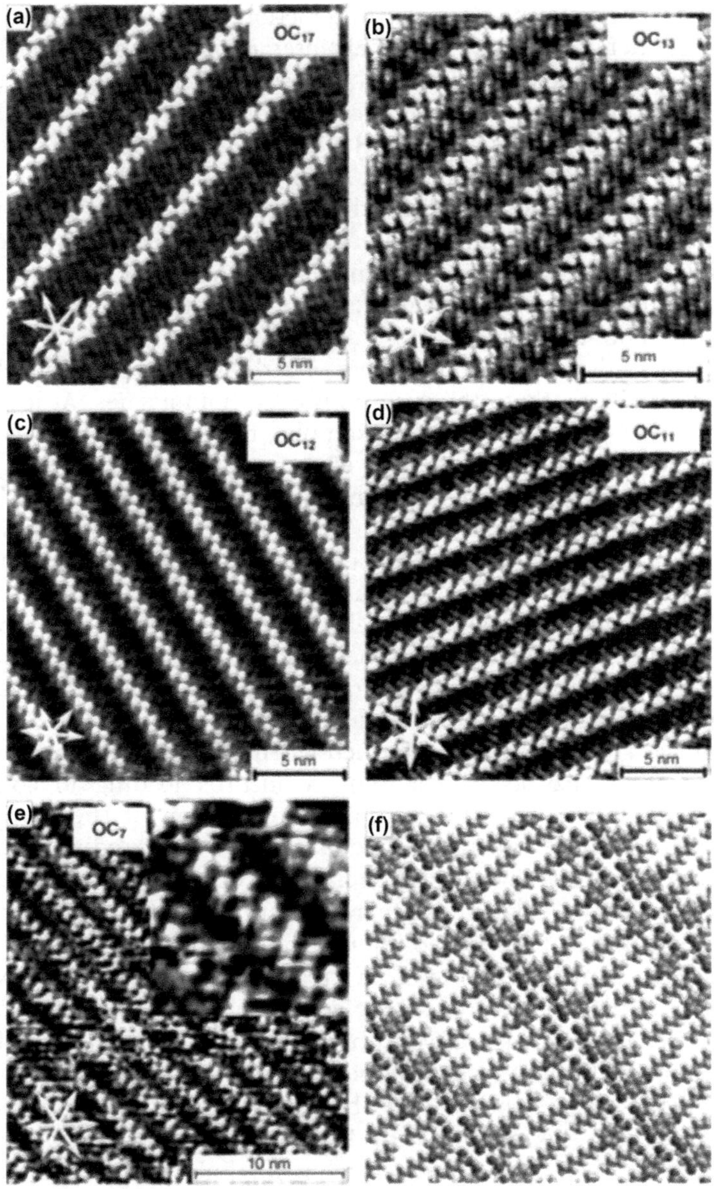

Figure 4.18 STM images of alkoxy benzoic acids with varying numbers of carbon atoms in their alkyl chains (indicated in the images) at the HOPG–1-octanoic acid interface. Notice that the lamellae are completely straight unlike the situation when 1-phenyloctane was used as the overlying medium. The model in (f) shows the proposed structure for the heptyl chain compound.

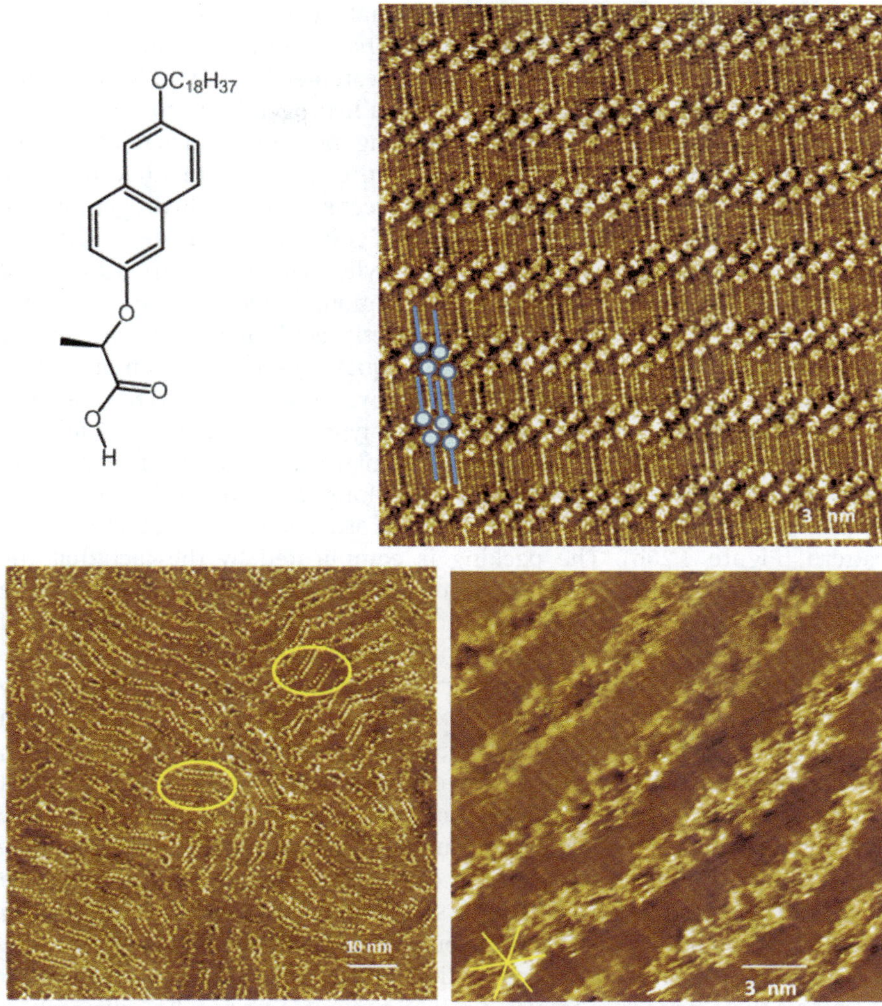

Figure 4.19 Room temperature STM image of monolayers of a chiral amphiphile with a carboxylic acid head group in its (*R*) form (structure shown) at the top right at the 1-phenyloctane–HOPG interface with a non-periodic but ordered structure, and below, larger and smaller scale images of the racemic modification where a disordered superstructure is observed, except in the circled small areas.

do apparently dominate the packing, because lamellae with interdigitated alkyl chains are formed, but individual naphthyl head groups can only be identified very infrequently. The multiple possible combinations of (*R*)- or (*S*)-isomers in the lamellae and their preference for a given conformation on the surface cause this poorly ordered situation.

Ferrocene carboxylic acid assembles into pentagonal and rectangular aggregate structures when sprayed onto a gold (111) surface from a benzene solution into a UHV chamber to give low coverage.[76] At high coverage, a previously unprecedented two-dimensional quasicrystal is formed comprising the supramolecular cyclic pentamers.[77] The five-membered supramolecular rings are believed to be a cyclic catemer (Figure 4.20) where the proton in the carboxylic acid moiety forms a hydrogen bond to the carbonyl group of the same moiety in the neighbouring molecule, and is reinforced by the formation of a hydrogen bond between the oxygen atom adjacent to the acidic hydrogen atom and one of the ferrocene aromatic hydrogen atoms. This arrangement is unique to the surface of gold, the solid state structure of the compound contains the classic carboxylic acid dimer structure[78] and even minor structural changes in the compound result in an on-surface dimer structure, as seen for ferroceneacetic acid that forms a dimeric structure under the same conditions. The quasicrystal—that contains long-range non-periodic order—that is formed was revealed by STM, and it is unique in being the only structure of this type (with 5,8,10 or 12-fold rotational symmetry) that results from molecular self-assembly. The STM images show definite local patterns but no long-range order is discernible (Figure 4.20a), and the intermediate range order matches a Penrose tiling pattern (Figure 4.20b). The packing is complicated by the fact that the supramolecular pentamer is chiral when adsorbed, and these isomers can be identified on the surface. It is not clear if this feature drives the formation of this unique superstructure. The result is remarkable because simply applying a molecule with five-fold symmetry onto a surface does not necessarily give rise to a quasicrystal, as seen for corannulene and its derivatives on copper (111),[79] where a methyl-substituted compound did display disorder, but mapped onto a hexagonal lattice presumably imposed by the metal adsorbent. The adsorpt penta-*tert*-butylcorannulene on copper (111) also finds an interesting packing arrangement so as to coincide with the symmetry of the surface, forming trimers that then surround a single molecule to give a hexagonal lattice.[80] Thus, it seems that a five-fold supramolecular structure is capable of quasi-crystal formation, while molecular systems, for now, are not. The role of the interaction between molecules or supramolecules and surfaces is sure to play a role in the outcome too. The interaction between gold and the ferrocene carboxylic acid is weak, imaging was performed at 77 K, and therefore the surface is probably less able to impose its symmetry on the long-range structure.

The structural dimensionality can be increased by adding carboxylic functions onto the molecules—the amphiphilic 3,5-dicarboxylic acid derivatives of benzene with alkyl chains at the 1-position lead to ribbon-like assemblies.[81,82] On the other hand, porphyrins bearing carboxylic acid functions on separate phenyl substituents form ring structures[83] and trimesic acid (1,3,5-benzene tricarboxylic acid) often forms an open superstructure on inert surfaces (as we shall see in the following section), but in the presence of a solvent that can interact with the carboxylic acid groups through hydrogen bonding—such as an aliphatic alcohol—leads to a lamellar packing where two of the acids form dimers, while the other two form a hydrogen-bonded chain incorporating the solvent into the network.[84]

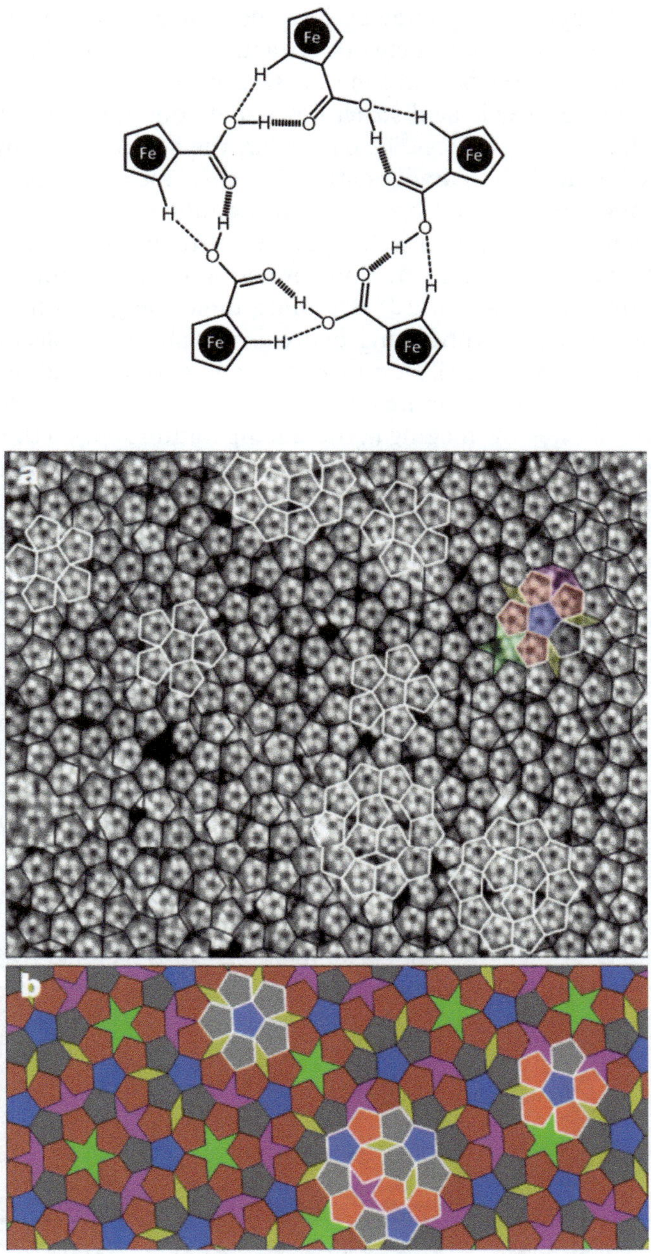

Figure 4.20 A chemical structure of the supramolecular cyclic pentamers formed by ferrocene carboxylic acid and (a) an STM image (40.5×36 nm) from the UHV deposited compound on gold (111) with pentagons showing long-range quasi-crystalline order, and (b) the Penrose P1 tiling where the white highlighted features coincide with the observed structure in (a). Reproduced with permission from Macmillan publishers Ltd., *Nature*, Copyright 2014, ref. 77.

A number of oligo-carboxylic acids have been used for the formation of principally two-dimensional network structures with voids in their plane, and these shall be discussed in the next section.

Beyond pure carboxylic acid structures, ionic bonding can be used to assemble molecules on surfaces. In particular, the case of biphenyl-3,3′,5,5′-tetracarboxylic acid is paradigmatic: for this molecule on Cu (100), adsorbate–adsorbate and adsorbate–substrate interactions were not of sufficient strength to generate a well-organised structure through interaction between the organic acids alone, but formation of the sodium carboxylate yielded highly stable networks.[85] The pure molecule gave only small and non-specific aggregates containing between two and six molecules, as witnessed by STM (Figure 4.21). Despite changes in coverage and temperature, no long-range ordered structures were formed. Deposition of NaCl in the UHV system followed by imaging at room temperature showed the formation of clusters of greater size than the organic molecule alone, but still with relatively short-range order. Annealing produced a very different situation, with large domains of a single kind of packing, with formation of the carboxylate form of the organic compound, clearly bound to the sodium ion as revealed by XPS of the sodium ion peaks. The stoichiometry of the system (also proved by XPS) showed that while a deficit of sodium per carboxylate yields small islands and many mobile molecules, at a ratio of 1.3 sodium to molecule, large ordered structures are formed. Further increment of the sodium content (to a 1.8 fraction) led to a smaller cluster formation presumably because of saturation of the outer edges of the clusters. The sodium carboxylate system at its optimum stoichiometry is in fact highly robust: STM imaging of the superstructure at 180 °C showed stable domains for several hours, something unthinkable in purely organic systems.

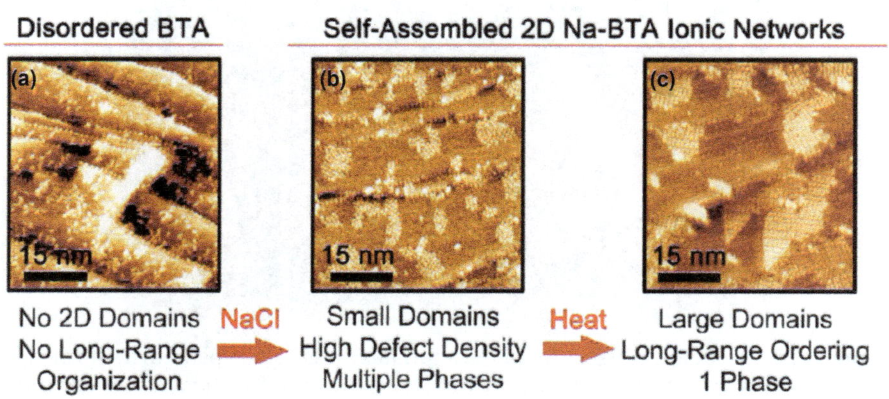

Figure 4.21 Scanning tunnelling microscope images of biphenyl-3,3′,5,5′-tetracarboxylic acid on copper (100) in UHV alone (left), after deposition of NaCl (middle) and then after thermal annealing.
Reprinted with permission from ref. 85. Copyright 2013 American Chemical Society.

Amides can also be used to influence the supramolecular structure of monolayers. Primary alkyl amides adsorbed onto graphite in UHV at sub-monolayer coverage formed lamellae in which the alkyl chains of neighbouring non-covalent chains abut (rather than the interpenetration that was seen for carboxylic acids).[86] The reason for this dense packing of the chains is that the adsorbates form head-to-head dimers at the centre of the lamellae that are linked into chains (Figure 4.22). The hydrogen bond between the formal dimers means that the intermolecular distance in the direction parallel to the lamellae is sufficient for only one alkyl chain in the unit cell. The study was performed on amides with different lengths of alkyl chain employing synchrotron X-ray diffraction and neutron diffraction to determine the structures, which were found to be more stable than non-amide alkyl derivatives. The strength of the amide bonds between head groups in this kind of amphiphilic molecule can also favour mixing of unlike structures, as seen for the case of binary mixtures of saturated and *cis*-unsaturated primary amides.[87] The bent shape that the *cis*-alkene has might be expected to force phase separation with linear molecules—as is the case of the unbranched saturated alkane—but in this case, the strength of the hydrogen bonding between the head groups is sufficient to overcome any thermodynamic penalty of mixing the unlike chains.

Secondary amides also form monolayers dominated by hydrogen bonds that form linear supramolecular polymers[88] that can be prone to polymorphism on graphite in the case of related carbamates.[89] The proximity between molecules within lamellae afforded by the amide bond can be used to influence the orientation of molecules on surfaces. Tetrathiafulvalene derivatives are interesting compounds because of their ability to act as charge carriers either in salts and complexes where they exist in their mixed valence state[90–92] or when they are incorporated into transistor devices.[93,94] To be able to perform this function, π stacking of the electron-rich units is necessary. When tetrathiafulvalene derivatives are substituted with simple linear alkyl chains and are physisorbed onto graphite from solution, the monolayers that form all pack with the aromatic system packed coplanar with the adsorbent's surface.[95] However, when amides are incorporated in the chains at either end of the aromatic unit, lamellae are formed where the tetrathiafulvalene residues are stacked on top of one another, in a quasi-perpendicular arrangement with respect to the surface plane (Figure 4.23).[96]

The compounds bearing two chains (Figure 4.24) can exist as *cis* and *trans* isomers that show isomer-dependent packing as their plain alkylated cousins do,[97] albeit in different ways. For the *cis* isomer bearing two amides, regular lines of high tunnelling current are observed by STM at a liquid–graphite interface. In larger scale images, these lines appear continuous, but occasionally sub-molecular resolution can be achieved showing the individual aromatic units. The alkyl chains are packed closely and do not interdigitate, and the distance between the centres of the lamellae is totally regular. Molecular modelling when compared with these images confirmed that the aromatic units are quasi-perpendicular to the plane of the surface

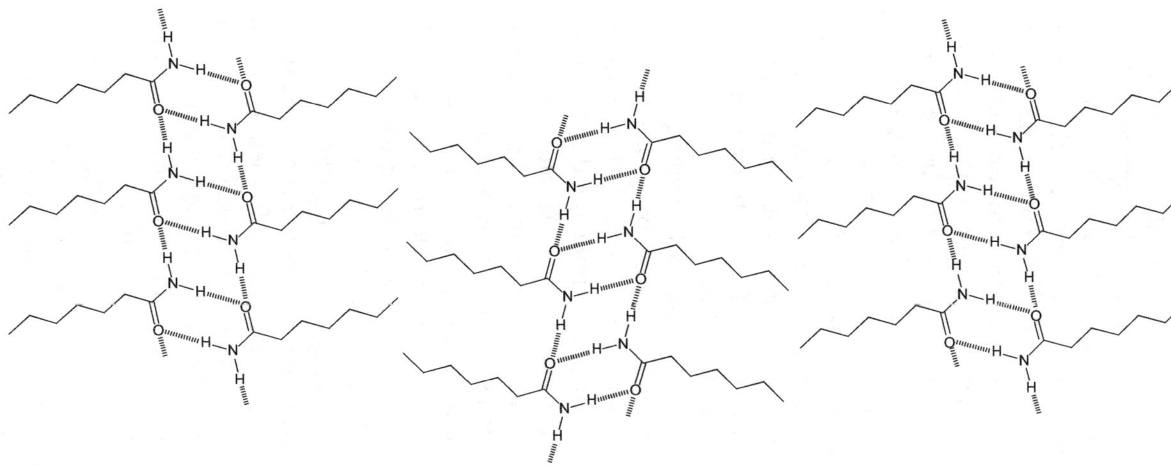

Figure 4.22 A representation of the crystal packing of hexanamide monolayers on graphite.

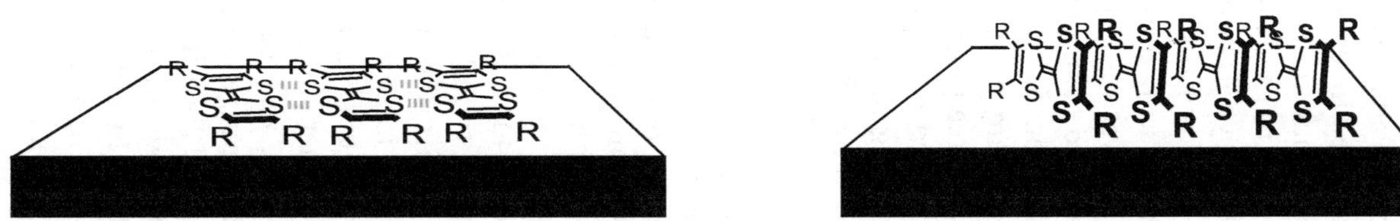

Figure 4.23 Cartoons showing the arrangement of tetrathiafulvalene derivatives on graphite when they are physisorbed. On the left is the arrangement coplanar to the surface that prevails when alkyl chains are attached to the electron-rich unit, and where sulfur–sulfur contacts exist when the alkyl chains do not occupy space around the aromatic preventing them. On the right is the perpendicular arrangement that exists when secondary amides are included in the chains and where the tetrathiafulvalene units enter into π–π stacking.

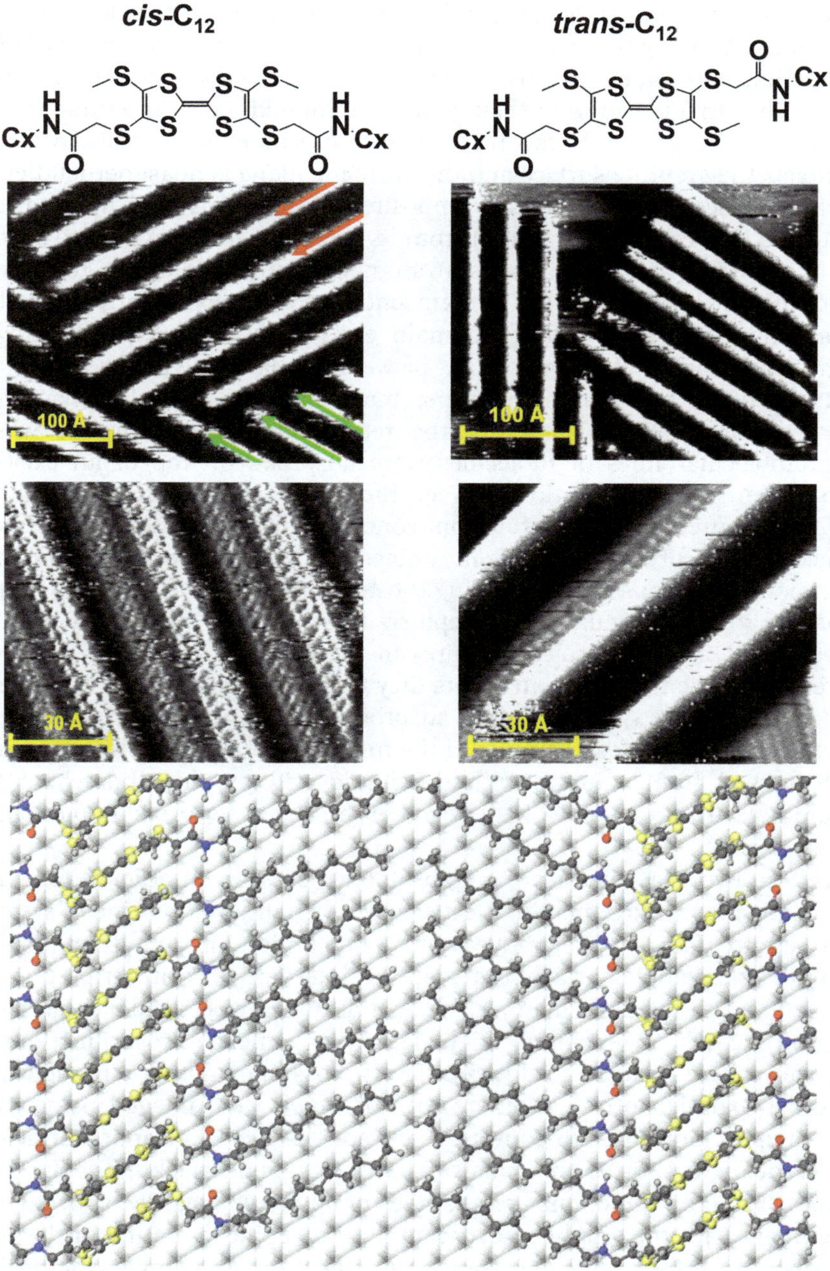

Figure 4.24 The structures of isomeric tetrathiafulvalene derivatives and STM images of the monolayers that they form spontaneously by physisorption on graphite and below a molecular model of the *cis* isomer packing showing the alignment of the alkyl chains with one of the surface's symmetry axes.

and that hydrogen bonds between the amide residues in the side chains are present. STM images of the *trans* isomer also show continuous lines of high tunnelling current, but the distance between the lamellae varies from area to area. This irregularity in packing was interpreted as a structural imperfection resulting from the fact that for the *trans* isomer, when one alkyl chain contacts the graphite surface and the aromatic plane is quasi-perpendicular to the surface, the diametrically opposite alkyl chain is held up from the graphite, and could either distort from its preferred conformation to adsorb to the surface or allow chains from neighbouring lamellae to adsorb underneath it. Occasionally, domain ends were observed early on in the adsorption where very abrupt domain ends were observed, indicating a significant non-covalent interaction between the lamellae (Figure 4.24).

In addition to these monolayers, multilayers of these lamellae were observed for the *cis* isomer of the tetrathiafulvalene derivative where supramolecular lines of molecules were adsorbed on top of an existing monolayer.[98] The interaction between the "sides" of the tetrathiafulvalene unit is favoured by inter-sulfur atom contacts. The orientation of the layers stacked on top of the monolayer in contact with the graphite follow the same symmetry directions strictly relaying those of the adsorbent, showing that at least up to the second layer the epitaxy is continued. In some areas, this means that the supramolecular wires in the second layer follow the same direction as the first, while in others they form an angle that is a multiple of 60° because of the symmetry of the adsorbent. The case is important because it is one of the very few examples of the formation of multilayers in a system physisorbed from solution with the liquid still present and where the supramolecular system is held together by relatively weak non-covalent interactions.

In fact, two alkyl chains are not absolutely necessary in order to force the perpendicular arrangement of the π-electron-rich unit in this kind of system. The tetrathiafulvalene derivative bearing one amide incorporating a lateral group self-assembles at a liquid–graphite interface to form strips that arrange in a head-to-head manner and that also form spectacular multilayers (Figure 4.25).[99] An extremely high density of lamellae cross junctions is present, where lamellae in different levels cross one another although they also run parallel to one another, both of which were imaged using STM where up to four different contrast levels are observed. The apparent height difference is 0.4 to 0.5 nm, consistent with a side-on stacking of the molecules. The cross junctions are at angles of 60 or 120° mainly, showing the alignment with the graphite adsorbent and the epitaxial nature of the physisorbed layer. Each lamella is composed of head-to-head single strands that occasionally split into two, each forming another full lamella row, as seen in the STM image, where slight imperfections in the row direction can also be seen. It is clear from the STM image shown in Figure 4.25 that the strands are not structurally perfect; many of these single molecule defects were attributed to tip removal of molecules during scanning and the defects actually aid in the assignment of the multilayer structure. The monolayer is

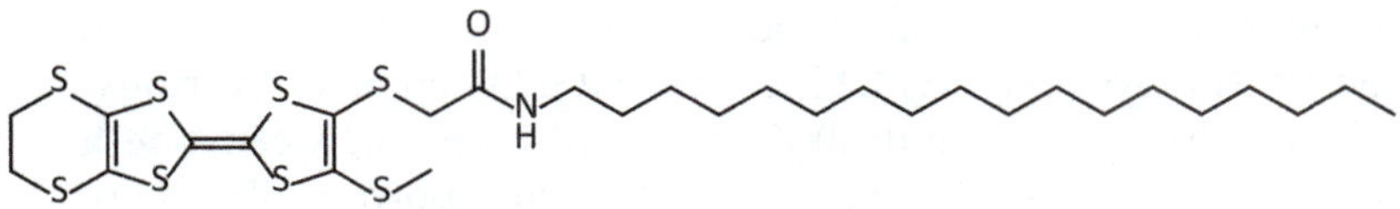

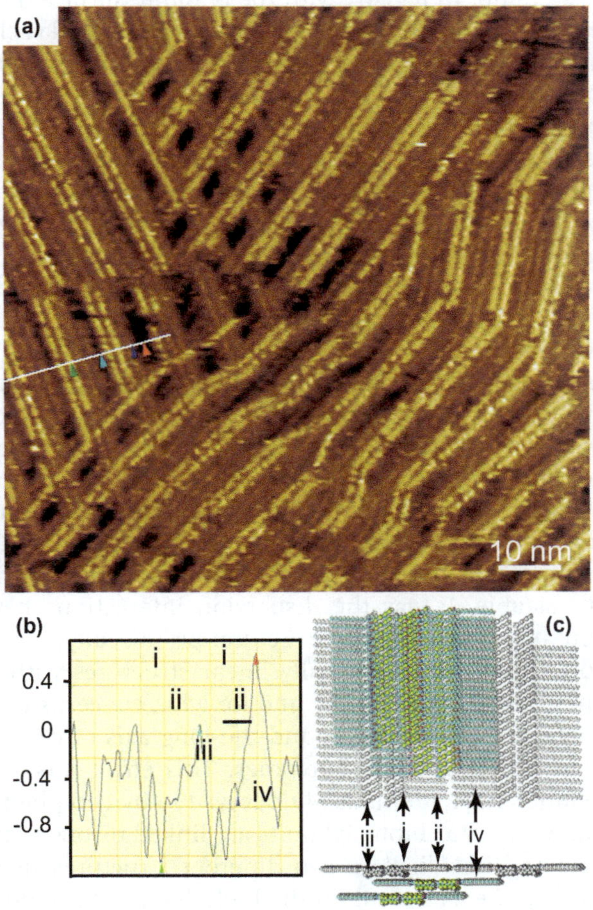

Figure 4.25 The structure of the tetrathiafulvalene derivative that forms multilayers at the graphite–octanoic acid interface. (a) shows the STM image where four levels of contrast can be discerned, picked out on the profile in (b). (c) shows a molecular model of the multilayer.
Reproduced from ref. 99 with permission from the Royal Society of Chemistry.

dynamic if maintained at the liquid–solid interface at room temperature: over time, the crossings tend to be replaced by co-parallel rows on top of one another. Inter-sulfur atom contacts between sheets as well as van der Waals interactions are surely responsible for the formation of these multilayers.

Long before this work, it had been shown that squarine dyes assemble through electrostatic interactions principally into multilayers on graphite from phenyloctane and the liquid crystalline solvent 4′-*n*-octyl-4-cyanobiphenyl.[100] The dyes (Figure 4.26) contained alkyl chains of various lengths, and those with R = OH and alkyl chains (*R*′) with between one and four carbon atoms, self-assembled into layers with polymorphic herringbone structures that were not in registry with the graphite surface on the basis of STM evidence. A low adsorption propensity of these short alkyl chains for the surface was inferred to explain the polymorphs, where the chains could be adsorbed or faced the overlying medium. For alkyl chains longer than eight carbon atoms, similar lamellae were formed. The binding of the molecules was strong, as adsorption of the dyes to the HOPG was always preferred to that of the liquid crystalline molecule, which can also form ordered layers. An interaction between the electron donor and acceptor regions of the molecules was inferred to explain the assembly. Quite extreme low concentrations (as low as 10^{-7} M) were required in order to be able to observe single layers of the dyes in the case of very long alkyl chains.[101] STM could be used to follow the formation of the layers: at low concentrations and short times after deposition (a few seconds), fibres less than 100 nm long were observed on the graphite surface. However, a while later, the deposition multilayer formation was observed unequivocally (Figure 4.26). Up to three layers were observed. Seven polymorphs were deciphered for the compound with octadecyl chains (and R = H), with different orientations of the alkyl chains, some of which were coincident with the HOPG substrate. The thermodynamic minimum was not reached, so many of these polymorphs are presumably metastable. It was the dispersion interactions between these chains that were used to justify the structural promiscuity.

The incorporation of topographically linear structures with an axle-like behaviour is interesting in general,[102] and for what concerns the discussion here in SAMs where the rotation around the long axis of certain units can provide different supramolecular outcomes. This effect was demonstrated for the compound 1,2,4,5-tetrakis-(4-dodecyl-4′-ethynylbiphenyl) benzene (Figure 4.27), in which four biphenyl units are linked through an alkyne to a central phenyl core, that displays two different structures in physisorbed monolayers on graphite as a result of the high degree of rotational freedom around the triple bond in the molecule.[103] The supramolecular rearrangement in the monolayer of the compound was observed using STM and the trigger to the change was the concentration of the compound in the solvent used in the experiment (hexadecane). When the adsorbate—whose alkyl chains surely help in attaining an ordered monolayer—was adsorbed on HOPG from a solution of the compound at a concentration of 32 mM, a lamellar structure was formed where alkyl chains form the main proportion of one part while the biphenyl groups parallel to this line come into close contact parallel to one another, but quasi-perpendicular to the plane of the interface. The STM images (Figure 4.27(a) and (c)) clearly show that the apparent area of these biphenyl groups is smaller than that of the equivalent moieties directed into the area containing the alkyl chains. The latter biphenyl groups are apparently coplanar with the interface in this monolayer.

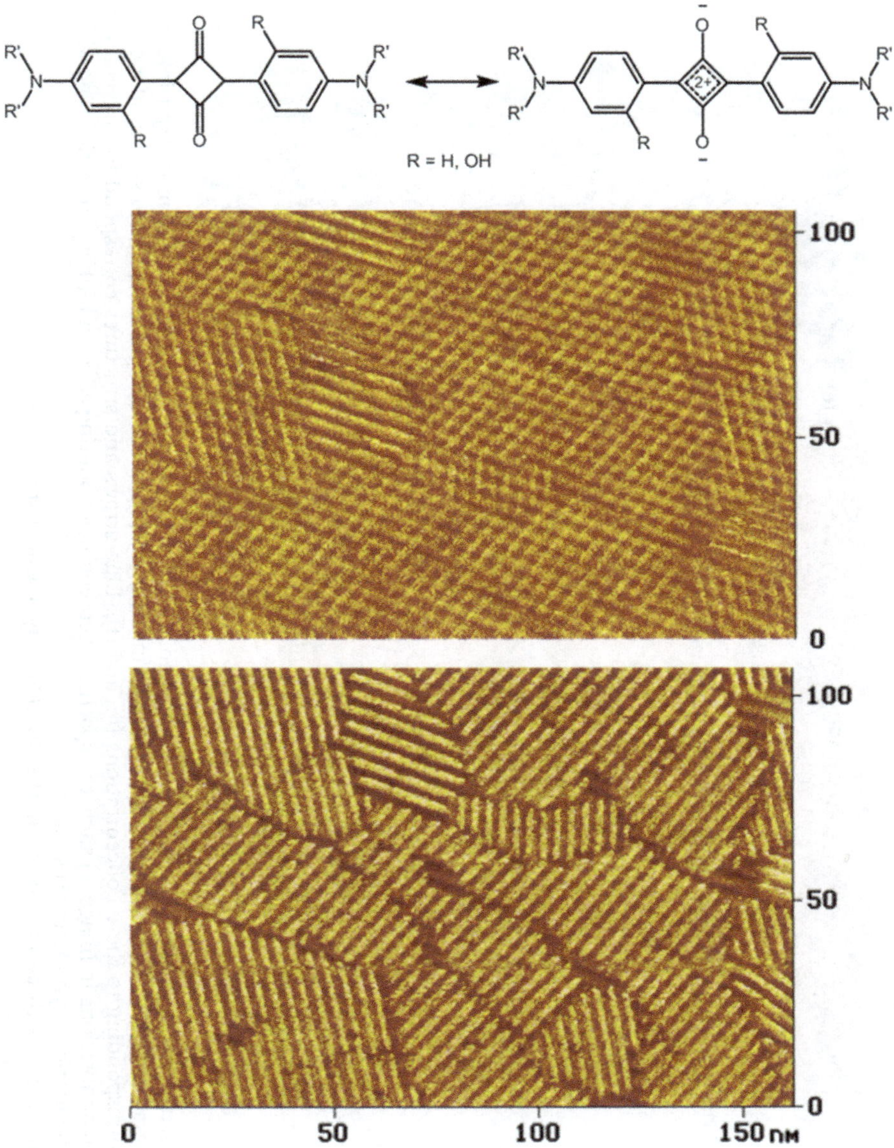

Figure 4.26 The structures of squarine dyes that form multilayers at graphite–solvent interfaces and two STM images of the compound with four octadecyl chains and R = H taken at different times during the growth of multilayers (the lower image 45 minutes after deposition, and the upper image 9 minutes earlier).
Reprinted with permission from ref. 101. Copyright 2003 American Chemical Society.

It was shown that a tilt angle of 37° of the biphenyl units from the normal to the interface allows a near ideal π–π stacking distance between the aromatic planes of 3.5 Å in the lamellar direction. Because of this dense packing,

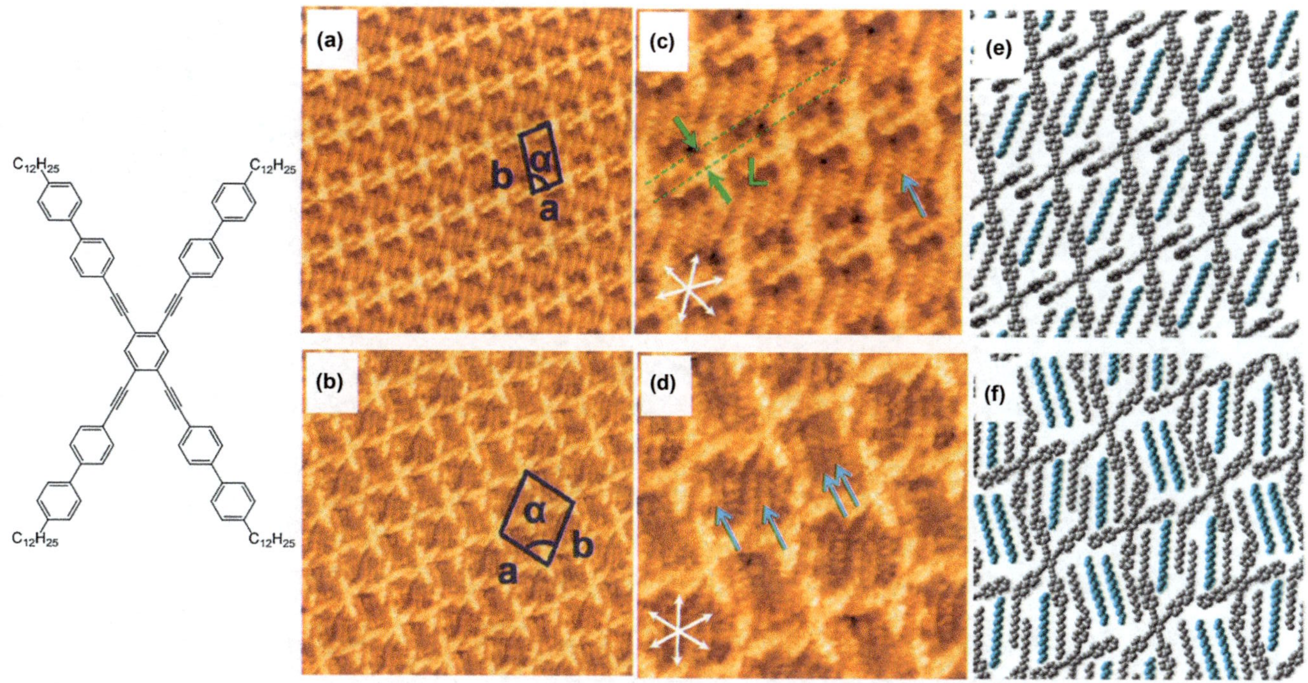

Figure 4.27 The chemical structure of 1,2,4,5-tetrakis-(4-dodecyl-4′-ethynylbiphenyl) benzene and STM images of the monolayer it forms at the graphite–hexadecane interface at concentrations of 0.32 mM ((a) and (c)) and 3.2 μM ((b) and (d)) and structural models of the packing corresponding to these concentrations ((e) and (f)). Blue arrows and structures correspond to molecules of solvent. The green annotations in image (c) show the spacing L between neighbouring edge-on biphenyl moieties. The white arrows indicate the graphite symmetry axes.
Reproduced from ref. 103 with permission from the Royal Society of Chemistry.

which is promoted by the concentration of the adsorbate in the overlying solution, the alkyl chains attached to the stacked biphenyl units are apparently not adsorbed onto the graphite surface, as seen by inspection of the packing model in Figure 4.27(c). On the other hand, one molecule of solvent is incorporated into spaces available in the frame that is formed by four of the adsorbate molecules. At a concentration of 3.2 μM, the situation is quite different, in that a more open monolayer structure is formed wherein the aromatic rings of all four biphenyl groups are parallel with the interface. A superstructure more similar to a parallelogram is formed, no lamellar structure is apparent. All the alkyl chains of the adsorbate interact with the graphite surface. The structures differ from one another in the SAM composition, the more open network incorporating an extra molecule of solvent in the loose frames formed by the target adsorbate compared with the more dense monolayer formed at higher concentrations. Scanning tunnelling spectroscopy provided further evidence of the difference between the structures, whereby higher tunnelling current was observed at negative bias, indicating tunnelling through the HOMO of the biphenyl units. The asymmetry in the curves was more pronounced for the structure containing the stacked molecules, indicating that this is consistent with splitting of degenerate energy levels in the proximal units so that the gap between HOMO and LUMO becomes narrower, in a Davydov type of splitting.[104] Apart from this example, the parallel stacking of dibenzo[*g,p*]chrysene[105] physisorbed on graphite has also been shown to be important for the electronic properties of the layers of the compound, and in general, the supramolecular arrangement of this kind of π-functional material can be closely correlated with its properties.[106]

STM under UHV conditions can be used to glean sub-molecular information concerning crystalline monolayer growth formation. A combination of the rod-like molecules shown in Figure 4.28—one of them chiral and the other achiral but prochiral—results in a complex structure on gold (111) whose detail can only be gleaned from knowledge of the conformations and how they influence the growth of the layer.[107] The achiral component can adopt interconvertible enantiomeric chiral conformations on the metal where the aldehyde groups are located diametrically opposite one another or in a "*meso*" structure where they point to the same side of the adsorbate. Beyond the molecule, these compounds form monolayers comprising "windmill" motifs where the four head groups are located close to the same group in a neighbouring molecule: these motifs are chiral on the surface (Figure 4.28). STM reveals that domains formed by co-adsorption of the achiral and a small fraction of chiral molecules are intrinsically chiral. Therefore, the chirality of the minority component is amplified into the monolayer domains. The chiral inducers can be seen in the monolayers as the head groups with the stereogenic centres appear with the brightest contrast,[108] allowing correlation of the chiral conformations of the molecules on the surface and the chirality of the domains that can be judged by the handedness of the windmill motifs. The chiral molecules can actually

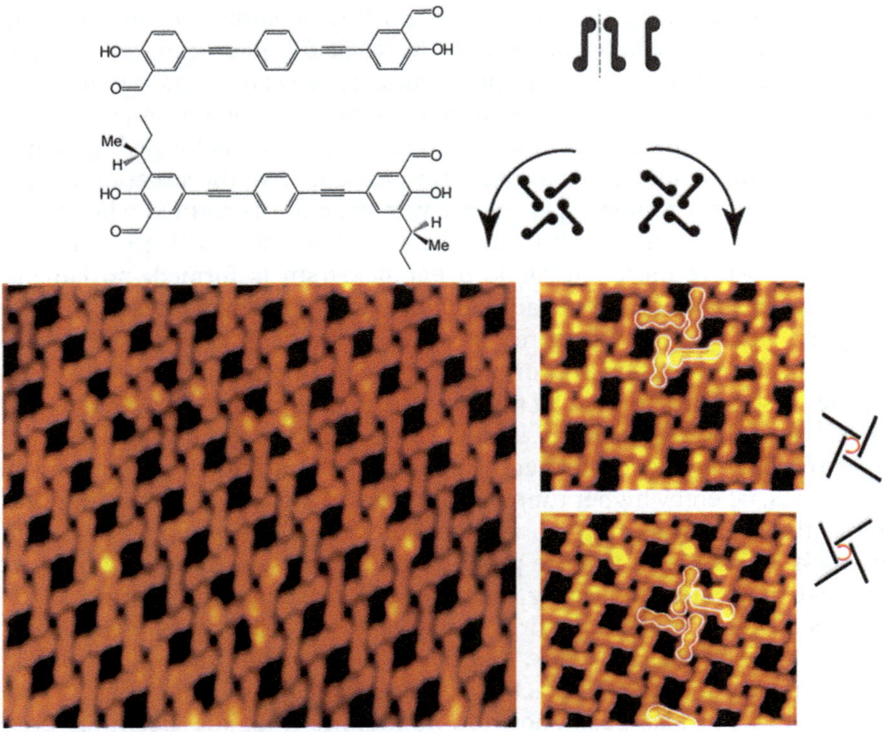

Figure 4.28 The structures of achiral and chiral conformationally flexible oligo(phenylene–ethynylene) rods with salicylaldehyde terminal groups and cartoons of the way they can appear with STM at Au (111). The chiral molecule can actually form diastereoisomeric windmill motifs. The STM images show a monolayer region where the molecules are perceived as bar-shaped objects comprising three lobes and the brightest protrusions correspond to the groups incorporating the stereogenic centres. The smaller STM images show areas with right (upper) and left (lower) handed windmill motifs.
Reprinted with permission from ref. 107. Copyright 2014 American Chemical Society.

exist as two diastereomeric conformations on the surface, leading to either right-handed or left-handed superstructures, as seen Figure 4.28.

Despite the fact that the chiral molecule can exist in two diastereomeric conformations, a clear preference is still seen for one of the two forms that leads to the left-handed windmill motif. Since the transfer of chirality to the monolayer domains is efficient, a study of individual domains was performed. Here, the UHV STM technique really comes into its own: as shown in Figure 4.29, the scanner for the microscope allowed imaging of separate domains that were generated over the gold surface. This possibility is particularly important in studies of chirality, because induction could in principle take place from domain to domain. In each area, several scans were performed so that the area's chirality could be probed from side to side (Figure 4.29).

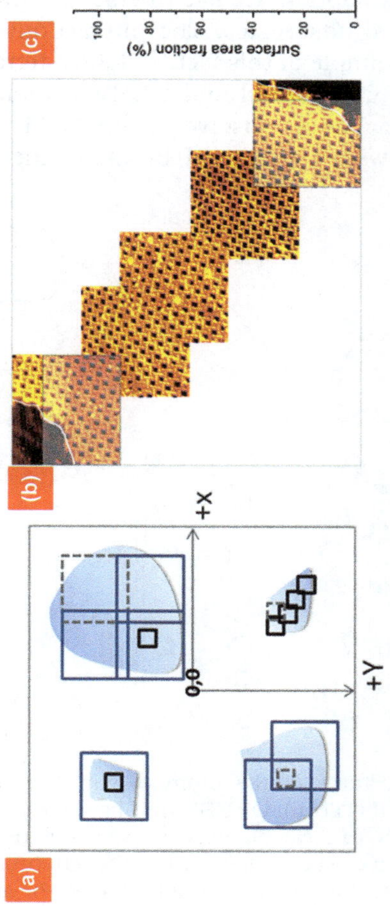

Figure 4.29 Following the chirality in monolayer islands of the molecules shown in Figure 4.28 by STM. (a) A cartoon showing the different monolayer areas (in shaded blue) that were imaged, (b) a sample of STM images taken across one of the monolayers and (c) the distribution of right- and left-handed windmill motif domains as a function of the percentage of chiral inducer. It can be seen that a higher proportion of chiral molecule does induce a more stereoselective outcome, but even at 3.4% of chiral compound, a 3 : 1 preference for one handedness is observed. Reprinted with permission from ref. 107. Copyright 2014 American Chemical Society.

The statistical element is critical in the thorough analysis of these kinds of monolayers. The result of this analysis showed that whether the chiral dopant was present in 3.4% or 11.6%, the majority domain was the left-handed one. This out of proportion influence of chiral induction is a suggestion that the sergeants and soldiers principle is at work. During the analysis, clusters of the chiral molecules were observed, indicating that they are responsible for nucleating the domains, and that because the chirality is fixed at this stage, the growth follows faithfully the handedness of the seed that is formed from the nucleus. This hypothesis was proven by performing heating and cooling cycles. The achiral molecules became very mobile and the islands dissolved at between 160 and 180 K, while the chiral islands were stable up to 200 K.[109] The work really emphasises the beauty of the STM for following these kinds of induction mechanisms molecule by molecule, information that would be unavailable directly through any other method.

Quite subtle differences in supramolecular interactions can give rise to packing differences in related adsorbates. The interactions between bromopyrene derivatives is an example of this fact.[110] 1,3,6,8-tetrabromopyrene when adsorbed on the (111) face of gold under UHV at room temperature forms two polymorphs that were observed with STM at monolayer coverage, whose structures were reproduced by modelling (Figure 4.30).

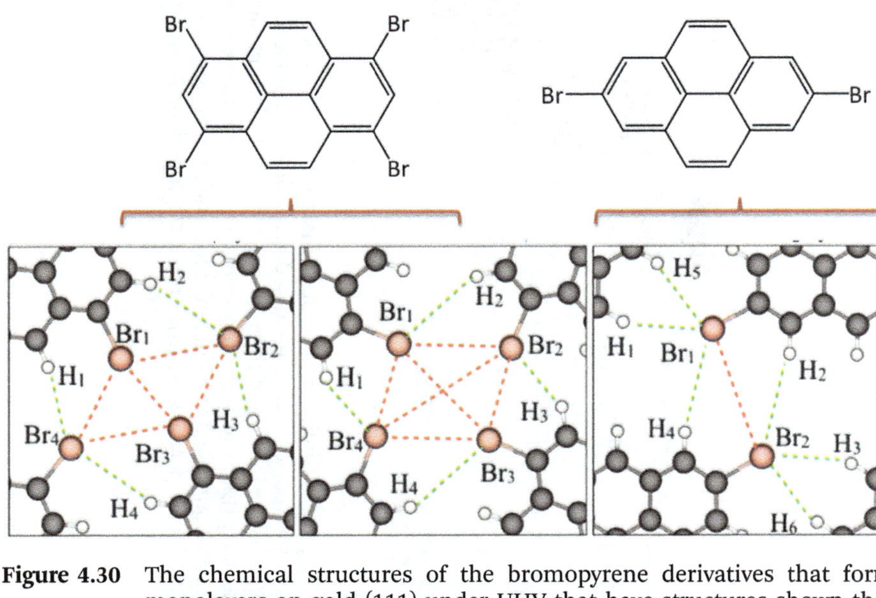

Figure 4.30 The chemical structures of the bromopyrene derivatives that form monolayers on gold (111) under UHV that have structures shown that were reproduced by DFT calculations. These simulations permit identification of the non-covalent interactions: the tetrabromopyrene exists as two polymorphs, both with intermolecular Br–Br and Br–H bonds (the red and green dashed lines, respectively), while the dibromo derivative forms a single dense-packed structure.
Reproduced from ref. 110 with permission from the Royal Society of Chemistry.

The micrographs revealed the herringbone reconstruction of the metal still present underneath the monolayers, proving that the gold was not modified or lifted when the molecules were adsorbed and confirming a weak interaction between molecule and substrate. The packing is dominated—in both polymorphs—by interactions between the bromine atoms (also seen in solid state structures from crystal engineering efforts[111]) as well as by Br–H interactions, forming triangles (Figure 4.30). The dibromo compound, on the other hand, forms a single apparent polymorph where Br–H interactions are the dominant ones. The density of packing of the dibromo compound was found to be lower than that of the tetrabromo compound, 0.87 molecules nm^{-2} rather than about 1 molecule nm^{-2} for both polymorphs, respectively. The differences in supramolecular structure are determined by the number and position of the relatively weakly interacting bromine substituents.

The formation of monolayers is also achieved by using multiple components to give solid solutions or ordered crystalline two-dimensional systems. A case of the formation of a solid solution is the self-assembly of bis-urea compounds (Figure 4.31)—readily prepared from commercial isocyanates and diamines. The bis-ureas adsorb leading to the formation of monolayers where the two N–H protons in the two urea moieties of one molecule hydrogen bond to the oxygen atoms in the same moiety in a neighbouring molecule, giving large lamellae where the long axis of the molecule is parallel to the interface plane.[112–114] Each molecule forms eight hydrogen bonds with its neighbours in a particular lamella, leading to an intermolecular distance of 4.6 Å. This arrangement was corroborated using molecular mechanics and molecular dynamics modelling. Mixing the thiophene derivative shown in Figure 4.31 and alkyl derivative with different chain lengths at the centre or extremities of the bis-ureas lead to exquisite control over the phase behaviour, from full mixing to full phase separation of the two components.[115] STM measurements of the mixtures at the HOPG–1-octanol interface showed that if the central chain in the aliphatic derivative is too short, compared with the dithiophene spacer, then phase separation takes place. For a fourteen carbon atom spacer in the aliphatic derivative, a perfect mixing takes place for a twelve-carbon chain at the "ends" of the molecule. Very interestingly, when only six carbon atoms are used as a chain at the ends, partial mixing takes place with isolated dithiophene derivatives observed in places, and also clusters of the same unit (Figure 4.31). The spacer between the urea units determines strongly if phase mixing occurs; if the spacer is too long or too short then optimal hydrogen bonding cannot take place. Molecules of identical overall length but with lateral and central chains of different lengths tend to phase separate. However, it is clear that the lengths of the side chains do play a role, and modifying this aspect can give rise to intriguing nanostructures employing essentially the same supramolecular synthon, emphasising the importance of interactions between chains of molecules at the interface. When bulkier groups are located on the 3 and 4 positions of the thiophene rings (and ethylenedioxy unit),

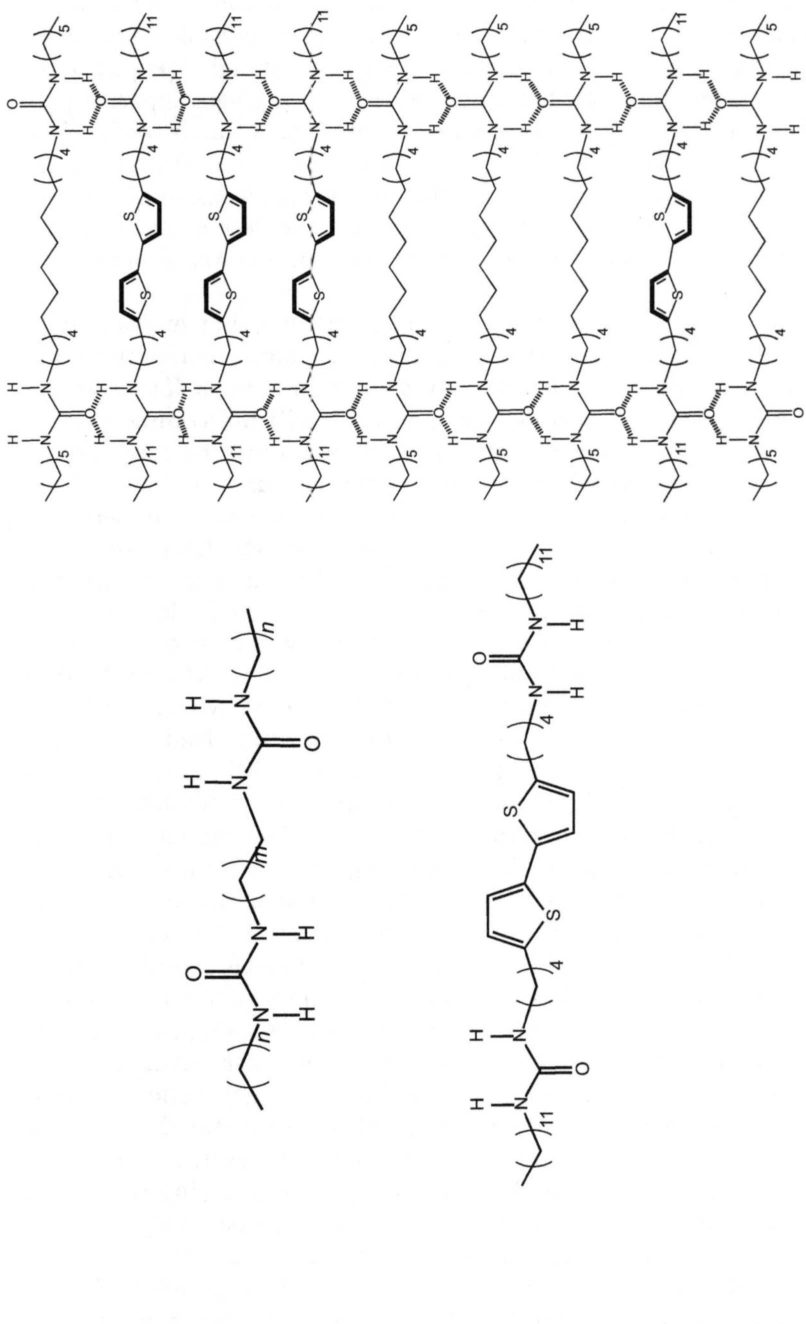

Figure 4.31 Chemical structures of bis-urea derivatives that self-assemble on HOPG at its interface with 1-octanol, and a representation of the packing in one of the combinations between a specific aliphatic derivative and one incorporating two thiophene units at its centre. This combination leads to the formation of mixing with short rows and individual thiophene-containing compounds in the aliphatic derivative rows.

ordered monolayers are not observed at the graphite–liquid interface, presumably because of the steric demands of forming the monolayer. In this case, co-deposition with a simple dialkyl urea affords mixed layers into which the bis-ureas are arranged apparently coplanar with the surface in lamellae with the other component.[116] This example demonstrates how co-adsorption can assist in the formation of stable monolayers that incorporate molecules that in their pure form would not adhere significantly at an interface.

Two-component systems that are formed through hydrogen bonding between carboxylic acids and pyridyl moieties are one of the more straightforward approaches to mixed monolayers, but can reveal intriguing behaviour as for the mixture of either (*S*)- or (*RS*)-16-methyloctadecanoic acid and bipyridine at the interface between graphite and phenyloctane.[117] STM images (Figure 4.32) show that for the racemic compound, chiral domains with opposite handedness exist, indicating that spontaneous resolution of enantiomers had taken place at the interface. The layer comprises lamellae that are formed from two aliphatic acids hydrogen-bonded to the nitrogen atoms of the pyridyl rings. These complexes line up giving homochiral tapes, the two halves being related by a C_2 axis (as we have seen, a common situation in these SAMs). Comparison of the images of the racemate with those of the enantiomerically pure sample was important to assign unambiguously the separation of enantiomers. Curiously, quasi-achiral domains are formed by the pure (*S*) enantiomer forms, with the alkyl chains oriented perpendicular to the tape axis, because the alkyl chain terminus' methyl or ethyl group are ambivalent as to which is adsorbed to the surface (as for quasi-enantiomorphous domains observed for an enantiopure liquid crystal[118]). Apparently, the complex formed with bipyridine forces a geometry with respect to the supramolecular tape axis whereby the alkyl group has to adopt a specific angle. A similar approach has been used for the assembly of 1-pyrenehexadecanoic acid with the same base, leading to bistable structures.[119]

Complementary hydrogen bonding leading to the formation of rosette-like structures can be applied to the solution–solid interface. This approach has been demonstrated for the assembly of oligo(*para*-phenylenevinylene)-functionalised melamines co-adsorbed with a perylene bisimide derivative.[120] Rosettes comprising donor and acceptor moieties are formed at the interface to give a twelve-membered supramolecular ring observed by STM. The assembly is not selective though, and to a degree reflects the promiscuity of the components in forming hydrogen-bonded complexes in solution. In this sense, the surface studies aid in the interpretation of solution-determined results, and certain combinations do lead to specific and well-defined two-dimensional patterns.[121] Rosette-type nanostructures have been prepared on graphite after evaporation of solvent and studied using AFM.[122] Two achiral components—a quadruple melamine and barbiturate derivative—form a stacked structure comprising four hydrogen-bonded hexamers. The twist between the rosette layers leads to enantiomeric

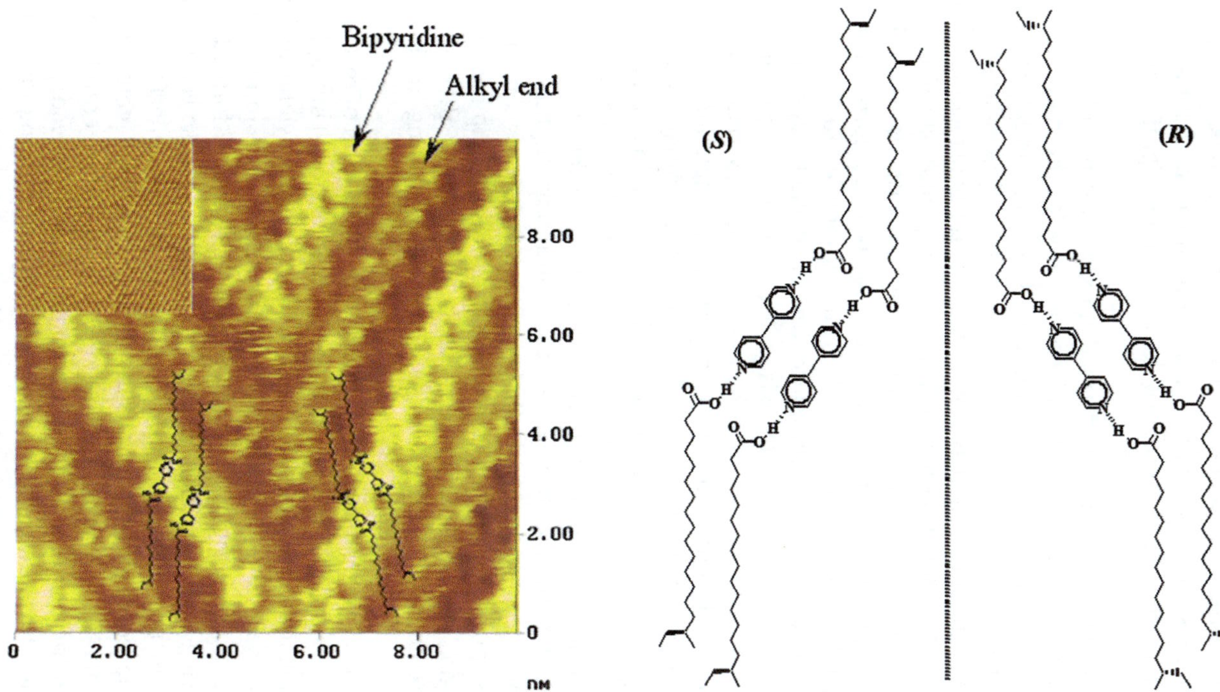

Figure 4.32 STM images (the main image shows lamellae up close and the inset a domain boundary) of the physisorbed layer formed on graphite from a mixture of (*RS*)-16-methyloctadecanoic acid (2 parts) and bipyridine (1 part) in phenyloctane, and a drawing of the supramolecular structure that is probably present in the conglomerate.
Reproduced from ref. 117 with permission from the Royal Society of Chemistry.

assemblies. Slow evaporation of a chloroform solution on graphite gives rods where rosettes are stacked, in domains with opposite chiralities.

Coordination chemistry at interfaces has also been used to generate multicomponent systems, from the complex rows formed on the copper (100) surface of one-dimensional copper–pyridyl coordination polymers linked with struts of terephthalic acid through hydrogen-bonding interactions to the side of the rod-like pyridyl system,[123] to larger iron coordination networks based on extended ligands of this type.[124]

4.4 Open Network Surface-based Systems by Physisorption

The formation of porous crystalline networks has been a subject of great interest for bulk materials for some time, and the self-assembly of mono- and multi-layers with spaces in the low dimensional supramolecular lattice[125] is seen as an opportunity both to understand the formation of this kind of structure as well as being architectures that can recognise or template other chemical objects. They are referred to here as "open" because spaces exist in the plane of the supramolecular networks, whereas the densely-packed systems considered up to here in this chapter are "closed".

Aromatic carboxylic acid derivatives are set up to form open network structures because of the high directionality of the carboxylic acid dimer (shown in the previous section) and the well-defined geometries of the cores. Trimesic acid—benzene-1,3,5-tricarboxylic acid—is the smallest molecule of this family, it is available commercially, and has provided remarkable discoveries. On copper (100) at a little over 200 K—under UHV conditions—trimesic acid forms the honeycomb structure that one might predict (Figure 4.33).[126] Here, the interactions between molecule and surface are relatively weak and the hydrogen bonds parallel to the surface lead to the fused rings formed by six molecules. It should be noted that if the temperature is raised, the carboxylic acid is deprotonated leading to a chemisorbed molecule that packs in a completely different way. The honeycomb network is also formed on graphite under UHV conditions, along with a second polymorph, more dense than the simple chicken wire structure, where complete trimesic acid hexagons come together through hydrogen bonds with the formation of a cyclic trimer of carboxylic acid moieties at the nodes (Figure 4.33).[127] STM was used to image the structures, that were also observed using LEED that showed that they are stable from 25 K up to room temperature. In both polymorphs, all the carbonyl and hydroxyl group are apparently involved in hydrogen bonds. At higher coverages, excess trimesic acid molecules enter the voids in the networks. It has been suggested, on the basis of theory, that this kind of polymorphism on graphene might be useful for influencing the electronic properties of the carbon allotrope.[128]

Under electrochemical STM conditions, potential controlled adsorption of trimesic acid onto a gold (111) surface that doubles as an electrode can be

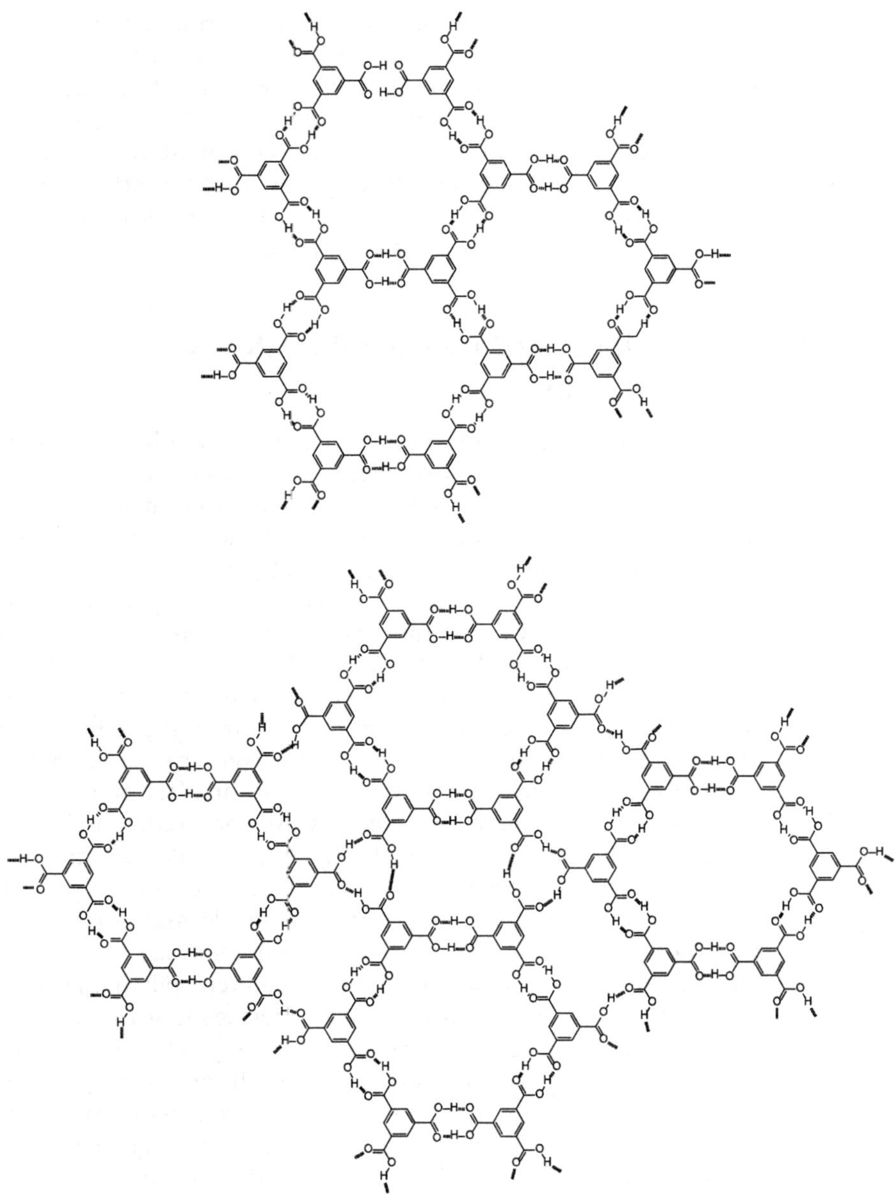

Figure 4.33 The polymorphic hexagonal networks formed by trimesic acid formed on graphite under UHV conditions.

achieved.[129] At relatively low potential in perchloric acid (0.1 M), a distorted chicken wire network structure is formed with two different void sizes, while higher potential causes a rearrangement into a close packed superstructure. Remarkably, the two phases are electrochemically reversible. The open structure has two different void sizes presumably because of the energetic drive towards a structure that is commensurate with the underlying surface. Dimers of the acid are formed with three of these coming together to form the smaller voids, while more open voids surrounded by six molecules are also present, and the superstructure has a remarkably large domain size (several tens of nanometres at least). This again shows how a delicate balance between adsorbent–adsorbate and adsorbate–adsorbate interactions can lead to a subtle variance in supramolecular structure in two-dimensional systems. In fact, water can be used as the solvent for deposition of the acid onto graphite and AFM and STM proved that this "green chemistry" approach to decorated surfaces is a viable one.[130] AFM proved that under these conditions, multilayers can be formed upon prolonged immersion times. The aqueous deposition of these acids does not necessarily lead to open networks, as shown for close-packed layers formed by isophthalate tetracarboxylic acids on the (111) face of gold.[131]

When the three carboxylic acids are kept in the same relative disposition but are separated from one another, as in 1,3,5-tris(4-carboxyphenyl)benzene, the voids in the monolayer of the hexagonal lattice become very large, and competition from lamellar structures that do not obviously have optimum packing are formed.[132] This system shows both solvent and temperature dependence in its packing. At room temperature, STM at the graphite–solvent interface showed that a hexagonal open network was formed in nonanoic acid; with heptanoic acid, a lamellar structure was formed and in octanoic acid, both phases co-existed at the interface. In the open network structure, the molecule is believed to be adsorbed totally flat and coplanar to the graphite adsorbent, with the classic hydrogen-bonded dimers of the carboxylic acid stabilising the structure as in trimesic acid. However, the row structure of the physisorbed 1,3,5-tris(4-carboxyphenyl)benzene has unit cell parameters from the STM images that are too small to be explained by a flat-lying molecule on the HOPG. In common with a derivative with slightly longer struts,[133] there is a competition between hydrogen bonding in the open structure and π-stacking in the row structure where one of the "arms" of the compound comes away from the adsorbent's surface and overlaps with aromatic units in a neighbouring molecule. The other two "arms" do form hydrogen bonds with neighbouring rows and help favour the lamellar type domains. Upon warming, the open framework system actually rearranges to give the monolayer comprising rows of partially-stacked molecules, and cooling results in the hexagonal lattice again. This effect is explained by the co-adsorption of solvent molecules in the low temperature phase, which results in a thermodynamically-favoured structure. However, as temperature is raised, the entropic gain of solvent molecule desorption to the bulk results in a collapse of the open framework to give the rows of adsorbate molecules.

The compound 4,4′-terphenyl-3,5,3′,5′-tetracarboxylic acid also forms hexagonal pores; in a way the middle ring could be considered a covalent equivalent (somewhat shorter) of the carboxylic acid dimer, and because of this similitude, five different rings can be envisaged resulting in the formation of a random tiling network (Figure 4.34).[134] The five rings

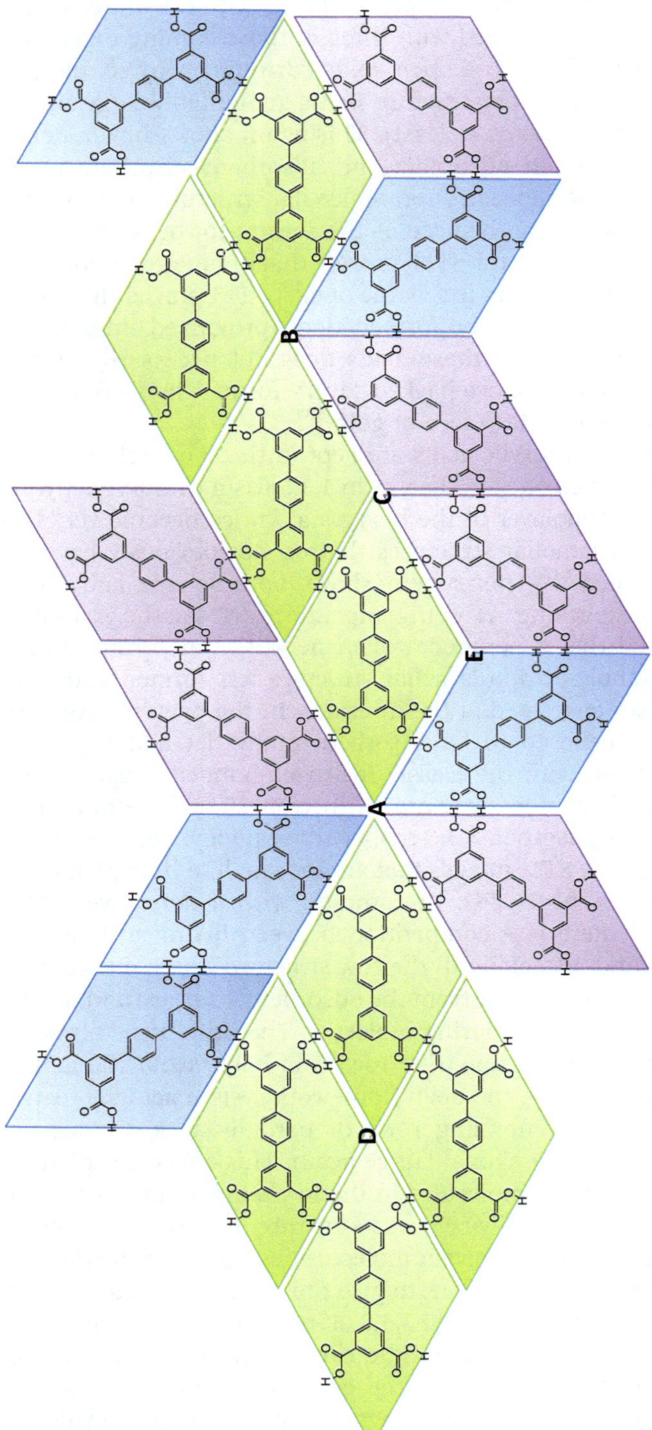

Figure 4.34 A drawing of the monolayer formed by the tetracarboxylic acid shown at the nonanoic acid–HOPG interface where random tiling is formed, with the formation of five different kinds of voids lettered A–E. The coloured rhombi are to indicate the relative orientation of each molecule in the hydrogen-bonded network.

incorporate three, four (two isomers), five or six molecules. The position of each of the three possible orientations of the molecule (shown in distinct colours in Figure 4.34) that coincide with the underlying graphite symmetry axes occurs at random, with small populations of defects that migrate in the dynamic system that is formed at the interface between HOPG and nonanoic acid. Thus, quasi-degenerate conformations and connectivities can interchange, yet despite this equilibrium the system remains random. The framework is totally non-periodic and therefore seems to be an entropically-stabilised phase. The system is very sensitive to the constitution of the tetracarboxylic acid; having an alkyne instead of a phenyl spacer at the centre causes a slight change in aspect ratio giving a tiled system with a greater degree of order.[135]

The use of a rigid linker between functional groups can lead to the spontaneous formation of complex and even chiral networks, as demonstrated for a family of rod-like molecules with nitrile groups at their ends and pointing along the long axis.[136] The oligo-*para*-phenylene dinitriles assemble on a silver (111) surface to give structures that depend on the length of the aromatic rod (Figure 4.35), but in which the weak hydrogen bond between the nitrogen atom in the functional group and hydrogen atoms attached to the phenyl ring apparently dictate the packing. When three phenyl rings space the functional groups, a chevron-like packing is observed, while increasing the number of rings by one leads to an open rhomb-like structure, where the non-symmetrical approach of the nitrile group to the edge of adjacent molecules leads to a chiral arrangement that is propagated through the domains of the monolayer in one handedness or the other. The compound with five phenyl groups in-between the nitrile moieties forms a remarkable Kagomé lattice where each node comprises a chiral arrangement of the organic molecules oriented end-to-edge. The dinitrile with six phenyl rings spacing the functional groups also leads to the formation of a dichotomous array of quantum corrals on the silver (111) surface, as a result of the coupling of the chiral Kagomé lattice with the surface electronic states of the metal,[137] but this compound also displays other chiral networks[138] as a result of the large supramolecular connectivity space available to it.

When alkyl chains interact in symmetrical ways with the core of the adsorpt as the centre of symmetry, open network systems can arise. This case is true for the nominally C_3 symmetric hexadehydrotribenzo[12]annulene derivatives and related compounds (Figure 4.36) whereby the alkyl chains interdigitate to give directional interdigitation and leave spaces between them.[139] The molecule with the smallest core size forms cyclic hexamers thanks to the interdigitation of the alkyl chains in two of the possible three directions. Apparently, the size of the core in this molecule is insufficient to allow effective packing of the chains in the third direction, but this becomes possible when the core is opened slightly by the incorporation of alkyne spacers between the benzene rings. In this case, there is perfect hexagonal packing and open spaces with exposed surfaces

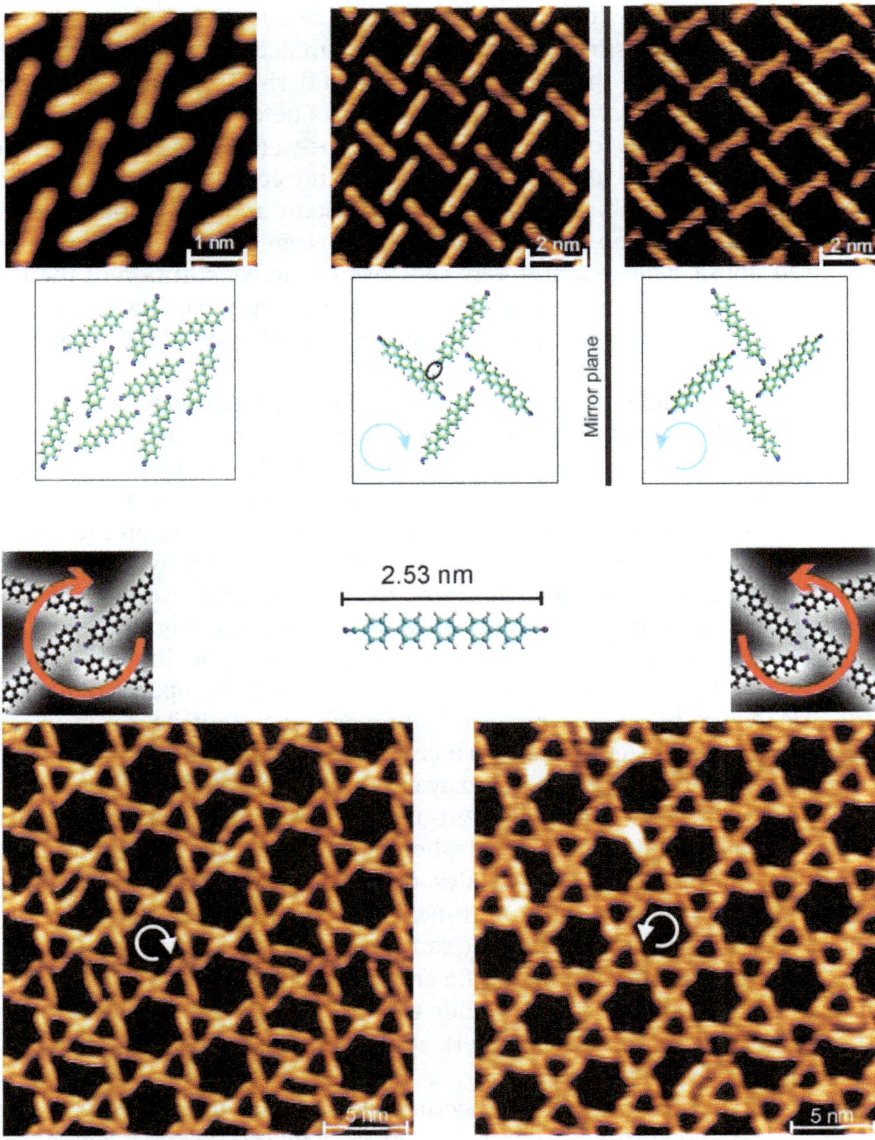

Figure 4.35 The structural variation of a series of oligophenyl dinitriles on a silver (111) surface as revealed by UHV STM. All patterns are consistent with the formation of weak hydrogen bonds between the nitrile and the hydrogen atoms attached to the phenyl rings. The molecule with three phenyl rings (top left) forms a chevron-like pattern, that with four (top middle and right) enantiomeric open rhomb structures, and the compound with five phenyl rings (bottom) a chiral Kagomé lattice where the two enantiomers are observed in separate domains. The curled arrows indicate the chirality of the domains. From the structural models, the proximity of the nitrile groups to the hydrogen atoms attached to the phenyl rings can be seen. Johannes Barth is warmly acknowledged for kindly providing the graphic material in this figure.

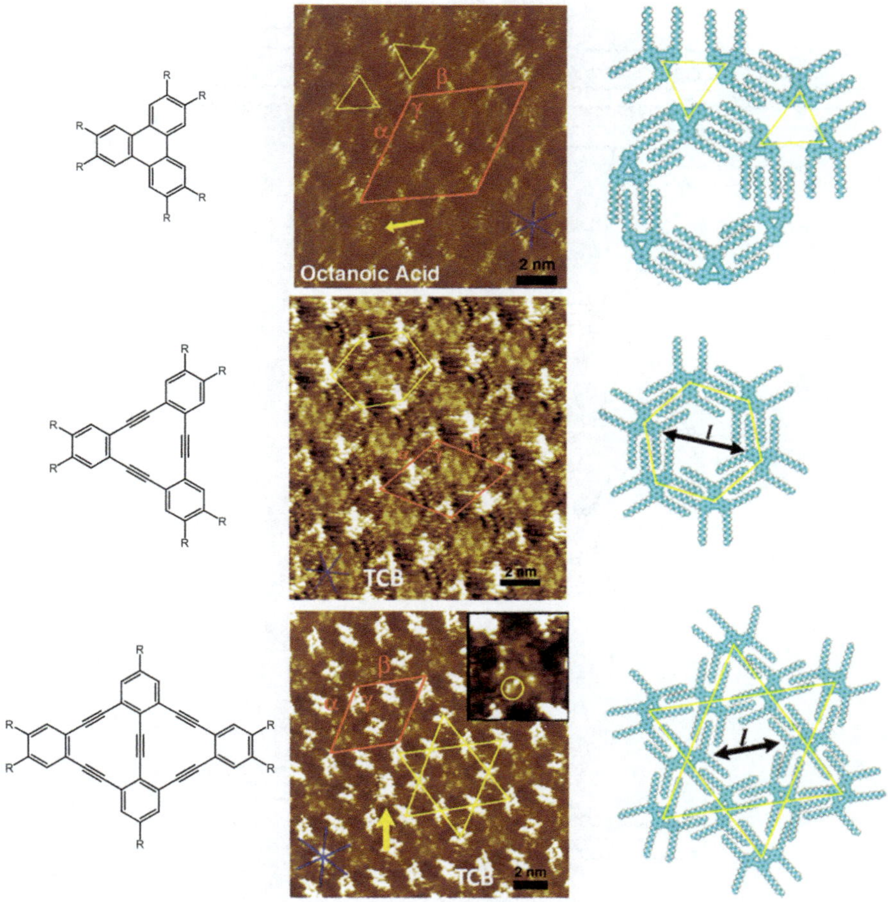

Figure 4.36 The chemical structures of a series of rigid core adsorpts that form surface networks on graphite with voids that were imaged using STM. The cartoons show packing models for the networks. The blue lines on the STM images correspond to the graphite axes, and the red lines the unit cells. TCB is trichlorobenzene. Illustration of STM and packing model kindly supplied by the authors of ref. 139.

(presumably filled by mobile solvent) surrounded by the edges of the hexagons made up of the alkyl chains. When the symmetry of the core is changed (as in the bottom row in Figure 4.36), different packing structures are observed as a result of the orientation of the alkyl chains, in the case shown leading to a Kagomé lattice (although in a different solvent, a polymorphic lamellar structure is observed). This and subsequent work shows the importance of the core size in alkyl chain interdigitation (itself shown influential in other cases[140]) and consequently in the formation of voids within the layer that are filled with the solvent used for deposition[141] as well as the temperature used for the experiments.[142] An entropy-driven

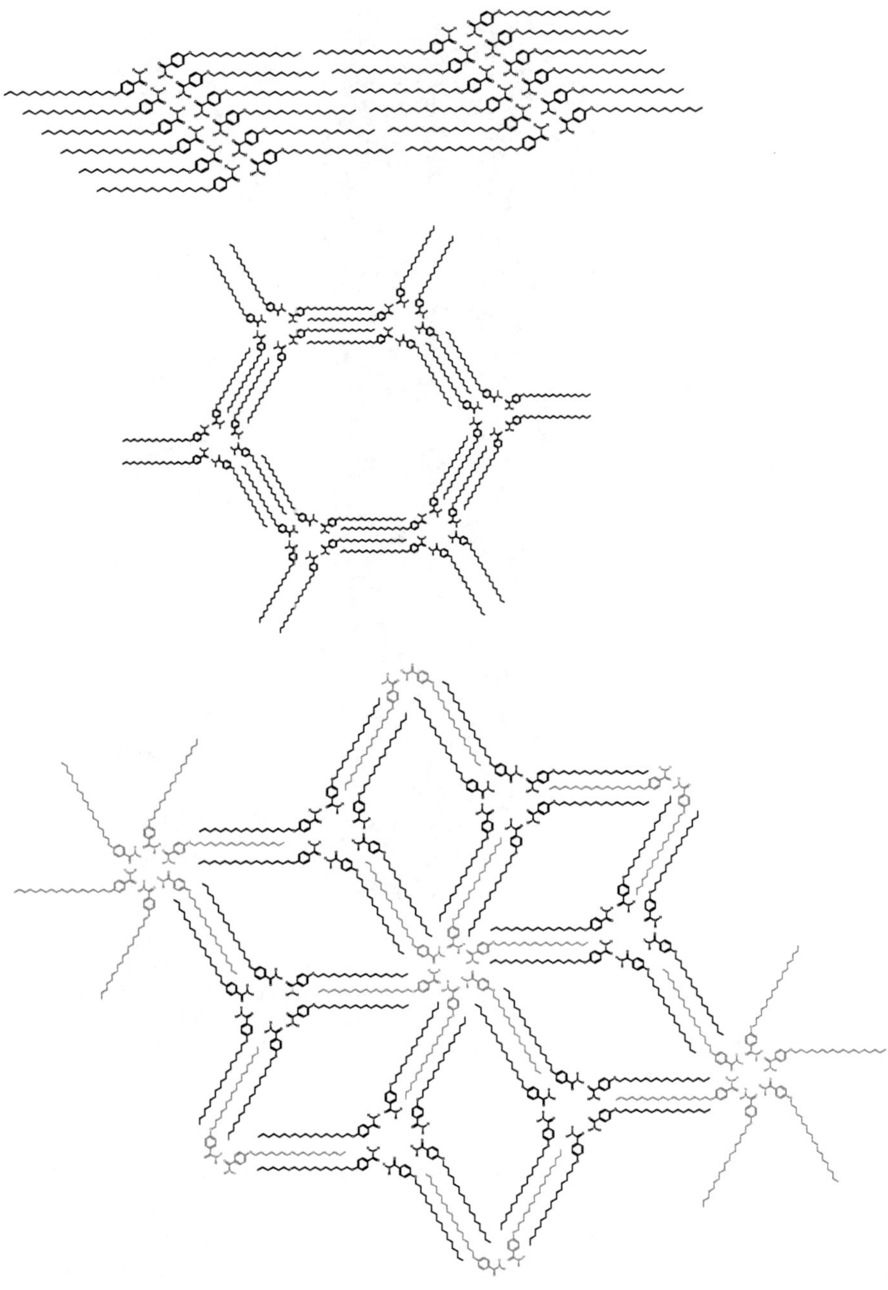

Figure 4.37 The chemical structures of polymorphic structures of simple 4-octadecyloxybenzamides that are formed at a graphite–liquid interface.

reversible phase transition from a densely-packed lamellar structure to an open hexagonal structure with interdigitated chains has been observed by STM.

The networks formed by dehydrobenzo[12]annulene derivatives have also been shown to be formed efficiently on graphene in a liquid medium.[143] The networks are sufficiently robust as to pass over defects in the graphene surface, and the system is stable in air. Hexagonal networks with periodicities of 3.9 and 7.0 nm were formed on the carbon that had been grown epitaxially on silicon. Proof of the formation of the hexagonal superstructure was provided by both STM and AFM, where the spaces in the monolayer were evident. These kinds of network-forming adsorbates thus provide a general route for the non-covalent functionalisation of graphitic carbons.

Concentration can play an important role in the formation of surface supramolecular structure, since the physisorption process can clearly be at a true thermodynamic equilibrium. Higher concentrations of simple trialkoxybenzenes lead to lamellar structures with interdigitated chains on graphite, while lowering the concentration results in the formation of open networks.[144] The length of the alkoxy chain also has a dramatic effect in this example, with long alkyl chains (16 carbons and more) tending to give only densely-packed structures because of their very high affinity to the adsorbent compared with the solvent. A truly remarkable example of this phenomenon was shown for the structurally simple 4-octadecyloxybenzamide, that displays a rich polymorphism at liquid–graphite interfaces.[145] In phenyloctane at a concentration of 100 μM, the lamellar polymorph shown in Figure 4.37 is formed, while at 33 μM and 25 μM, a hexagonal open network and a rhombus-shaped open structure are formed, respectively. The rhombus-shaped network comprises alternating modes of the hexagonal network with a cyclic hexamer of the adsorbate. The diversity in structure and flexibility in the supramolecular arrangement of the head groups can be understood when one considers the multiple ways of arranging the amide group—the only one in the molecule with strong hydrogen-bonding capacity—in orientations that align the alkyl chains along the graphite symmetry axes. In addition to these factors, additives can play an absolutely determining role in the structures that are formed at the graphite–liquid interface. Additives that are adsorbed strongly at the interface can drive the kinetic formation of less favourable polymorphs than under thermodynamic conditions, whereby the additive apparently stabilises an otherwise metastable situation that is "frozen" rather than transforming into a lamellar dense packed structure.[146]

Open networks in three dimensions have become widespread in the form of metal–organic frameworks in which coordination nodes are connected by rigid (generally aromatic) spacer units. In two-dimensional systems, a variety of structures have also been observed that could be classified under this banner. While the coordination chemistry is probably not directly comparable to bulk crystals where metal ions are used, the metal atoms on metallic surfaces are surely electropositive compared with the bulk metal and for that reason form specific complexes with ligands typical of solution chemistry. On a silver (111) surface, the rod-shaped ligands 4,4′-biphenol and 1,4′,4′,1″-terphenyl-4,4″-dicarbonitrile form hexagonal lattices by coordination of the phenolic oxygen atoms to iron adatoms and the nitrile group to cobalt adatoms, respectively.[147] The networks are different on a copper (100) surface, which has lower symmetry than the (111) and shows this effect. Tetra-4-pyridyl porphyrin forms a remarkable open network when deposited onto a gold (111) surface and annealed.[148] The molecule was evaporated onto the gold and the purity of the sample was checked using Auger spectroscopy. Annealing at 250 °C was followed by STM imaging at room temperature that revealed a network with a clear hexagonal-shaped ring, built up of triangular units (Figure 4.38). This framework is termed a Kagomé lattice, one comprising interconnected hexagons and triangles. The spaces between the centres of the molecules that were determined by STM imply that there is a gold adatom located between each pyridyl group, implying a pyridyl–gold bond with a distance of 2.7 Å. The islands are large, with lateral dimensions of up to 200 nm. The orientation of the network could not be established because of the lack of imaging of the gold adsorbent. It was clear, however, that the gold substrate in its herringbone reconstruction was lifted up, which was previously known to result in an increase of adatom concentration.[149] A study of the same porphyrin on the gold surface under electrochemical conditions showed that a close-packed structure was formed,[150] prompting the authors of the present study to infer the importance of adatoms in the formation of the open network.

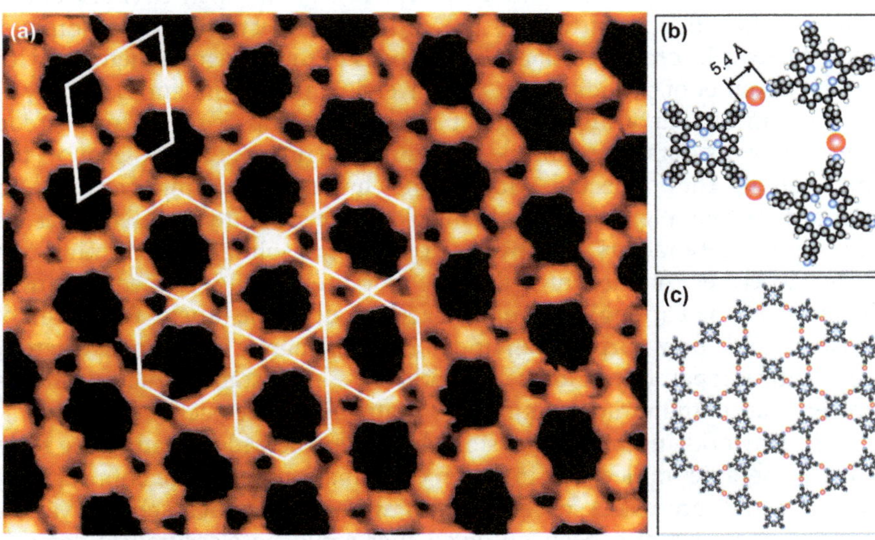

Figure 4.38 An STM image (a) of the Kagomé network formed by tetra-4-pyridyl porphyrin on a gold (111) surface after annealing at 250 °C. The model in (b) shows the trimeric unit held together by pyridyl–gold–pyridyl bonds that in turn link through the same interaction to give a surface-supported network with hexagonal spaces in the plane (c).

4.5 Molecular Recognition in Physisorbed Monolayers

Interactions between molecules adhered at an interface and solution-borne molecules are possible provided that the guests are strongly bound above the monolayer, or in the plane of the layer in contact with the substrate. The non-covalent interactions between physisorbed molecules that are confined to monolayers with guests in spaces in the monolayers can be aided by interactions between the substrate and the guest. These situations can give unique insight into forces that can lead to the segregation of molecules, as shall be shown in the specific cases that follow.

One of the landmark pieces of research into supramolecular chemistry on surfaces was the formation of a non-covalent honeycomb network on a silver-terminated Si (111) surface and its control in the organisation of C_{60} fullerene in the spaces.[151] The surface was chosen because molecules were known to diffuse freely over the surface under UHV conditions,[152–154] and diffusion is important for the formation of well-ordered domains close to a thermodynamic outcome of the adsorption. The network, which comprises melamine and perylene tetra-carboxylic diimide, was formed under UHV (with a base pressure of approximately 5×10^{-11} torr) with the molecules sublimed from effusion cells at 100 and 360 °C, respectively, to give a coverage of up to 0.3 monolayers and the surface was annealed at 100 °C. This procedure afforded a honeycomb network, as proven by *in situ* STM, where the melamine–melamine distance is close to commensurate with the surface lattice. The bicomponent mixture affords spaces big enough to incorporate C_{60} fullerene: deposition of 0.03 monolayers worth of the molecular ball afforded the three-component supramolecular structure drawn in Figure 4.39. At this coverage, some voids contained fewer fullerenes, while others were completely empty. Further deposition of fullerene led to filling of the pores and location of fullerene over the larger component of the honeycomb network. The work proved how a non-covalent network on a surface could stabilise small clusters of a third species in the surface spaces. When C_{84} fullerene was introduced into the bicomponent melamine and perylene tetra-carboxylic diimide hexagonal framework, tight-packed clusters were also formed, but they showed a remarkable reorientation when going from three to four guests in the host network spaces.[155] The delicate balance between guest–substrate and guest–network non-covalent forces results in a variety of outcomes—with a trimer structure favoured—and leads to reorientation of molecules as these aggregates grow.

C_{60} fullerene has also been shown to organise selectively on top of a monolayer formed by a porphyrin derivative, with remarkably regular dimer-like structures being observed by STM,[156] onto a densely-packed monolayer of 4,4′-dimethylbianthrone on Cu (111) (under UHV, where the fullerene can be moved with the STM tip)[157] as well as into spaces in open frames formed by coordination networks.[158] The latter case showed nicely how the

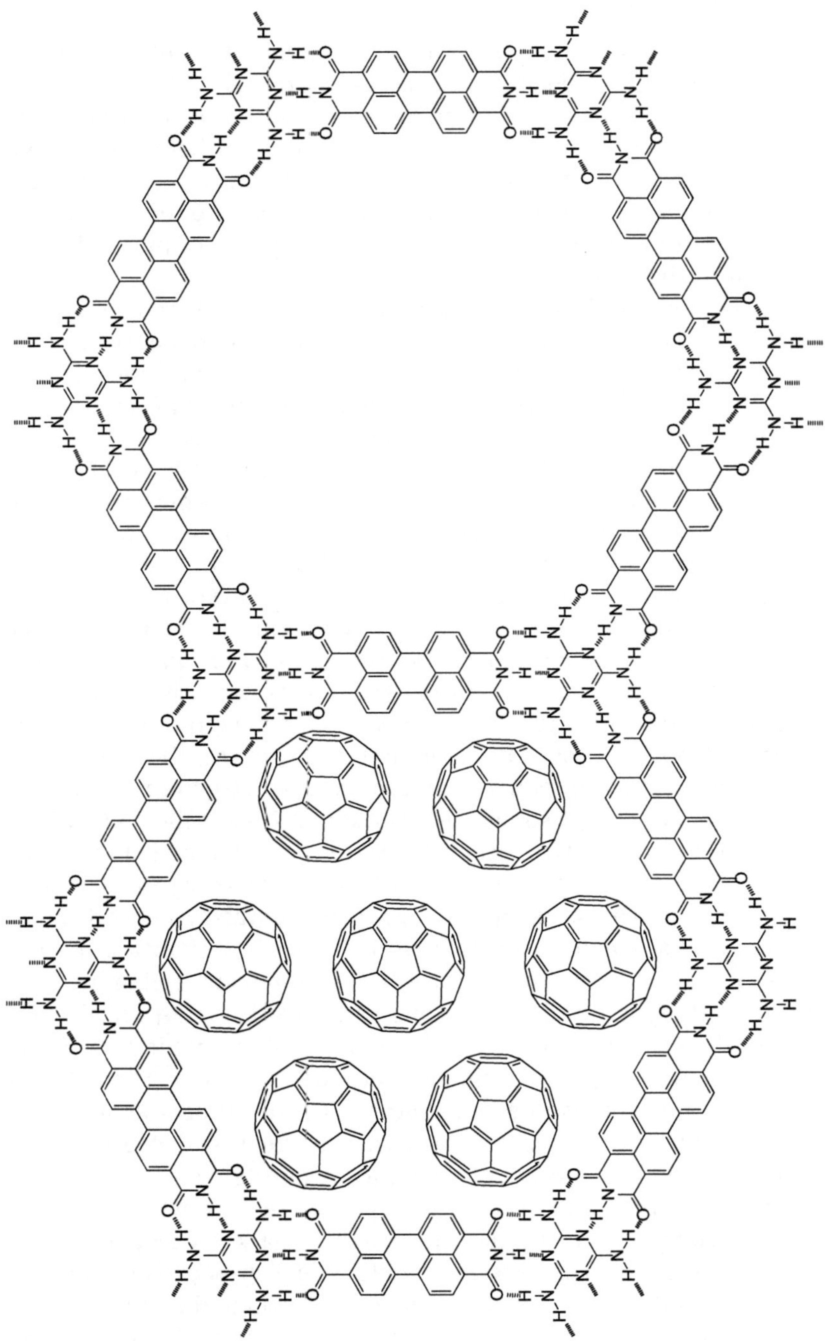

Figure 4.39 A hydrogen-bonded honeycomb network formed by melamine and perylene tetra-carboxylic diimide on silver-terminated silicon (111) (not shown) that is able to trap heptamer clusters of C_{60} at low coverages of the fullerene under UHV conditions. The drawing shows a filled and adjacent empty space, as observed by STM on the surface.

temperature onset of fullerene mobility was modified by the size of the spaces in the coordination networks and the type of interaction between the host and guest system. C_{60} fullerene has also been shown to adsorb as a single guest in the smaller pores of a bicomponent supramolecular system grown on a silicon adatom substrate (Si(111)–B) under UHV.[159] This substrate has single silicon adatoms (6.6 Å apart) with "dangling bonds" because of sub-surface boron atoms. The adsorbate 1,3,5-tri(4″-bromo-4,4′-biphenyl)benzene forms a stable network on the substrate where the arms appear to brace each other in a non-covalent manner, giving hexagonal (1.1 nm sides) and triangular (2.0 nm sides) spaces. The phenyl rings are twisted up from the surface, causing the authors to infer π–π interactions, an idea supported by modelling. When C_{60} was deposited on this layer, the fullerene was observed by STM as small clusters (66%) or on the host network in small triangular spaces (6%) or hexagonal spaces (28%). In order to provide a more specific organisation, the triangular spaces were selectively filled with 1,3,5-tri(4′-bromophenyl)benzene, thanks to the formation of C–H–π and C–Br–π interactions. This two-dimensional co-crystal is highly stable because of the combination of attractive interactions between adsorbates themselves and adsorbate–adsorbent. When the fullerene is evaporated onto this lattice, it selectively occupies the hexagonal dimples in a periodic fashion, with only 5% of the fullerenes being observed as clusters. It is an inspiring example of how self-assembly of small molecules leads to a precise array on a semiconductor surface, which from a technological viewpoint, is more relevant than metallic surfaces.

A truly remarkable three-component self-assembled monolayer was formed using a network-forming alkoxylated dehydrobenzo[12]annulene as the structural frame, an intermediate ring from isophthalic acid and a core guest of coronene.[160] The three components (Figure 4.40) form a single superstructure on gold (111) where the largest structure forms a continuous hexagonal superstructure through interdigitation of the alkyl chains, the bis-acid forms a cyclic hexamer within the spaces in the frame, and the coronene plugs the gap in the hexagonal supramolecular macrocycle. It should be noted that the frame formed on gold by the annulene is somewhat more expanded than that formed on graphite, and this slight difference means that the HOPG only has space for the coronene comfortably. Although the framework does form on HOPG,[161] that generated on gold is qualitatively much more easily formed (it happens at lower concentration of the adsorpts). This very subtle change in supramolecular arrangement caused by the election of surface thus has a significant effect on adsorption strength. It was hypothesised that diffusion barriers on gold were relatively higher than on graphite, leading to a guest-nucleated formation of the surrounding supramolecular macrocycle and framework. A related multi-component system can also self-assemble with the star-shaped hexakis-(phenylethynyl)benzene and unsubstituted dehydrobenzo[12]annulene to give a remarkable and complex crystalline layer without interdigitation of the alkyl chains between the dehydrobenzo[12]annulene derivative's alkyl chains.[162]

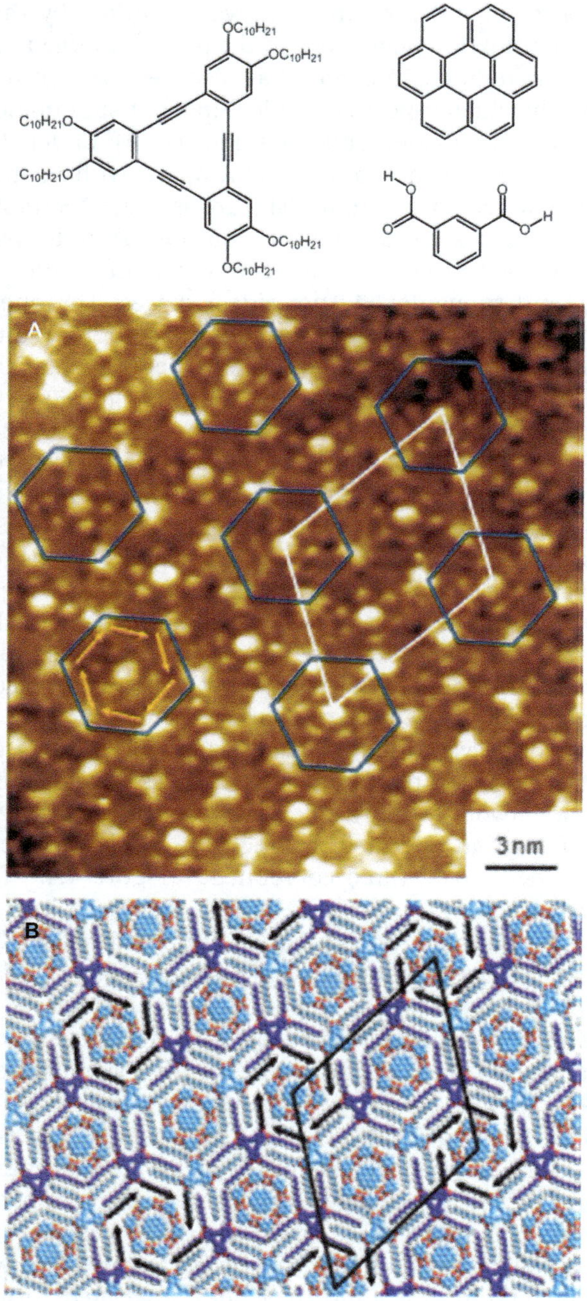

Figure 4.40 The structures of the three components of the monolayer and an STM image of the self-assembled two-dimensional system at the gold (111)–octanoic acid interface. B shows a molecular model of the superstructure. Reprinted with permission from ref. 160. Copyright 2012 American Chemical Society.

The hallmarks of self-assembly in these supramolecular systems are plain: selectivity, efficiency, complexity and error-checking.

The incorporation of molecules in the voids in the monolayers of the hexadehydrotribenzo[12]annulenes are determined by their size that can be set with the length of the alkyl chains attached to them (Figure 4.41). When six twenty-carbon chains are attached to the hub, the molecule alone physisorbs at the interface of 1,2,4-trichlorobenzene forming a dense packed structure but addition of a guest such as a "nanographene" causes a re-organisation into a framework with this second component in the centre of the rings.[163] Smaller lengths of alkyl chain result in the incorporation of five, four, two and one nanographene in the open part of the network on graphite. STM gave clear indications of the orientation of the host in the networks, which are able to accommodate guests because of their supramolecular flexibility arising from the weak interactions—van der Waals interactions between alkyl chains—holding the hubs together in the honeycomb topology.

The incorporation in an open framework of guests of sufficient height can actually promote the formation of a bilayer.[164] When a nonanoic acid solution of C_{60} was added to a monolayer of 4-terphenyl-3,5,3″,5″-tetra-carboxylic acid–that was shown earlier to display a random tiled monolayer–under the same solvent, STM revealed that the fullerene was incorporated into the pores of the network but also a second network of the framework was templated on top of the monolayer in contact with the graphite ad-sorbent. The nature of the superstructure is shown in Figure 4.42, where the fullerene is surrounded by the void type of greatest stability, both theoretically and experimentally. It is this kind of void that has the largest area of aromatic groups at the edges facing the fullerene guest. The supramolecular

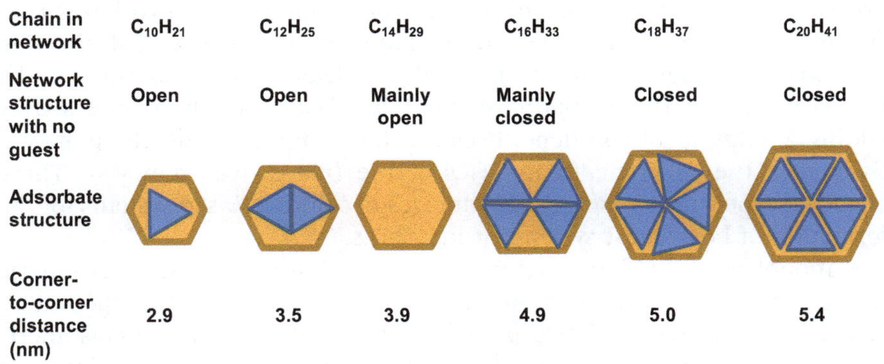

Figure 4.41 The structures of the components of the monolayers where honeycomb networks incorporate nanographene guests on graphite under a trichlorobenzene solution. The table indicates the number of carbon atoms attached to the hexadehydrotribenzo[12]annulene, the nature of the host network prior to complexation, a cartoon showing the complex formed in the monolayer, and the corner-to-corner distance of the voids in the host network.

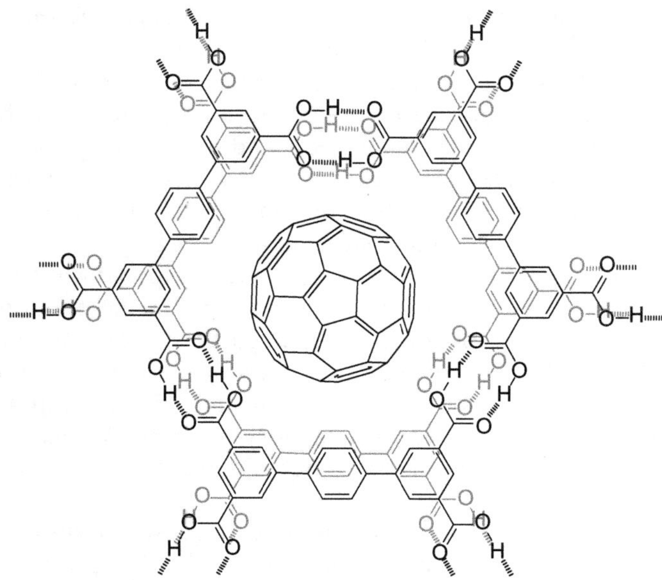

Figure 4.42 The bilayer formed by 4-terphenyl-3,5,3″,5″-tetracarboxylic acid and the complex it forms with C_{60} was added to it at the nonanoic acid–graphite interface.

structure was co-dependent, meaning that the fullerene was not observed significantly in single-layer networks, and the bilayer structure free of fullerene was only present in areas very close to fullerene inclusion. STM showed that the minimum distance between C_{60} guests was approximately 16.5 Å, coincident with the pore spacing, and once incorporated, the guests did not change location (with very few exceptions). Topographic analysis proved that the apparent height of the hydrogen-bonded network was consistent with a bilayer. The growth of the layer was followed by STM, showing how the host–guest complex effectively nucleated the formation of the bilayer structure. The formation of the bilayer can actually be reversed by adding a solution of coronene: because this compound adsorbs preferentially over the fullerene to the HOPG, the bilayer was removed. These beautiful experiments show the advantages of solution-based systems for the formation of host–guest systems at interfaces.

A monolayer of chiral acids has been used for the diastereoisomeric resolution of a mixture of enantiomers from solution.[165] This experiment, based on the principle first explored by Pasteur for the diastereoisomeric salts of tartaric acid and quinine,[166] involved binding a single enantiomer of a chiral amine from a mixture of enantiomers in a solution using a single enantiomer of an amphiphilic acid that had been previously shown to form chiral monolayers.[167] Determining the monolayer structure of the pure acid on graphite was important for the subsequent binding studies: a chiral layer comprised of lamellae with interdigitated alkyl chains is formed (Figure 4.43). The aromatic head groups—that appear as bright dots in the

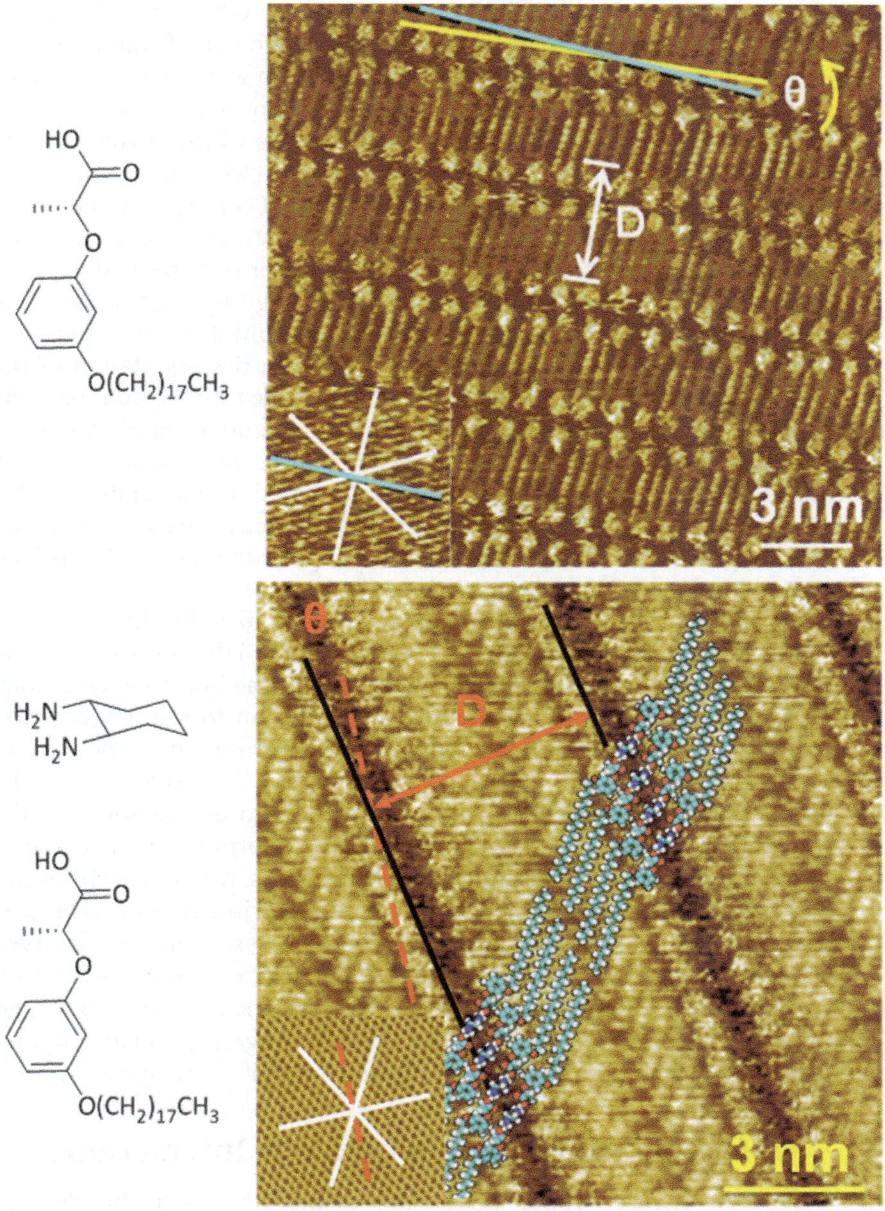

Figure 4.43 STM images of the amphiphilic acid resolving agent (top) and the mixture of this compound with (*R*,*R*)-diaminocyclohexane (bottom), both at the graphite–phenyloctane interface. The angle θ is that between the graphite reference axis and the lamellae direction, and D shows the width of the lamellae. The insets show images of the graphite surface with the symmetry axes indicated and the reference axis. Models of the molecules are overlaid on the lower image.

STM images—are in close proximity to one another. When the (*R*)-enantiomer of the acid is combined with the (*R,R*) enantiomer of diaminocyclohexane, a completely different monolayer structure is observed. Most strikingly, the alkyl chains of the amphiphile are no longer interdigitated, but abut, leading to a corresponding increase in the width of the lamellae and the unit cell dimensions. A row that appears with dark contrast is apparent in the STM images (Figure 4.43), that would correspond to the area where the diaminocyclohexane is bound between two acid moieties. Understanding the precise structure of the monolayer requires insight from molecular modelling, which shows how the amine groups of the guest are bound to the acidic head group of the adsorbed amphiphile. The change in the packing of the acid is explained by a change in the orientation of the phenyl ring upon binding. Because it is slightly lifted in the complex, the amphiphiles can pack more densely, leading to the abutting alkyl chains. The chirality of the layer—indicated by the angle formed between the lamellae direction and a reference axis of the graphite substrate—is also influenced by this structural arrangement; the complex forms a much higher angle (around 10°) than the pure amphiphile (approximately 3°). Therefore, the free and bound acid are identified readily by STM.

The complex forms for the acid and diamine of like chirality, however, when the (*R*) configuration acid is mixed with the (*S,S*)-diaminocyclohexane at a 2 : 1 ratio, only the pure acid is seen adsorbed at the interface, and much higher proportions of amine are required in solution to see any complex formation expressed in supramolecular surface structure. This difference in adsorption of the diastereoisomeric complexes can be exploited for the resolution experiment. Mixing the acid and the two enantiomers of the diamine in a 4 : 1 : 1 ratio leads exclusively to the adsorption of the complex that is the favoured one in the individual experiments, where like chiralities form an adsorbed complex. This remarkable effect witnessed by STM (it is hard to think of another way this kind of complex could be observed unequivocally) reveals that although the acids form complexes with both amines in solution, the adsorption adds a fillip to the selectivity, and perhaps points the way to the design of effect-resolving agents, specifically, and perhaps more generally to the design of selective host–guest systems.

4.6 Observing Areas of Non-covalent Interactions

In the majority of cases where specific non-covalent bonds in the plane of the monolayer are responsible for the self-assembly of well-defined aggregates, interactions between molecules are inferred from intermolecular distances from STM measurements, often accompanied by molecular modelling support for the superstructures.

Scanning tunnelling spectroscopy (STS, where current is monitored while potential is swept) can actually be used to probe the electronic states of molecules on metallic surfaces and to infer non-covalent interactions. This method was employed to study bis-urea derivatives (prepared from racemic

2-ethylhexylamine and 2,4-toluene diisocyanate[168]) where either isolated molecules or hydrogen-bonded chains of them were measured.[169] The chains of molecules linked by hydrogen bonds between the oxygen atoms and two NH groups of the urea unit show an unusual contrast in the STM images of the layer on gold (111), where one "end" of the molecule is far brighter than the other (Figure 4.44). While this behaviour might be expected for the non-C_2 symmetric molecule, the large difference is very likely caused by a very different adsorption geometry as a result of the methyl substituent on the phenyl ring between the two urea units. According to the authors, this situation could cause a charging effect on the molecule (as seen for copper-phthalocyanines in similar experiments[170]), that has a varying strength of molecule–gold interaction. This effect gives rises to very different STS curves, because the LUMO energy level (through which the electron could tunnel on its way between tip and substrate) is different at each position. This situation is seen in the curve in Figure 4.44, where it can be appreciated that the LUMO energy is raised as the molecule separates from the surface.

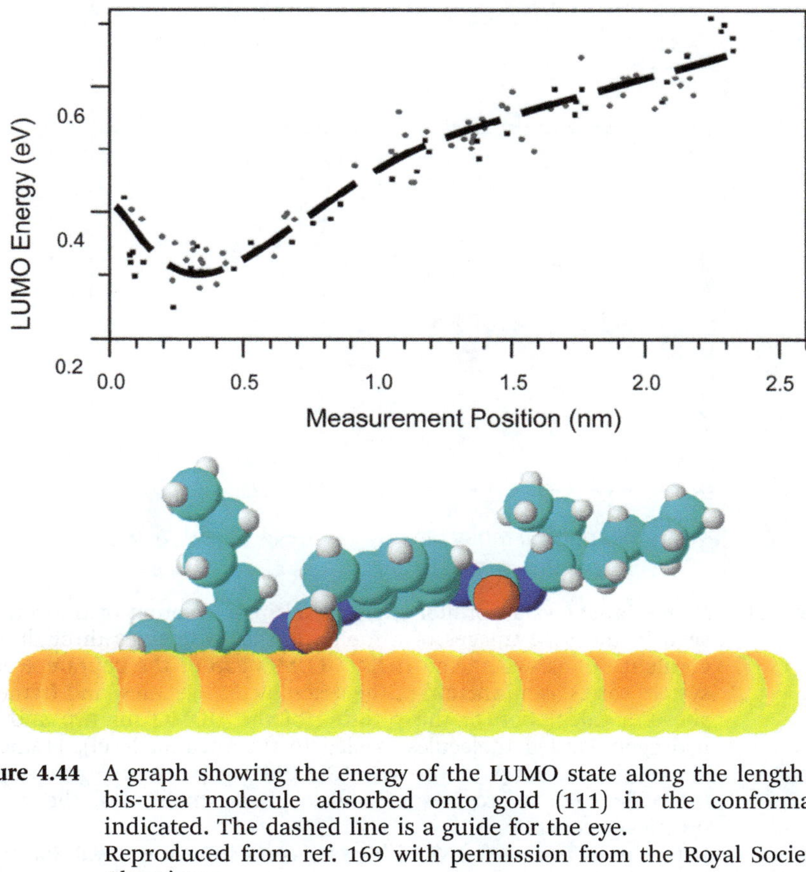

Figure 4.44 A graph showing the energy of the LUMO state along the length of a bis-urea molecule adsorbed onto gold (111) in the conformation indicated. The dashed line is a guide for the eye.
Reproduced from ref. 169 with permission from the Royal Society of Chemistry.

The situation is quite different for isolated molecules or small groups of molecules that were generated by applying a voltage pulse to the ordered monolayer. This experiment desorbs the molecules temporarily, and re-adsorption takes place but into a metastable situation because free diffusion is not possible in this experiment carried out in vacuum. Comparison of the images and STS curves of isolated molecules, small groups of them and the full monolayer implied energy level splitting as a result of non-covalent bond formation, as shown in the energy level diagrams in Figure 4.45. Single molecules have electronic states whose degeneracy is split upon linking through hydrogen bonds or π-interactions. The molecules have their sub-levels split as a result either of the hydrogen bonding between molecules and/or a distortion-induced Jahn–Teller splitting.[171] The more molecules

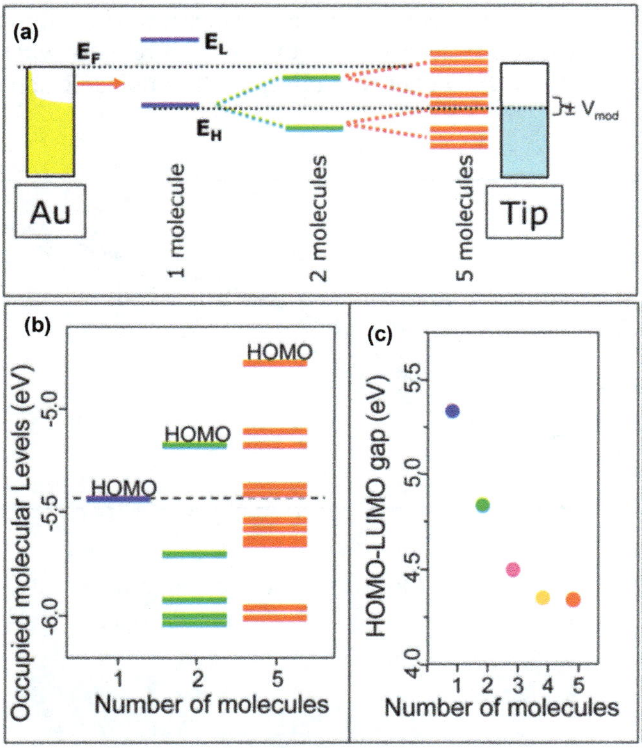

Figure 4.45 Frame (a) shows an orbital energy level interpretation of the contrast seen in the STM images of a molecule that assembles through non-covalent bonds, where the HOMO–LUMO gap of the supramolecular system varies as a function of the assembled molecules, and frame (b) shows a calculation of the splitting of the HOMO for two and five hydrogen-bonded molecules (scaled to the vacuum level). Frame (c) shows the lowering in the HOMO–LUMO gap for the grouping of molecules. For aggregates of four or more molecules, the gap is practically constant.
Reproduced from ref. 169 with permission from the Royal Society of Chemistry.

that are linked in the aggregate, the greater number of states is theoretically possible, until a limiting value of the gap between HOMO and LUMO is reached. The results show the cooperative nature of hydrogen bonds in this kind of aggregate in an electronic sense, that might be of significance for charge transport. This observation would be of great interest in molecules in which electrons could be delocalised through π stacking, where the in-vacuum measurements are somewhat unique in allowing the isolation of aggregates of different sizes on the surface.

Hydrogen-bonded assemblies can also be probed using dynamic force microscopy (also known as non-contact AFM), which shows remarkable sub-molecular resolution thanks to functionalisation of the AFM tip that in principle probes electron density, as shown for 8-hydroxyquinoline on copper (111)[172] and the naphthalene tetracarboxylic diimide monolayer.[173] Using the technique, the tip–sample force field is mapped quantitatively, leading to contrast from repulsive tip–sample interactions. While it is intuitively tempting to judge the bonding pattern directly from the images, it was shown that the coupled tip–molecule system has to be analysed with care and with the aid of modelling. That includes using a tip-adsorbed molecule—the source of the high resolution—in the calculations. Ignoring the influence of the tip on the system is not realistic. When this is done, however, the level of contrast obtained and the detail gleaned is far more well-resolved than most STM experiments, and, although still controversial,[174] it seems likely that much more will be heard of this technique for the analysis of supramolecular structure.

4.7 Chemical Reactivity in Physisorbed Monolayers

The specific geometry of molecules adsorbed to surfaces and the proximity of the underlying substrate provide physisorbed layers with a unique environment for both topochemical reactions and metal ion reactivity.[175–177] This area is a burgeoning one because of the ability to observe the outcomes of the reactions using scanning probe microscopes that is allowing the discovery of new chemistry, as well as supramolecular phenomena.

Incorporation of reactive groups into specific locations of molecules that form physisorbed layers can lead to surface-adhered polymers, as shown for a terephthalic acid derivative bearing a chain that incorporates a diacetylene function (Figure 4.46).[178] The molecule adsorbs to graphite from 1-undecanol solution to give a monolayer where the chains coming from each head group interdigitate leading to lamellae. In-between these strips are solvent lamellae. The carboxylic acid groups can in principle form hydrogen bonds both within the like-molecule lamellae as well as with the solvent areas. The diacetylene groups are very proximal (approximately 4.7 Å). Irradiation of the system—with the solvent present—for about half an hour at 254 nm followed by reintroduction of the STM tip gave images where the initial supramolecular structure was present, but also where polymerised compound was present, as indicated by the change in contrast, direction with respect to the

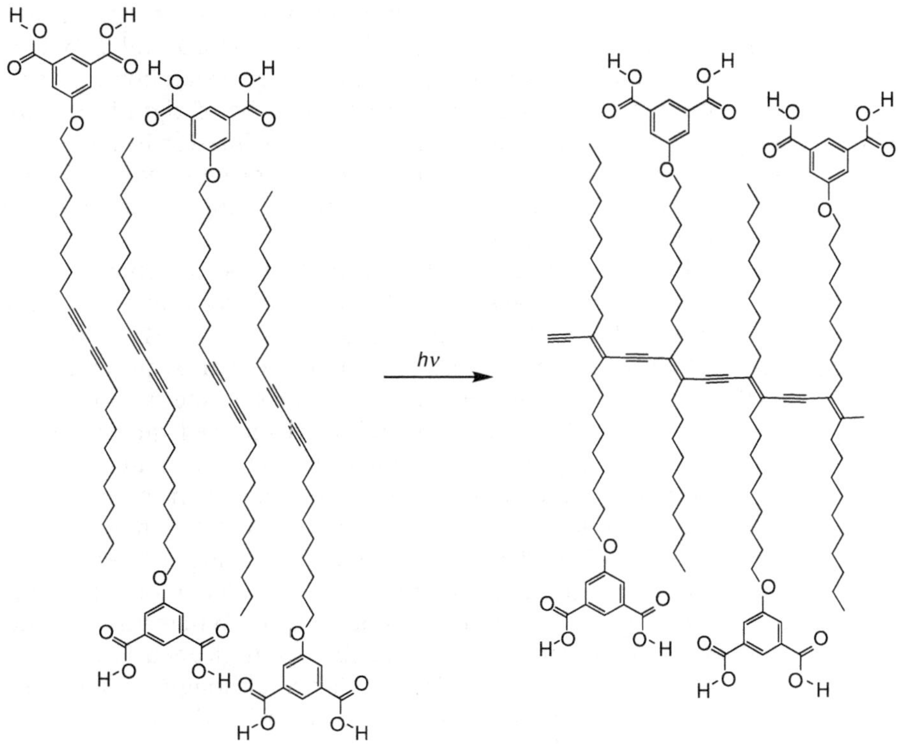

Figure 4.46 The packing structure that a diacetylene-containing amphiphile adopts at the interface between HOPG and 1-undecanol, and the polydiacetylene that is formed after irradiation, and was observed on the surface by STM.

graphite substrate and unit cell dimensions. The very straight polymer chains, the fact that a solvent lamella was present after irradiation and that the photochemical reaction is topochemical[179] indicate strongly that the reaction is taking place at the interface. In this two-dimensional crystal engineering, one can appreciate how the surface can determine the organisation of adsorpts, their reactivity, and the organisation of the products. Graphene has also proven to be an effective template,[180] and, rather than with light, the polymerisation can also be initiated by the STM tip at specific locations.[181,182] Subsequent work has shown how this kind of polymer formed on graphite, either under ambient or UHV conditions, can be desorbed and manipulated.[183] These results are of great promise in the area of single molecule electronics.[184]

In situ coordination chemistry in the plane parallel to the interface can also be exploited to construct supramolecular objects. Redox reactions are a way to perform this, a case being that of a dipyridyl tetrazine that reacts with platinum to give a coordination polymer on gold where the platinum is oxidised to the divalent state and the tetrazine moiety is reduced

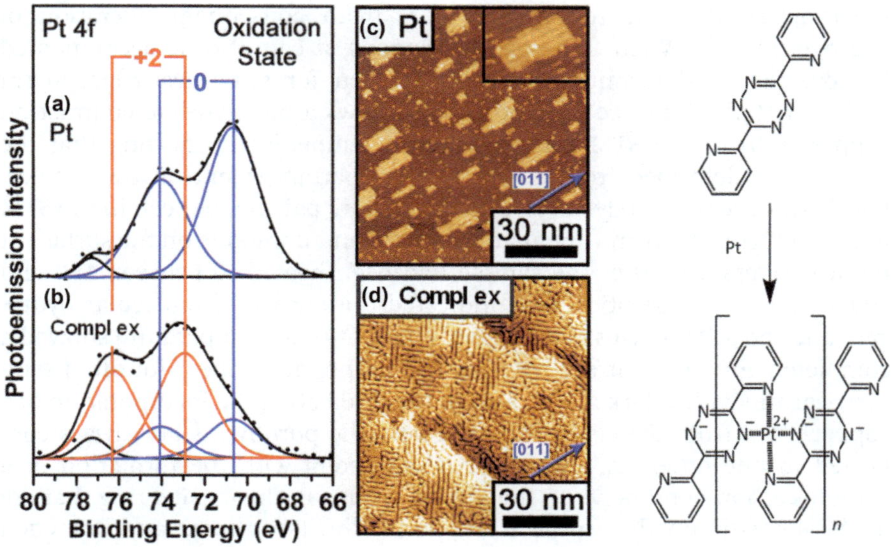

Figure 4.47 XPS spectra of the 4f platinum region showing (a) pure platinum and (b) platinum co-deposited with the dipyridyl tetrazine ligand (shown to the right) both on the gold (100) reconstructed surface. STM images of the two samples (both annealed for 20 minutes at 170 and 150 °C, respectively) evidence the formation of a new compound. While the metal forms rectangular islands (c), the coordination polymer exists as chains up to 30 nm long, approximately.
Parts (a)–(d) adapted with permission from ref. 185. Copyright 2014 American Chemical Society.

(Figure 4.47).[185] When platinum is deposited on the gold (100) surface, rectangular islands are observed by STM and XPS measurements indicate that it is present as the zero oxidation state. The dipyridyl tetrazine compound also forms islands with a small amount of one-dimensional chains (oriented with the reconstruction rows of the surface). When the metal was deposited on a submonolayer quantity of the dipyridyl tetrazine ligand and annealed, a quite different structure was seen at the interface. The STM images of the layer show straight rods up to 30 nanometres long that lie at an angle of 45° with respect to the surface rows of the gold adsorbent. Non-contact AFM showed that these rods are made up of the ligands that are twisted at an angle to the long axis, as shown in the chemical structure in the figure (the ligands were not resolved by STM). XPS shows that in these complexes, there is a complete charge transfer from the metal to the ligand, so that the divalent metal ion is formed and observed by this spectroscopy, it is held in-between the ligands that are reduced. The ligand provides a nearly square-planar coordination pocket that can align parallel to the gold surface, as well as having redox-activity in the tetrazine moiety that permits charge transfer within the layer of adsorbed materials.

The oxidation of metal ions also takes place in molecules adsorbed on surfaces when annealing porphyrins on metal surfaces where ions are incorporated into the tetrapyrrolic core (with consequent liberation of hydrogen).[186–189] When this process happens, it can also be accompanied by covalent bond forming reactions, as seen for tetra(mesityl)porphyrin (Figure 4.48).[190] The free base porphyrin shows a butterfly-like contrast on copper (110) in the STM images at room temperature, showing that the molecule is deformed from a strictly planar arrangement on the surface. Porphyrins are very rarely in close contact (a rare pair can be seen in the STM image in Figure 4.48) and are mainly isolated and immobile on the surface at room temperature. After the sample had been heated to between 150 and 200 °C, thermal desorption data indicated the release of molecular hydrogen, and the STM images of the sample back at room temperature showed a completely different surface pattern. The porphyrins appear four-fold symmetrical and with a dark dot at their centre, indicating the incorporation of a copper atom from the surface. In addition, the porphyrins are always connected to one other, in a way that is consistent with the formation of a carbon–carbon bond between the methyl groups in the mesityl ring that are in the 4-position with respect to the porphyrin. Both linear and branched oligomers are formed in a way characteristic of a condensation-type polymerisation, with a variety of topologies and sizes. The physisorbed nature of these oligomers was proven by performing STM at 120 °C and showing how a linear trimer (corresponding to $n = 2$ in Figure 4.48) moved slowly across the metal's surface, while the monomers were far more mobile. The experiment also demonstrated that the three units were held together covalently, since any non-covalent linkage would be expected to dissociate at this temperature. The results are complementary to the more familiar couplings carried out on surfaces to produce macromolecules.[191,192] The incorporation of the metal ions is not necessarily through coordination-type bonds, but can be through organometallic linkers,[193] although these tend to be directly adhered to the metal and are classified here under chemisorbed monolayers.

The reaction of a manganese(III) porphyrin with oxygen can be monitored in real time and space using STM at the HOPG–1-octanoic acid interface.[194] The porphyrin shown in Figure 4.49 adsorbs on the surface of graphite at its interface with 1-octanoic acid at room temperature. STM of the monolayers showed a characteristic lamellar structure, but occasionally brighter features were observed corresponding to the cores of the molecules. In fact, at least four types of brightness are present in the images of the compound under these conditions. The STM image in Figure 4.49 shows these four types of contrast where each one has been assigned to one of the oxygen-manganese compounds. The profile in the inset provides a vertical scale, bearing in mind that the STM does not measure height accurately. The darkest contrast would appear to be a manganese(II) porphyrin that results from reduction of the adsorbent (that has intermediate contrast) by the surface, with the corresponding loss of the chloride ion from the metal ion. This low valent state of manganese is oxidised readily by ambient oxygen to give a manganese(IV) oxo species, where a single oxygen atom is attached from a split dioxygen molecule. This manganese(IV) adsorpt can react with the initial

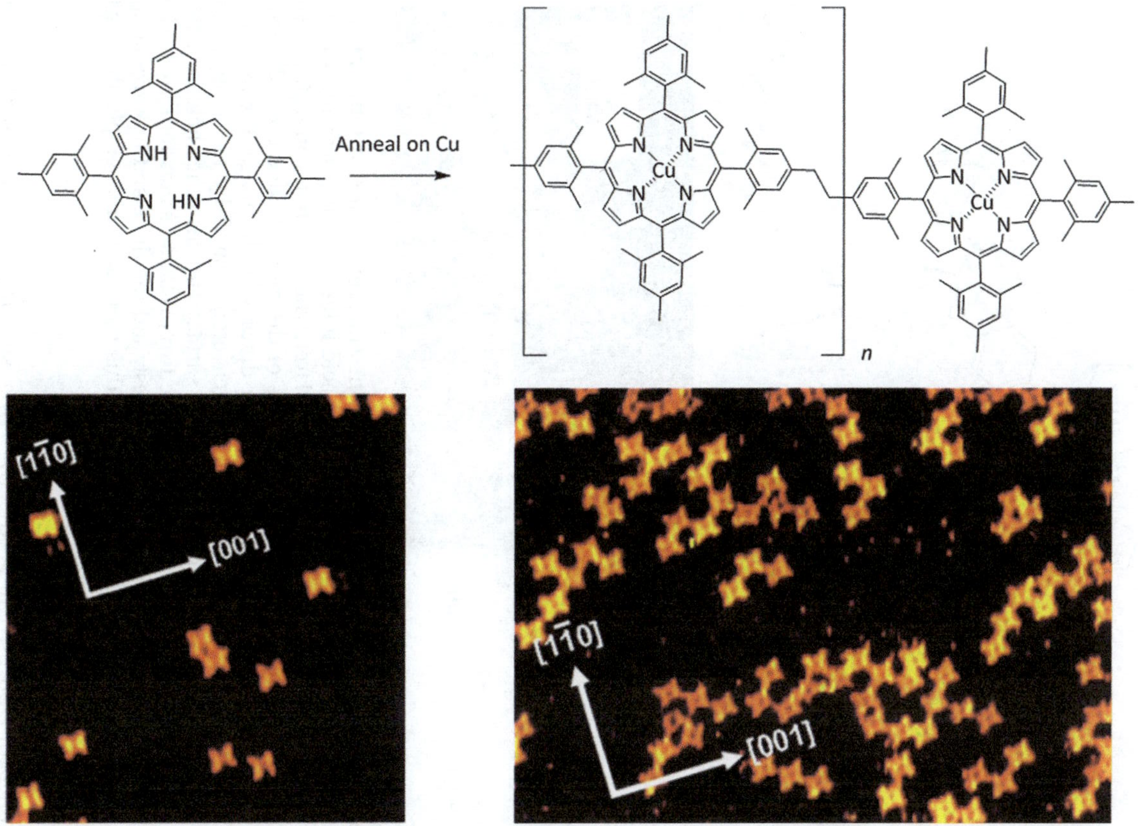

Figure 4.48 The reaction of tetra(mesityl)porphyrin on the copper (110) surface and STM images of the porphyrin prior to (left) and after (right) reaction above 150 °C on the surface.

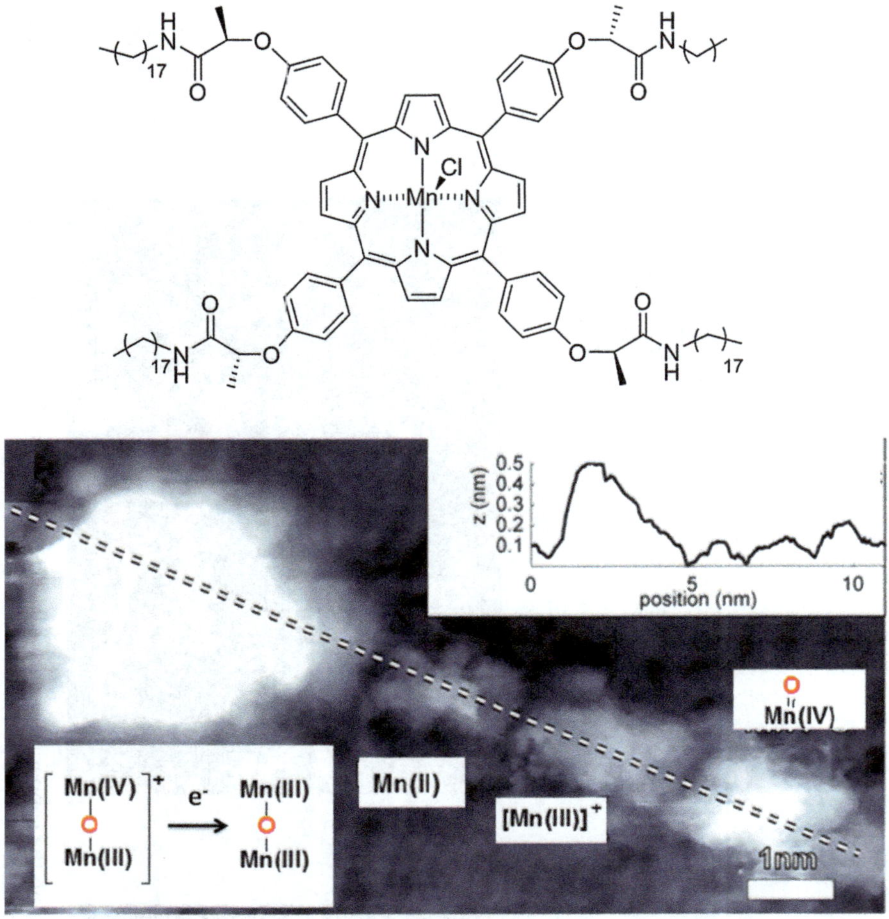

Figure 4.49 The chemical structure of manganese(III) porphyrin that forms monolayers at the graphite–1-octanoic acid interface that react with oxygen, as revealed in the STM image of a layer, where the four contrasts shown probably correspond to the states of the molecule that are indicated. The profile corresponds to the line drawn over the image. The brightest contrast corresponds to an oxygen atom bound by two manganese porphyrins (one adsorbed to the surface and the other held over the monolayer) that could exist in the two oxidation states indicated. The manganese(II) species is the darkest, and the three and four oxidation states with increasing clarity. In this particular micrograph, the alkyl chains are not imaged.

manganese(III) porphyrin in the overlying solution to give complexes in which a single oxygen atom is located between the manganese ions in separate porphyrins, one adsorbed to the surface and the other held above the monolayer. These complexes, of which two oxidation states are likely (Figure 4.49), shine out from the micrographs as very bright protrusions. Determining the precise structure of the single molecule reaction products is beyond the state-of-the-art—spectroscopy on ensembles is usually necessary—but these images do show how the chemistry of the monolayer develops over time with temporal resolution on the order of minutes. Indeed, quantitative analysis of the population of manganese(IV) species over time either under argon or oxygen shows how the production of this compound is dependent on the atmospheric reagent, reaching up to 20% of the adsorbed molecules. It also shows how, over four hours, it becomes the dominant oxidised specie after an initial build-up of the manganese(II) compound. Importantly, the surface is essential for this chemistry that does not occur in a homogeneous solution of the porphyrin in 1-octanoic acid.

Frequently, pairs of manganese(IV) are present at the interface. The relative occurrence of neighbouring manganese(IV) porphyrin in the same row of molecules is roughly twice that expected for a random process, and points to its correlated formation by reaction of dioxygen with adjacent adsorpts. This correlated oxygen adsorption was also observed during reaction of a different manganese porphyrin adsorbed on Au (111) and used to oxidise an alkene.[195] A concerted reaction for oxygen dissociation by very proximal iron adatoms coordinated by carboxyl ligands on copper in UHV has also been described, where molecular modelling using DFT methods demonstrated the role of the interface configuration in the process at the atomic level.[196] In the case of these manganese porphyrins, the remarkable feature is to be able to follow in real time under ambient conditions, and in addition to probe solution-borne oxidants. After dioxygen fission, the non-adsorbed oxygen atom (adsorpt molecules would not be sufficiently proximal to bind both) would bind to the site with highest effective molarity, in this case the neighbouring adsorpt porphyrin, to generate a second manganese(IV)–oxo adsorpt. When the single oxygen donor pentafluoroiodosylbenzene was added to monolayers of the manganese(III) porphyrin, these neighbouring oxo species were not observed, instead a fifth noisy type of response was observed, that transformed into the dimanganese porphyrin oxygen sandwich complex. This signal is consistent with the generation of a manganese(V) porphyrin oxo complex. Thus, STM shows how different reagents convert the initial adsorpt at the single molecule level, and how the distribution of compounds changes over time in this stable supramolecular surface structure.

The possibility of using the tip of the microscope to influence the reaction was also explored.[197] Above a certain potential, a voltage pulse between the platinum–iridium STM tip and the graphite surface through the monolayers of the manganese(III) porphyrin (that was discussed earlier) leads to the formation of bright protrusions that are consistent with the bisporphyrin oxo complex (Figure 4.50). This observation shows how the local potential can influence the reactivity of the molecules in the monolayer. In the present case, the position where the pulse was made was devoid of bright protrusions. The distance between the centre of the pulse and the highest population of bisporphyrin μ-oxo complexes was approximately 15 nm, as measured over 48 pulse experiments (corresponding to approximately 200 bright objects). This can be interpreted as a tip-induced removal of the

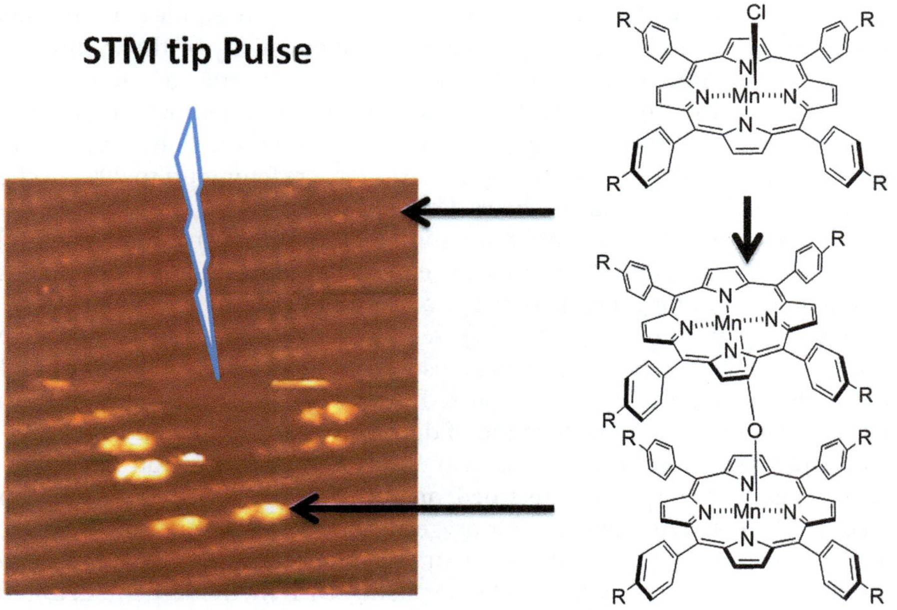

Figure 4.50 An STM image showing the region of a monolayer of the manganese(III)chloride porphyrin shown with a pulse event at the point indicated on the micrograph, from which bright protrusions appear consistent with the formation of the bisporphyrin μ-oxo complex whose structure is shown.

molecules from the surface and quasi-simultaneous formation of an oxo-intermediate that diffuses through the solution until adsorbing to a part of the monolayer that is occupied by the initial adsorpt. This local chemical reaction, induced by the tip in solution, is unique in that the physisorbed molecule is removed from its original location, which is then filled with molecules in the solution.

The oxidation process is not limited to metal-containing compounds. Adlayers of the oligopyridine compound 2-phenyl-4,6-bis(6-(pyridine-2-yl)-4-(pyridine-4-yl)pyridine-2-yl)pyrimidine adsorbed onto the (111) surface of both gold and silver are ordered,[198] but undergo a reaction only on silver to afford a selective oxidation that can be followed using video STM.[199] The molecule forms a network sustained by hydrogen bonds between C–H groups and the pyridyl nitrogen atoms. Exposure of the monolayer on silver to oxygen leads to the appearance of bright protrusions in STM images at very specific locations. It is believed that the N-oxide is formed after oxygen fission has taken place at the interface by dissociative adsorption, and the oxygen atoms diffuse to the most reactive site. The adsorpt is totally unreactive in its pure state to oxygen under ambient conditions, thus the interface generates unique conditions. Theoretical simulations supported the hypothesis that the reaction is energetically favourable and catalysed by silver. The result is considered important because in many devices involving organic molecules there are interfaces with metals, and silver can clearly facilitate the reaction of oxygen with certain molecules in this heterogeneous system.

The incorporation of these covalent bonding transformations into systems where non-covalent bonds influence directly the strong bonds that are being formed is perhaps still a challenge for physisorbed layers because of their dynamic nature. Nonetheless, it is foreseeable that with the non-covalent tools at our disposal and the control that can be attained over molecular orientation on surfaces that these and other related challenges become attainable in the not too distant future.

References

1. S. De Feyter and F. C. De Schryver, *Chem. Soc. Rev.*, 2003, **32**, 139.
2. O. Ivasenko and D. F. Perepichka, *Chem. Soc. Rev.*, 2011, **40**, 191.
3. N. R. Champness and S. De Feyter, *CrystEngComm*, 2011, **13**, 5531.
4. S. De Feyter and F. C. De Schryver, *J. Phys. Chem. B*, 2005, **109**, 4290.
5. K. Ishida, S. Taki, H. Okabe and K. Matsushige, *Jpn. J. Appl. Phys.*, 1995, **34**(Pt. 1), 3846.
6. G. H. Findenegg, *J. Chem. Soc., Faraday Trans 1*, 1973, **69**, 1069.
7. X. Qiu, C. Wang, Q. Zeng, B. Xu, S. Yin, H. Wang, S. Xu and C. Bai, *J. Am. Chem. Soc.*, 2000, **122**, 5550.
8. S. Mohnani and D. Bonifazi, *Coord. Chem. Rev.*, 2010, **254**, 2342.
9. K. S. Mali and S. De Feyter, *Philos. Trans. R. Soc., A*, 2013, **371**, 20130263.
10. W. Ostwald, *Zeitschrift fur Physikalische Chemie*, 1897, **22**, 289.
11. T. Threlfall, *Org. Proc. Res. Dev.*, 2003, **7**, 1017.
12. A. Stabel, R. Heinz, F. C. De Schryver and J. P. Rabe, *J. Phys. Chem.*, 1995, **99**, 505.
13. F. S. Tautz, *Prog. Surf. Sci.*, 2007, **82**, 479.
14. W. A. English and K. W. Hipps, *J. Phys. Chem. C*, 2008, **112**, 2026.
15. W. Song, N. Martsinovich, W. M. Heckl and M. Lackinger, *J. Am. Chem. Soc.*, 2013, **135**, 14854.
16. S. De Feyter, A. Gesquière, M. M. Abdel-Mottaleb, P. C. M. Grim and F. C. De Schryver, *Acc. Chem. Res.*, 2000, **33**, 520.
17. W. Song, N. Martsinovich, W. M. Heckl and M. Lackinger, *Phys. Chem. Chem. Phys.*, 2014, **16**, 13239.
18. U. Mazur and K. W. Hipps, *Chem. Commun.*, 2015, **51**, 4737.
19. A. Bhattaral, U. Mazur and K. W. Hipps, *J. Am. Chem. Soc.*, 2014, **136**, 2142.
20. D. S. Karpovich and G. J. Blanchard, *Langmuir*, 1994, **10**, 3315.
21. A. Bellec, C. Arrigoni, G. Schull, L. Douillard, C. Fiorini-Debuisschert, F. Mathevet, D. Kreher, A.-J. Attias and F. Charra, *J. Chem. Phys.*, 2011, **134**, 124702.
22. R. T. Baker, J. D. Mougous, A. Brackley and D. L. Patrick, *Langmuir*, 1999, **15**, 4884.

23. T. Yang, S. Berber, J.-F. Liu, G. P. Miller and D. Tománek, *J. Chem. Phys.*, 2008, **128**, 124709.
24. M. Böhringer, K. Morgenstern, W.-D. Schneider, R. Berndt, F. Mauri, A. De Vita and R. Car, *Phys. Rev. Lett.*, 1999, **83**, 324.
25. M. Böhringer, W.-D. Schneider and R. Berndt, *Surf. Rev. Lett*, 2000, **7**, 661.
26. M. Bohringer, K. Morgenstern, W. D. Schneider and R. Berndt, *Angew. Chem., Int. Ed.*, 1999, **38**, 821.
27. M. Böhringer, K. Morgenstern, W. D. Schneider, M. Wühn, C. Wöll and R. Berndt, *Surf. Sci.*, 2000, **444**, 199.
28. K. Morgenstern, E. Lægsgaard, I. Stensgaard, F. Besenbacher, M. Böhringer, W.-D. Schneider, R. Berndt, F. Mauri, A. De Vita and R. Car, *Appl. Phys. A: Mater. Sci. Process.*, 1999, **69**, 559.
29. T. Yokoyama, S. Yokoyama, T. Kamikado, Y. Okuno and S. Mashiko, *Nature*, 2001, **413**, 619.
30. N. Wintjes, J. Hornung, J. Lobo-Checa, T. Voigt, T. Samuely, C. Thilgen, M. Stçhr, F. Diederich and T. A. Jung, *Chem. – Eur. J.*, 2008, **14**, 5794.
31. D. Heim, K. Seufert, W. Auwarter, C. Aurisicchio, C. Fabbro, D. Bonifazi and J. V. Barth, *Nano Lett.*, 2009, **10**, 122.
32. W. Auwärter, F. Klappenberger, A. Weber-Bargioni, A. Schiffrin, T. Strunskus, C. Wöll, Y. Pennec, A. Riemann and J. V. Barth, *J. Am. Chem. Soc.*, 2007, **129**, 11279.
33. J. Weckesser, A. De Vita, J. V. Barth, C. Cai and K. Kern, *Phys. Rev. Lett.*, 2001, **87**, 096101.
34. J. V. Barth, J. Weckesser, G. Trimarchi, M. Vladimirova, A. De Vita, C. Cai, H. Brune, P. Günter and K. Kern, *J. Am. Chem. Soc.*, 2002, **124**, 7991.
35. J. Lipton-Duffin, J. A. Miwa, S. G. Urquhart, G. Contini, A. Cossaro, L. Casalis, J. V. Barth, L. Floreano, A. Morgante and F. Rosei, *Langmuir*, 2012, **28**, 14291.
36. F. Vidal, E. Delvigne, S. Stepanow, N. Lin, J. V. Barth and K. Kern, *J. Am. Chem. Soc.*, 2005, **127**, 10101.
37. B.-I. Kim, C. Cai, X. Deng and S. S. Perry, *Surf. Sci.*, 2003, **538**, 45.
38. M.-C. Blüm, E. Ćavar, M. Pivetta, F. Patthey and W.-D. Schneider, *Angew. Chem., Int. Ed.*, 2005, **44**, 5334.
39. J. V. Barth, H. Brune, G. Ertl and R. J. Behm, *Phys. Rev. B: Condens. Matter Mater. Phys.*, 1990, **42**, 9307.
40. D. D. Chambliss, R. J. Wilson and S. Chiang, *Phys. Rev. Lett.*, 1991, **66**, 1721.
41. M. Pivetta, M.-C. Blüm, F. Patthey and W.-D. Schneider, *Angew. Chem., Int. Ed.*, 2008, **47**, 1076.
42. M. Pivetta, M.-C. Blüm, F. Patthey and W.-D. Schneider, *ChemPhysChem*, 2010, **11**, 1558.
43. F. Cicoira, J. A. Miwa, D. F. Perepichka and F. Rosei, *J. Phys. Chem. A*, 2007, **111**, 12674.

44. J. A. Miwa, F. Cicoira, S. Bedwani, J. Lipton-Duffin, D. F. Perepichka, A. Rochefort and F. Rosei, *J. Phys. Chem. C*, 2008, **112**, 10214.
45. J. A. Miwa, F. Cicoira, J. Lipton-Duffin, D. F. Perepichka, C. Santato and F. Rosei, *Nanotechnology*, 2008, **19**, 424021.
46. A. Gourdon, *Eur. J. Org. Chem.*, 1998, 2797.
47. T. Zambelli, H. Tang, J. Lagoute, S. Gauthier, A. Gourdon and C. Joachim, *Chem. Phys. Lett.*, 2001, **348**, 1.
48. M. Yu, N. Kalashnyk, R. Barattin, Y. Benjalal, M. Hliwa, X. Bouju, A. Gourdon, C. Joachim, E. Lægsgaard, F. Besenbacher and T. R. Linderoth, *Chem. Commun.*, 2010, **46**, 5545.
49. C. Ludwig, B. Gompf, J. Petersen, R. Strohmaier and W. Eisenmenger, *Z. Phys. B: Condens. Matter*, 1994, **93**, 365.
50. M. Mura, F. Silly, G. A. D. Briggs, M. R. Castell and L. N. Kantorovich, *J. Phys. Chem. C*, 2009, **113**, 21840.
51. N. Kalashnyk, M. Yu, R. Barattin, Y. Benjalal, M. Hliwa, C. Joachim, E. Lægsgaard, F. Besenbacher, A. Gourdon, X. Bouju and T. R. Linderoth, *Chem. Commun.*, 2014, **50**, 10619.
52. F. Rosei, M. Schunack, P. Jiang, A. Gourdon, E. Lægsgaard, I. Stensgaard, C. Joachim and F. Besenbacher, *Science*, 2002, **296**, 328.
53. K. E. Plass, A. L. Grzesiak and A. J. Matzger, *Acc. Chem. Res.*, 2007, **40**, 287.
54. A. I. Kitaigorodskii, *Organic Chemical Crystallography*, Consultants Bureau, New York, 1959.
55. M. J. J. Coenen, D. den Boer, F. J. van den Bruele, T. Habets, K. A. A. M. Timmers, M. van der Maas, T. Khoury, D. Panduwinata, M. J. Crossley, J. R. Reimers, W. J. P. van Enckevort, B. L. M. Hendriksen, J. A. A. W. Elemans and S. Speller, *Phys. Chem. Chem. Phys.*, 2013, **15**, 12451.
56. J. S. Foster and J. E. Frommer, *Nature*, 1988, **333**, 542.
57. Y. Iwakabe, M. Hara, K. Kondo, K. Tochigi, A. Mukoh, A. Yamada, A. F. Garito and H. Sasabe, *Jpn. J. Appl. Phys*, 1991, **30**(Pt. 1), 2542.
58. D. M. Walba, F. Stevens, D. C. Parks, N. A. Clark and M. D. Wand, *Science*, 1995, **267**, 1144.
59. F. Stevens, D. J. Dyer and D. M. Walba, *Angew. Chem., Int. Ed.*, 1996, **35**, 900.
60. F. Charra and J. Cousty, *Phys. Rev. Lett.*, 1998, **80**, 1682.
61. S. De Feyter, H. Uji-i, W. Mamdouh, A. Miura, J. Zhang, P. Jonkheijm, A. P. H. J. Schenning, E. W. Meijer, Z. Chen, F. Wurthner, N. Schuurmans, J. Esch, B. L. Feringa, A. E. Dulcey, V. Percec and F. C. Schryver, *Int. J. Nanotechnol.*, 2006, **3**, 462.
62. L. Xu, L. Yang and S. Lei, *Nanoscale*, 2012, **4**, 4399.
63. A. Gesquière, P. Jonkheijm, A. P. H. J. Schenning, E. Mena-Osteritz, P. Bäuerle, S. De Feyter, F. C. De Schryver and E. W. Meijer, *J. Mater. Chem.*, 2003, **13**, 2164.
64. A. P. H. J. Schenning, P. Jonkheijm, E. Peeters and E. W. Meijer, *J. Am. Chem. Soc.*, 2001, **123**, 409.

65. A. Gesquière, P. Jonkheijm, F. J. M. Hoeben, A. P. H. J. Schenning, S. De Feyter, F. C. De Schryver and E. W. Meijer, *Nano Lett.*, 2004, **4**, 1175.
66. P. Jonkheijm, A. Miura, M. Zdanowska, F. J. M. Hoeben, S. De Feyter, A. P. H. J. Schenning, F. C. De Schryver and E. W. Meijer, *Angew. Chem., Int. Ed.*, 2004, **43**, 74.
67. A. Miura, P. Jonkheijm, S. De Feyter, A. P. H. J. Schenning, E. W. Meijer and F. C. De Schryver, *Small*, 2005, **1**, 131.
68. A. Minoia, Z. Guo, H. Xu, S. J. George, A. P. H. J. Schenning, S. De Feyter and R. Lazzaroni, *Chem. Commun.*, 2011, **47**, 10924.
69. M. Lackinger and W. M. Heckl, *Langmuir*, 2009, **25**, 11307.
70. G. R. Desiraju, *Angew. Chem., Int. Ed.*, 1995, **34**, 2311.
71. G. R. Desiraju, *Chem. Commun.*, 1997, 1475.
72. A. K. Bickerstaffe, N. P. Cheah, S. M. Clarke, J. E. Parker, A. Perdigon, L. Messe and A. Inaba, *J. Phys. Chem. B*, 2006, **110**, 5570.
73. K. S. Mali, K. Lava, K. Binnemans and S. De Feyter, *Chem. – Eur. J.*, 2010, **16**, 14447.
74. F. Tao and S. L. Bernasek, *Chem. Rev.*, 2007, **107**, 1408.
75. W. J. Saletra, H. Xu, T. Vosch, S. De Feyter and D. B. Amabilino, *CrystEngComm*, 2011, **13**, 5578.
76. R. C. Quardokus, N. A. Wasio, J. A. Christie, K. W. Henderson, R. P. Forrest, C. S. Lent, S. A. Corcelli and S. A. Kandel, *Chem. Commun.*, 2014, **50**, 10229.
77. N. A. Wasio, R. C. Quardokus, R. P. Forrest, C. S. Lent, S. A. Corcelli, J. A. Christie, K. W. Henderson and S. A. Kandel, *Nature*, 2014, **507**, 86.
78. F. A. Cotton and A. H. Reid Jr., *Acta Cryst. C*, 1985, **41**, 686.
79. T. Bauert, L. Merz, D. Bandera, M. Parschau, J. S. Siegel and K.-H. Ernst, *J. Am. Chem. Soc.*, 2009, **131**, 3460.
80. O. Guillermet, E. Niemi, S. Nagarajan, X. Bouju, D. Martrou, A. Gourdon and S. Gauthier, *Angew. Chem., Int. Ed.*, 2009, **48**, 1970.
81. K. Eichhorst-Gerner, A. Stabel, G. Moessner, D. Declerq, S. Valiyaveettil, V. Enkelmann, K. Müllen and J. P. Rabe, *Angew. Chem., Int. Ed.Engl.*, 1996, **35**, 1492.
82. S. De Feyter, A. Gesquière, M. M. Abdel-Mottaleb, P. C. M. Grim, F. C. De Schryver, C. Meiners, M. Sieffert, S. Valiyaveettil and K. Müllen, *Acc. Chem. Res.*, 2000, **33**, 520.
83. T. Yokoyama, T. Kamikado, S. Yokoyama and S. Mashiko, *J. Chem. Phys.*, 2004, **121**, 11993.
84. K. G. Nath, O. Ivasenko, J. A. Miwa, H. Dang, J. D. Wuest, A. Nanci, D. F. Perepichka and F. Rosei, *J. Am. Chem. Soc.*, 2006, **128**, 4212.
85. D. Skomski and S. L. Tait, *J. Phys. Chem. C*, 2013, **117**, 2959.
86. T. Bhinde, S. M. Clarke, T. K. Phillips, T. Arnold and J. E. Parker, *Langmuir*, 2010, **26**, 8201.
87. K. S. Mali, B. Van Averbeke, T. Bhinde, Ad.Y. Brewer, T. Arnold, R. Lazzaroni, S. M. Clarke and S. De Feyter, *ACS Nano*, 2011, **5**, 9122.
88. H. Nanjo, P. Qian, T. Yokoyama and T. M. Suzuki, *Jpn. J. Appl. Phys.*, 2003, **42**, 6560.

89. K. Kim, K. E. Plass and A. J. Matzger, *Langmuir*, 2005, **21**, 647.
90. C. Dourthe, M. Izumi, C. Garrigou-Lagrange, T. Buffeteau, B. Desbat and P. Delhaes, *J. Phys. Chem.*, 1992, **96**, 2812.
91. J. Puigmartí-Luis, V. Laukhin, A. Pérez del Pino, J. Vidal-Gancedo, C. Rovira, E. Laukhina and D. B. Amabilino, *Angew. Chem., Int. Ed.*, 2007, **46**, 238.
92. Q. Wang, M. Zecchini, J. D. Wallis, Y. Wu, J. M. Rawson and M. Pilkington, *RSC Adv.*, 2015, **5**, 40205.
93. M. Mas-Torrent and C. Rovira, *Chem. Soc. Rev.*, 2008, **37**, 827.
94. J. S. Brooks, *Chem. Soc. Rev.*, 2010, **39**, 2667.
95. M. M. S. Abdel-Mottaleb, E. Gomar-Nadal, M. Surin, H. Uji-I, W. Mamdouh, J. Veciana, V. Lemaur, C. Rovira, J. Cornil, R. Lazzaroni, D. B. Amabilino, S. De Feyter and F. C. De Schryver, *J. Mater. Chem.*, 2005, **15**, 4601.
96. J. Puigmartí-Luis, A. Minoia, H. Uji-i, C. Rovira, J. Cornil, S. De Feyter, R. Lazzaroni and D. B. Amabilino, *J. Am. Chem. Soc.*, 2006, **128**, 12602.
97. E. Gomar-Nadal, M. M. S. Abdel-Mottaleb, S. De Feyter, J. Veciana, C. Rovira, D. B. Amabilino and F. C. De Schryver, *Chem. Commun.*, 2003, 906.
98. J. Puigmartí-Luis, A. Minoia, S. Lei, V. Geskin, B. Li, R. Lazzaroni, S. De Feyter and D. B. Amabilino, *Chem. Sci.*, 2011, **2**, 1945.
99. S. Lei, J. Puigmartí-Luis, A. Minoia, M. Van der Auweraer, C. Rovira, R. Lazzaroni, D. B. Amabilino and S. De Feyter, *Chem. Commun.*, 2008, 703.
100. M. E. Stawasz, D. L. Sampson and B. A. Parkinson, *Langmuir*, 2000, **16**, 2326.
101. M. E. Stawasz and B. A. Parkinson, *Langmuir*, 2003, **19**, 10139.
102. P. F. H. Schwab, M. D. Levin and J. Michl, *Chem. Rev.*, 1999, **99**, 1863.
103. S.-L. Lee, H.-J. Wu, Y.-J. Hsu, H.-H. Chen, H.-F. Hsu and C.-h. Chen, *Chem. Commun.*, 2014, **50**, 14093.
104. J. Cornil, D. Beljonne, J. P. Calbert and J. L. Bredas, *Adv. Mater.*, 2001, **13**, 1053.
105. S.-L. Lee, M.-J. Huang, C.-h. Chen, C.-I. Wang and R.-S. Liu, *Chem. – Asian J.*, 2011, **6**, 1181.
106. R. Shokri, M.-A. Lacour, T. Jarrosson, J.-P. Lère-Porte, F. Serein-Spirau, K. Miqueu, J.-M. Sotiropoulos, F. Vonau, D. Aubel, M. Cranney, G. Reiter and L. Simon, *J. Am. Chem. Soc.*, 2013, **135**, 5693.
107. A. Nuermaimaiti, C. Bombis, M. M. Knudsen, J. R. Cramer, E. Lægsgaard, F. Besenbacher, K. V. Gothelf and T. R. Linderoth, *ACS Nano*, 2014, **8**, 8074.
108. S. Weigelt, C. Busse, L. Petersen, E. Rauls, B. Hammer, K. V. Gothelf, F. Besenbacher and T. R. Linderoth, *Nat. Mater.*, 2006, **5**, 112.
109. F. Masini, N. Kalashnyk, M. M. Knudsen, J. R. Cramer, E. Lægsgaard, F. Besenbacher, K. V. Gothelf and T. R. Linderoth, *J. Am. Chem. Soc.*, 2011, **133**, 13910.

110. T. A. Pham, F. Song, M.-T. Nguyen and M. Stöhr, *Chem. Commun.*, 2014, **50**, 14089.
111. T. T. T. Bui, S. Dahaoui, C. Lecomte, G. R. Desiraju and E. Espinosa, *Angew. Chem., Int. Ed.*, 2009, **48**, 3838.
112. J. van Esch, S. De Feyter, R. M. Kellogg, F. De Schryver and B. L. Feringa, *Chem. – Eur. J.*, 1997, **3**, 1238.
113. A. Gesquière, S. De Feyter, F. C. De Schryver, F. Schoonbeek, J. van Esch, R. M. Kellogg and B. L. Feringa, *Nano Lett.*, 2001, **1**, 201.
114. S. De Feyter, P. C. M. Grim, J. van Esch, R. M. Kellogg, B. L. Feringa and F. C. De Schryver, *J. Phys. Chem. B*, 1998, **102**, 8981.
115. S. De Feyter, M. Larsson, N. Schuurmans, B. Verkuijl, G. Zoriniants, A. Gesquière, M. M. Abdel-Mottaleb, J. van Esch, B. L. Feringa, J. van Stam and F. C. De Schryver, *Chem. – Eur. J.*, 2003, **9**, 1198.
116. S. De Feyter, M. Larsson, A. Gesquière, H. Verheyen, F. Louwet, B. Groenendaal, J. van Esch, B. L. Feringa and F. C. De Schryver, *ChemPhysChem*, 2002, **11**, 966.
117. P. Qian, H. Nanjo, T. Yokoyama, T. M. Suzuki, K. Akasaka and H. Orhui, *Chem. Commun.*, 2000, 2021.
118. D. M. Walba, F. Stevens, N. A. Clark and D. C. Parks, *Acc. Chem. Res.*, 1996, **29**, 591.
119. H. Uji-i, M. Yoshidome, J. Hobley, K. Hatanaka and H. Fukumura, *Phys. Chem. Chem. Phys.*, 2003, **5**, 4231.
120. F. J. M. Hoeben, J. Zhang, C. C. Lee, M. J. Pouderoijen, M. Wolffs, F. Wurthner, A. P. H. J. Schenning, E. W. Meijer and S. De Feyter, *Chem. – Eur. J.*, 2008, **14**, 8579.
121. S. De Feyter, A. Miura, S. Yao, Z. Chen, F. Würthner, P. Jonkheijm, A. P. H. J. Schenning, E. W. Meijer and F. C. De Schryver, *Nano Lett.*, 2005, **5**, 77.
122. H. Schönherr, M. Crego-Calama, G. J. Vancso and D. N. Reinhoudt, *Adv. Mater.*, 2004, **16**, 1416.
123. A. Langner, S. L. Tait, N. Lin, R. Chandrasekar, M. Ruben and K. Kern, *Angew. Chem., Int. Ed.*, 2008, **47**, 8835.
124. A. Langner, S. L. Tait, N. Lin, C. Rajadurai, M. Ruben and K. Kern, *Proc. Natl. Acad. Sci. U. S. A.*, 2007, **104**, 17927.
125. H. Liang, Y. He, Y. Ye, X. Xu, F. Cheng, W. Sun, X. Shao, Y. Wang, J. Li and K. Wu, *Coord. Chem. Rev.*, 2009, **253**, 2959.
126. A. Dmitriev, N. Lin, J. Weckesser, J. V. Barth and K. Kern, *J. Phys. Chem B*, 2002, **106**, 6907.
127. S. Griessl, M. Lackinger, M. Edelwirth, M. Hietschold and W. M. Heckl, *Single Mol.*, 2002, **3**, 25.
128. F. Shayeganfar and A. Rochefort, *Langmuir*, 2014, **30**, 9707.
129. Y. Ishikawa, A. Ohira, M. Sakata, C. Hirayama and M. Kunitake, *Chem. Commun.*, 2002, 2652.
130. V. V. Korolkov, S. Allen, C. J. Roberts and S. J. B. Tendler, *J. Phys. Chem. C*, 2012, **116**, 11519.
131. I. Cebula, E. F. Smith, M. C. Gimenez-Lopez, S. Yang, M. Schröder, N. R. Champness and P. H. Beton, *Phys. Chem. C*, 2013, **117**, 18381.

132. R. Gutzler, T. Sirtl, J. F. Dienstmaier, K. Mahata, W. M. Heckl, M. Schmittel and M. Lackinger, *J. Am Chem. Soc.*, 2010, **132**, 5084.
133. R. Gutzler, S. Lappe, K. Mahata, M. Schmittel, W. M. Heckl and M. Lackinger, *Chem. Commun.*, 2009, 680.
134. M. O. Blunt, J. C. Russell, M. C. Giménez-López, J. P. Garrahan, X. Lin, M. Schröder, N. R. Champness and P. H. Beton, *Science*, 2008, **322**, 1077.
135. A. Stannard, J. C. Russell, M. O. Blunt, C. Salesiotis, M. C. Giménez-López, N. Taleb, M. Schröder, N. R. Champness, J. P. Garrahan and P. H. Beton, *Nat. Chem*, 2012, **4**, 112.
136. U. Schlickum, R. Decker, F. Klappenberger, G. Zoppellaro, S. Klyatskaya, W. Auwärter, S. Neppl, K. Kern, H. Brune, M. Ruben and J. V. Barth, *J. Am. Chem. Soc.*, 2008, **130**, 11778.
137. F. Klappenberger, D. Kühne, W. Krenner, I. Silanes, A. Arnau, F. J. García de Abajo, S. Klyatskaya, M. Ruben and J. V. Barth, *Nano Lett.*, 2009, **9**, 3509.
138. D. Kühne, F. Klappenberger, R. Decker, U. Schlickum, H. Brune, S. Klyatskaya, M. Ruben and J. V. Barth, *J. Phys. Chem. C*, 2009, **113**, 17851.
139. K. Tahara, S. Furukawa, H. Uji-i, T. Uchino, T. Ichikawa, J. Zhang, W. Mamdouh, M. Sonoda, F. C. De Schryver, S. De Feyter and Y. Tobe, *J. Am. Chem. Soc.*, 2006, **128**, 16613.
140. D. Bléger, D. Kreher, F. Mathevet, A.-J. Attias, G. Schull, A. Huard, L. Douillard, C. Fiorini-Debuischert and F. Charra, *Angew. Chem., Int. Ed.*, 2007, **46**, 7404.
141. K. Tahara, C. A. Johnson II, T. Fujita, M. Sonoda, F. C. De Schryver, S. De Feyter, M. M. Haley and Y. Tobe, *Langmuir*, 2007, **23**, 10190.
142. M. O. Blunt, J. Adisoejoso, K. Tahara, K. Katayama, M. Van der Auweraer, Y. Tobe and S. De Feyter, *J. Am. Chem. Soc.*, 2013, **135**, 12068.
143. B. Li, K. Tahara, J. Adisoejoso, W. Vanderlinden, K. S. Mali, S. De Gendt, Y. Tobe and S. De Feyter, *ACS Nano*, 2013, 7, 10764.
144. X. Shen, X. Wei, P. Tan, Y. Yu, B. Yang, Z. Gong, H. Zhang, H. Lin, Y. Li, Q. Li, Y. Xie and L. Chi, *Small*, 2015, **11**, 2284.
145. S. Ahn and A. J. Matzger, *J. Am. Chem. Soc.*, 2010, **132**, 11364.
146. S. Ahn and A. J. Matzger, *J. Am. Chem. Soc.*, 2012, **134**, 3208.
147. S. Stepanow, N. Lin, D. Payer, U. Schlickum, F. Klappenberger, G. Zoppellaro, M. Ruben, H. Brune, J. V. Barth and K. Kern, *Angew. Chem., Int. Ed.*, 2007, **46**, 710.
148. Z. Shi and N. Lin, *J. Am. Chem. Soc.*, 2009, **131**, 5376.
149. P. Maksymovych, D. C. Sorescu and J. T. Yates, *Phys. Rev. Lett.*, 2006, **97**, 146103.
150. Y. He and E. Borguet, *Angew. Chem., Int. Ed.*, 2007, **46**, 6098.
151. J. A. Theobald, N. S. Oxtoby, M. A. Phillips, N. R. Champness and P. H. Beton, *Nature*, 2003, **424**, 1029.
152. M. D. Upward, P. Moriarty and P. H. Beton, *Phys. Rev. B: Condens. Matter Mater. Phys.*, 1997, **56**, R1704.
153. T. Nakayama, J. Onoe, K. Takeuchi and M. Aono, *Phys. Rev. B: Condens. Matter Mater. Phys.*, 1999, **59**, 12627.

154. M. D. Upward, P. H. Beton and P. Moriarty, *Surf. Sci.*, 1999, **441**, 21.
155. J. A. Theobald, N. S. Oxtoby, N. R. Champness, P. H. Beton and T. J. S. Dennis, *Langmuir*, 2005, **21**, 2038.
156. D. Bonifazi, H. Spillmann, A. Kiebele, M. de Wild, P. Seiler, F. Cheng, H.-J. Guntherodt, T. Jung and F. Diederich, *Angew. Chem., Int. Ed.*, 2004, **43**, 4759.
157. M. T. Cuberes, R. R. Schlittler and J. K. Gimzewski, *Appl. Phys. A: Mater. Sci. Process.*, 1998, **66**, S745.
158. S. Stepanow, M. Lingenfelder, A. Dmitriev, H. Spillmann, E. Delvigne, N. Lin, X. Deng, C. Cai, J. V. Barth and K. Kern, *Nat. Mater*, 2004, **3**, 229.
159. B. Baris, J. Jeannoutot, V. Luzet, F. Palmino, A. Rochefort and F. Chérioux, *ACS Nano*, 2012, **6**, 6905.
160. T. Balandina, K. Tahara, N. Sandig, M. O. Blunt, J. Adisoejoso, S. Lei, F. Zerbetto, Y. Tobe and S. De Feyter, *ACS Nano*, 2012, **6**, 8381.
161. S. Lei, M. Surin, K. Tahara, J. Adisoejoso, R. Lazzaroni, Y. Tobe and S. De Feyter, *Nano Lett.*, 2008, **8**, 2541.
162. K. Tahara, K. Kaneko, K. Katayama, S. Itano, C. H. Nguyen, D. D. D. Amorim, S. De Feyter and Y. Tobe, *Langmuir*, 2015, **31**, 7032.
163. S. Lei, K. Tahara, X. Feng, S. Furukawa, F. C. De Schryver, K. Müllen, Y. Tobe and S. De Feyter, *J. Am. Chem. Soc.*, 2008, **130**, 7119.
164. M. O. Blunt, J. C. Russell, M. C. Gimenez-Lopez, N. Taleb, X. Lin, M. Schroeder, N. R. Champness and P. H. Beton, *Nat. Chem*, 2011, **3**, 74.
165. H. Xu, W. J. Saletra, P. Iavicoli, B. Van Averbeke, E. Ghijsens, K. S. Mali, A. P. H. J. Schenning, D. Beljonne, R. Lazzaroni, D. B. Amabilino and S. De Feyter, *Angew Chem., Int. Ed.*, 2012, **51**, 11981.
166. L. Pasteur, *Ann. Chim.*, 1853, **38**, 437.
167. P. Iavicoli, H. Xu, T. Keszthelyi, J. Telegdi, K. Wurst, B. Van Averbekee, W. J. Saletra, A. Minoia, D. Beljonne, R. Lazzaroni, S. De Feyter and D. B. Amabilino, *Chirality*, 2012, **24**, 155.
168. F. Lortie, S. Boileau, L. Bouteiller, C. Chassenieux, B. Demé, G. Ducouret, M. Jalabert, F. Lauprêtre and P. T. Erech, *Langmuir*, 2002, **18**, 7218.
169. F. Vonau, R. Shokri, D. Aubel, L. Bouteiller, O. Guskova, J.-U. Sommer, G. Reiter and L. Simon, *Nanoscale*, 2014, **6**, 8250.
170. M. Toader, T. G. Gopakumar, P. Shukrynau and M. Hietschold, *J. Phys. Chem. C*, 2010, **114**, 21548.
171. F. Vonau, M. Linares, B. Isare, D. Aubel, M. Habar, L. Bouteiller, G. Reiter, V. Geskin, F. Zerbetto, R. Lazzaroni and L. Simon, *J. Phys. Chem. C*, 2009, **113**, 4955.
172. J. Zhang, P. Chen, B. Yuan, W. Ji, Z. Cheng and X. Qiu, *Science*, 2013, **342**, 611.
173. A. M. Sweetman, S. P. Jarvis, H. Sang, I. Lekkas, P. Rahe, Y. Wang, J. Wang, N. R. Champness, L. Kantorovich and P. Moriarty, *Nat. Commun*, 2014, **5**, 3931.
174. S. K. Hämäläinen, N. van der Heijden, J. van der Lit, S. den Hartog, P. Liljeroth and I. Swart, *Phys. Rev. Lett.*, 2014, **113**, 186102.

175. A. Gourdon, *Angew. Chem., Int. Ed.*, 2008, **47**, 6950.
176. G. Franc and A. Gourdon, *Phys. Chem. Chem. Phys.*, 2011, **13**, 14283.
177. R. Lindner and A. Kühnle, *ChemPhysChem*, 2015, **16**, 1582.
178. P. C. M. Grim, S. De Feyter, A. Gesquiere, P. Vanoppen, M. Riicker, S. Valiyaveettil, G. Moessner, K. Müllen and F. C. De Schryver, *Angew. Chem., In. Ed.*, 1997, **36**, 2601.
179. G. Wegner, *Narurforsch. B*, 1969, **24**, 824.
180. A. Deshpande, C.-H. Sham, J. M. P. Alaboson, J. M. Mullin, G. C. Schatz and M. C. Hersam, *J. Am. Chem. Soc.*, 2012, **134**, 16759.
181. Y. Okawa and M. Aono, *Nature*, 2001, **409**, 683.
182. Y. Okawa and M. Aono, *J. Chem. Phys.*, 2001, **115**, 2317.
183. R. Giridharagopal and K. F. Kelly, *J. Phys. Chem. C*, 2007, **111**, 6161.
184. Y. Okawa, M. Akai-Kasaya, Y. Kuwahara, S. K. Mandal and M. Aono, *Nanoscale*, 2012, **4**, 3013.
185. D. Skomski, C. D. Tempas, K. A. Smith and S. L. Tait, *J. Am. Chem. Soc.*, 2014, **136**, 9862.
186. K. Diller, F. Klappenberger, M. Marschall, K. Hermann, A. Nefedov, C. Wöll and J. V. Barth, *J. Chem. Phys.*, 2012, **136**, 014705.
187. Y. Li, J. Xiao, T. E. Shubina, M. Chen, Z. Shi, M. Schmid, H. P. Steinrück, J. M. Gottfried and N. Lin, *J. Am. Chem. Soc.*, 2012, **134**, 6401.
188. J. Nowakowski, C. Wackerlin, J. Girovsky, D. Siewert, T. A. Jung and N. Ballav, *Chem. Commun.*, 2013, **49**, 2347.
189. H. Marbach, *Acc. Chem. Res.*, 2015, **48**, 2649.
190. M. In't Veld, P. Iavicoli, S. Haq, D. B. Amabilino and R. Raval, *Chem. Commun.*, 2008, 1536.
191. L. Grill, M. Dyer, L. Lafferentz, M. Persson, M. V. Peters and S. Hecht, *Nat. Nanotechnol.*, 2007, **2**, 687.
192. L. Dong, P. N. Liu and N. Lin, *Acc. Chem. Res.*, 2015, **48**, 2765.
193. S. Haq, F. Hanke, M. S. Dyer, M. Persson, P. Iavicoli, D. B. Amabilino and R. Raval, *J. Am. Chem. Soc.*, 2011, **133**, 12031.
194. D. den Boer, M. Li, T. Habets, P. Iavicoli, A. E. Rowan, R. J. M. Nolte, S. Speller, D. B. Amabilino, S. De Feyter and J. A. A. W. Elemans, *Nat. Chem.*, 2013, **5**, 621.
195. B. Hulsken, R. Van Hameren, J. W. Gerritsen, T. Khoury, P. Thordarson, M. J. Crossley, A. E. Rowan, R. J. M. Nolte, J. A. A. W. Elemans and S. Speller, *Nat. Nanotechnol.*, 2007, **2**, 285.
196. S. Fabris, S. Stepanow, N. Lin, P. Gambardella, A. Dmitriev, J. Honolka, S. Baroni and K. Kern, *Nano Lett.*, 2011, **11**, 5414.
197. M. Li, D. den Boer, P. Iavicoli, J. Adisoejoso, H. Uji-i, M. Van der Auweraer, D. B. Amabilino, J. A. A. W. Elemans and S. De Feyter, *J. Am. Chem. Soc.*, 2014, **136**, 17418.
198. H. E. Hoster, M. Roos, A. Breitruck, C. Meier, K. Tonigold, T. Waldmann, U. Ziener and R. J. Behm, *Langmuir*, 2007, **23**, 11570.
199. T. Waldmann, D. Künzel, H. E. Hoster, A. Groß and R. J. Behm, *J. Am. Chem. Soc.*, 2012, **134**, 8817.

CHAPTER 5

Chemisorbed Layers at Interfaces

5.1 Introduction

The formation of a strong chemical bond in-between a surface and a molecule's reactive head group leads to robust monolayer systems (Figure 5.1). The surface—which is certainly reactive in order to enter into bond forming—is often passivated by the adsorbate. The chemical bonds that are formed are more often than not covalent, but the family of chemisorbed self-assembled monolayers often includes systems where an ionic bond exists between the surface and the head group of the adsorbate.[1] Perhaps the most habitual orientation of the adsorbates in the layers is with the long axis of the molecules quasi-perpendicular to the initial surface plane. This arrangement and the exact angle to the perpendicular are determined by the geometry around the adsorbate–adsorbent bond(s). Despite this strong (and often essentially irreversible) connection, supramolecular chemistry plays a big role in the formation of the layers: the interaction between the sides of the molecules can influence dramatically the structure and properties of these ultrathin films. This interaction (shown in Figure 5.1 in cartoon form for a specific non-covalent bond) would propagate in one or two dimensions in a plane parallel to the surface.

Kinetic effects are important in determining the structures of chemisorbed layers, and the rate of chemisorption—which can be influenced using the temperature of the deposition—controls film quality. Also, in terms of their construction, chemisorbed monolayers exhibit self-limiting growth: the functional head group of the adsorbate will react with the surface until a complete layer is formed. In the initial step, this means that after non-specific physisorption, the system must overcome the activation energy for the bond formation between the surface and the adsorbate

Monographs in Supramolecular Chemistry No. 19
Supramolecular Chemistry at Surfaces
By David B. Amabilino

Published by the Royal Society of Chemistry, www.rsc.org

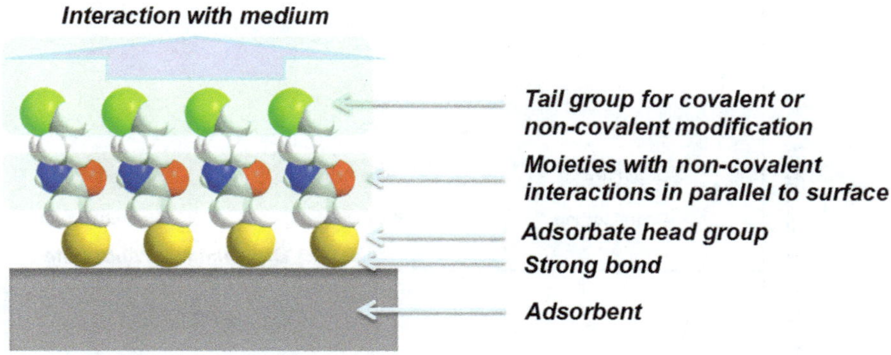

Figure 5.1 A cartoon view of a chemisorbed layer, where non-covalent interactions can be important parallel to the plane of the surface or perpendicular to it.

head group ($E^{\ddagger}_{\text{phys-chem}}$, Figure 5.2). Clearly, chemisorption energy is more favourable than the lowest possible energy of physisorption. No matter how much more of the same material arrives at the newly formed surface, if the tail group does not react covalently with the head group, a robust layer will not grow. Any spuriously physisorbed material can simply be removed by cleaning (sublimation, washing, *etc.*). On the other hand, a crucial advantage of chemisorbed systems is that it is easy to create multicomponent monolayers, because of the dominant role of the bond linking substrate to adsorbate and therefore the kinetics in the formation of the film, particularly at high concentration of adsorbates and fast reaction rate of formation for the covalent-like bond.

In all SAMs formed through chemisorption, there are many processes acting during film growth.[2] In the case of a solution process, the adsorpts move to the substrate–liquid interface through both convective and diffusive transport. The subsequent attachment to the surface may be controlled by diffusion, adsorption rate, or a collusion of the two. Up to here, the phenomena are similar to adsorption from a vapour; although the surface tension (lowered by SAM formation) is somewhat more important than the surface coverage, the two are linked by the equation of state for the system.[3] The phase diagram can contain the following states: (i) a surface "gas" phase where adsorbates are randomly distributed over the interface and have high mobility and low density; (ii) a disordered phase of medium density containing disordered molecules or ones that lie parallel to the interface, and (iii) a phase in which the molecules are conformationally ordered, densely packed, and quasi-perpendicular to the interface plane. Other phases can exist and are largely determined by the chemical structure of the adsorpt. In addition, the phases that are accessible to a system depend on transport on the surface, the rate of bond formation, the mobility of surface atoms and the relative importance of intermolecular bonding, the main factor that concerns us here. All these parameters will in turn depend on the characteristics of the molecule, the medium (solvent or vacuum), the surface, and

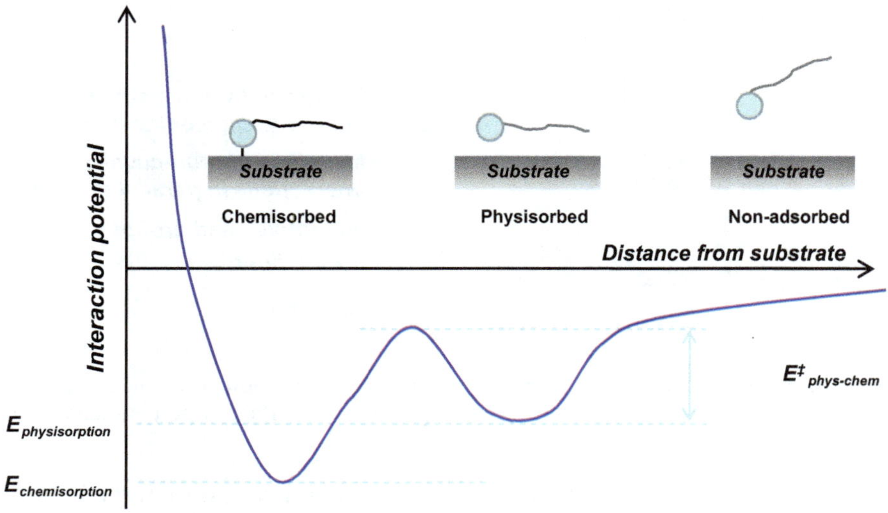

Figure 5.2 A representative plot of the chemical interaction potential between an adsorpt and a surface as a function of the distance between the two. Two minima are represented corresponding to optimal physisorption, when most of the area of the adsorbate is in contact with the substrate, and chemisorption, when a covalent-like bond is formed between the two. Notice that an activation barrier must be passed to go from a physisorbed state to the chemisorbed one, which corresponds to any bond breaking in either component. Importantly, the level of this energy barrier can be affected by non-covalent interactions with neighbouring physisorbed molecules, leading to cooperative processes characteristic of self-assembly.

the temperature at which the adsorption takes place. In this sense, the monolayers can be looked at in terms of a phase diagram (Figure 5.3) where at low coverages and temperatures the molecules can organise into striped phases.[4] This state, largely thermodynamic, arises from the side-on interactions between molecules that can lead to the formation of lamellae and the head groups react with the surface to anchor the situation. At the other extreme, at high coverage and low temperature, dense packing is favoured with the head groups in a closely-packed arrangement, with ordered chains coming off the head group, and a low proportion of defects between domains. The ordered phases would show a melting point,[5] just as any molecular solid phase, where chains become mobile even if the head groups are chemisorbed.

The importance of the rate of bond forming to produce the chemisorbed molecule means that the position of the phase diagram on the temperature scale, and also the progression of one point in the phase diagram to another over time depend dramatically on deposition conditions. Short exposure times of adsorbate to substrate result in low coverage and low overall order (though local order may be present).

The formation of SAMs in general has based its foundation on the study of the adsorption of simple aliphatic compounds comprising a chemisorbing

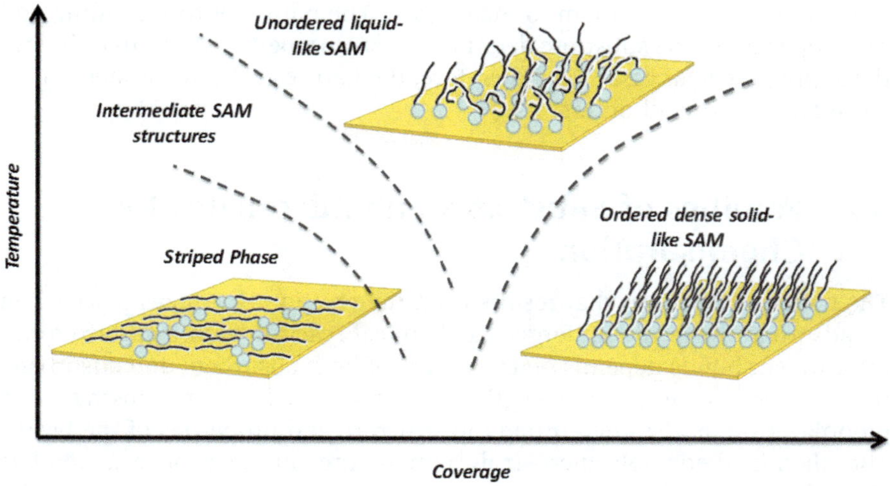

Figure 5.3 A hypothetical phase diagram for SAMs on a flat surface, showing the relationship between coverage and temperature. At low temperature and coverage, a striped phase can be formed, while at high coverage and low temperature, ordered phases are possible with the long axes of the molecules perpendicular to the interface. At higher temperatures and intermediate coverage, the layers are liquid-like, with disorder and mobility of the adsorbate's chains possible.

head group bearing alkyl chains, where the packing of the oligomethylene segments that are oriented at a high angle to the interface plane favours order that depends on the length of these units.[6] Changing this alkyl group to others clearly affects the two-dimensional packing in the layer, and many approaches rely on aliphatic spacers between the head group and functional unit to achieve dense packing in layers. An important alternative supramolecular effect that can influence the packing of all kinds of chemisorbed layers is the use of hydrogen bonding to increase adsorbate–adsorbate interactions—prior to adsorption, during adsorption, and after adsorption—whereby order can be enhanced. This effect was demonstrated for SAMs on gold where the thiol was spaced by an ethylene group to the amide function that was N-substituted with alkyl chains of varying length in order to probe the relative importance of each interaction.[7] The amide mid-layers are ordered in every case, as demonstrated through thorough XPS, contact angle goniometry and reflective IR spectroscopy. They are the dominant structure in the phase diagram, with a striped phase only being accessible through electrochemical manipulation of the monolayer.[8] Clearly, the position of the amide group has a profound effect on monolayer structure, and inclusion of the unit near the farthest point of an alkyl chain from the anchoring thiol group also results in hydrogen bonding between adsorbates with a modified angle of their long axis with respect to the interface.[9]

The synthetic routes to molecules provides huge possibilities for adsorbate design, but for chemisorption all are limited by requiring a functional group that will anchor them to the surface. The nature of this bonding and the properties of the substrate determine many properties that can influence the supramolecular assembly. Therefore, the nature of the substrates will be considered before all others.

5.2 Families of Substrates and Adsorbates for Chemisorption

The nature of the layer that results from the reaction that takes place when an adsorbate molecule undergoes a chemical reaction with a surface forming covalent bonds depends on the nature of both adsorbent and adsorbate: the relationship and the strength of interactions between substrate and monolayer themselves determine the structure and properties of the layers. The chemisorbed molecules—truly part of the interface once a bond is formed—change the surface properties because of their characteristics (hydrophilic or hydrophobic, for example). The resulting monolayers are robust because of the strength of the chemical bonds between the surface atoms of the substrate and the adsorbate. The general stability of these systems means that their applications are potentially multiple. But one has to choose the right adsorbates for the right surfaces for these applications to be realised. The synthetic chemistry of SAM growth has to be controlled.

In general, the relatively flat and ordered surfaces that are the subject of this monograph can be split up into four groups for the more common chemisorbed systems (Figure 5.4). Metals comprise the group of substrates that permit the most precise studies of chemisorbed structures. Monocrystalline faces can be purchased and cleaned in a more or less routine manner. They form bonds with soft atoms in general, although the harder metals also bind oxygen-headed adsorbates. The mobility of the adsorbates on the surface can be appreciable, as one could envisage that either the organic–metal bond could be mobile, or the softer metals could provide motion with their own atoms. This mobility can lead to extremely ordered monolayers. Inorganic semiconductors—silicon being the most common—form very specific bonds with carbon compounds through addition to alkenes to the hydrogenated surface, for example. This chemistry has been reviewed,[10,11] and while of huge interest because of the substrate, the supramolecular exploration of these films is not widespread. The instability of the surface to ambient conditions is surely the reason for this, and the oxide layer that is formed on the element is a far more common surface to be used for SAMs, where silyl ethers and chlorides are most usually used to chemisorb molecules. These same adsorbates are also used for glass, obviously. Finally, the other oxides, many extremely important for electronics and catalysis, where linking through carboxylates, phosphonates, sulfonates, and catecholates, bind well.[12] On all these inorganic materials, the surfaces comprise very specific and directional bonding, meaning that once a bond is formed to them, it can be essentially irreversible, and the kinetics of the SAM

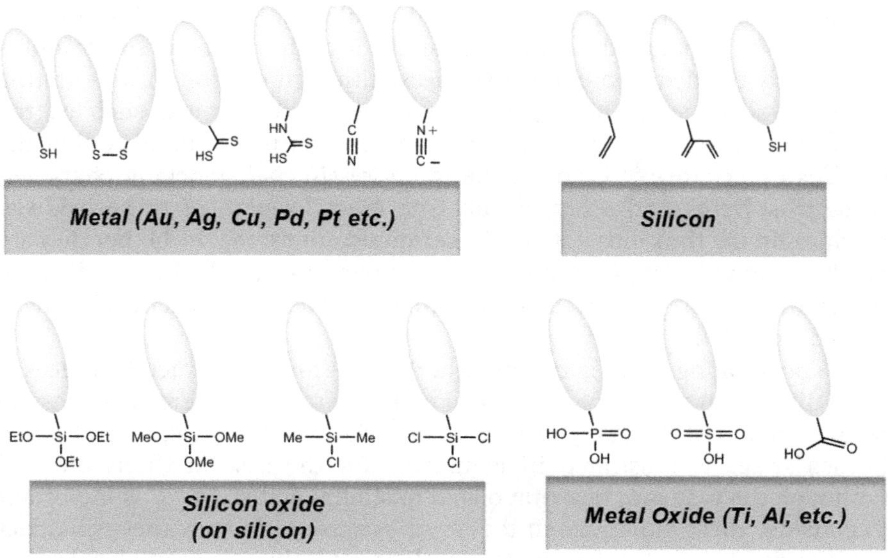

Figure 5.4 Pairing adsorbents and adsorpts for the more common chemisorbed monolayers. Silicon oxide is separated from the other oxides because its use as a covering to elemental silicon makes it stand apart in terms of chemistry and applications.

growth truly dominate. For this reason, the nature of the films that are formed is quite different, and thus each group of surfaces shall be discussed in turn, beginning with the metals.

5.3 Self-assembled Monolayers on Metals

Metal surfaces are fascinating places to look at the formation of self-assembled monolayers and the formation of these systems can provide applications: the adsorption of molecules to metals is perhaps historically of greatest importance because the layers formed can prevent the corrosion of the surfaces if they are sensitive to the environmental conditions under which they are used,[13–15] and can also prevent biofouling (where biological material grows off a surface).[16] From a purely scientific perspective, metals are useful adsorbents because they can be bought and treated to generate flat crystal planes on their surface, allowing precise determination of surface structure and properties.[17–20] They are conducting, a feature that makes them ideal for study by scanning tunnelling microscopy, which, as we have seen, is the optimum imaging tool for this kind of setting. The formation of metal complexes—either in the formation of coordination-type complexes with metal–heteroatom bonds, or organometallic complexes with carbon–metal bonds—shows remarkable effects of the surfaces not necessarily perceived in homogeneous media.[21]

The binding of thiols, thioethers and disulfides and related adsorbates has been widely studied, and it is generally accepted that the sulfur atoms are directly bonded to the metal surfaces, although caution in interpreting chemical structure is necessary. Under certain conditions, such as the presence of ozone in the laboratory, Raman experiments indicate that the sulfides can oxidise.[22] Even so, the layers retain their blocking characteristics to solvents and other agents. The exact location of these oxidised products in the films has not been determined, for example whether they are at phase boundaries. Routine XPS measurements should be sufficient to determine any impurities of this nature, or even non-specifically physisorbed material.

The positioning of thiolate (and other) head group(s) on metal surfaces is often difficult to ascertain with precision. Even assuming there is no massive reconstruction of the (111) surface, to take the simplest case, several possibilities arise, as illustrated in Figure 5.5 for the case of thiols on gold (although the case may be extrapolated to similar systems[23]). It is important to consider these situations in the layer, because ultimately they will affect the packing of molecules on the surface and their supramolecular behaviour. The most frequently observed situation for thiolates on gold (111) is that where the sulfur atom occupies the "hollow" in-between three gold atoms.[24] It could also sit on top of a single gold atom, either in the monolayer or as an adatom, or in a hollow defect left by an ejected gold atom. Another possibility could be the adsorption of a dithiol (arising from oxidation of the thiol precursor) across two adjacent gold atoms. Finally, a gold adatom could be "clamped" by two thiolates in a kind of "gold(I) complex" (though this cannot be "purely" gold(I) because of its intimate contact with the underlying metal). This kind of situation has actually been observed by STM.[25]

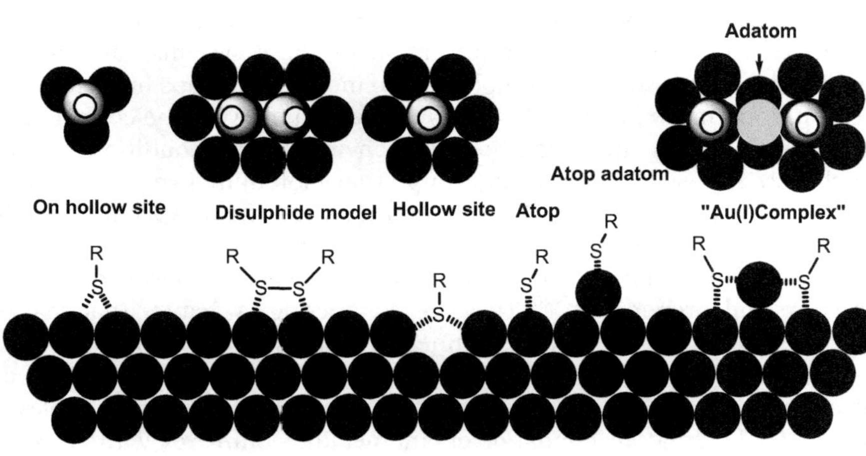

Figure 5.5 Hypothetically possible adsorption sites of thiolates on gold, also applicable to other metal–adsorbate systems.

The supramolecular chemistry of these chemisorbed layers is at play in their formation as a result of the side-on interactions between the adsorbates, usually in a plane parallel to the surface of the adsorbent. Even all the fundamental work on alkyl-derived adsorbates—the only other functionality being the reactive head group—shows that they owe their structure to van der Waals interactions for the order in the monolayers. The van der Waals interactions are largely responsible for the ''self-assembly''. Be that as it may, the monolayers can be very sensitive to the medium in which they are deposited and the concentration, characteristics of a chemical bond forming process. For example, the solvent plays an important role. When 1,4-benzenedimethanethiol is chemisorbed onto Au (111) from hexane or water, very different outcomes are observed (Figure 5.6).[26] At low concentration in water, scanning tunnelling microscopy indicates that the dithiols adsorb in a lying down arrangement with both sulfur atoms adsorbed to the metal surface. The situation in hexane is completely different: initially, the solvent physisorbs to the surface, and then the dithiol replaces it. At high concentration in hexane, a complete monolayer is formed quickly on the gold and XPS measurements indicated that there were free thiol groups demonstrating a standing-up arrangement of the molecule on the metal. Interestingly, the lying down arrangement in the SAM formed in water can be modified by exposure of the pre-dried layer to a hexane solution of the dithiol, where an upright monolayer is achieved. It is clear that if this kind of layer is to be used for further adsorption through the thiol groups for whatever purpose, the standing-up arrangement is required and this fundamental observation is important and perhaps general.

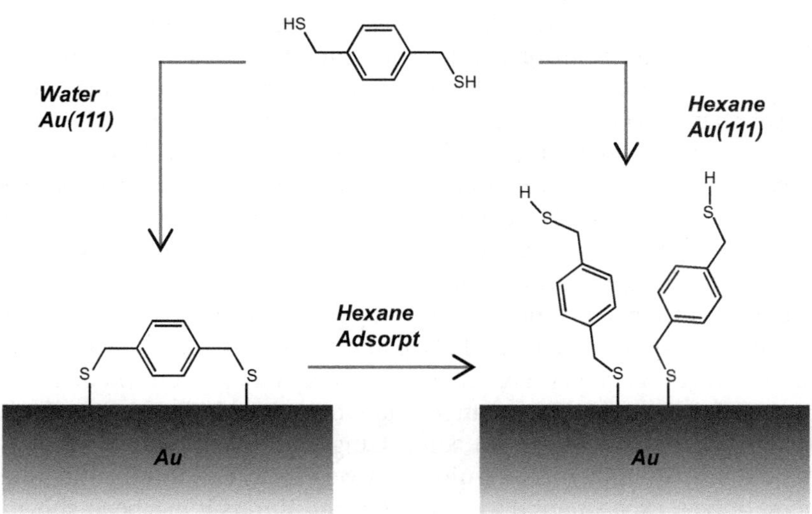

Figure 5.6 A schematic view of chemisorption of 1,4-benzenedimethanethiol onto gold (111) from either water or hexane, and the conversion of the lying down structure generated in water to the standing-up arrangement.

The surface atoms of many metals are not immobile: the structure of metals at surfaces can change upon formation of layers on top of them. This phenomenon is termed *reconstruction.*[27] These reconstructions have quite profound effects when attempting to explain the packing of the molecules in monolayers on the surfaces.[28] Clearly if one assumes that a metal surface is unmodified upon chemisorption of a molecule, and reconstruction does actually take place, the explanation of structure at the atomic level could well be incorrect. The density of surface metal atoms can change dramatically, as seen for crystalline surfaces of the so called "coinage" metals (copper, gold and silver) that have been proven to reconstruct upon the adsorption of alkane thiolates.[29] Through a combination of low energy electron diffraction, scanning tunnelling microscopy and synchrotron radiation-based normal incidence X-ray standing wave measurements it has been shown that the three metals' surfaces rearrange in different ways when a sulfur atom chemisorbs. It was pointed out in this work that this rearrangement is bound to affect or be affected by the inter-adsorbate interactions in the layer.

Given the quite different behaviour of different metals in SAM formation and characteristics of the films,[30] the subsequent sub-sections will discuss chemisorbed supramolecular systems on the different elemental surfaces, starting with the most widely used surface of all: gold.

5.3.1 Self-assembled Monolayers on Gold

Gold is perhaps THE most popular surface for the study of self-assembly and supramolecular chemistry at surfaces in general.[31] But why gold? The reason for this popularity is that the gold surface—particularly the (111) terminated surface—(1) can be purchased and prepared easily, (2) is essentially inert to the atmosphere (though this factor is not entirely innocent) and (3) is not toxic, giving it potential for applications such as sensors in biological systems. Furthermore, the SAMs that form are generally quite stable[32] and the physical characteristics of the gold surface make it very amenable to scanning probe microscopy in general[33] but also to optical techniques such as surface plasmon resonance measurements. The metal can also be used to coat the quartz in a microbalance. Despite their widespread use, there remains great debate and doubt concerning precise details of the structure in all but the most simple and deeply studied cases. The reason for this debate is the great number of possible rearrangements that even the most symmetrical surface of gold (the (111) plane) can undergo.[34] Some of these possible rearrangements are shown as snapshots from models[35–42] in Figure 5.7 for the case of methanethiolate on gold. This plethora of models for the relatively simple system comprising methanethiolate on gold demonstrates the difficulty in discerning how we expect this family of ligands to adhere. In fact, experimentally under UHV conditions it has been shown (on the same surface) by normal incidence X-ray standing waves and S 1s scanned-energy mode photoelectron diffraction that this particular thiolate is adsorbed atop a gold adatom in the hollow of the (111) surface, and less

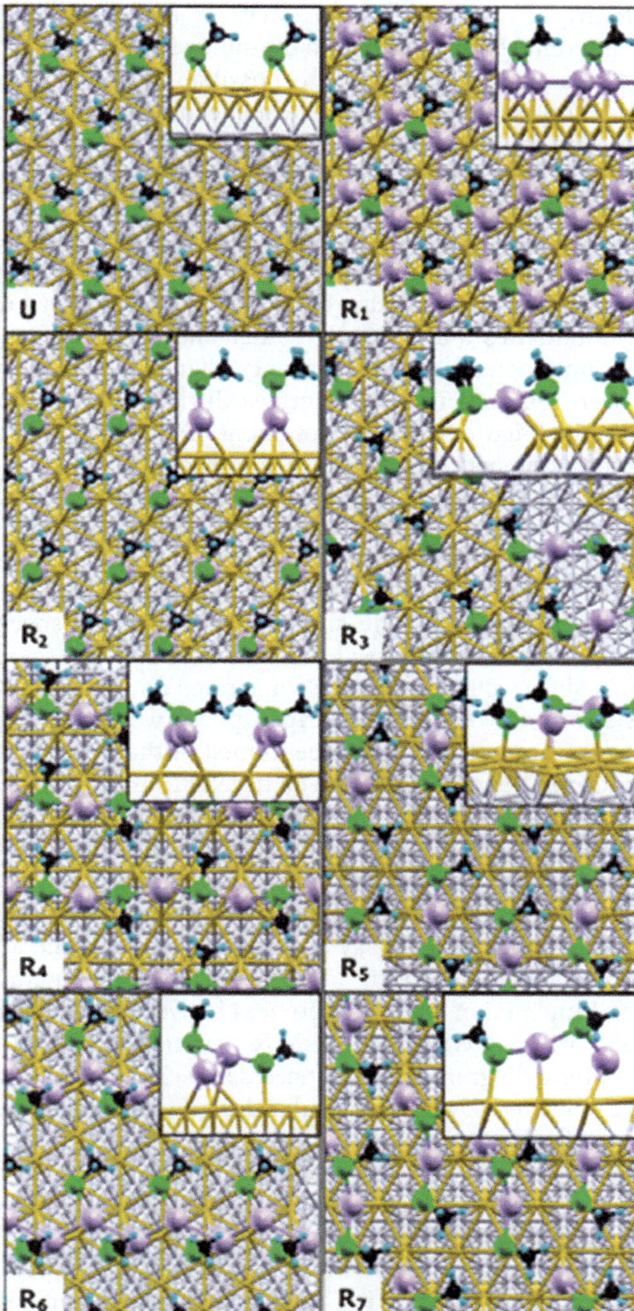

Figure 5.7 Models showing $\sqrt{3}\times\sqrt{3}$ R30° and c(4×2) methanethiolate lattices for the unreconstructed (U) gold (111) surface, and for seven possible reconstructions where the thiolates bridge adjacent atoms (R_1), stand atop metal adatoms (R_2), bridge a removed surface atom (R_3), locate around two adatoms (R_4, R_5, R_6 and R_7). The insets show views along the plane of the interface. Reproduced from ref. 34 with permission from the Royal Society of Chemistry.

than 10% (if any) of other species.[43] However, for other experimental conditions and different chains attached to the sulfur atom, these reconstructed surface structures may be, and surely are, possible. It has been proposed that intermolecular forces between adsorbate molecules move the sulfur from this atop site, and cause reconstruction of the metal's top layer.[44]

Monolayers on gold are mainly constructed through the formation of gold–sulfur bonds, when either thiols, thioethers or disulfides are exposed to the surface of the metal, either in vacuum or in different solutions mainly with comparable results.[45] For the longer alkyl chains at high coverage, standing-up phases of ordered adsorbate molecules are observed,[46] with all the domain boundaries and step edges that one might expect on such a surface, occasional defects, but also vacancy islands, where a layer of gold atoms has been removed from the surface during the adsorption of the sulfur amphiphile (Figure 5.8). The movement of gold atoms is apparent in these and other systems (styrene as an adsorbant at 80 K) as shown by STM,[47] and in the vacancy islands on gold, Ostwald ripening takes place at the edges showing the dynamics of these layers over time.[48] Apart from these defects, the layers can contain non-chemisorbed adsorpts, because of the difficulty of rinsing with poor solvents, such as ethanol, especially in the case of very hydrophobic materials. This situation can be probed specifically with XPS, focussing on the sulfur 2p peak.[49]

Alkanethiols show well defined phase diagrams on the gold (111) surface when deposited under UHV.[50] The striped phase where the molecules lie parallel to the surface in lamellae can be formed by thermal treatment of the more favoured upright monolayer.[51] For decanethiolate, four ordered and one disordered phase have been identified by STM.[52] In-between the striped phases and the quasi-upright monolayer, phases exist where the tilt angle is intermediate in value,[53] and even where one layer apparently lies flat on the gold while the other is bent over a neighbouring alkyl row.[54] The ability of the alkyl thiolates to tilt and overlay is partially a result of the ability of the aliphatic part of the molecule to adopt several side-on interactions with relatively small interaction energy differences between them, because of the non-directional van der Waals interactions that compact them. This complexity in the phase diagram does not necessarily exist for other adsorbates, but has to be considered. Supramolecular design can lead to more specific outcomes.

Polymorphism can be observed in ordered SAMs of thiolates on gold, as demonstrated for terphenylthiol.[55] After adsorption of the thiol to the gold (111) surface in ethanol for 18 hours and then annealing the sample in pure ethanol at 40 °C for around half an hour, STM shows monolayers with quite large ordered domains (over 20 nm in width at least), and two different packing patterns (Figure 5.9). The polymorphs are most easily recognised by their different contrast in STM and the directions that their rows follow on the surface, but also the size of the features of the α phase are smaller than those of the β phase, which are more elliptical. The packing density of the β phase is about 30% of the α. These patterns had actually been inferred

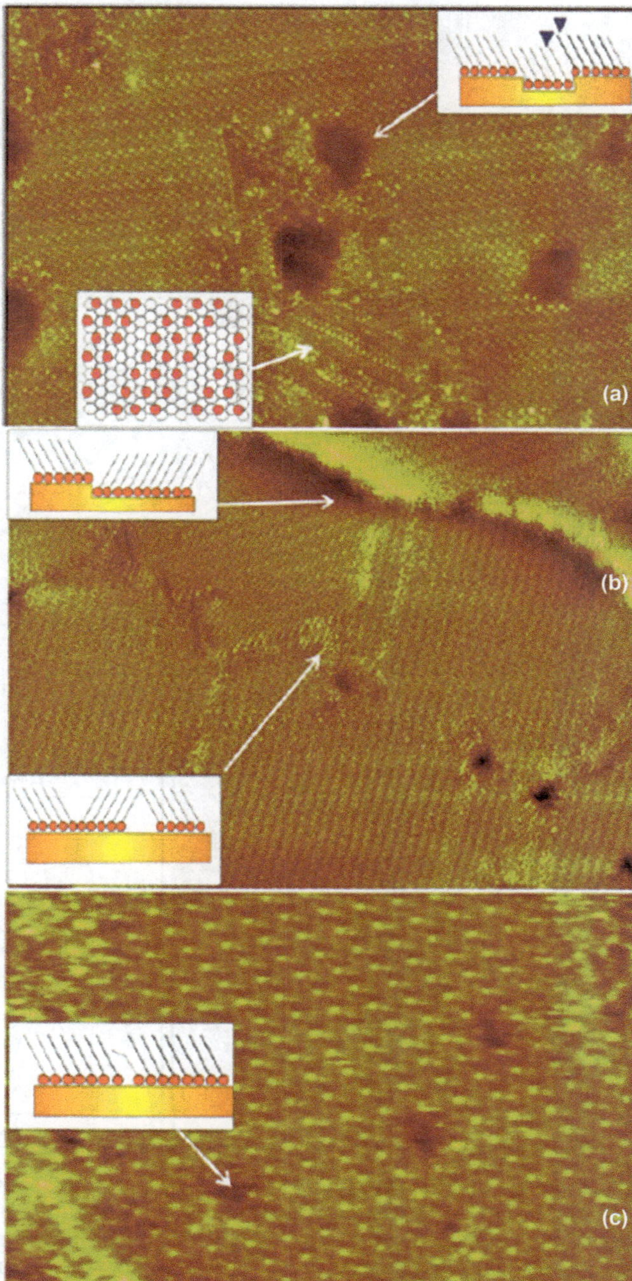

Figure 5.8 Images from an STM illustrating some of the defect types in alkanethiolate SAMs on gold (111), each with a cartoon showing the type in the inset: (a) shows missing rows and vacancy islands (48 nm×32 nm), (b) step edges and frontiers between domains (46 nm×33 nm) (c) adsorbate defects (so-called "pinholes") (19 nm×12 nm).
Reproduced from ref. 34 with permission from the Royal Society of Chemistry.

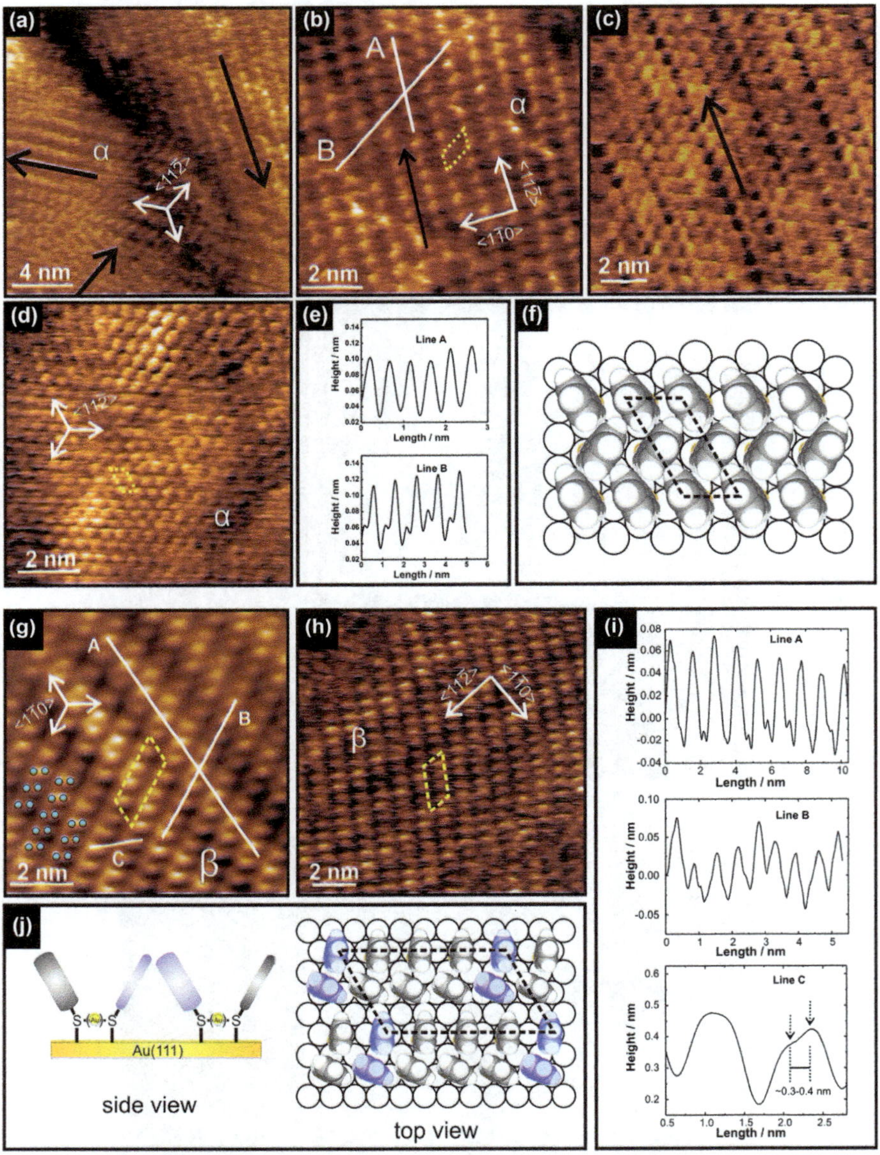

Figure 5.9 Polymorphic phases of terphenylthiol on gold (111) assembled in ethanol and annealed in the same solvent: STM images (a)–(d) show the α phase, with black arrows showing the domain directions and white arrows showing crystallographic directions and yellow parallelograms the unit cells. Graph (e) shows the line profiles indicated in (b), and (f) shows a model of this close-packed structure. STM images (g) and (h) show the β phase, (i) the line profiles indicated in (h), and (j) a model of the adsorption geometry of the thiolate head groups around a gold adatom and a top view of the packing.
Reprinted with permission from ref. 55. Copyright 2013 American Chemical Society.

(albeit with less precision) in studies on the same compound deposited from either ethanol or dichloromethane,[56] but in general the initial reports on non-annealed samples had domains of very small size and order.[57,58] The tilt angles for the two phases, based on modelling the packing, were 13° and 33–49° with respect to the perpendicular to the interface plane, respectively. The high tilt angle of the terphenyl was interpreted as being either the formation of a disulfide link on the surface or the binding of a gold adatom by two thiolates as shown in Figure 5.9. The STM images do infer the tilt of the unit, but do not allow a definitive assignment of this structure though. The coexistence of the phases makes a spectroscopic picture of the surface difficult to achieve though, because these methods measure a far larger area than the STM and therefore observe both the coexisting phases.

For the connection of functional units to the surface of gold, it has been shown that both aliphatic and aromatic groups can give a degree of order in the SAMs. However, some units do not form ordered layers, either because of lack of a suitable commensurate layer for matching the gold lattice or because the interactions between groups in the adsorbate are not appropriate to favour two-dimensional packing. The spacer between the functional group and the metal (or other surface indeed) determines to great measure the order that is observed. For example, thiophene units are of interest for their conducting properties and a control over their orientation is of interest. A series of dithiols studied on gold (111) under UHV conditions showed that the two with mercaptomethylthienyl moieties at their extremes formed hexagonal phases where one thiol was attached to the metal surface while the other extended away from it.[59] The origin of this structure apparently awaits an explanation, but the empirical observation that the mercaptomethylthienyl group favours an upright conformation is an interesting one. On the other hand, thiophene groups spaced from a thiol attachment group by either two, six or twelve methylene units all form stable monolayers on gold with the thiophene moiety directed away from the interface.[60] The compound with the longer alkyl chain is the most densely-packed and well-ordered SAM, thanks to the ordering of the aliphatic segment. The thickness of the SAMs determined by ellipsometry is 6.7, 10.7, and 17.9 Å for the compounds with two, six and twelve methylene groups, respectively, implying long-axis tilt angles of 42, 41, and 35° assuming an all-*trans* alkyl conformation. These layers are stable up to 100 °C, but decompose by desorption at about 150 °C. A subsequent exploration of the electrochemical stability of this kind of monolayer showed that for reversibility it was necessary to have all the carbon positions of the thiophene units substituted.[61]

The use of hydrogen bonding in the plane of the gold surface has proven effective for the formation of structurally-ordered SAMs.[62] Ethanol solutions of the thiol that is a precursor to the thiolate in the monolayer (Figure 5.10) were used to form the films over a period of 24 hours. Infrared spectroscopy showed that the alkyl chains ordered and that there is inter-adsorbate hydrogen bonding through the amide fragments, and capacitance measurements proved a densely-packed film. The thickness of the film was

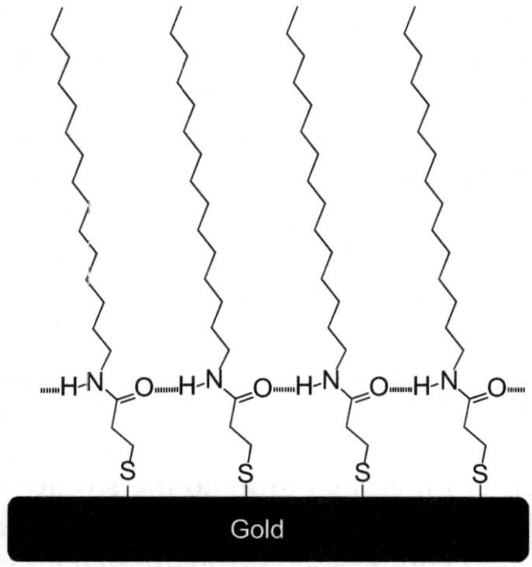

Figure 5.10 The hydrogen-bonded chains that form on gold upon adsorption of an amide-containing aliphatic thiol.

measured with XPS (the Au($4f_{7/2}$) photoelectrons are attenuated by the organic film) giving a value of about 28 Å, corresponding to a quite upright orientation in contrast to the higher tilt shown by purely alkyl thiols.[63] This approach to forming ordered and stable layers has become quite general.[64,65] The incorporation of oligo(ethylene oxide) segments has proven of particular interest for the use of SAMs in preventing non-specific biomolecule adsorption, and it has been shown that an amide group in the adsorbates leads to an increase of the helical conformer of this segment in the monolayer.[66]

Under UHV conditions, the assembly of molecules on gold can yield great insight into a number of matters. The adsorption of the amino acid cysteine onto the gold (110) surface is a wonderful example of chiral self-assembly at an interface. Pairs of molecules self-assemble when they are sublimed onto the gold.[67] When the racemic modification of cysteine was adsorbed onto the surface, homochiral dimers were generated, in part because a restructuring of the metal takes place. Density functional theory simulations indicated that the selective creation of three bonds between each cysteine molecule and the surface are responsible for the homochiral dimer's stability. This feature made the authors appeal to the three-point model for stereoselective binding in biological systems as an explanation for the remarkable selectivity here. Deposition of the racemic modification also generated an excess of one of the mirror image dimers,[68] although the reasons for this effect are as yet unclear. The amount of material on the surface is minute, and an undetectable scalemic mixture cannot be ruled out, although in principle the molecule-by-molecule counting available through STM could resolve this conundrum. When more material is deposited, high resolution STM indicates that there is an apparent spontaneous resolution of the bigger clusters as well.[69]

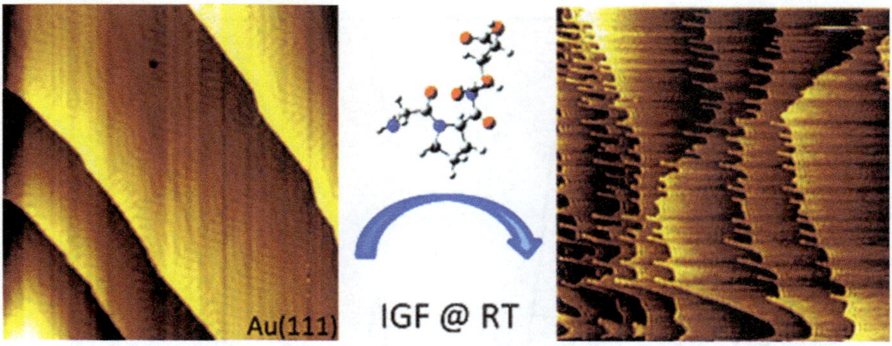

Figure 5.11 STM images showing a gold (111) surface before (left) and after (right) adsorption of a tripeptide (insulin growth factor) and reconstruction of the surface at room temperature.
Reprinted with permission from ref. 70. Copyright 2012 American Chemical Society.

One of the most remarkable restructuring effects seen on the gold (111) surface arises from the adsorption of a heterotripeptide insulin growth factor (IGF, gly-pro-glu).[70] When this compound adsorbs on the metal at room temperature, reconstruction initiates not only at kinks and steps edges, but even within defect-free terraces of the surface, generating ten nanometre-wide fingers extending across the surface (Figure 5.11). The directionality of the reconstruction is apparently a result of very weak interactions of the STM tip with the surface, indicating a very weak cohesion of the metal's top layer. The tripeptide exists mainly as a zwitterion on the surface (as evidenced by RAIRS and XPS), and this feature enables a very high adsorption energy: its interaction with a gold atom from the surface was calculated to be stronger than the cohesion energy of the surface metal atoms. This situation leads to the reconstruction, but the origins of the precise mechanism are yet to be detailed. It is intriguing to think of how one might devise systems where surfaces could be reconstructed specifically in this manner as a result of adsorbate-facilitated mass transport.

The previous example of reconstruction implied a strong non-covalent interaction (aided electrostatically) between the carboxylate function of the peptide and the gold surface, and the bonding between hard electronegative atoms and the gold surface does indeed give rise to SAMs for cucurbiturils, cyclic disk-shaped molecules whose faces are covered with oxygen atoms from carbonyl groups.[71] After immersion of gold into aqueous saturated solutions of the cycles, there was clear evidence for the formation of monolayers, perhaps the most impelling data coming from the infrared reflection absorption spectroscopy (IRRAS) data (Figure 5.12). The strong signal seen for the carbonyl peaks in the ring and the selection rules of the IRRAS technique allow affirmation that one face of the macrocycle is in contact with the gold surface. XPS data confirmed this idea, and a comparison with the data obtained from deposition experiments on silicon

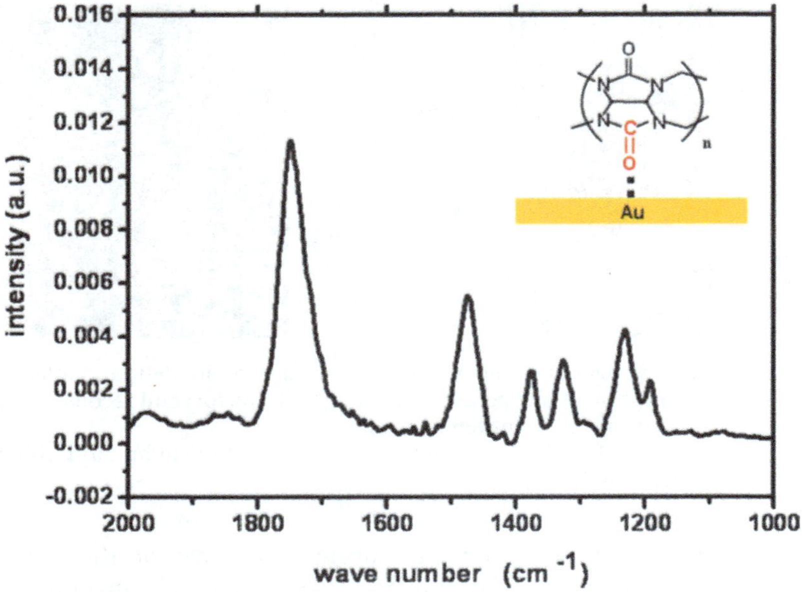

Figure 5.12 Infrared reflection absorption spectroscopy of a self-assembled monolayer or cucurbit[6]uril on gold, showing the flat orientation of the disk-shaped molecules on the surface.
Reproduced from ref. 71 with permission from the Royal Society of Chemistry.

confirmed a shift of the carbonyl peak in this spectroscopy. The formation of the films was also confirmed by measuring their capacitance electrochemically using the oxidation of a solution of ferrocene. The larger macrocycle (formed of eight monomers) gave a capacitance slightly less than the bare metal, because the analyte was able to penetrate into the core of the adsorbant into contact with the electrode surface. The capacitance of the seven- and six-membered macrocycles is sequentially less, because the smaller cavity sizes restrict the access of the ferrocene to the gold. The number of ferrocene molecules contacting the metal surface was calculated from the area of the redox peak. For the seven-membered host, the surface density was estimated at a minimum of 2.4×10^{13} molecules cm^{-2}, while the density of an alkanethiol is about 4.5×10^{14} molecules cm^{-2} (partially because the surface area occupied by the thiol is far less than the macrocycle, but also because of the low coverage). The surface binding by cucurbiturils has been used for the supramolecular electrochemical analysis of analytes.[72] The interaction of a monolayer of the cucurbit[7]uril with adamantyl moieties was done by atomic force spectroscopy by attaching a chain with this group to the tip of the AFM and using dynamic force microscopy to measure the binding.[73] The dissociation rate constants of the non-specific adamantyl–gold interaction was found to be 0.3 s^{-1} and the host–guest complex was 0.03 s^{-1}, indicating a strong binding that could be used to assemble multilayers (despite the incomplete coverage observed previously), as shall be shown. The strong binding often observed in an aqueous environment is a direct result of the role of water in the binding.[74]

A single molecule can incorporate multiple chemisorption sites for the gold surface.[75] Bespoke chelating spiroalkanedithiols are effective adsorbates for gold, giving similar order to the straight chain analogues,[76] and tridentate trithiols derived from cyclohexane prove to be more stable than straight chain compounds.[77] For ease of access, the commercial lipoic acid (aka thioctic acid) is perhaps the most frequently used divalent tether,[78] although the monolayer of the compound is quite disordered and dynamic.[79] Aromatic derivatives have also been explored, including 2-mercaptomethylbenzenethiol, which forms a disordered monolayer at room temperature but after annealing does give striped organised monolayers in which the aromatic rings are apparently in close contact with one another.[80] This kind of dithiolate is prone to desorption though, through the formation of a dithiolane-type compound. The use of dithiolates where the sulfur atoms are far apart from one another and cannot form an intramolecular disulfide is potentially interesting.[81]

Exchange of adsorbates on gold can be carried out in both gaseous and solution conditions. A monolayer of methylthiolate on gold (111) when exposed to dibutyl disulfide gas at room temperature resulted in desorption of the initial adsorbate.[82,83] Observation of the species leaving the surface by line-of-sight mass spectroscopy indicated the presence of both dimethyl disulfide and the cross product methylbutyl disulfide. An equilibrium is established quickly between the chemisorbed, immobile, thiolates and physisorbed disulfides that can diffuse across the surface. In solution, perhaps the most precise studies derive from nanoparticles in solution, and although not identical to the situation on a crystalline face of a block of metal, do give an idea of the processes that can take place.[84] As a system, the exchange of thiolates from gold and other metal substrates (including silver and copper) has been shown to be a result of first-order dissociation of the thiolate–metal bond at a rate that depends on the overlying solvent.[85] It has also been shown by using STM that the exchange of thiolates with selenol is fast and complete, and the kinetics depend on a perimeter growth model.[86] The replacement is falsest at the defects in the film, be they domain boundaries, step edges, or other areas where the thiolate is not stabilised by a surrounding like-layer holding it in place. When STM was used (in positive bias) to study the SAMs, it was found that the selenolate–gold bond is labile: they are replaced by thiolates. This, the scanning probe method, can be used to both observe and instigate these exchange processes. In fact, even pressing monolayers of decanethiol leads to the formation of SAM structures not observed under thermal annealing.[87] The tip's electronic and mechanical influence on structure in monolayers is therefore a possible variable in the formation of chemisorbed layers on gold.

Defects in layers can be exploited deliberately or the formation of mixed monolayers with more or less controlled isolation of molecules in a

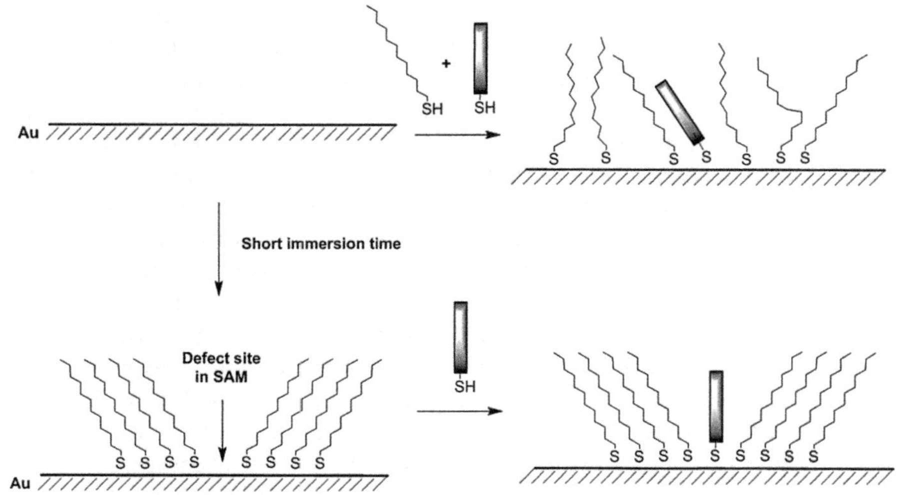

Figure 5.13 A representation of the formation of mixed monolayers of alkanethiols and some other thiol either by co-deposition to give a disordered mixed monolayer (top) and the sequential deposition of alkanethiol during a short adsorption period to give a partially ordered layer in which domain boundaries are abundant followed by adsorption of a second thiol in the spaces left in this first layer to give a mixed layer comprising partially ordered alkanethiol with inserted thiolate at defect sites.

surrounding matrix.[88] When alkanethiols are incubated for short periods of time in solution, the monolayers that form are of low density and are poorly ordered, they contain abundant areas of domain boundaries where the surface is not saturated with adsorbate (Figure 5.13). Mixed monolayers with different thiols inserted at these boundaries and defects are produced by immersion into solutions of this second substrate. The route is completely different to the co-deposition of the two thiols from the same solution, where disordered monolayers are formed with a random distribution of both components. The insertion of thiols into the defects is a useful route for the incorporation of any kind of functional unit into a monolayer so that its properties can be probed there without neighbouring like-adsorbates interfering in the measurements.

An ingenious approach to the preparation of low density SAM phases involves the threading of the adsorbate in a host that obviates the close packing in the layer followed by removal of this negative template. This idea was proven for cyclodextrin (CyD)-encapsulated 16-mercaptohexadecanoic acid, where the macrocyclic sugar was washed off after deposition.[89] The complete formation of the SAM and subsequent removal of the CyD was measured using cyclic voltammetry (CV). The capacitance of the gold changes as the SAM is formed, so that essentially no response is seen at complete SAM formation while removal of the CyD leaves space around the thiol and therefore the CV response increases. This process of SAM formation and CyD removal was also followed using a gold surface in a QCM set-up. The low density monolayer showed electrical potential-dependent wetting, because the chain is able to re-orient as a function of the field so that either the carboxylate group or alkyl chain are exposed to the solvent medium. Binding of proteins to the low density layers was proven, which depended on the isoelectric points of the molecules. Thus, streptavidin or avidin could be bound, applying different electrical potentials to the film.

5.3.2 Self-assembled Monolayers on Silver

The nature of the chemical bond in-between any given adsorbate and silver is quite similar to that with gold, and therefore the type of reactive group compatible with the formation of self-assembled structures is similar. Thiolates are typically used, and at a macroscopic level, the characteristics are similar to the ones formed on gold.[90] As a substrate, silver is of interest because it has the highest thermal and electrical conductivity of all metals. However, an oxide layer forms immediately in air. Indeed, an important challenge, and perhaps opportunity, in the chemistry of SAMs on silver is their reactivity. Benzyl and phenyl thiolates are quite readily cleaved to give mercaptides, indicating the rupture of C–S and S–S bonds, at least under irradiation, as these observations were made under surface-enhanced Raman scattering.[91] The lattice size of alkanethiolates on silver is somewhat larger than that of the gold analogues, leading to a change in the orientation of the adsorbate molecules and the surface is highly reconstructed upon adsorption.[92]

A beautiful example of the difference between the orientation of molecules at the surface is that of SAMs formed by a series of ω-(4′-methyl-biphenyl-4-yl)-alkanethiols (CH_3–$(C_6H_4)_2$–$(CH_2)_n$–SH, $n = 1$–6).[93] The biphenyl unit located at varying distances from the metal surface acted as a probe to test the orientation of the last methylene unit in the chain. The results of IRRAS and NEXAFS indicated that this orientation followed an odd–even effect.[94] For an even number of methylene units, the long axis of the biphenyl unit was directed quasi-perpendicularly to the interface, and implying an sp hybridisation of the sulfur atom on silver. For the molecules with an odd number of methylene units, the angle of the biphenyl unit was much less with respect to the surface plane, and as a result the coverage of the surface was lower when compared with the even carbon number analogues. When compared with SAMs of the same compounds on gold, an inverse relationship with carbon number was observed that was interpreted as an sp^3 hybridisation of the sulfur atom on this metal. Similar effects were observed when the adsorbing group was selenium[95] (see Figure 5.14) and when the head group was azobenzene on both metal surfaces.[96]

5.3.3 Self-assembled Monolayers on Copper

The self-assembly of structures on copper, as has been mentioned, is important for inhibiting corrosion of the metal.[97,98] Like gold and silver,

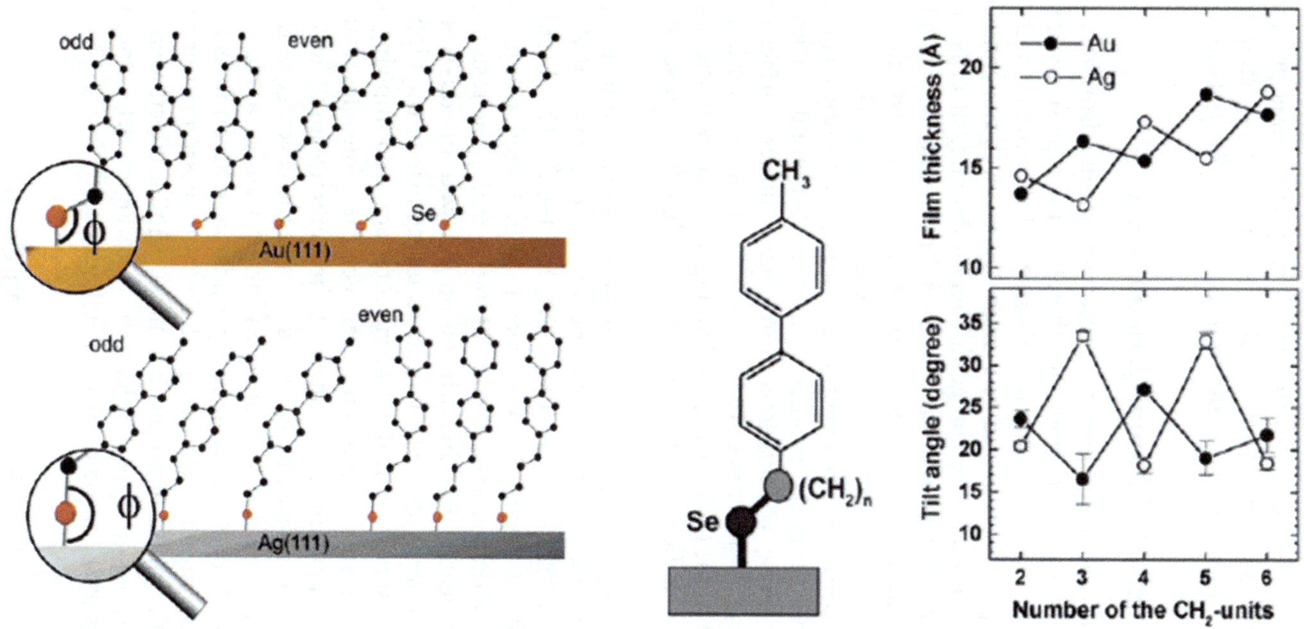

Figure 5.14 SAMs formed by ω-(4′-methyl-biphenyl-4-yl)-alkaneselenols where odd–even effects are seen for the head group orientations, with an opposite dependence for gold and silver surfaces.
Reprinted with permission from ref. 95. Copyright 2007 American Chemical Society.

at high coverage thiols form self-assembled monolayers on copper.[99] The picture is more complex at low coverage, where disorder reigns in the adsorbate film for alkyl thiolates.[100,101] A series of halo thiophenols do form ordered structures on the copper (111) surface at low coverage,[102] whereas the non-halogenated compound has a disordered phase at low temperature and stands upright at higher coverage.[103] An important difference between copper and gold (our reference here) is that the S–Cu bond is stronger than the S–Au bond, and the S–Cu potential corrugation amplitude as a function of coordinates on the surface has a much larger amplitude. For alkanethiols, this has the interesting effect that it is actually easier to organise shorter-chain compounds rather than longer-chain ones.[104] When adsorption takes place, a very significant reconstruction of the surface takes place.[105]

The supramolecular chemistry of chemisorbed structures on copper under UHV conditions is remarkable, and avoids the difficulties of establishing a clean copper surface under solution conditions, where the oxidation of the metal is very easy. Remarkably, though, this is not a deterrent to SAM formation, as octadecylthiol does self-assemble to give high quality films on a partially-oxidised copper surface.[106] Because of the nature of the chemical bonding of copper with adsorbates, the ligating groups most frequently used contain oxygen and nitrogen.

The evaporation of tartaric acid onto copper (110) results in the formation of a family of monolayers whose structure and bonding depend on coverage and chirality of the adsorbate. When the racemic modification of tartaric acid is deposited, spontaneous resolution of the enantiomers takes place, with coverage dependence.[107] Two enantiomorphic domains are indicated by LEED at low surface coverage of the adsorbate which has both acid groups deprotonated and in contact with the metal surface, as do the pure enantiomers individually.[108] (Alcohols have also been shown to deprotonate on copper upon adsorption.[109]) When the monolayers are made with a higher flux to yield greater coverage, the adsorbates are forced to make a denser phase so that one carboxylate per molecule is chemisorbed. In this surface phase, the symmetry of the surface dominates the packing, racemic two-dimensional crystalline domains are formed. This racemic compound phase has greater adsorbate density than the enantiopure "standing-up" phase.

Spontaneous resolution[110] of adsorbate molecules is observed when copper (110) is used as the substrate for the pro-chiral *meso*-tartaric acid[111] or succinic acid[112] (where the hydroxyl groups of tartaric acid are replaced with hydrogen atoms) under UHV conditions. In both cases, the origin of the separation of conformational enantiomers is apparently the twist in the ethylene link between the two carboxylate groups, which are both in contact with the metal. For succinic acid, the twisted conformation favours the contact of all oxygen atoms of the adsorbate with the copper (Figure 5.15). The chances of forming either chiral conformer upon adsorption are equal, but once the twist starts to go in one chiral direction—either because of the

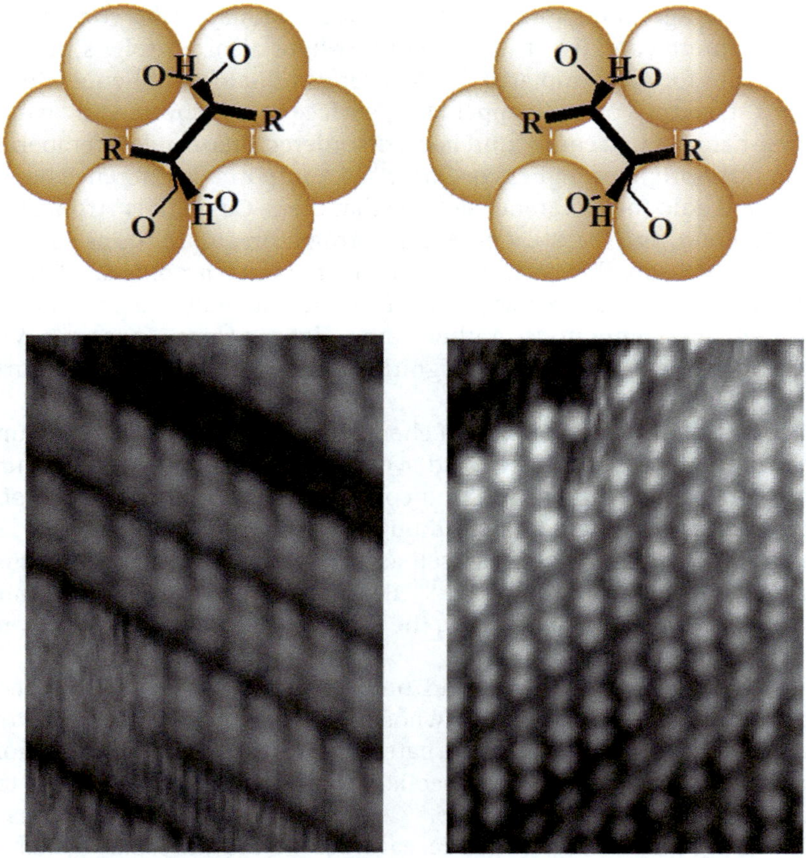

Figure 5.15 A cartoon showing the twists generated in the carbon skeleton of succinic acid (R = H) and tartaric acid (R = OH) upon chemisorption on Cu (110) leading to chiral interface structures that are observed by STM images (below) of the enaniomorphous succinic acid domains. Reproduced from ref. 110 with permission from the Royal Society of Chemistry.

molecular distortion or metal reconstruction—it is propagated through the domains. Both LEED and STM were used to prove the existence of enantiomorphous domains.

The hydroxyl groups attached to the same stereogenic centres as the carboxylate moieties in tartrate force a twist of the organic framework in a given and opposite sense for the two enantiomers, and this arrangement is propagated over the surface. The achiral molecule of *meso*-tartaric acid, which has a plane of symmetry running through its central bond off any surface, is curious when considering chiral induction, since when small amounts of either (*R*,*R*)- or (*S*,*S*)-tartaric acid are evaporated with it onto copper, one of the enantiomorphs of the achiral compound is suppressed and the monolayer's ordered domains become homochiral.

The supramolecular surface chemistry of amino acids has received considerable attention because of their availability and relevance to the formation of proteins, where surfaces may play an important role.[113] The adsorption of tryptophan onto copper led to the formation of chiral monolayers, proof coming from LEED measurements.[114] The enantiomers of the adsorbate produce mirror image domains, providing a proof of the authenticity of the observations. Studies of the compound on the metal's (111) surface by STM show that the adsorbate forms clusters at low coverage, followed by chains and then a homochiral film whose packing is determined by intermolecular hydrogen bonding, interactions between the indole groups of adjacent molecules and the copper.[115] The simplest chiral amino acid, gives a kind of "nematic phase" at room temperature on copper (110),[116] but when annealed at 400 K gives monolayers with large chiral domains, and STM shows that the amino acids pack in groups of six or eight molecules.[117] It has been shown that the carboxylate group that binds the molecule to the surface has a chiral footprint in the metal, and the groups do not coincide with the major symmetry directions of the copper surface. When the sample is further annealed to 470 K, an achiral packing results, which is apparently the thermodynamic state. The chiral packing for the amino acids is a common trait, at least at realtively low temperatures, as phenyl glycine also forms a chiral monolayer with a chiral two-dimensional space group.[118] Theory demonstrated that the phenyl group favours a chiral arrangement because of its position and the volume it occupies in the layer.[119]

Adsorption of proline onto the copper (110) surface generates the prolate adsorbate whose oxygen atoms are bound to atoms in the same surface row.[120] STM images of the monolayers formed by the prolate adsorbate exhibit molecules with different contrast. This effect is a result of the adsorption configuration on the surface: one enantiomer of the compound has two adsorbate structures accessible as a result of formation of a bond between the nitrogen atom with copper atoms on neighbouring rows. This situation affords two diastereoisomeric conformations (Figure 5.16). Adsorption of the racemic modification of proline to the copper (110) surface gives a monolayer that is hugely complex. No long-range order is present, yet the enantiomers and their diastereoisomeric surface conformations can be picked out in the STM images of the monolayer.[121] The handedness—the intrinsic chirality from the stereogenic centre—and footedness—the chiral adsorption pattern on the metal—causes four distinct contrasts. There is a random arrangement of handedness (the position of the enantiomers), but the chiral adsorption pattern (the footedness) is organised in lines (double—carboxylate—and single—nitrogen—coordination alternate on a copper row). The sample is essentially a solid solution of the four adsorption possibilities.

Rigidification of the heterocyclic ring in 3-pyrroline-2-carboxylic acid gives a completely random packing on the same metal face.[122] Therefore the "handedness" of an adsorpt (its intrinsic chirality) does not necessarily

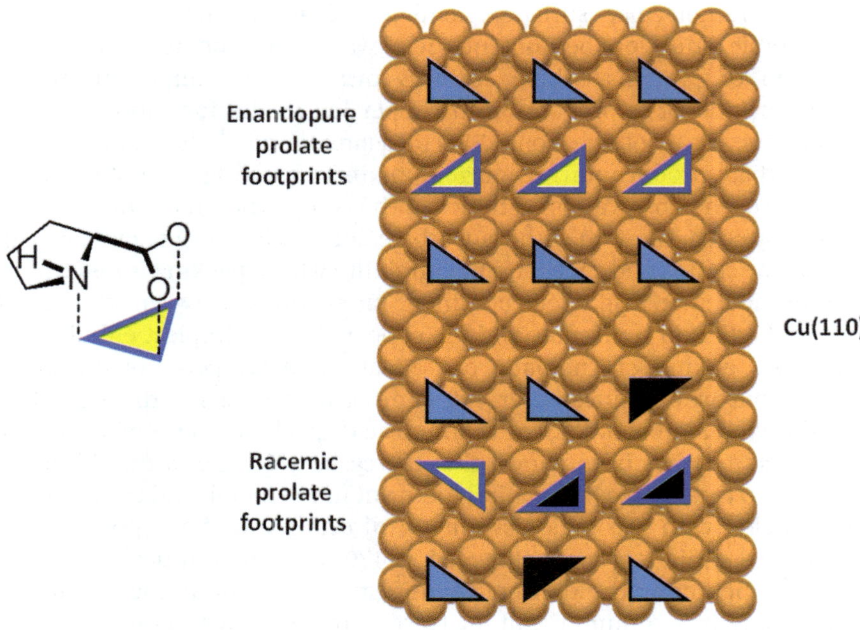

Figure 5.16 A representation of one of the adsorption footprints of prolate on the copper (110) and the packing on the surface that result from adsorption of enantiomerically pure or racemic samples.

correlate with its "footedness" (adsorption pattern) when attached to a surface.[123] The small structural dissimilarity between proline and 3-pyrroline-2-carboxylic acid leads to a homochiral and ordered two-dimensional footprint on copper (110).[124]

A huge combinatorial space is generated when considering the adsorption possibilities of oligomers comprising different amino acids, and for now studies have been mainly focussed on homopeptides (made of the same amino acid in each monomer unit). Adsorption of di-L-alanine onto copper (110) leads to the generation of chiral rows of molecules.[125] Homochiral chains are also observed when diphenylalanine enantiomers are adsorbed onto the same surface.[126] The enantiomers diffuse over the surface before forming homochiral chains, and video STM reveals that like-enantiomers unite to give chiral chains, while opposite enantiomers do not interact significantly.[127] Co-deposition of terephthalic acid improved order in the chains allowing self-assembly of a two-dimensional co-crystal.[128] Deposition of the heterodipeptide glycine-proline also results in adsorption through the carboxylate group as well as the amino group, as shown clearly through XPS and PM-RAIRS, giving a three-point-type interaction.[129] The linking tertiary amide group is lifted up from the surface. STM reveals ordered and large domains—which are chiral, as also shown by LEED—are formed comprising rows of molecules, apparently favoured thanks to intermolecular hydrogen bonding between the amino hydrogen atoms and carbonyl groups of the carbonyl moieties. Tripeptides—tri-L-alanine and tri-L-leucine—adsorb to copper (110) through the now familiar carboxylate–copper adsorption mode as well as through the amino group at the termini.[130] RAIR spectroscopy showed that the amide groups hydrogen bond between one another in the plane of the interface. The phase space of this monolayer is very sensitive to deposition conditions.

Copper can be a reactive metal for the synthesis of molecules in heterogeneous solution, and has also been shown to be effective when adsorbates are heated on its surfaces. The reaction of hydrocarbons on metallic copper leads to the formation of self-assembled systems, where organometallic bonds are formed between the metal and the adsorpt, as shown for porphyrins on a copper (110) surface.[131] Heating free-base porphyrins on the surface first leads to incorporation of copper from the surface into the centre of the porphyrin ring[132,133] in what appears a habitual metallation of a porphyrin, with consequent release of molecular hydrogen (as revealed by temperature programmed desorption, Figure 5.17). Then, for both previously metallated and *in situ* metallated systems, the porphyrin rings link up selectively along the (001) direction of the copper surface, because of the match of molecular geometry to the surface dimensions, so that aligned macromolecular strands are generated. STM reveals that these strands can incorporate between one and three copper atoms from the surface. Modelling of the connectivity confirmed this hypothesis (Figure 5.17). Interestingly, on silver (111), a related process leads to the formation of carbon–carbon covalently linked structures.[134]

5.4 Self-assembled Monolayers on Oxides

The surfaces of oxides differ greatly from metals and other elements, in that surface mobility of atoms is much lower and the surfaces are generally much less perfect. The covalent bonds formed between adsorbate and the surfaces are stronger and irreversible on habitual temperature ranges and for most adsorbates. The oxide layer on elemental silicon is a very convenient place to study self-assembled monolayers. When freshly etched or polished silicon (or germanium) is exposed to air at room temperature, the surface reacts quickly to give a thin oxide layer.[135] The thickness of this layer depends on a logarithmic growth rate that gives a thickness of around 0.5 to 2.5 nm over a time period of the order 1 minute to 1 year.[136]

The self-assembly of organosilanes on surfaces of elemental silicon was initially done to modify the wettability of the surfaces,[137] as well as to explore the knowledge that this kind of chemical system could provide about the interactions between the electronic states of different organic adsorbates and the semiconductor.[138] The oxide layer can be so thin that it allows tunnelling electrons to pass through it. This feature means that the study of electronic properties can be achieved and also that the samples do not charge excessively if electron microscopy is performed (whereas pure silica is an insulator that would have charging problems with these techniques).

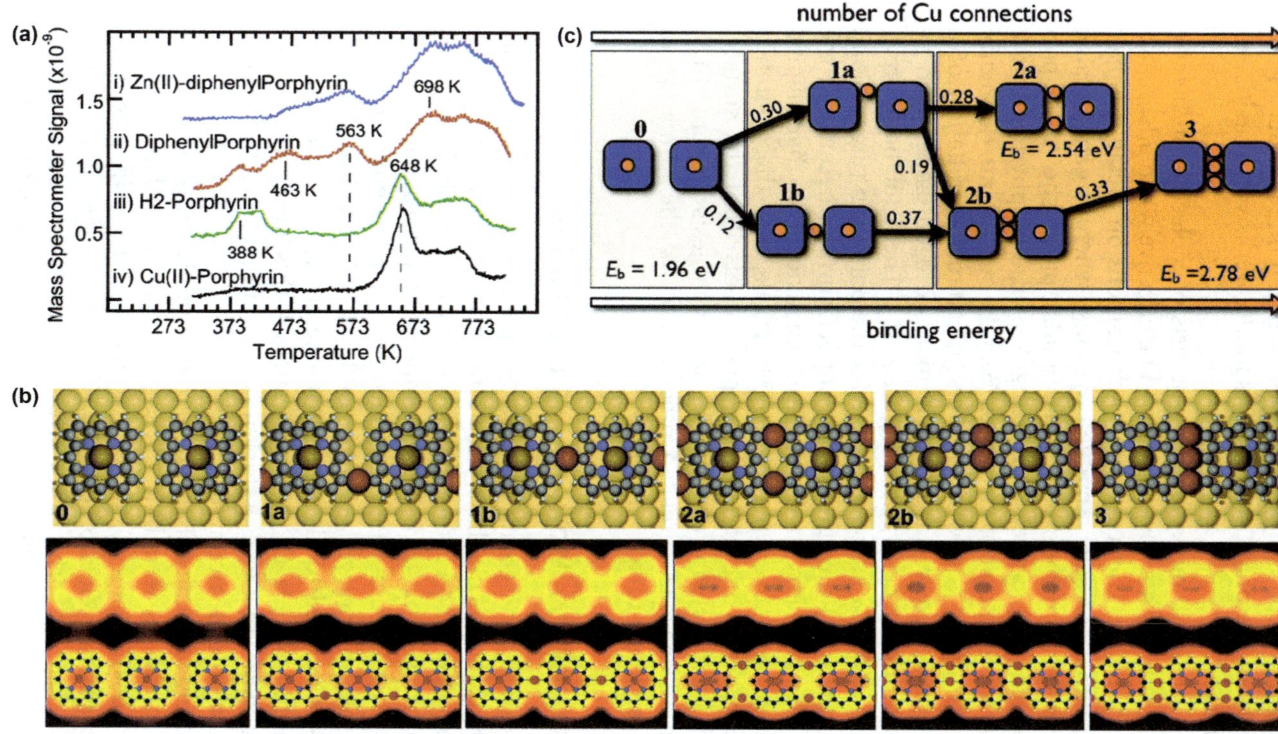

Figure 5.17 The generation of organometallic porphyrin strands on copper (110). (a) The temperature-programmed desorption curves for H_2 measured for different porphyrin compounds. The process at 388 K corresponds to centre porphyrin ring metallation and above 420 and 600 K to the organometallic linkages. (b) The calculated geometries (top) and simulated STM images (bottom) for six different porphyrin oligomers with zero, one, two, or three connecting copper atoms. (c) The computed reaction energies for the porphyrin linking demonstrating the exothermic nature of each individual bond-forming reaction. The binding energy increases from left to right. The net adsorption energies E_b are shown for three representative cases. Reprinted with permission from ref. 131. Copyright 2011 American Chemical Society.

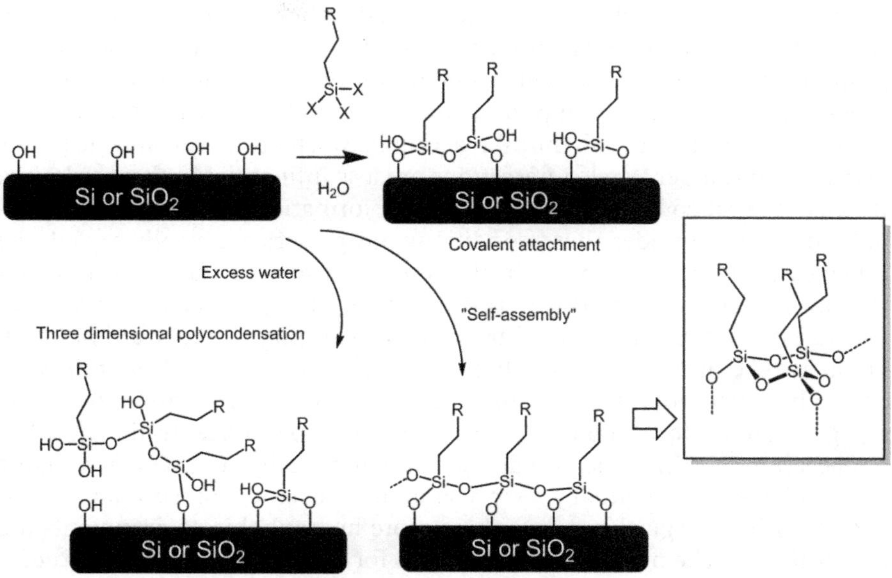

Figure 5.18 The possible outcomes of the reaction of silanes (where X is a leaving group, generally chlorine, methoxy or ethoxy) and a hydroxylated silicon surface.

In the same work, it was realised that the reaction of organosilanes on this flat substrate would also give a model system for assembly on non-flat surfaces. Particulate silicates can arise during solution treatments, but can be avoided with refluxing toluene or xylene solutions of the siloxane in an inert atmosphere. Smooth monolayers are formed because the adsorpt has a low vapour pressure in the mixture. The possible outcomes of the treatment of a silica surface with silanes are shown in Figure 5.18. Simple attachment to the surface corresponds to immobilisation of the silane, but monolayer formation is possible, with an optimised silicon dioxide network and the alkyl groups emanating perpendicularly away from the interface.[139] If too much water is present, polycondensation can occur, leading to rough surfaces and even particulate material in the system.

The self-assembly of *n*-octadecyltrichlorosilane from 2.5 mM solutions in hexadecane–carbon tetrachloride (70 : 30, for 60 minutes) on freshly hydrated silicon oxide on silicon surfaces shows a marked and well-behaved temperature effect on deposition.[140] Below approximately 25 °C, the surface energy is practically constant—the contact angle of a water drop was approximately 114°—and consistent with a purely methyl group terminus. When the chemisorption was performed above this temperature, the surface energy increased (lower contact angle), and in addition, the coverage determined by ellipsometry was lower, decreasingly practically linearly with temperature. Infrared spectroscopy also showed the differences, where the films prepared at low temperature had well-formed quasi-crystalline-like methylene signals while the monolayers prepared at higher temperature revealed liquid-like hydrocarbon chain signals. These results confirmed and clarified prior experiments with similar findings.[141] It seems reasonable to propose that below the critical temperature, non-covalent interactions between chains play a very important role in monolayer formation, while at higher temperature, the chemical reaction between adsorbate and surface is fast and the van der Waals interactions are less influential in their ordering. Because the silicon–oxygen covalent bond formation is strong and essentially irreversible under these conditions, no posterior annealing would be expected to improve the order in the film because no diffusion or reorganisation can take place readily in this interface. An ingenious recent method used a diffusion barrier formed by a spin-coated polymer to avoid build-up of particulate material in the self-assembled monolayers.[142] It is somewhat contradictory to note that *N*-(2-aminoethyl)(3-aminopropyl)trimethoxysilane can generate uniform monolayers from water at low concentration and low temperature.[143] This kind of adsorpt, of which many commercial variants are available, is potentially very useful in the study of supramolecular chemistry in biological systems where more hydrophobic environments are less desirable. The use of this kind of SAM for molecular recognition studies involves the immobilisation of classic recognition units used for supramolecular chemistry in solution,[144] discussed in the next section.

The reliable formation of SAMs on surfaces to act as dielectric materials in transistors is highly important as they determine device performance in grand measure. The treatment of the surface in question prior to chemisorption plays a hugely important role in efficient attachment, as demonstrated for SAMs of phosphonic acids on a silicon wafer with a mixed silicon–aluminium oxide surface layer.[145] Before the adsorpt was attached, the surface of the silicon wafer was treated with air plasma containing aluminium ions, which leaves a film with a thickness of the order of nanometres on the surface, terminated with hydroxyl groups. The phosphonic acids (with alkyl or fluoroalkyl chains) were attached by spin coating and then annealing at 140 °C to give hydrophobic SAMs (contact angles of the order of 100° for the alkyl derivatives) that were resistant to solvent and microwave treatment for long periods. The absence of the aluminium on the surface leads to layers with a much lower contact angle (approximately 60°), and relatively low proportions of phosphorus were observed by XPS. TOF-SIMs data showed that the adsorbate was attached to the aluminium ions, highlighting their importance in the adsorption.

The attachment of carboxylates to metal oxide surfaces is an effective means of SAM formation in certain cases. The attachment of *n*-alkanoic acids onto oxidised aluminium leads to monolayers; although they are dynamic and complex in nature, ellipsometry did prove the formation of monolayers—most clearly above a twelve-carbon chain.[146] Several days were required to form the equilibrium structures from solution, where the carboxyl groups are in contact with the surface although the order in the films was not high according to IR data.[147] Precise studies on titanium dioxide[148] reveal that acids bind to the surface with the carboxylate adsorbate bridge-bonded at the five-fold coordinated Ti(IV) centres.[149] The molecules apparently are directed away from the plane of the surface. Terephthalic acid shows a transition from flat lying to perpendicular arrangement on the same material, as shown by NEXAFS.[150] Non-contact AFM indicated the presence of two polymorphs at high coverage, one of which involved hydrogen bonding between the carboxylic acid groups directed away from the interface. This kind of layer has not found real use in supramolecular chemistry to the present. It has been shown that terminal alkenes bearing amide groups at the opposite end of an alkyl chain can be attached to titanium dioxide through a photochemical reaction,[151] although the generality of this approach is questionable because of the photocatalytic nature of the substrate.

The chemisorbed layers can be used for subsequent covalent chemistry, and the robustness of the surface–adsorbate bond allows modification of the newly-formed monolayer. The attachment of thiophene adsorbates to oxides as adhesion promoters for grafting polythiophene has been achieved.[152] The adsorpts were ω-(thiophen-3-yl) alkyl phosphonic acids, and the films were characterised (on several substrates) through IR spectroscopy and contact angle measurements. A precise picture of the bonding between the phosphonate group and the oxide has been developed through XPS analysis of related compounds, showing that all three possible P–O–metal bonds are formed with titanium while for aluminium and tantalum, only the P–O(H) bond is present.[153] On the other hand, silanation of a hydroxylated silicon surface with either a trichlorosilane[154] or a triethoxysilane[155] bearing a thiophene unit was used as a precursor for the formation of poly(thiophene) and graphene, respectively.

5.5 Supramolecular Functions for Self-assembled Chemisorbed Monolayers

The study of interactions between molecules or ions in the atmosphere or in solution and surface self-assembled systems reveals the nature of the environment of the moieties involved in non-covalent interactions. It can show how the interface assembly can affect binding interactions that might have been established in bulk solution, and make these interactions useful because the surface-based systems can act as sensors in devices whereas homogeneous systems are less obviously applicable.

Electrochemistry was widely used early on in the study of supramolecular systems at interfaces because of the easy assembly of components onto gold and the use of this metal as one of the electrodes.[156–158] The electrochemical sensing of cations by self-assembled monolayers on metals has proven very effective. The self-assembled monolayers of a tetrathiafulvalene derivative bearing a dithia-18-crown-6 annelated at one end, the π-electron rich unit, and an (oligo)methylene thiol at the other end showed small but detectable electrochemical sensing of silver.[159] The response to the silver cation was recorded in acetonitrile, giving a similar response to that in solution, where the first oxidation of the π-electron rich moiety was shifted by approximately 70 mV. The reversibility of this first oxidation process for this kind of compound meant that this kind of system is very applicable generally, although applying very positive potentials causes a loss of reversibility.

This approach has been developed using crown-annelated tetrathiafulvalene derivatives anchored to gold through lipoic acid to sense the presence of alkali cations through shifts in the cyclic voltammetry waves.[160] A "lateral" crown ring in a surface-mounted system leads to selective binding of barium cations.[161] The electrochemical detection of alkali metals can be performed both with electroactive adsorpts as well as adsorpts with no electroactive component incorporated.[162] The two tetrathiafulvalene-containing adsorpts in Figure 5.19 both form very stable self-assembled monolayers on gold and show shifts in their oxidation waves when sodium or potassium (but not lithium) are introduced into the electrolyte. The adsorpts derived from penta(ethylene glycol) and tetra(ethylene glycol) with lipoic acid termini also bind sodium and potassium in the self-assembled monolayers that they form, despite the fact that RAIRS showed that the layers were disordered and essentially "liquid-like". Impedance spectroscopy (which shows the reorganisation of dipoles near the electrode surface, in this case associated with the ions) showed that the longer ethylene glycol unit favoured potassium complexation and the shorter one sodium. It was found that the selectivity in binding was greatly improved in the monolayer when compared with cyclic crown structures either in monolayers or in solution. The same piece of research showed that the recognition of metal ions (apart from the intrinsic affinity of certain hosts for

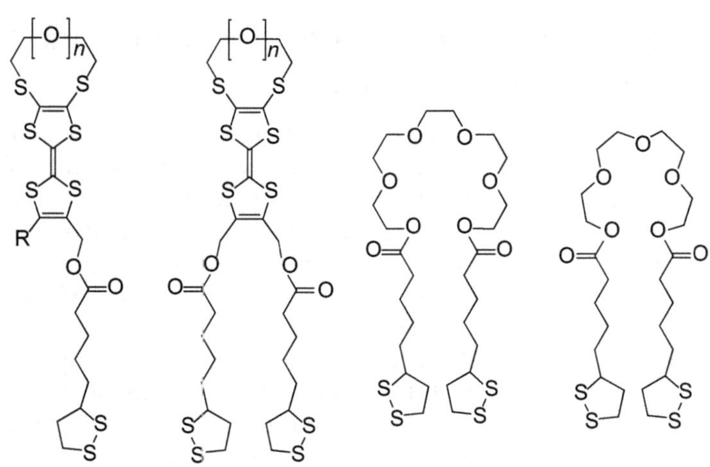

Figure 5.19 Two adsorpts that are electrochemically active and are capable of selective recognition of alkali metal ions (left) and two podand-like receptors that form monolayers and bind potassium and sodium ions, respectively.

guests) can be influenced by the medium when the monolayer is formed, through an "imprinting" method whereby the target ions were present during the formation of the monolayers.

The incorporation of redox-active units in the monolayers can also facilitate the recognition of non-redox-active organic hosts in solution. A mixed monolayer of a ferrocene amphiphile on gold (Figure 5.20) was used as an electrochemical probe for the binding of an amphiphilic calix[6]arene to the interface.[163] The mixed monolayer of alkyl thiol and ferrocenyl was designed so that the ferrocenyl moiety was "standing outside" the monolayer to ensure electrochemical reversibility[164] as well as accessibility by the host. The coverage of ferrocene moieties was determined by integration of the anodic wave in the voltammogram of the covered gold in electrolyte. The calix[6]arene in the figure is sensed by this monolayer at concentrations as low as 0.05 μM in aqueous electrolyte, as evidenced by a shift in the oxidation potential of the ferrocene moiety. The presence of the alkyl chains on one rim of the calix[6]arene is essential to the response, as with hydroxyl moieties the host is entirely soluble in the electrolyte and not adsorbed to the electrode surface. The dodecyl chains are apparently interacting with the filling alkyl chain at the interface that facilitates binding assisted through van der Waals interactions and hydrophobicity along with the intrinsic binding of the ferrocene to the macrocycle. Similar results were also seen for calix[6]arenes binding to physisorbed monolayers containing 4,4′-bipyridinium derivatives.[165]

The boot can be put on the other foot: solution-borne hosts with appended electroactive units can be recognised by appropriate guests bound to the surface. The benzo-18-crown-6 appended fullerene shown in Figure 5.21 forms a complex with a self-assembled monolayer with an ammonium terminus to the thiol chemisorbed on gold, which acts as an electrode.[166] The use of the ammonium–crown interaction[167] allows the formation of a supramolecular monolayer (in dichloromethane containing an electrolyte) that was detected using Osteryoung square wave voltammetry, observing the reduction of the fullerene moiety, which indicated a coverage of approximately a third the maximum theoretically possible. The assembly of totally unfunctionalised fullerene in its anionic form has been achieved more recently using a gold surface covered with per-6-thio-β-cyclodextrin, where the anion (which is held stronger than neutral C_{60} in solution) binds to the host through dispersion interactions.[168] Once bound, the anion was allowed to re-oxidise to the neutral form, and the film was characterised by electrochemistry (where the reduction of the fullerene was observed clearly), which gave a very high coverage value (actually larger than that expected considering a full monolayer of cyclodextrin, perhaps indicating interstitial binding as well). Also, contact angle measurements indicated the inclusion of the fullerene, as the contact angle of the host–guest surface complex (77°) was higher than that of the cyclodextrin monolayer (36°).

Defect sites in SAMs of alkanethiols that have not reached their optimum packing can be used for trapping molecular wires so as to be able to measure

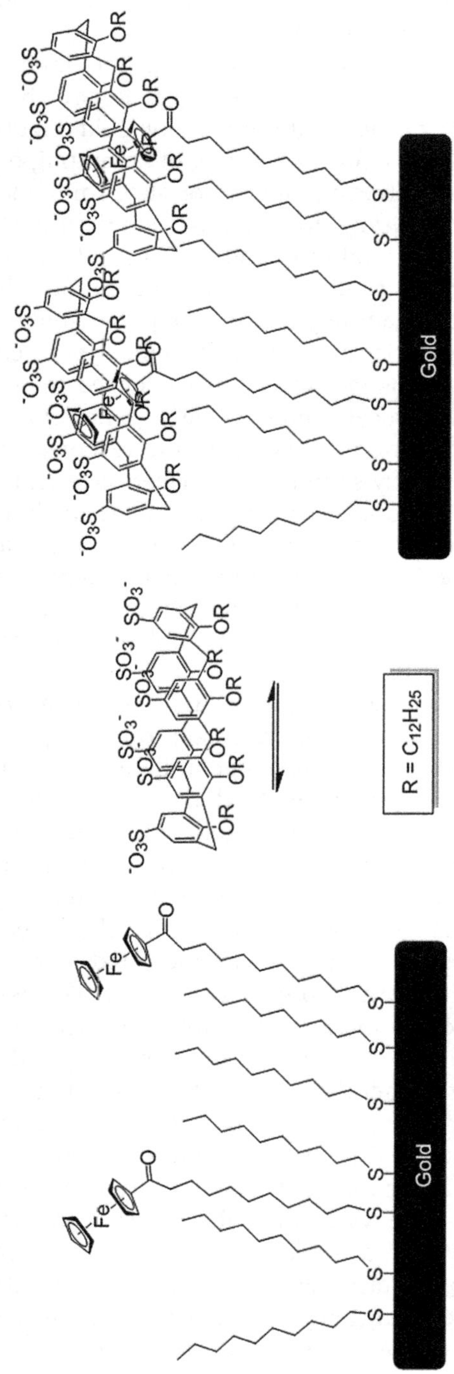

Figure 5.20 A depiction of the binding in an aqueous electrolyte of an amphiphilic calix[6]arene to a mixed monolayer incorporating ferrocene moieties that are used to sense the binding electrochemically.

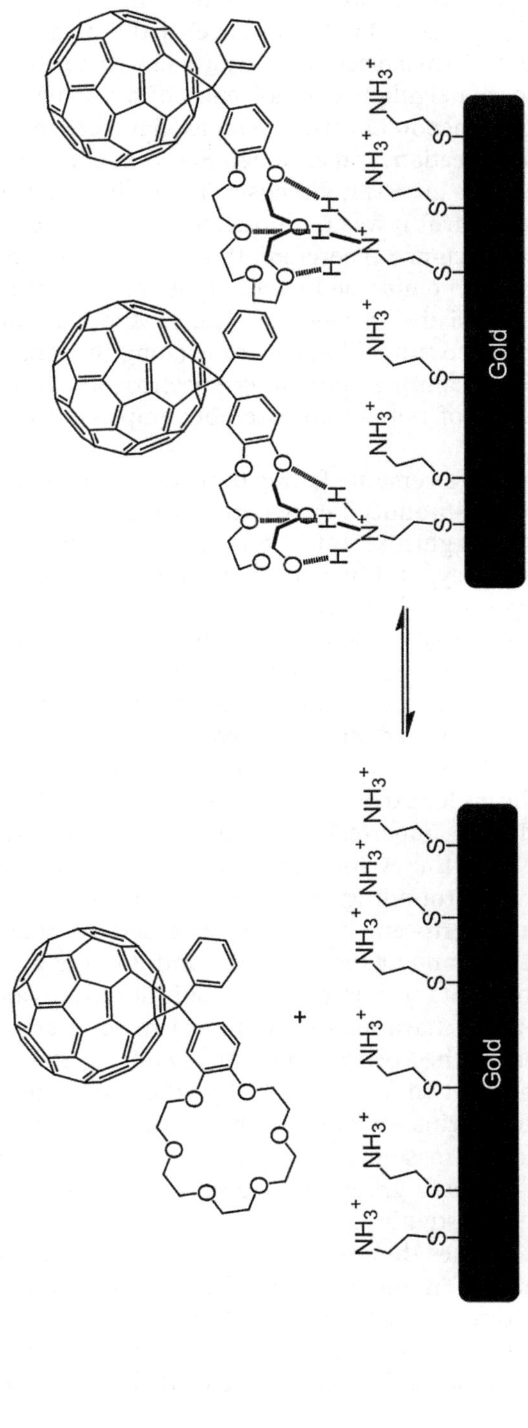

Figure 5.21 A depiction of the binding in dichloromethane and an electrolyte of a crown-appended fullerene to a chemisorbed ammonium-terminated SAM on gold where binding was observed electrochemically.

their conductance.[169] The exposed gold substrate in SAMs formed during short immersion times in alkanethiol was reacted with conjugated dithiols so that the latter were forced to attach to the metal through only one of their reactive groups (Figure 5.22). In the absence of the alkanethiol blocking layer, the dithiol could form a face-down adsorbate structure on the gold and in any case would have neighbouring molecules of the same type that could influence the charge transport effects. The molecular wires in the defect sites have a very small cross section, and although the wires could be seen by their bright appearance with an STM, because of the limited stability of the microscope to thermal drift it was not possible to reliably contact the outer thiol to measure conductance. Therefore, this free thiol was reacted with a gold nanoparticle that chemisorbed to the upper surface of the layer. The attachment was clearly visible in topographical STM measurements and the molecular conductivity to tunnelling electrons along the length of the wire was measured using scanning tunnelling spectroscopy. The route is an elegant demonstration of bottom-up assembly employing multiple supramolecular concepts.

The cyclodextrins are a versatile family of hosts, in many ways. The rigid and deep macrocycles—produced enzymatically form starch[170]—comprise six, seven or eight α-D-glucose units for the α-, β- and γ-cyclodextrins (Figure 5.23), respectively, and have been exploited in monolayers for the chemisorbed recognition of solution-borne guests. The β-cyclodextrins, where all primary hydroxyl moieties are converted to thiol groups, self-assemble on gold forming imperfect monolayers.[171] Electrochemical measurements (reductive desorption) showed that approximately two thirds of the gold was covered with adsorbate. A supramolecular route to the modification of the surface was effective in generating a layer where only the bases of the cyclodextrin cavities had gold exposed to the solution (Figure 5.23): firstly, the imperfect layer was exposed to ferrocene, which formed a complex with the cyclodextrin. Subsequently, pentanethiol was chemisorbed in the surrounding area that was exposed to the solution. Removal of the bound ferrocene by washing afforded a surface that was an effective binding environment for ferrocene and 3-toluic acid, with comparable binding constants to homogeneous solution. Inorganic anions[172] and the anion of C_{60}[173] have also been shown to bind effectively to the cavities. Subsequently, the role of different anchoring groups, including amine and isocyanide compared with thioether, for the assembly of monolayers of cyclodextrins on gold, has been studied.[174] Each anchoring group provides unique assembly kinetics and structure. Interestingly, cyanide and isothiocyanate groups provided films of the greatest quality and charge transfer resistivities.

In order to overcome the difficulties associated with packing the cyclodextrins effectively in the plane of the gold–solution interface, dialkylsulfides were introduced to the primary face of the hosts (Figure 5.24).[175] When a layer of these compounds is assembled on gold, the alkyl chains are present to fill the space between the adsorbed thiol groups and the

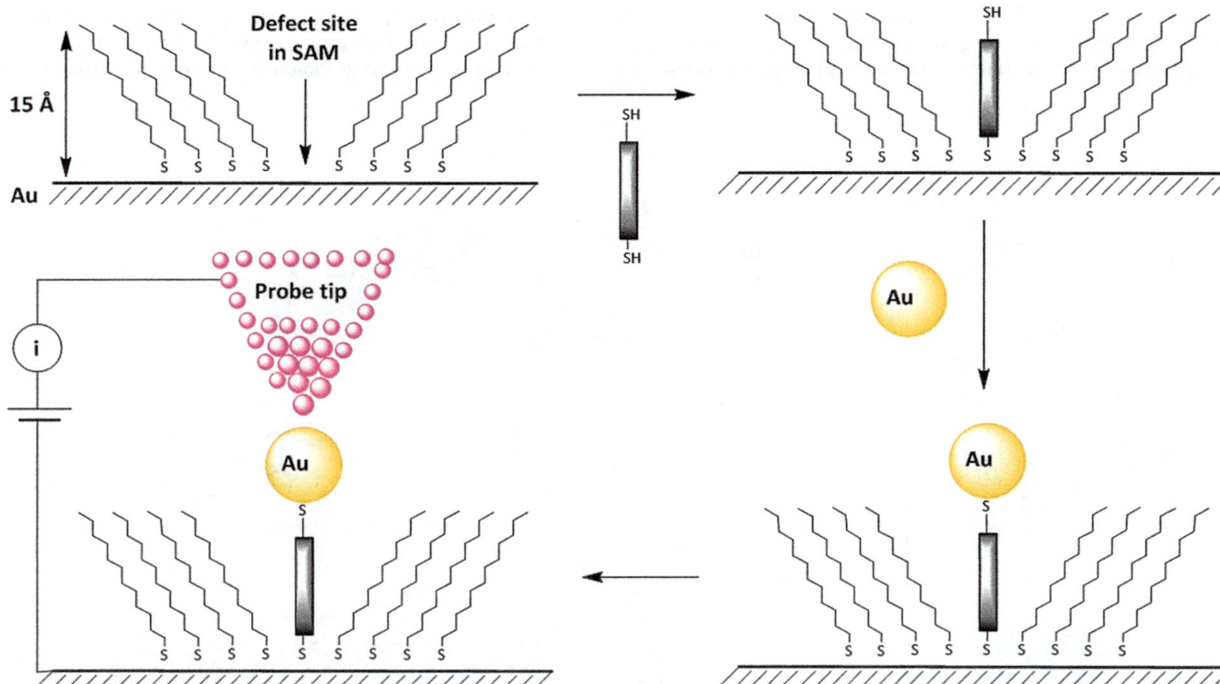

Figure 5.22 The route molecular wire chemisorbed to a surface and a nanoparticle for measurement of conductivity. The key steps are the insertion of a dithiol in defect sites in a non-optimum SAM to ensure upright arrangement followed by attachment of a gold nanoparticle with a labile surfactant covering. This approach allows relatively easy positioning of an STM tip over the wire (that otherwise has a very small cross section).

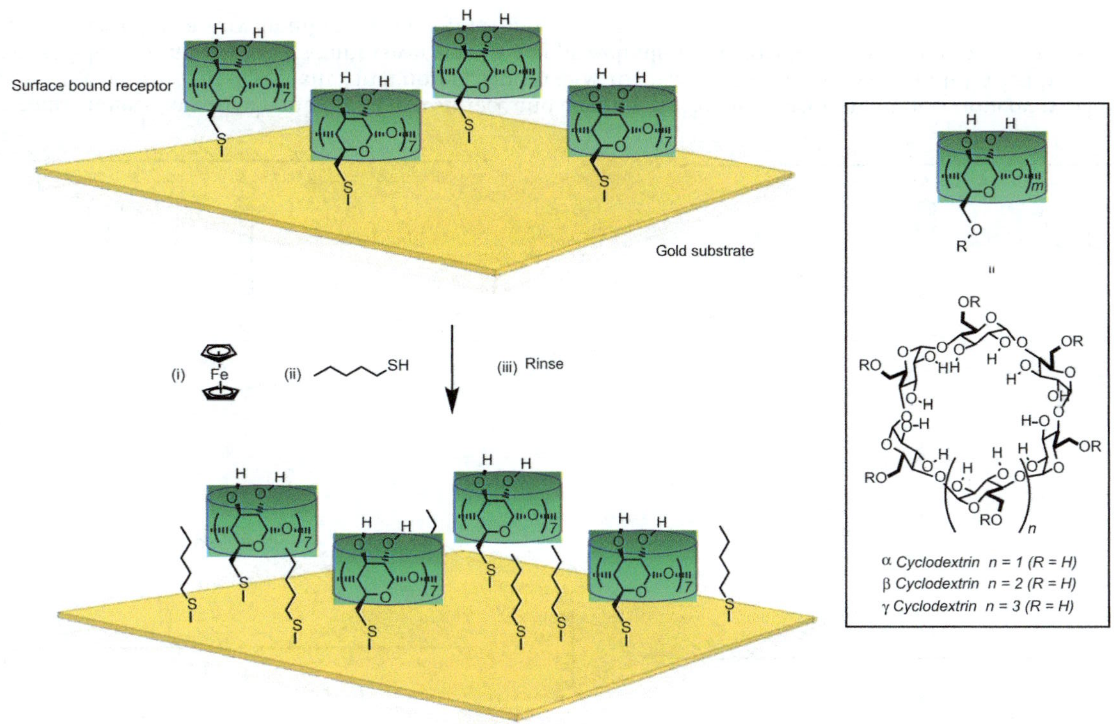

Figure 5.23 A representation of a layer of per-6-thio-β-cyclodextrin (see key for the different cyclodextrins on the right), which has an imperfect layer structure that can be tuned by binding ferrocene in the cavities of the hosts, reacting pentanethiol with the gold in-between the hosts, and then removing the guest, leaving a monolayer with vacant cyclodextrin cavities with "bare" gold at their base.

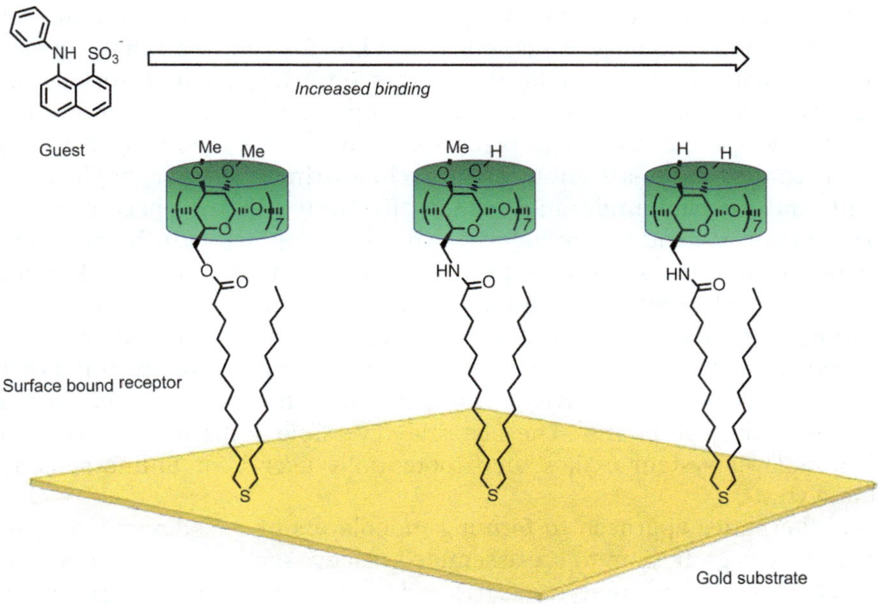

Figure 5.24 A cartoon showing three different gold surface-anchored β-cyclodextrin hosts where binding of the depicted guest increases as methoxy groups are replaced with hydroxyl groups from left to right.

cyclodextrin moieties. The monolayers were prepared by exposing solutions of the compounds to the gold for sixteen hours at sixty degrees centigrade. Interestingly, atomic force microscopy measurements on the monolayers prepared from derivatives of α-, β- and γ-cyclodextrins showed that only those from the ring incorporating seven glucose units had local two-dimensional crystalline order. Positive TOF-SIMs spectra showed peaks corresponding to the entire adsorbate bound to a gold atom (in addition to sulfur–gold clusters at lower molecular weight), and XPS was used to determine the amount of sulfur atoms adhered to the gold. In the case of the permethylated β-cyclodextrin adsorbate (Figure 5.24), this proportion was only 39%, perhaps because of the large conformational flexibility of this particular compound. The hydroxyl groups aid in rigidifying the ring and it is known that the preorganisation afforded by the intramolecular hydrogen bonds can maximise binding. This effect is also seen in the surface-bound hosts (Figure 5.24) where the adsorbate with both secondary alcohol functions is the most effective binder of aromatic guests from solution. The β cyclodextrin SAMs respond through both SPR measurements as well as electrochemically (impedance spectroscopy) to detect low concentrations of guests. The complex formation is negligible with two methoxy groups on the secondary side of the β-cyclodextrin-based adsorbate, while one methoxy and one hydroxyl group do give a response, but the association constant of the SAM based on the cyclodextrin with two hydroxyl groups binds a guest with an association constant of approximately 300 000 M^{-1}, a value far greater than in a homogeneous solution. While for simple aromatic guests the selectivity of the monolayers mirrors observations in solution, for steroidal bile salts it does not, and an approach based on a more disordered system that can adapt is more effective.[176] This monolayer system was prepared by co-assembling a β-cyclodextrin bearing a single thiol chain and mercaptoundecanol, which effectively fills the space between the gold surface and the cyclodextrin that is in contact with the medium. Therefore, selectivity might be probed by co-assembling host blocks with different co-adsorbates that might influence the relative selectivity in the binding of guests. Beyond gold, a layer of thiolated β-cyclodextrin on silver was proven to bind cytochrome c functionalised with adamantyl units that enter the cyclodextrin's cavity.[177] Addition of competitive guests caused release of the compound. The fast and reversible response of this kind of monolayer system makes them potentially useful for online sensing of analytes.

An alternative approach to forming monolayers of cyclodextrins to gold surfaces is to graft them to pre-assembled reactive monolayers, as shown in the reaction of a β-amino-cyclodextrin with a SAM formed by a thiopropionic acid-activated ester.[178] This film was used for the selective binding of the *trans* form of *N*-methyl-*N*′-(1-phenylazobenzyl)-4,4′-bipyridinium, while photoisomerisation to the *cis* form resulted in reversible decomplexation. This effect was observed by both amperometric electrochemical methods and a microgravimetric quartz crystal microbalance. It shows how binding at an interface can be modified through isomerisation, in this case releasing a bound guest.

A number of spectacular surface-bound synthetic receptors have been reported based on combinations of resorcarenes and calixarenes, among them the system shown in Figure 5.25 that comprises two calix[4]arenes attached to a central resorcin[4]arene with thioether anchor groups on one face of the latter.[179] The SAMs of the receptor were prepared dissolving the adsorbate in a mixture of chloroform and ethanol (1 : 1) and heating to 60 °C for sixteen hours to ensure high coverage.[180] The monolayer was characterised by XPS (where a quantitative measurement indicated the correct elemental composition), water contact angle determination (that shows a hydrophobic surface, angle approximately 103°) and polarised infrared external reflectance spectroscopy (PIERS) that showed the bands expected for alkyl chains and amide groups, principally, along with signals corresponding to the aromatic groups. The compactness of the layer was judged by electrochemical capacitance and resistance measurements, which showed a relatively poorly-ordered SAM. The binding of a variety of steroids from aqueous solutions to the monolayers was monitored using SPR spectroscopy. This technique showed binding of the guests principally through hydrophobic interactions (the more water-soluble steroids were bound less strongly). Comparison with a SAM of octadecylthiol confirmed this hypothesis, as a significant signal was observed here also, although with

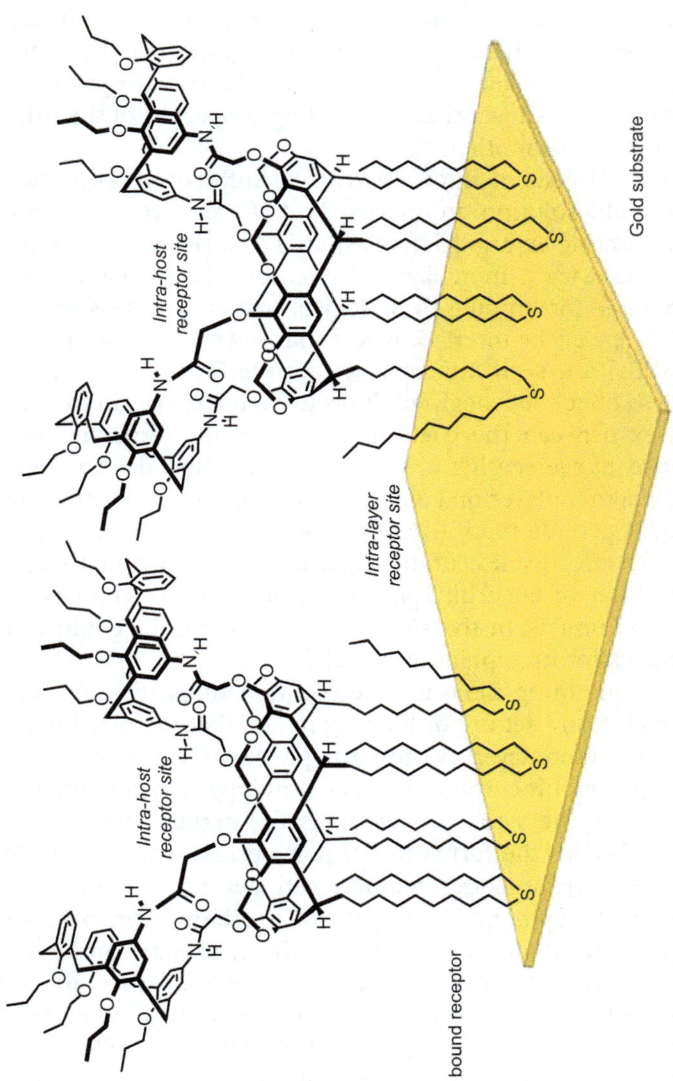

Figure 5.25 The structure of a steroid-binding surface-anchored receptor, comprising two calix[4]arenes attached to a central resorcin[4]arene with thioether anchor groups. The desired intra-host receptor and serendipitous intra-layer receptor sites are indicated.

different selectivities to the supramolecular host. Filling the host molecular layer with 11-mercaptouncecan-1-ol reduced the binding of steroids, showing the importance of a principally hydrophobic environment on the binding. This result indicated the presence of two binding sites for the steroids in the monolayer of the host (Figure 5.25), the designed ones in the cavities of the adsorbate molecules—the intra-host receptor site—and the spaces between the molecules—the intra-layer receptor spaces—in the relatively low density monolayer. The attachment of multivalent adsorpts to different surface substrates has shown that the binding properties of the printboards are very similar to one another.[181–184]

The adhesion of cells at interfaces can be influenced using host–guest chemistry at gold–solution interfaces.[185] A ferrocenyl-substituted cyclic peptide incorporating an arginylglycylaspartic acid (RGD) cell binding motif forms a complex with a monolayer of cucurbit[7]uril leaving the peptide sequence available for binding to biological matter, and this was observed during microscopy of the interface where many cells (human umbilical vein endothelial cells) are seen attached at the interface (Figure 5.26). It was necessary to use oligoethyleneglycol thiols to fill the spaces in the monolayer of the complex between the cucurbit[7]uril and the ferrocenyl-RGD compound, because (as seen earlier in this chapter) the host does not apparently form a complete monolayer on gold. Control experiments on the bare metal or with the RAD peptide motif (that does not bind cells as effectively as the RGD) proved the effectiveness of the approach, and clearly neither bare gold nor the monolayer of cucurbit[7]uril effect adhesion of the cells. Release of the ferrocenyl unit from the surface electrochemically could lead to responsive surfaces for manipulation of cells.

Monolayers containing molecular recognition units directed away from the initial interface are set up for binding material at the new interface. For example, a mixed monolayer incorporating a thymine-terminated SAM was able to bind a polymer incorporating 2,6-diamidopyridine moieties.[186] Apart from adhering polymers to a functionalised interface, the surface functionality can influence the formation of polymers. Chemisorbed SAMs prepared from 12-mercaptododecanamide can act as a template for the formation of hydrogen-bonded adlayers of a different primary amide.[187] When a diacetylene-containing monomer with a primary amide hydrogen was physisorbed from decalin and decalin–toluene mixtures, an organised adlayer with similar organisation to the templating SAM film was formed (Figure 5.27), as shown by grazing angle IR spectroscopy, ellipsometry and contact angle measurements. Exposure of this hetero-bi-layer system to ultraviolet light led to polymerisation of the diacetylenes to produce a polymer in the layer. Resonance Raman spectroscopy proved the change in the conjugation in the adlayer. The polymerisation of diacetylenes is highly dependent on their packing, and the adsorption of thiols of these kinds of compounds on gold does not lead to efficient reaction between the moieties and it has been demonstrated that a layer of silver on the gold modifies the packing of the chemisorbed diacetylene thiolate and allows

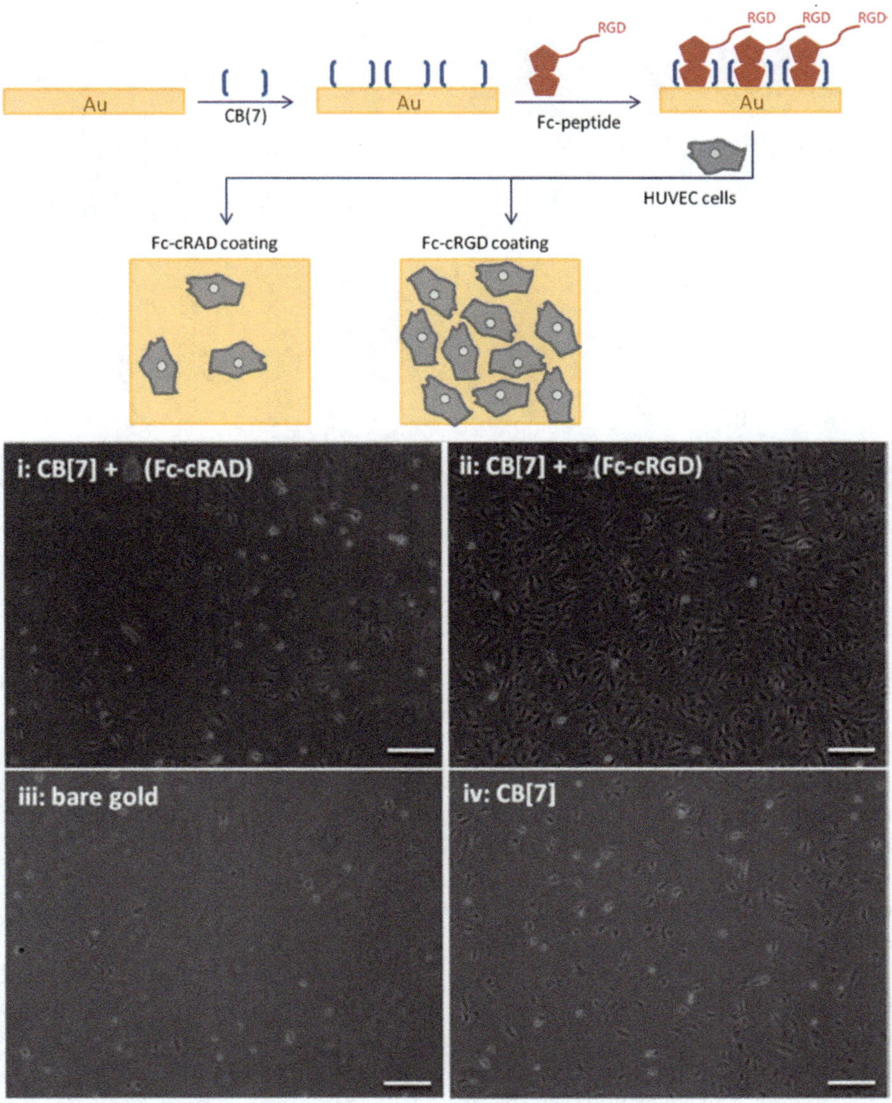

Figure 5.26 The use of a cucurbit[7]uril (CB[7]) monolayer on gold for the promotion of cell adhesion. The cartoons at the top show the binding of ferrocenyl-substituted peptides to the monolayer of cyclic hosts and the schematic view of the growth of cells on RAD (poor cell binding) and RGD (strong cell binding) peptide layers. Below are optical micrographs showing the growth of a kind of cell (HUVEC) on the monolayers with different functional groups as well as bare gold (scale bar: 200 μm). Reproduced from ref. 185 with permission from the Royal Society of Chemistry.

polymerisation.[188] Therefore, molecular alignment for a specific function can be achieved through hydrogen bonding on adsorbed layers or through the differences in adsorbate orientation as a function of substrate.

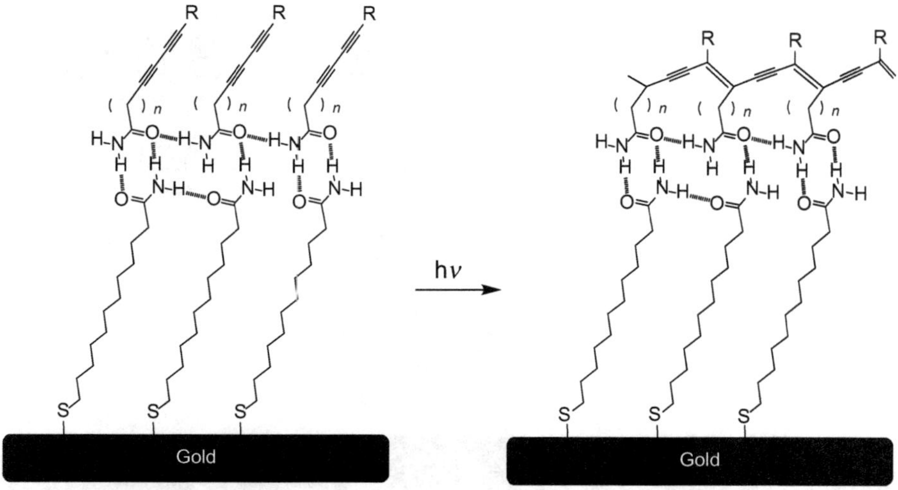

Figure 5.27 The surface-templated synthesis of a poly(diacetylene) by specific hydrogen bonding between a chemisorbed primary amide and an amide-derived diacetylene that adsorbs to it.

The amide functionality present in the side-chains of triethoxy silanes has also proven to be of great advantage in the preparation of spontaneously aligned films on indium tin oxide (ITO) that can be used as alignment layers for liquid crystals.[189] The size of the grooves in the substrate is amplified over three orders of magnitude by this anisotropic grafting procedure that apparently generates rigid siloxane oligomers.[190] In addition, the presence of a 4-amidopyridyl group at the opposite extreme to the triethoxy silanes' functionality affords surfaces that can be used to bind Lewis acid type complexes for the alignment of functional materials on the transparent electrical conductor.[191]

The interactions between surface-bound hosts and solution-dissolved guests clearly give the chance for multivalent binding, where several groups from the two components interact and increase significantly the molar binding constant and therefore the adsorption ability. This avenue has been explored for the layer-by-layer formation of thin films, which shall be discussed in the next chapter.

References

1. A. Ulman, *Chem. Rev.*, 1996, **96**, 1533.
2. D. K. Schwartz, *Annu. Rev. Phys. Chem.*, 2001, **52**, 107.
3. C.-H. Chang and E. I. Franses, *Colloids Surf.*, 1995, **100**, 1.
4. F. Schreiber, *J. Phys.: Condens. Matter*, 2004, **16**, R881.
5. F. Schreiber, A. Eberhardt, T. Y. B. Leung, P. Schwartz, S. M. Wetterer, D. J. Lavrich, L. Berman, P. Fenter, P. Eisenberger and G. Scoles, *Phys. Rev. B: Condens. Matter Mater. Phys.*, 1998, **57**, 12476.

6. F. P. Zamborini and R. M. Crooks, *Langmuir*, 1998, **14**, 3279.
7. R. S. Clegg and J. E. Hutchison, *J. Am. Chem. Soc.*, 1999, **121**, 5319.
8. M. Kim, J. N. Hohman, A. C. Serino and P. S. Weiss, *J. Phys. Chem. C*, 2010, **114**, 19744.
9. P. N. Angelova, K. Hinrichs, I. L. Philipova, K. V. Kostova and D. T. Tsankov, *J. Phys. Chem. C*, 2010, **114**, 1253.
10. D. D. M. Wayner and R. A. Wolkow, *J. Chem. Soc., Perkin Trans. 2*, 2002, 23.
11. J. M. Buriak, *Chem. Rev.*, 2002, **102**, 1271.
12. S. P. Pujari, L. Scheres, A. T. M. Marcelis and H. Zuilhof, *Angew. Chem., Int. Ed.*, 2014, **53**, 6322.
13. A. Edwards, C. Osborne, S. Webster, D. Klenerman, M. Joseph, P. Ostovar and M. Doyle, *Corros. Sci.*, 1994, **36**, 315.
14. G. K. Jennings and P. E. Laibinis, *Colloids Surf., A*, 1996, **116**, 105.
15. S. Ramachandran, B.-L. Tsai, M. Blanco, H. Chen, Y. Tang and W. A. Goddard III, *Langmuir*, 1996, **12**, 6419.
16. L. Y. Li, S. F. Chen, J. Zheng, B. D. Ratner and S. Y. Jiang, *J. Phys. Chem. B*, 2005, **109**, 2934.
17. N. Sheppard, *Annu. Rev. Phys. Chem.*, 1988, **39**, 589.
18. N. Sheppard and C. de la Cruz, *Adv. Catal.*, 1998, **42**, 181.
19. S. M. Barlow and R. Raval, *Surf. Sci. Rep.*, 2003, **50**, 201.
20. F. Rosei, M. Schunack, Y. Naitoh, P. Jiang, A. Gourdon, E. Laegsgaard, I. Stensgaard, C. Joachim and F. Besenbacher, *Prog. Surf. Sci.*, 2003, **71**, 95.
21. J. M. Basset, F. Lefebvre and C. Santini, *Coord. Chem. Rev.*, 1998, **178**, 1703.
22. M. H. Schoenfisch and J. E. Pemberton, *J. Am. Chem. Soc.*, 1998, **120**, 4502.
23. C. Vericat, M. E. Vela, G. Corthey, E. Pensa, E. Cortés, M. H. Fonticelli, F. Ibañez, G. E. Benitez, P. Carro and R. C. Salvarezza, *RSC Adv.*, 2014, **4**, 27730.
24. H. Häkkinen, *Nat. Chem.*, 2012, **4**, 443.
25. O. Voznyy, J. J. Dubowski, J. T. Yates and P. Maksymovych, *J. Am. Chem. Soc.*, 2009, **131**, 12989.
26. A. M. Sharif, F. R. Laffir, D. N. Buckley and C. Silien, *Chem. Phys.*, 2014, **441**, 77.
27. H. Over, *Prog. Surf. Sci.*, 1998, **58**, 249.
28. S. Titmuss, A. Wander and D. A. King, *Chem. Rev.*, 1996, **96**, 1291.
29. D. P. Woodruff, *Appl. Surf. Sci.*, 2007, **254**, 76.
30. P. E. Laibinis, G. M. Whitesides, D. L. Allara, Y. Tai Tao, A. N. Parikh and R. G. Nuzzo, *J. Am. Chem. Soc.*, 1991, **113**, 7152.
31. J. C. Love, L. A. Estroff, J. K. Kriebel, R. G. Nuzzo and G. M. Whitesides, *Chem. Rev.*, 2005, **105**, 1103.
32. L. Srisombata, A. C. Jamison and T. R. Lee, *Colloids Surf., A*, 2011, **390**, 1.
33. G. E. Poirier, *Chem. Rev.*, 1997, **97**, 1117.

34. C. Vericat, M. E. Vela, G. Benitez, P. Carro and R. C. Salvarezza, *Chem. Soc. Rev.*, 2010, **39**, 1805.
35. L. M. Molina and B. Hammer, *Chem. Phys. Lett.*, 2002, **360**, 264.
36. Y. Morikawa, C. Liew and H. Nozoye, *Surf. Sci.*, 2002, **514**, 389.
37. M. Yu, N. Bovet, C. J. Satterley, S. Bengio, K. R. J. Lovelock, P. K. Milligan, R. G. Jones, D. P. Woodruff and V. Dhanak, *Phys. Rev. Lett.*, 2006, **97**, 166102.
38. R. Mazzarello, A. Cossaro, A. Verdini, R. Rousseau, L. Casalis, M. F. Danisman, L. Floreano, S. Scandolo, A. Morgante and G. Scoles, *Phys. Rev. Lett.*, 2007, **98**, 016102.
39. H. Gronbeck and H. Hakkinen, *J. Phys. Chem. B*, 2007, **111**, 3325.
40. H. Gronbeck, H. Hakkinen and R. L. Whetten, *J. Phys. Chem. C*, 2008, **112**, 15940.
41. D.-e. Jiang, *Chem. Phys. Lett.*, 2009, **477**, 90.
42. O. Voznyy and J. J. Dubowski, *Langmuir*, 2009, **25**, 7353.
43. D. C. Jackson, A. Chaudhuri, T. J. Lerotholi, D. P. Woodruff, R. G. Jones and V. R. Dhanak, *Surf. Sci.*, 2009, **603**, 807.
44. M. G. Roper, M. P. Skegg, C. J. Fisher, J. J. Lee, V. R. Dhanak, D. P. Woodruff and R. G. Jones, *Chem. Phys. Lett.*, 2004, **389**, 87.
45. S. Xu, S. J. N. Cruchon-Dupeyrat, J. C. Garno, G.-Y. Liu, G. K. Jennings, T.-H. Yong and P. E. Laibinis, *J. Chem. Phys.*, 1998, **108**, 5002.
46. A. Badia, R. B. Lennox and L. Reven, *Acc. Chem. Res.*, 2000, **33**, 475.
47. A. E. Baber, S. C. Jensen, E. V. Iski and E. C. H. Sykes, *J. Am. Chem. Soc.*, 2006, **128**, 15384.
48. O. Cavalleri, A. Hirstein and K. Kern, *Surf. Sci. Lett.*, 1995, **340**, L960.
49. D. Castner, K. Hinds and D. W. Grainger, *Langmuir*, 1996, **12**, 5083.
50. G. E. Poirier, W. P. Fitts and J. M. White, *Langmuir*, 2001, **17**, 1176.
51. R. Staub, M. Toerker, T. Fritz, T. Schmitz-Hübsch, F. Sellam and K. Leo, *Langmuir*, 1998, **14**, 6693.
52. K. Sotthewes, H. Wu, A. Kumar, G. J. Vancso, P. M. Schön and H. J. W. Zandvliet, *Langmuir*, 2013, **29**, 3662.
53. E. Barrena, C. Ocal and M. Salmeron, *J. Chem. Phys.*, 2001, **114**, 4210.
54. M. Toerker, R. Staub, T. Fritz, T. Schmitz-Hübsch, F. Sellam and K. Leo, *Surf. Sci.*, 2000, **445**, 100.
55. A. Bashir, W. Azzam, M. Rohwerder and A. Terfort, *Langmuir*, 2013, **29**, 13449.
56. T. Ishida, W. Mizutani, H. Azehara, F. Sato, N. Choi, U. Akiba, M. Fujihira and H. Tokumoto, *Langmuir*, 2001, **17**, 7459.
57. C. Fuxen, W. Azzam, R. Arnold, A. Terfort, G. Witte and C. Wöll, *Langmuir*, 2001, **17**, 3689.
58. T. Ishida, W. Mizutani, H. Azehara, K. Miyake, Y. Aya, S. Sasaki and H. Tokumoto, *Surf. Sci.*, 2002, **514**, 187.
59. T. Nakamura, H. Kondoh, M. Matsumoto and H. Nozoye, *Langmuir*, 1996, **12**, 5977.
60. H. Ahn, M. Kim, D. J. Sandman and J. E. Whitten, *Langmuir*, 2003, **19**, 5303.

61. T. Ikeda, Y. Nagata, Y. Zheng, D. Liu, H.-J. Butt and M. Shimoda, *Langmuir*, 2014, **30**, 1536.
62. R. S. Clegg and J. E. Hutchison, *Langmuir*, 1996, **12**, 5239.
63. A. Ulman, J. E. Eilers and N. Tillman, *Langmuir*, 1989, **5**, 1147.
64. S. Sek, B. Palys and R. Bilewicz, *J. Phys. Chem. B*, 2002, **106**, 5907.
65. H.-H. Lee, Ž. Ruželė, L. Malysheva, A. Onipko, A. Gutés, F. Björefors, R. Valiokas and B. Liedberg, *Langmuir*, 2009, **25**, 13959.
66. R. Valiokas, S. Svedhem, S. C. T. Svensson and B. Liedberg, *Langmuir*, 1999, **15**, 3390.
67. A. Kuhnle, T. R. Linderoth, B. Hammer and F. Besenbacher, *Nature*, 2002, **415**, 891.
68. A. Kuhnle, T. R. Linderoth and F. Besenbacher, *Top. Catal.*, 2011, **54**, 1384.
69. A. Kuhnle, T. R. Linderoth and F. Besenbacher, *J. Am. Chem. Soc.*, 2003, **125**, 14680.
70. V. Humblot, A. Vallée, A. Naitabdi, F. Tielens and C.-M. Pradier, *J. Am. Chem. Soc.*, 2012, **134**, 6579.
71. Q. i. An, G. Li, C. Tao, Y. Li, Y. Wu and W. Zhang, *Chem. Commun.*, 2008, 1989.
72. C. S. H. Dominguez, M. C. Quintana and P. Hernandez, *Electroanalysis*, 2013, **25**, 1217.
73. A. Gomez-Casado, P. Jonkheijm and J. Huskens, *Langmuir*, 2011, **27**, 11508.
74. S. J. C. Lee, J. W. Lee, H. H. Lee, J. Seo, D. H. Noh, Y. H. Ko, K. Kim and H. I. Kim, *J. Phys. Chem. B*, 2013, **117**, 8855.
75. P. Chinwangso, A. C. Jamison and T. R. Lee, *Acc. Chem. Res.*, 2011, **44**, 511.
76. Y.-S. Shon and T. R. Lee, *Langmuir*, 1999, **15**, 1136.
77. B. Singhana, A. C. Jamison, J. Hoang and T. R. Lee, *Langmuir*, 2013, **29**, 14108.
78. Q. Cheng and A. Brajter-Toth, *Anal. Chem.*, 1992, **64**, 1998.
79. T. M. Willey, A. L. Vance, C. Bostedt, T. van Buuren, R. W. Meulenberg, L. J. Terminello and C. S. Fadley, *Langmuir*, 2004, **20**, 4939.
80. X. Stammer, K. Tonigold, A. Bashir, D. Käfer, O. Shekhah, C. Hülsbusch, M. Kind, Axel Groß and Christof Wöll, *Phys. Chem. Chem. Phys.*, 2010, **12**, 6445.
81. E. Gomar-Nadal, G. K. Ramachandran, F. Chen, T. Burgin, C. Rovira, D. B. Amabilino and S. M. Lindsay, *J. Phys. Chem. B*, 2004, **108**, 7213.
82. M. G. Roper and R. G. Jones, *Langmuir*, 2005, **21**, 11684.
83. M. G. Roper and R. G. Jones, *Phys. Chem. Chem. Phys.*, 2008, **10**, 1336.
84. T. Bürgi, *Nanoscale*, 2015, **7**, 15553.
85. J. B. Schlenoff, M. Li and H. Ly, *J. Am. Chem. Soc.*, 1995, **117**, 12528.
86. J. N. Hohman, J. C. Thomas, Y. Zhao, H. Auluck, M. Kim, W. Vijselaar, S. Kommeren, A. Terfort and P. S. Weiss, *J. Am. Chem. Soc.*, 2014, **136**, 8110.
87. I. Touzov and C. B. Gorman, *J. Phys. Chem. B*, 1997, **101**, 5263.

88. M. T. Cygan, T. D. Dunbar, J. J. Arnold, L. A. Bumm, N. F. Shedlock, T. P. Burgin, L. Jones, D. L. Allara, J. M. Tour and P. S. Weiss, *J. Am. Chem. Soc.*, 1998, **120**, 2721.
89. L. Mu, Y. Liu, S. Zhang, B. Liu and J. Kong, *New J. Chem.*, 2005, **29**, 847.
90. P. E. Laibinis, M. A. Fox, J. P. Folkers and G. M. Whitesides, *Langmuir*, 1991, 7, 3167.
91. C. J. Sandroff and D. R. Herschbach, *J. Phys. Chem.*, 1982, **86**, 3277.
92. M. Yu, D. P. Woodruff, C. J. Satterley, R. G. Jones and V. R. Dhanak, *J. Phys. Chem. C*, 2007, **111**, 10040.
93. H.-T. Rong, S. Frey, Y.-J. Yang, M. Zharnikov, M. Buck, M. Wühn, C. Wöll and G. Helmchen, *Langmuir*, 2001, **17**, 1582.
94. F. Tao and S. L. Bernasek, *Chem. Rev.*, 2007, **107**, 1408.
95. A. Shaporenko, J. Müller, T. Weidner, A. Terfort and M. Zharnikov, *J. Am. Chem. Soc.*, 2007, **129**, 2232.
96. D. Gnatek, S. Schuster, J. Ossowski, M. Khan, J. Rysz, S. Krakert, A. Terfort, M. Zharnikov and P. Cyganik, *J. Phys. Chem. C*, 2015, **119**, 25929.
97. D. F. Lim, S. K. Goulet, M. Bergkvist, J. Wei, K. C. Leong and C. S. Tan, *J. Electrochem. Soc.*, 2011, **158**, H1057.
98. O. Azzaroni, M. Cipollone, M. E. Vela and R. C. Salvarezza, *Langmuir*, 2001, **17**, 1483.
99. P. E. Laibinis and G. M. Whitesides, *J. Am. Chem. Soc.*, 1992, **114**, 1990.
100. K. Wong, K.-Y. Kwon, B. V. Rao, A. Liu and L. Bartels, *J. Am. Chem. Soc.*, 2004, **126**, 7762.
101. H. Kondo, N. Saito, F. Matsui, T. Tsuzuki, T. Ota and H. Yoem, *Nippon Kagakkai Koen Yokoshu*, 1999, **76**, 158.
102. K. L. Wong, X. Lin, K.-Y. Kwon, G. Pawin, B. V. Rao, A. Liu, L. Bartels, S. Stolbov and T. S. Rahman, *Langmuir*, 2004, **20**, 10928.
103. S.-Y. Hong, P.-C. Yeh, J. I. Dadap and R. M. Osgood, *ACS Nano*, 2012, **6**, 10622.
104. S. M. Driver and D. P. Woodruff, *Langmuir*, 2000, **16**, 6693.
105. N. P. Prince, D. L. Seymour, D. P. Woodruff, R. G. Jones and W. Walter, *Surf. Sci.*, 1989, **215**, 566.
106. H. Ron, H. Cohen, S. Matlis, M. Rappaport and I. Rubinstein, *J. Phys. Chem. B*, 1998, **102**, 9861.
107. S. Romer, B. Behzadi, R. Fasel and K.-H. Ernst, *Chem. – Eur. J.*, 2005, **11**, 4149.
108. M. Ortega-Lorenzo, C. J. Baddeley, C. Muryn and R. Raval, *Nature*, 2000, **404**, 376.
109. G. Contini, P. Gori, M. Di Giovannantonio, N. Zema, S. Turchini, D. Catone, T. Prosperi and A. Palma, *J. Phys. Chem. C*, 2013, **117**, 10545.
110. L. Pérez-García and D. B. Amabilino, *Chem. Soc. Rev.*, 2007, **36**, 941.
111. M. Parschau, T. Kampen and K.-H. Ernst, *Chem. Phys. Lett.*, 2005, **407**, 433.
112. V. Humbolt, M. Ortega-Lorenzo, C. J. Baddeley, S. Haq and R. Raval, *J. Am. Chem. Soc.*, 2004, **126**, 6460.

113. J. F. Lambert, *Origins Life Evol. Biospheres*, 2008, **38**, 211.
114. L. L. Atanasoska, J. C. Buchholz and G. A. Somorjai, *Surf. Sci.*, 1978, **72**, 189.
115. E. N. Yitamben, A. Clayborne, S. B. Darling and N. P. Guisinger, *Nanotechnology*, 2015, **26**, 235604.
116. S. M. Barlow, S. Louafi, D. Le Roux, J. Williams, C. Muryn, S. Haq and R. Raval, *Surf. Sci.*, 2005, **590**, 243.
117. S. M. Barlow, S. Louafi, D. Le Roux, J. Williams, C. Muryn, S. Haq and R. Raval, *Langmuir*, 2004, **20**, 7171.
118. Q. Chen, C. W. Lee, D. J. Frankel and N. V. Richardson, *PhysChem-Comm*, 1999, **2**, 41.
119. S. Blankenburg and W. G. Schmidt, *Nanotechnology*, 2007, **18**, 424030.
120. M. Forster, M. S. Dyer, M. Persson and R. Raval, *J. Am. Chem. Soc.*, 2009, **131**, 10173.
121. M. Forster, M. S. Dyer, M. Persson and R. Raval, *Angew. Chem., Int. Ed.*, 2010, **49**, 2344.
122. M. Forster, M. S. Dyer, S. D. Barrett, M. Persson and R. Raval, *J. Phys. Chem. C*, 2011, **115**, 1180.
123. A. G. Mark, M. Forster and R. Raval, *ChemPhysChem*, 2011, **12**, 1474.
124. M. Forster, M. S. Dyer, M. Persson and R. Raval, *J. Am. Chem. Soc.*, 2011, **133**, 15992.
125. I. Stensgaard, *Surf. Sci.*, 2003, **545**, L747.
126. M. Lingenfelder, G. Tomba, G. Costantini, L. Colombi-Ciacchi, A. De Vita and K. Kern, *Angew. Chem., Int. Ed.*, 2007, **46**, 4492.
127. G. Tomba, M. Lingenfelder, G. Costantini, K. Kern, F. Klappenberger, J. V. Barth, L. C. Ciacchi and A. De Vita, *J. Phys. Chem. A*, 2007, **111**, 12740.
128. Y. Wang, M. Lingenfelder, T. Classen, G. Costantini and K. Kern, *J. Am. Chem. Soc.*, 2007, **129**, 15742.
129. H. Cruguel, C. Methivier, C. M. Pradier and V. Humblot, *Chirality*, 2015, **27**, 411.
130. S. M. Barlow, S. Haq and R. Raval, *Langmuir*, 2001, **17**, 3292.
131. S. Haq, F. Hanke, M. S. Dyer, M. Persson, P. Iavicoli, D. B. Amabilino and R. Raval, *J. Am. Chem. Soc.*, 2011, **133**, 12031.
132. K. Diller, F. Klappenberger, M. Marschall, K. Hermann, A. Nefedov, C. Wöll and J. V. Barth, *J. Chem. Phys.*, 2012, **136**, 14705.
133. S. Ditze, M. Stark, M. Drost, F. Buchner, H.-P. Steinrück and H. Marbach, *Angew. Chem., Int. Ed.*, 2012, **51**, 10898.
134. A. Wiengarten, K. Seufert, W. Auwärter, D. Ecija, K. Diller, F. Allegretti, F. Bischoff, S. Fischer, D. A. Duncan, A. C. Papageorgiou, F. Klappenberger, R. G. Acres, T. H. Ngo and J. V. Barth, *J. Am. Chem. Soc.*, 2014, **136**, 9346.
135. R. J. Archer, *J. Electrochem. Soc.*, 1957, **104**, 619.
136. S. I. Raider, R. Flitsch and M. J. Palmer, *J. Electrochem. Soc.*, 1975, **122**, 413.
137. W. D. Bascom, *J. Colloid Interface Sci.*, 1968, **27**, 789.

138. I. Haller, *J. Am. Chem. Soc.*, 1978, **100**, 8050.
139. A. Ulman, *An Introduction to Ultrathin Organic Films*, Academic Press, Boston, 1991.
140. N. L. Parikh, D. L. Allara, I. B. Azouz and F. Rondelez, *J. Phys. Chem.*, 1994, **98**, 7577.
141. P. Silberzan, L. Léger, D. Ausserré and J. J. Benattar, *Langmuir*, 1991, 7, 1647.
142. R. M. Diebold and D. R. Clarke, *Langmuir*, 2012, **28**, 15513.
143. J. Zhang, J. Hoogboom, P. H. J. Kouwer, A. E. Rowan and T. Rasing, *J. Phys. Chem. C*, 2008, **112**, 20105.
144. S. Onclin, B. J. Ravoo and D. N. Reinhoudt, *Angew. Chem., Int. Ed.*, 2005, **44**, 6282.
145. O. Acton, D. Hutchins, L. Árnadóttir, T. Weidner, N. Cernetic, G. G. Ting, T.-W. Kim, D. G. Castner, H. Ma and A. K.-Y. Jen, *Adv. Mater.*, 2011, **23**, 1899.
146. D. L. Allara and R. G. Nuzzo, *Langmuir*, 1985, **1**, 45.
147. D. L. Allara and R. G. Nuzzo, *Langmuir*, 1985, **1**, 52.
148. C. L. Pang, R. Lindsay and G. Thornton, *Chem. Rev.*, 2013, **113**, 3887.
149. Q. Guo and E. M. Williams, *Surf. Sci.*, 1999, **433–435**, 322.
150. P. Rahe, M. Nimmrich, A. Nefedov, M. Naboka, C. Woll and A. Kuhnle, *J. Phys. Chem. C*, 2009, **113**, 17471.
151. B. Li, R. Franking, E. C. Landis, H. Kim and R. J. Hamers, *ACS Appl. Mater. Interfaces*, 2009, **1**, 1013.
152. E. Jaehne, D. Ferse, G. Busch, H.-J. P. Adler, A. Singh and I. K. Varma, *Des. Monomers Polym.*, 2002, **5**, 427.
153. B. Adolphi, E. Jähne, G. Busch and X. Cai, *Anal. Bioanal. Chem.*, 2004, **379**, 646.
154. D. Appelhans, D. Ferse, H.-J. P. Adler, W. Plieth, A. Fikus, K. Grundke, F.-J. Schmitt, T. Bayer and B. Adolphi, *Colloids Surf., A*, 2000, **161**, 203.
155. H. Jing, M. Min, S. Seo, B. Lu, Y. Yoon, S. M. Lee, E. Hwang and H. Lee, *Carbon*, 2015, **86**, 272.
156. H. O. Finklea, *Electroanal. Chem.*, 1996, **19**, 109.
157. P. L. Boulas, M. Gómez-Kaifer and L. Echegoyen, *Angew. Chem., Int. Ed.*, 1998, **37**, 216.
158. A. Kaifer and M. Gómez-Kaifer, *Supramolecular Electrochemistry*, Wiley-VCH, 1999.
159. A. J. Moore, L. M. Goldenberg, M. R. Bryce, M. C. Petty, A. P. Monkman, C. Marenco, J. Yarwood, M. J. Joyce and S. N. Port, *Adv. Mater.*, 1998, **10**, 395.
160. H. Liu, S. Liu and L. Echegoyen, *Chem. Commun.*, 1999, 1493.
161. G. Trippé, M. Oçafrain, M. Besbes, V. Monroche, J. Lyskawa, F. Le Derf, M. Sallé, J. Becher, B. Colonna and L. Echegoyen, *New J. Chem.*, 2002, **26**, 1320.
162. M. A. Herranz, B. Colonna and L. Echegoyen, *Proc. Natl. Acad. Sci. U. S. A.*, 2002, **99**, 5040.

163. L. Zhang, L. A. Godinez, T. Lu, G. W. Gokel and A. E. Kaifer, *Angew. Chem., Int. Ed. Engl.*, 1995, **34**, 235.
164. G. K. Rowe and S. E. Creager, *Langmuir*, 1991, 7, 2307.
165. A. R. Bernardo, T. Lu, E. Cordova, L. Zhang, G. W. Gokel and A. E. Kaifer, *J. Chem. Soc., Chem. Commun.*, 1994, 529.
166. F. Arias, L. A. Godinez, S. R. Wilson, A. E. Kaifer and L. Echegoyen, *J. Am. Chem. Soc.*, 1996, **118**, 6086.
167. R. M. Izatt, K. Pawlak, J. S. Bradshaw and R. L. Bruening, *Chem. Rev.*, 1991, **91**, 1721.
168. W. Liu, Y. Zhang and X. Gao, *J. Am. Chem. Soc.*, 2007, **129**, 4973.
169. X. D. Cui, A. Primak, X. Zarate, J. Tomfohr, O. F. Sankey, A. L. Moore, T. A. Moore, D. Gust, G. Harris and S. M. Lindsay, *Science*, 2001, **294**, 571.
170. G. Wenz, *Angew. Chem., Int. Ed. Engl.*, 1994, **33**, 803.
171. M. T. Rojas, R. Königer, J. F. Stoddart and A. E. Kaifer, *J. Am. Chem. Soc.*, 1995, **117**, 336.
172. Y. Domi, K. Ikeura, K. Okamura, K. Shimazu and M. D. Porter, *Langmuir*, 2011, **27**, 10580.
173. W. Liu, Y. Zhang and X. Gao, *J. Am. Chem. Soc.*, 2007, **129**, 4973.
174. A. Mendez-Ardoy, T. Steentjes, T. Kudernac and J. Huskens, *Langmuir*, 2014, **30**, 3467.
175. M. W. J. Beulen, J. Bügler, M. R. de Jong, B. Lammerink, J. Huskens, H. Schönherr, G. J. Vancso, B. A. Boukamp, H. Wieder, A. Offenhäuser, W. Knoll, F. C. J. M. van Veggel and D. N. Reinhoudt, *Chem. - Eur. J.*, 2000, **6**, 1176.
176. M. R. de Jong, J. Huskens and D. N. Reinhoudt, *Chem. - Eur. J.*, 2001, 7, 4164.
177. A. Fragoso, J. Caballero, E. Almirall, R. Villalonga and R. Cao, *Langmuir*, 2002, **18**, 5051.
178. M. Lahav, K. T. Ranjit, E. Katz and I. Willner, *Chem. Commun.*, 1997, 259.
179. A. Friggeri, F. C. J. M. van Veggel and D. N. Reinhoudt, *Chem. - Eur. J.*, 1999, **5**, 3595.
180. E. U. Thoden van Velzen, J. F. J. Engbersen, P. J. de Lange, J. W. G. Mahy and D. N. Reinhoudt, *J. Am. Chem. Soc.*, 1995, **117**, 6853.
181. S. Onclin, A. Mulder, J. Huskens, B. J. Ravoo and D. N. Reinhoudt, *Langmuir*, 2004, **20**, 5460.
182. T. Auletta, B. Dordi, A. Mulder, A. Sartori, S. Onclin, C. M. Bruinink, M. Peter, C. A. Nijhuis, H. Beijleveld, H. Schönherr, G. J. Vancso, A. Casnati, R. Ungaro, B. J. Ravoo, J. Huskens and D. N. Reinhoudt, *Angew. Chem., Int. Ed.*, 2004, **43**, 369.
183. J. Huskens, A. Mulder, T. Auletta, C. A. Nijhuis, M. J. W. Ludden and D. N. Reinhoudt, *J. Am. Chem. Soc.*, 2004, **126**, 6784.
184. A. Mulder, T. Auletta, A. Sartori, S. Del Ciotto, A. Casnati, R. Ungaro, J. Huskens and D. N. Reinhoudt, *J. Am. Chem. Soc.*, 2004, **126**, 6627.

185. P. Neirynck, J. Brinkmann, Q. An, D. W. J. van der Schaft, L.-G. Milroy, P. Jonkheijm and L. Brunsveld, *Chem. Commun.*, 2013, **49**, 3679.
186. T. B. Norsten, E. Jeoung, R. J. Thibault and V. M. Rotello, *Langmuir*, 2003, **19**, 7089.
187. D. W. Mosley, M. A. Sellmyer, E. J. Daida and J. M. Jacobson, *J. Am. Chem. Soc.*, 2003, **125**, 10532.
188. Y.-H. Chan, J.-T. Lin, I. P. Chen and C.-h. Chen, *J. Phys. Chem. B*, 2005, **109**, 19161.
189. J. Hoogboom, T. Rasing, A. E. Rowan and R. J. M. Nolte, *J. Mater. Chem.*, 2006, **16**, 1305.
190. J. Hoogboom, M. Behdani, J. A. A. W. Elemans, M. A. C. Devillers, R. De Gelder, A. E. Rowan, T. Rasing and R. J. M. Nolte, *Angew. Chem., Int. Ed.*, 2003, **42**, 1812.
191. J. Hoogboom, P. M. L. Garcia, M. B. J. Otten, J. A. A. W. Elemans, J. Sly, S. V. Lazarenko, T. Rasing, A. E. Rowan and R. J. M. Nolte, *J. Am. Chem. Soc.*, 2005, **127**, 11047.

CHAPTER 6

Layer-by-layer Growth

6.1 Introduction

In general, layer-by-layer approaches aim to produce films of a more or less defined thickness on a surface in a sequential manner by formation of non-covalent complexes. The thickness of the film will depend on the number of deposition repetitions that are performed. In this sense, the result might seem similar to the Langmuir–Blodgett (LB) technique for the preparation of multilayers (Chapter 3). However, layer-by-layer (LbL) approaches require a specific interaction between an adsorbent and an adsorbate. In the LB technique, the consecutive layers are forced to be in contact with one another, with potential implications for long-term stability. LbL films rely on spontaneous processes driven thermodynamically, and therefore have intrinsically greater stability. They are self-limiting structures because of the saturation of a surface in a determined moiety that can interact with a different chemical in a subsequent step. LbL approaches where growth is self-limited are distinctive from more general thin film formation where multiple layers can form at any particular point in the multilayer system because growth is equally likely at any level of the growing film. The specificity required for self-limiting growth in LbL methods also makes the choice of components more restrictive.

6.2 From Surface Layer to Sequential Deposition Strategies

Two cartoons showing LbL growth are shown in Figure 6.1. In the first, a surface is treated with a string-like material that interacts strongly with the adsorbent, but not with itself because of the lack of any self-complimentary chemical functionality (step (i)). Because the adsorpt cannot interact strongly with itself, one layer only is formed. If this film is treated with a

Monographs in Supramolecular Chemistry No. 19
Supramolecular Chemistry at Surfaces
By David B. Amabilino

Published by the Royal Society of Chemistry, www.rsc.org

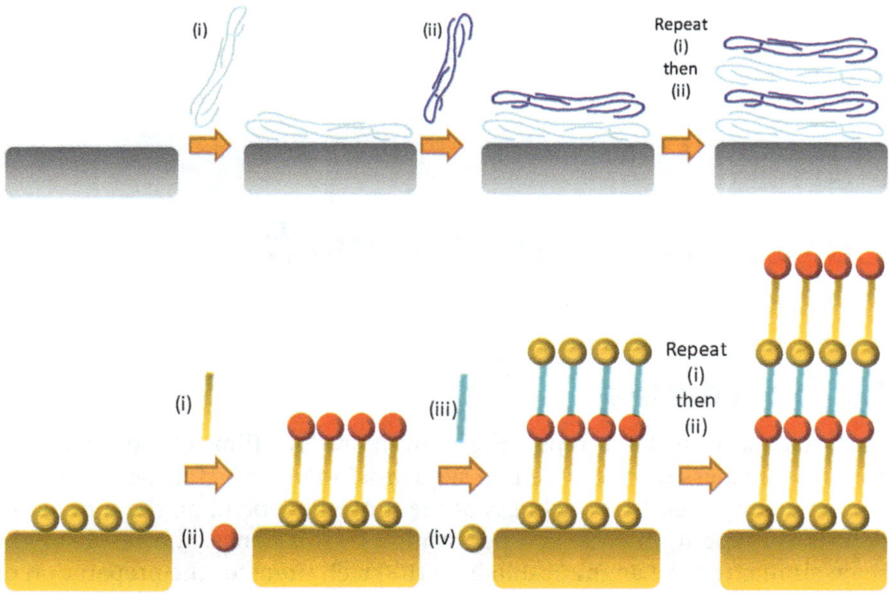

Figure 6.1 Two contrasting types of layer-by-layer assembly where both are self-limiting at each step. At the top, in solution a material (i) is adhered to a surface through multiple interactions, but because the material does not interact significantly with itself a single layer is formed. Subsequently, a second material (ii) is introduced that interacts with the newly-formed surface, but, again, not with itself so that a single layer is formed. Subsequently, new materials can be deposited in a sequential manner. At the bottom, an alternative strategy involves the use of specific nodes on a surface that adsorb a rigid component (i) in a specific orientation so that both "ends" do not interact with the surface. Then a new node (ii) is generated through a specific interaction with the free "end". Subsequent and sequential introduction of rigid linkers and nodes results in layer-by-layer growth. In both cases, excess reagents are removed between steps to avoid defect multilayers.

second string-like material that interacts strongly with the first string-like material, and not with itself, a different second layer is formed (step (ii)). Repeating steps (i) and (ii) results in an alternating film of the two string-like substances. Each layer is self-limiting, because like does not interact with like. Of course, the sequence is not limited to two materials, as long as the different substances obey the self-limiting interaction rules. Each layer would be expected to bind strongly at the interface, essentially irreversibly, and diffusion and rearrangement at the interface in this step would be very limited. Once any particular monolayer has been formed, the character of the interface is changed, alternating after each step.

In the second sequence in Figure 6.1, a surface comprising specific nodes (that might be ions or other chemical functionalities) is exposed first to a rod-like object (step (i)) and then with another node (step (ii)) with different character to the first. In common with the first sequence we discussed, neither rod nor node can interact with themselves. Now, in steps (iii) and (iv), a second rod is bound to the second node, and then the initial node is reintroduced. Because nodes and rods cannot interact with themselves, the process is self-limiting at each step. The sequence can be carried on, in principle, indefinitely. The only limit in principle is the time taken to perform each step.

In both examples, an excess of reagent can be used in every stage, ensuring complete coverage and fast and efficient monolayer formation. The excess adsorpt in the solution could be recycled. At each step it will be essential to remove any excess reagent from the interface. If any layer forming material from step (i) in either example is present when step (ii) is carried out, more than one layer could be formed in certain places. If this were a vacuum process, this sequence would mean removing all of the excess reagent by pumping thoroughly, and in a solution process by washing thoroughly.

These two cartoon examples correspond to polyelectrolyte LbL formation,[1] and other kinds of coordination or non-covalent bond forming LbL film preparation,[2] which will be discussed in the following sections. The approach has certain parallels with atomic layer deposition methods,[3,4] and conceptually similar covalent bonding approaches to layers, which will not be discussed here because of the supramolecular focus of the book, but which are very interesting and efficient in the preparation of films using bottom-up ideas.[5]

6.3 Polyelectrolyte Films at Surfaces Through Layer-by-layer Deposition

The binding of polyelectrolytes—molecules, macromolecules, clusters or colloids bearing a large net charge in their structure with associated low molecular weight counter-ions—to oppositely charged surfaces, of any shape, with the release of the counter-ions of each component of the electrostatic complex is a powerful assembly method because of the strength and lack of directional dependence of electrostatic interactions.[6] This area developed initially from the study of polymeric polyelectrolytes, especially from macromolecules of biological origin.[7] The preparation of multilayers of cationic and anionic amphiphiles was shown to proceed in a sequential manner through charge reversal,[8] and this was subsequently applied to synthetic polymer polyelectrolytes.[9] In principle, the amount of order in these systems is low, in that the polymer chains are usually assumed to be distributed randomly and they are flexible, and so can adapt to local structure in the substrate to which they adsorb.

6.3.1 Preparation of Polyelectrolyte Films

The formation of polyelectrolyte multilayers (PEMs) first requires a charged surface, the initial adsorbent (in fact this initial adsorption is a way of fractionating polyelectrolytes[10]). The surface charge of a material depends on its character, but for perhaps most of the materials relevant in this section it involves a dissociation of groups to give a dissolved cation and negatively-charged surface, as in the case of the surfaces of inorganic oxides for example, or *vice versa*, as is the surface charge. A case of adsorption of a polymer electrolyte to an inorganic surface is that of poly(4-styrene sulfonate) on platinum from aqueous sodium chloride solution.[11] Ellipsometry proved that the adsorption depended on the salt concentration, but was independent of the molecular weight of the polymer. This kind of adsorption is the first step in the formation of multilayers. Figure 6.2 shows the principle of formation of these PEM films and representative polymers that can be used. As described in a more abstract and general way, the process starts with a charged substrate (in the example negatively charged) and is exposed to a solution containing an oppositely charged polyelectrolyte that adheres to the surface. The polymer adsorbs so that the charges at the surface of the substrate are compensated but much of the polymer remains separate from the initial substrate interface and a new interface is generated comprised of part of the adsorbed polymer and its small counter-ions. It is clear that a driving force for adsorption is the entropic gain of liberating small counter-ions to the medium, both from the initial interface and from the polymer.[12] This entropic effect was confirmed by a thorough analysis of the thermodynamics of the assembly using isothermal calorimetry.[13] The figure shows examples of polymers of this type, such as protonated poly(ethylene imine) as the poly(cation) or poly(4-styrenesulfonate) for the poly(anion). The solution used for the adsorption process contains a large excess of polyelectrolyte that is necessary to ensure charge inversion, but this excess should not adsorb at the interface because it has like charge, and thus the structure is self-limiting. Nevertheless, it is necessary to ensure that no excess of the polymer in solution remains, otherwise local multilayers may form, and thus washing the surface with pure solvent is required. After this thorough rinsing of the surface, exposure to a solution of an oppositely charged polymer results in the formation of a hetero-bilayer where once again the newly formed surface has inverted charge compared with the starting substrate. Again the growth is self-limiting. Rinsing, and repartition of these steps with the same two polymers leads to the formation of multilayers with alternating character of the individual components. Apart from the examples given here, there is a huge variety of polymers, from those with an extremely high density of charge—up to one cation or anion per carbon atom in the polymer backbone in the case of poly(fumaric acid) and poly(methyleneamine)[14]—to others where large and complex structures are combined, as shall be presented in the last section of this chapter. The process does not rely on having a smooth surface from the outset, and defects on the initial substrate can be smoothed over by the polymers, although obviously the films are not molecularly smooth.

Alternate treatment of a charged surface with polyelectrolytes of different types leads to a wide variety of unique films.[15] The versatility of the

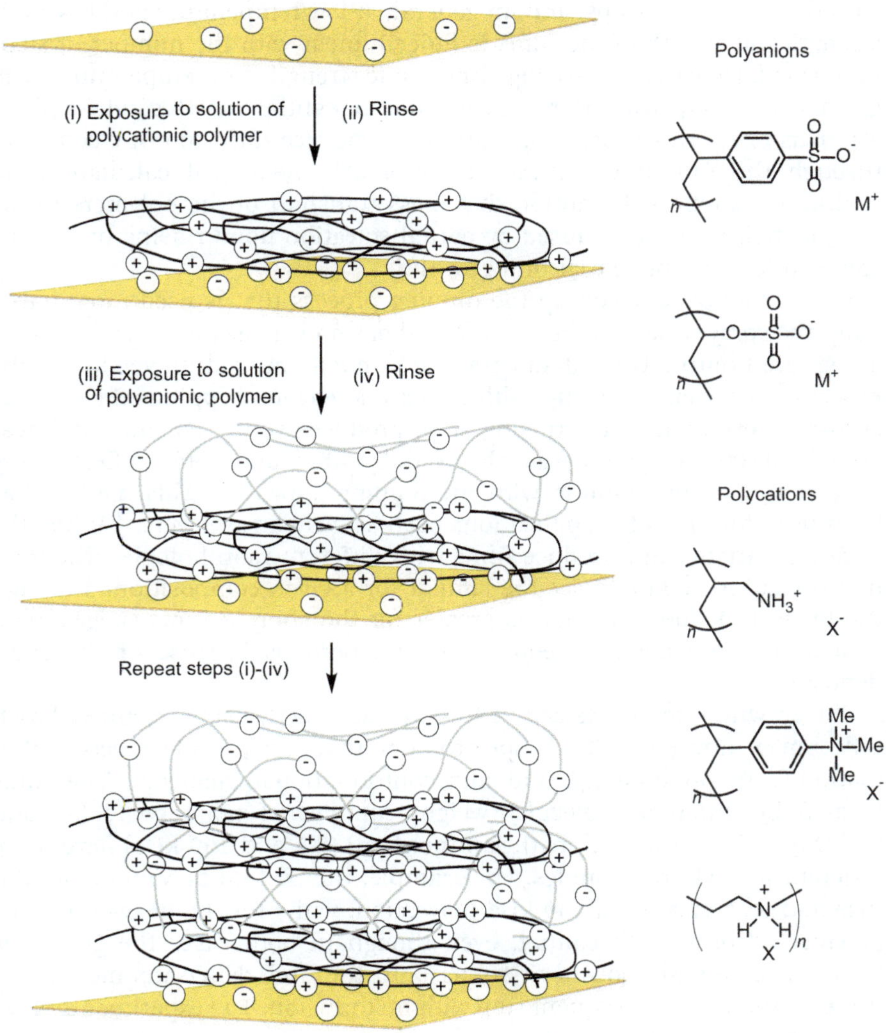

Figure 6.2 A cartoon showing the steps in the formation of polyelectrolyte thin films by the alternating exposure of a surface with net negative charge to polymers with positive and negative charge, respectively. The structures on the right are typical charged synthetic polymers that can be used in the process.

technique is tremendous when one considers that any charged component materials can be employed. While the general picture of assembly of the layers is relatively uncontroversial, many authors assert that the exact mechanism of film assembly remains available for an exact explanation. This picture of PEMs constructed using the LbL method—while wholly logical and under certain conditions largely correct—is ideal and can be inaccurate. For example, in principle a linear growth in thickness is expected with the number of steps, and, in general, at low temperature and low ionic strength the growth of the films is indeed linear with the number of steps performed. However, increasing either ionic strength[16] or temperature[17] can give rise to a "superlinear" regime where exponential growth can take place. The reason for this process appears to be the free diffusion of electrolytes through the film under these conditions.[18] Theoretical calculations of multilayers support the notion that mass transport of the polymers in the film, as well as defects caused by partial solvation upon rinsing or defects, can lead to the exponential film growth.[19]

In an attempt to speed up the dipping process that is used more habitually, the spray variant of the LbL method has proven useful, where solutions are sprayed onto a vertical substrate in a much speeded up version of the process.[20] Alternant spraying, with the excess reagent dripping off the initial substrate, even without a rinsing step, produces very even films. A linear growth of the thickness was observed by AFM and X-ray reflectometry measurements, in common with the dipping approach. This method has been used for PEMs for applications such as gas barrier layers.[21] While the different variants can produce films with different morphologies, the thermal transitions taking place are similar for a given composition, although the breadth of the transition is greater for the sprayed films suggesting a somewhat more heterogeneous structure, perhaps because of the short deposition time.[22]

The growth of the PEMs can be followed accurately using attenuated total reflectance infrared (ATR-IR) spectroscopy that can provide precise information both on coverage and any conformational change.[23] The films formed by different molecular weight samples of polyacrylate (30 and 60 kDa) and poly(diallyldimethylammonium) (8.5 and 150 kDa) have quite different formation dynamics. For example, the 30 kDa polyanion can diffuse and rearrange on a 150 kDa polycation, while for the heavier anionic polymer adsorption, it cannot, even though in both cases the growth is linear. The cationic polymer flattens onto the underlying anionic surface from solution in a subsequent step with no diffusion. The experiments show that diffusion does not necessarily give rise to exponential growth, and show how real-time spectroscopy can provide valuable information concerning adsorption mechanisms.

The growth behaviour of the PEMs is not necessarily either linear or exponential, and can show two linear regimes where one corresponds to the near-interface growth and the other bulk growth.[24] The spray variant of the LbL method was used[25] for the preparation of films incorporating the inorganic polyelectrolyte poly(sodium phosphate) (inorganic, water-soluble, and displaying unusual interactions with charged species[26]) and considerably larger poly(allylamine hydrochloride), which shows very unusual concentration-dependent growth characteristics (Figure 6.3). At the lowest concentrations studied (10^{-5} and 10^{-4} M), the growth is apparently linear, while at 10^{-3} M, exponential growth prevails. At 10^{-2} M, the growth has to be fitted to a combination of exponential then linear models. While optical interference in

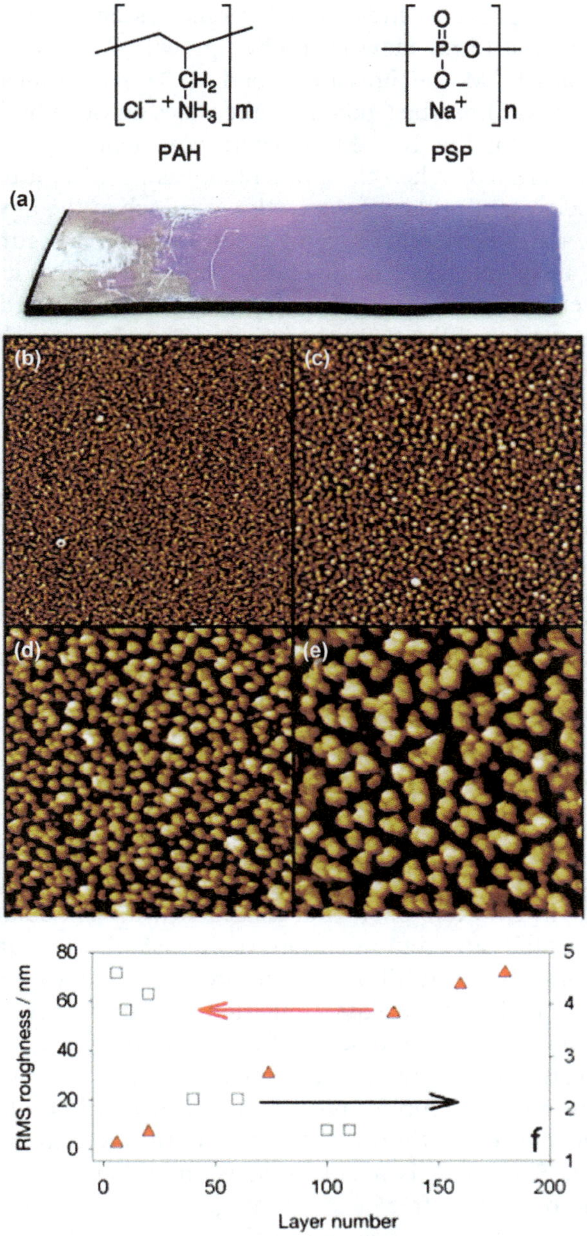

Figure 6.3 The chemical structures of polyelectrolytes used for PEM formation (top) followed by a photograph of a 150-layer film (with handling marks on the left) (a) and AFM images ((b)–(e), all 2 μm×2 μm) of PEMs made by spray deposition at a concentration of each polymer of 10^{-4} M (with 0.15 M NaCl at pH 6.7). The images show the topography of 6, 20, 60 and 110 layers. The evident increase in roughness is plotted in the graph (f) for polymer concentrations of 10^{-4} M (left-hand axis) and 10^{-3} M (right-hand axis). Reprinted with permission from ref. 25. Copyright 2010 American Chemical Society.

the films would appear to indicate a layer-like assembly, the topographic AFM images of the films show island-like growth (5–75 nm roughness) at 10^{-4} M, while at 10^{-3} M, the films are smooth (1–5 nm roughness). Also, while in more usual PEM films, the ζ-potential alternates between high positive and negative values,[27] for the 10^{-4} M concentration deposition no alternating character was observed. When the ζ-potential was positive, it decreased with increasing thickness. Regular film growth ceased when the ζ-potential came close to zero. Nanometre-scale heterogeneities or perhaps strong hydrogen bonding might explain these characteristics, but nonetheless it is unusual, if not unprecedented,[28,29] that a positively (or negatively) charged polymer can adsorb onto a surface of the same charge sign. Therefore, purely electrostatic interactions do not appear to be responsible for the assembly, or perhaps dynamic rearrangement takes place during growth at the newly-forming interface. Whatever the case, it is a stunning example of "self-patterning growth" where nanometre-scale supramolecular objects of quite well-defined dimensions and shape are formed spontaneously upon deposition. The authors of this work pointed out that most PEMs have been prepared using more commonly available polyelectrolytes, and the discovery of interesting growth modes might be uncovered should novel polymer compositions and topologies be tested.

The polymers in the films are not necessarily precisely charge balanced, and any local imbalance once made results in the inclusion of low molecular weight ions. While in certain instances this situation can be taken advantage of—for the preparation of composite films as shall be presented later—the preparation of PEMs without these counter-ions is desirable from both a fundamental point of view and in terms of making full potential of the materials properties of the films, such as mechanical strength.[30] The PEM formed by poly(diallyldimethylammonium) and poly(styrenesulfonate) was shown to contain chloride ions,[31] but treatment with salt solution followed by poly(styrenesulfonate) removes more than 95% of the counter-ions present in the as-prepared multilayer.[32] The process works by plasticising the multilayer beyond its room glass transition by exposure to 2 M sodium chloride solution[33] that increases diffusion in the film,[34] "scrambling" the excess cationic polymer. Subsequent treatment with a sodium poly(styrenesulfonate) solution results in incorporation of the anionic polymer[35] and consequently the remnant chloride ions from preparation leave the film. The result is a PEM that is smoother than the as-formed multilayer and practically free of small inorganic ions. The measurement of the films using AFM *in situ* has shown that exposure of PEM films to salt solutions leads to swelling of the layers followed by a smoothing when compared to the as-made films.[36]

The anions present as counter-ions to the polycation affect the thickness of the layers that are formed (the preceding discussion considered chloride only): dry PEMs comprising poly(4-styrenesulfonate) and poly(diallyldimethylammonium) have thickness in a wide range (20–90 nm) whose value depends approximately on the position of the counter-ion to the polycation in the Hofmeister series.[37] When the same PEMs were wet, they showed dramatically different stiffness depending on the same series of anions.[38] The storage shear modulus of the films and their mass density depends on the anion's hydration entropy. This characteristic of the anion is also responsible for the dramatic differences in swelling of the same type of PEM incorporating either bromide or fluoride ions.[39] The multilayers formed in the presence of fluoride showed essentially no swelling while those incorporating bromide ions displayed large swelling (particularly at low concentration of the salt), and the films show swelling that is largely correlated with the Hofmeister series for solutions of HCOONa, $NaBrO_3$, NaCl, $NaClO_3$ and $NaNO_3$. In addition to these monovalent anions, divalent anions can also affect the structure and properties of the films of polyelectrolytes.[40] Relatively low concentrations ($\leq$50 mM) of sulfate, hydrogen phosphate or organic dicarboxylates in aqueous solutions make multilayer films more UV absorbent, thicker, and rougher than the equivalent ones made in the presence of chloride, although linear growth was observed in all cases. A model for layer growth in which the polycation component formed aggregates with bridging divalent cations in solution prior to adsorption explains the observed film characteristics. The nature of the dianion—from the more spherical inorganic ones to the linear organic carboxylates—also influence the nature of the aggregates that the polycations form and therefore the multilayer properties. These observations demonstrate how charge-assisted supramolecular interactions can influence material properties, and offer many opportunities for exploring related avenues.

There are certain tangible correlations between the growth characteristics of the polyelectrolyte multilayers and the phase behaviour of the complexes of the charged macromolecules in solution.[41] Consideration of the pH and the ionic strength of the deposition solutions and the intrinsic characteristics of the macromolecules—the ratio of the lengths of the component moieties in the polyelectrolyte chains and their nature—allows a certain control and even predictability of the quality and properties of the films. The erosion of multilayers and their stability can be correlated with the nature of the polyelectrolyte complexes in solution.

One would expect that molecules with low charge would not be favoured in their adsorption at charged surfaces, as the entropic gain in counter-ion exchange is lost. This hypothesis holds in some cases, where adsorption and stability are greater for more highly charged systems.[42,43] There are optimum values for a given system. An illustrative case is that of the random copolymers formed from the cationic monomer diallyldimethylammonium chloride and charge neutral *N*-methyl-*N*-vinylacetamide (which have similar molecular weights) and between 24 and 100% of the charged monomer forming LbL films with poly(styrene sulfonate).[44] The formation of the PEM was witnessed by UV-visible and IR absorption spectroscopies as well as quartz crystal microbalance (QCM). When the multilayers were prepared using solutions with salt added, a clear limit of cation composition was identified, resulting in a change from layer growth to an essentially insignificant growth. For the present system this "charge density limit" was between 75 and 53%. When the charge density was below this value, deposited cationic polymer was actually stripped from the film upon addition of poly(styrenesulfonate). When the concentration of cations in the macromolecule was above the critical charge density, the presence of a greater salt concentration during adsorption created thicker films with a quite rough morphology, as shown by AFM. Thus, there is a range of charge content in the macromolecules that is apparently optimum for the formation of smooth and dense layers. Apart from modifying this proportion in the covalent chemistry of the component polymers, the assembly can also be tuned by adjusting the pH of the solutions used to deposit the samples, as shown for multilayer formation for films comprising the natural polymers alginate and chitosan (that have carboxyl and sulfonate groups, respectively).[45] It is the charge on the polymers that leads to these effects in the cited systems.

The presence of other non-covalently interacting groups can reduce the expected charge density in the polymers, one exception being the copolymer (related to the one described earlier) formed between diallyldimethylammonium chloride and acrylamide that forms PEMs with poly(styrenesulfonate).[46] The methods used previously to monitor the growth of the PEMs showed that as little as 8% cationic monomer incorporation facilitated multilayer formation. Rather than the more usual increase of adsorbed material with charge density above the critical charge density seen in other systems, the thickness of the layers in films incorporating the cationic component between 8 and 100%—as shown by small-angle X-ray scattering and ellipsometry—was essentially the same. This remarkable effect—recalling that in the related system described earlier the lower charge limit was 53%—was attributed to the synergic action of hydrogen bonds between the primary amide groups in the acrylamide monomeric units. Therefore, non-covalent forces can aid and abet the electrostatic interactions that are the main driving force in the assembly of these kinds of multilayers.

Metal ion binding moieties—bipyridyl—have been incorporated into a polyelectrolyte in which permanently charged quaternary ammonium moieties (that are less prone to a change in structure as a function of pH than primary or secondary ammonium ions) drive the formation of the PEMs with poly(4-styrenesulfonate) or through metal ion coordination.[47] The polyelectrolyte type LbL showed linear growth of films that were homogeneous and smooth: this morphology was proven by X-ray reflectometry, ellipsometry and AFM. This film can complex divalent metal salts of iron, nickel and zinc at the bipyridyl moieties, and the same ions can be released from the film by introducing a solution of the metal ion chelant ethylenediaminetetraacetic acid. The films prepared by metal ion coordination were less robust because of the reversible nature of the metal ion coordination used in this case; addition of the metal ion sequestering agent used in the PEM films caused a disassembly of the film, a feature that was proposed for use in stimulus-triggered release functions from the materials.

6.3.2 Structure of Polyelectrolyte Films

The zone model is a good approximation for the structure of PEMs, which explains a lot of the observations made in these systems. It explains why the film structure in the layer-by-layer method is distinct from other films. Three zones based on charge and composition are defined in this model (Figure 6.4). In zone 1, a few polyelectrolyte layers close to the substrate have unique interactions compared with the rest of the film because of the interface of the "primer" polyelectrolyte layer with the substrate, and clearly the nature of this interface depends on the charge and charge density of the substrate. Zone 3 comprises the outermost layer (or few outermost layers) where the polyelectrolyte that is deposited last has a much higher content of small counter-ions than the rest of the multilayer. Clearly, the charge is defined by the last adsorbed material. The rest of the film, all that lies between the interface with the substrate and the last formed outer interface, is zone 2, the "bulk" of the material that is not influenced by the interfaces with substrate or surroundings. Charges are largely compensated between neighbouring polyelectrolytes of opposite charge, each one being "sandwiched" with identical layers. The model incorporates observations concerning the surface charge, the counter-ion content at any position, the charge stoichiometry and the layer interpenetration. The strong interactions mean that the layers are quite flat, depending on the polymer, as little as two chain widths.[48] In this model, there are no sharp borders, and it can only really be applied to relatively thick films. The model also assumes that deposition conditions are identical and that the polyelectrolytes have strictly alternating materials. It is clear that more zones could be generated by using different materials or different deposition conditions.

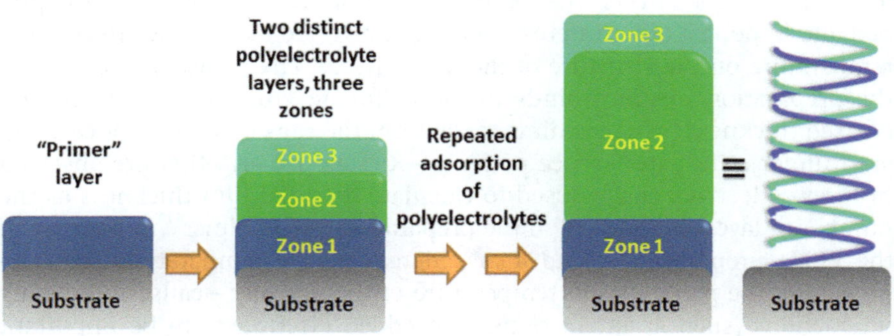

Figure 6.4 The zone model of layer-by-layer assembly of polyelectrolytes, where the three zones correspond to the "primer" contact interface between the first polyelectrolyte adsorption to the substrate (zone 1), the "bulk" zone 2 where all charges are in principle compensated between the component polymers, and the outside interface where the charge is defined by the last polyelectrolyte to be adsorbed and contains the counter-ions that accompany it. The overlaying curves on the right indicate the concentrations of each polymer in each part of the multilayer, where the zones can be appreciated next to the bar representation.

The nature of the surface of the films is routinely characterised by techniques such as AFM for their morphology and XPS for their chemical composition, but the zone model has its origins in scattering techniques. Neutron and X-ray reflectivity measurements were used to measure the internal structure of PEMs formed from partially deuterated samples of poly(4-styrenesulfonate) and poly(allylamine) prepared on a silicon wafer substrate.[49] The PEM comprising 48 "strata", eight repetitions of five perprotonated layers followed by one perdeuterated layer, had an overall thickness of approximately 120 nm, and internally it was established that the layers in contact with the silicon substrate were thinner than those farther from the substrate by a not insignificant amount. The effect of the substrate is also evident in the measures of surface roughness, approximately 4 Å at the silicon–PEM interface, and 13 Å at the PEM–air interface. Nonetheless, the exterior surface of the PEM is obviously very smooth. As the film was prepared in 2 M NaCl, a significant number of counter-ions (0.5–0.8 per monomer unit) were incorporated into the layers, as revealed by the X-ray absorption. The neutron reflectivity measurements showed clear Bragg reflections. From these peaks, it was determined that the polycation and polyanion layers were approximately two and three and a half nanometres thick, respectively, in their equilibrium state, while at the surface, the polyanion had a thickness of approximately 1.3 nm. The approximate interpenetration distance of the layers in their equilibrium thickness was estimated at approximately 12 Å. The equilibrium thickness was reached after approximately four layers in this system.

The structure of PEMs prepared from the now familiar model system poly(allylamine) hydrochloride and poly(4-styrenesulfonate) but under conditions of high potassium chloride concentration—to generate a thicker film to accentuate the effects when studying with X-ray reflectometry—was probed and (amongst other interesting effects) revealed a dramatic effect of the temperature on the structure of the multilayer.[50] The potassium present in the preparation medium binds to the sulfonate group leading to the increased thickness (whereas films made from the salts of sodium or caesium are thinner). The interference patterns—Kiessig fringes—that are observed in X-ray reflectivity can be used to calculate the total film thickness as the number of layers is changed in the preparation of the films.[51] In the case of the PEMs prepared at 5 and 40 °C, increasing the temperature—and approaching the precipitation temperature of the polymer—causes secondary interactions (such as interactions with added electrolyte) to be important and therefore the film is thicker, by up to 40%, depending on salt concentration. Figure 6.5 shows schematically the nature of the PEMs and the electron density profiles that are obtained from the X-ray reflectivity measurements. The roughness of the films also increases significantly under these higher temperatures. Evidently, if more compact and smoother layers are required for these PEMs, lower temperature adsorption processes seem to be preferred.

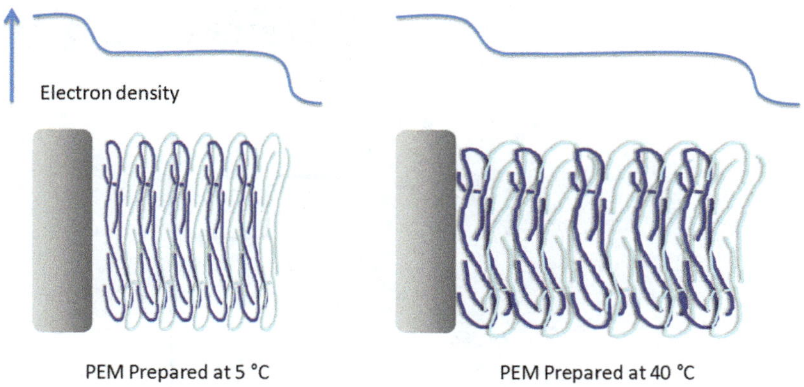

Figure 6.5 A representation of the thickness of PEMs prepared at different temperatures; those prepared at higher temperatures are more expanded and those at lower temperature are relatively more compact, and (above) a cartoon indicating approximately the electron density profiles that are observed from X-ray reflectivity measurements.

6.3.3 Properties of Polyelectrolyte Films

The principle property of the PEMs is the possibility for changing the porosity of the films in a response to pH changes in the medium. Typically, protonation, say, of a carboxylate group will result in the weakening of the cohesion between layers and entry of solvent and swelling of the film. The porosity of the films is also increased dramatically. The treatment with acid or base does not necessarily produce identical effects, as shown for LbL films of polyelectrolytes formed from linear poly(ethyleneimine) and poly(acrylic acid) (PAA) and poly(allylamine hydrochloride),[52] where treatment with base affords more porous films. The application of an electric field to the materials has been shown to increase porosity, apparently through local changes in pH although other mechanisms involving trapped counter-ions might be plausible, and this effect may be useful for the release of entrapped guest molecules.[53] In fact, the entrapment of polymers through hydrogen bonding followed by their release under pH modification can be used for the controlled preparation of polymer thin films.[54]

A fascinating use of the LbL polymer films arises when conducting polymers are used as one of the polyelectrolytes. The formation of PEMs employing either poly(aniline) or a colloid of the material as one component and an anionic polymer as the other has led to films that when used as electrodes show potentially interesting charge storage properties.[55] The PEM film was grown on a (3-aminopropyl)triethoxysilane-modified ITO surface by sequential deposition of the two components (Figure 6.6). This process was performed using an automated slide stainer apparatus. The linear growth of the PEM film was followed by absorption spectroscopy where a charge transfer (resulting from hopping of electrical charge in the material) was

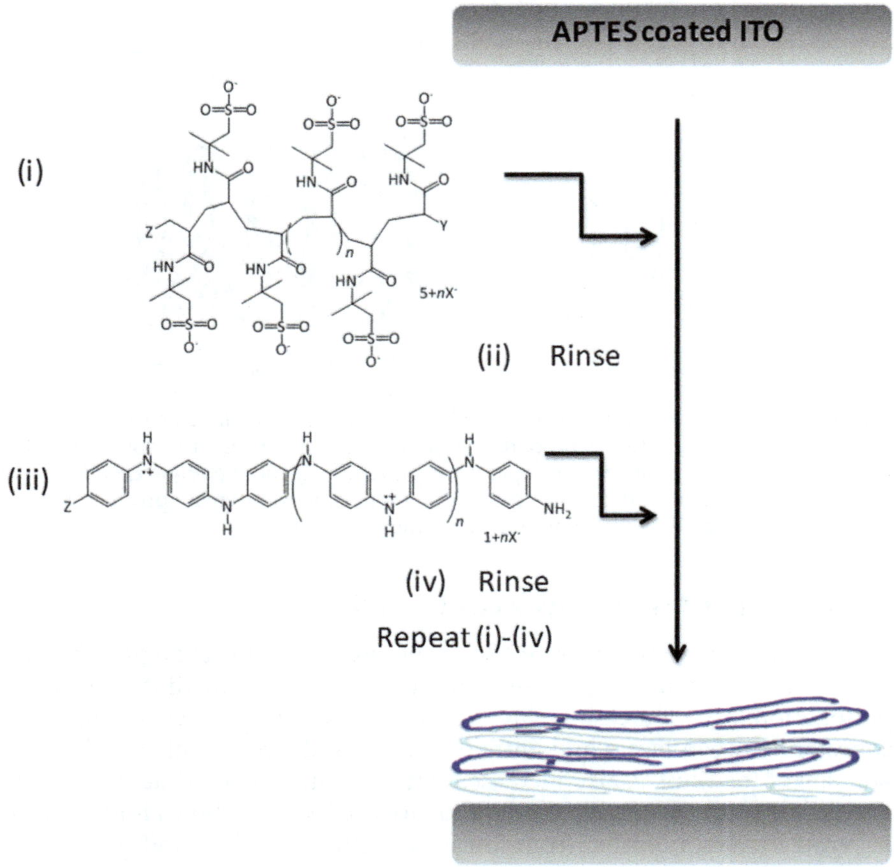

Figure 6.6 The LbL preparation of a PEM that acts as an electrode showing potentially interesting charge storage properties.

observed from 600 nm and well into the infrared. In addition, quartz crystal microbalance measurements showed an equimolar proportion of anionic and cationic components in the material. The roughness of the PEM measured with profilometry showed a very smooth surface (approximately 4.4 nm). Although the films showed high initial performance with good charging because of the high proportion of poly(aniline), they had poor ability for cycling, probably because of irreversible formation of the pernigraniline base. In fact, an analogous film prepared in the same piece of work where a colloid of poly(aniline) and the anionic polymer was layered with a purely structural polycation (the PEM film does not undergo linear growth and is much rougher than the other cases) showed very good performance over 1000 cycles between 1.5 and 4.5 V. A similar approach to colloidal polymer LbL assembly has proven beneficial in electrochromic materials with poly(aniline) as the active component.[56] Subsequently, the poly(aniline)-based materials have been deposited by the spray-on LbL method (see earlier) and used as electrodes.[57,58] The preparation route used to make the colloid in this system has a significant impact on performance, and an understanding of the supramolecular chemistry in these systems would surely lead to greater understanding and potentially improved properties. The properties of these films and other electrode materials based on PEM ideas[59] could be interesting for studies involving scanning probe techniques to study the morphological and electrical nature of the interfaces in them.

The preparation of multilayer films that combine both ionic and hydrogen bonds as non-covalent forces between their components leads to stretchable materials that have a high gas barrier.[60] So-called quadlayers constructed with a polyethylenimine, poly(acrylic acid), poly(ethylene oxide), and poly(acrylic acid) sequence at pH 3 on 1 mm thick polyurethane rubber gave films with a quite dense morphology and no significant porosity. A twenty quadlayer film (that was around 1 μm in thickness) showed a transmission rate for oxygen that was fifteen times less than the polymeric substrate. This remarkable improvement in barrier properties—useful for materials packaged under an inert atmosphere—is believed to be a result of the synergy between the interdigitated layers of electrolyte and hydrogen bond-forming polymers. Stretching this film by 10% caused the barrier to be a factor of 28 more effective to oxygen permeation resistance when compared to the substrate polyurethane. The combination of effective gas barrier and stretchy nature is believed to make these self-assembled layers promising as high barrier elastomers. The pH plays a critical role in determining the structure of the film. When the pH values used for each component deposition were 10 (polyethylenimine, unprotonated), 4 (poly(acrylic acid)), 2.5 (poly(ethylene oxide)), and 2.5 (poly(acrylic acid)), completely different morphology and properties were observed. The film was much thicker than the one prepared using pH 3 throughout the process, with a high degree of porosity, as a result of high chain mobility causing significant in-and-out diffusion of the partially-charged polyelectrolytes (PEI at pH 10 and PAA at pH 4). This material has a very poor gas barrier. The results are a demonstration of the control in material properties that can be achieved by determining the non-covalent interactions that take place between the component polymers at each step.

6.4 Layer-by-layer Assembly in Neutral Supramolecular Systems

There are a great many complementary supramolecular pairs that one might envisage for the assembly of multilayers in a layer-by-layer approach. Single point interactions between adsorbents and adsorbates must be specific and either to be carried out in vacuum or to be extremely strong in solution to allow selective growth of monolayers one at a time. The desorption process from the film must be disfavoured, with binding of each successive layer being irreversible on the timescale of the adsorption in order to exert control over the assembly and have a true layer-by-layer assembly.

Multivalency—the binding of several complementary recognition units between two chemical entities—can lead to high association constants compatible with the LbL assembly.[61] In principle, this effect is responsible when different polymers, one bearing a chemical functionality that can interact non-covalently with a complementary unit on the other, are combined. The assembly is thermodynamically distinct to polyelectrolyte LbL assembly, because entropy becomes somewhat less important because of the absence of released counter-ions. Therefore, the enthalpy of binding should, in principle, be high. These conditions are met by polymers bearing 9-carbazolyl and 3,5-dinitrobenzoyl units as π-electron rich and poor groups, respectively.[62] The polymers, based on methacrylate backbones, were adsorbed alternately from methylene chloride onto both gold and quartz substrates, by exposure of 0.1 mM solutions (with the exception of the first layer of the electron-rich component on gold where 1 mM concentration was used) for five minutes. Rinsing with the same solvent and repetition of the steps, switching the polymer used in alternate manner, afforded the multilayers that rely on a charge-transfer interaction between π-electron poor and rich groups at the interfaces that is witnessed by a long tail in the absorption spectra above around 400 nm and up to 600 nm (there is no complete charge transfer of an electron that would lead to absorption at much higher wavelength). The effect of multivalency was confirmed by the higher association constants in solution for the polymers compared with the monomers. Surface plasmon measurements of the films after each layer adsorption on the gold substrate showed that the approximate thickness per layer was 3.5 nm. Therefore, the interactions between electron rich and poor units is a viable strategy for film build-up, and generation of colour through a supramolecular interaction in the films.[63–66] Hydrogen bond donating and accepting polymers, such as poly(4-vinylphenol) and with poly(4-vinylpyridine), have also been exploited for the LbL approach.[67–72] For the cited example, a smooth (the surface roughness was approximately 1.9 nm) and homogeneous film was proven through AFM and XRD measurements, and UV and IR absorption spectroscopies proved the linear build-up of the film through sequential exposure to hydrogen bond accepting then donating polymer.

A very elegant approach involving the complexation of hydrophobic moieties in the cavities of cyclodextrins (CyDs) on spherical nano-objects, namely dendrimers and nanoparticles, led to LbL-type assembly thanks to multivalent supramolecular interactions (Figure 6.7).[73] SAMs of CyDs had been previously shown to bind dendrimers through multivalent interactions to give kinetically stable complexes,[74,75] exemplified by a dendrimer coated with sixty-four ferrocene moieties that binds through around seven moieties to SAMs of CyDs on gold.[76] Clearly, this leaves many ferrocene groups at the new surface of the film, ready for binding. This feature was exploited in a pair of complementary nanostructures for the formation of films in an LbL approach, outlined in Figure 6.7. A poly(propylene imine) dendrimer coated with between four and sixty-four adamantyl groups (sixteen are shown) was adhered to the SAM of a CyD derivative. In order to achieve this binding,

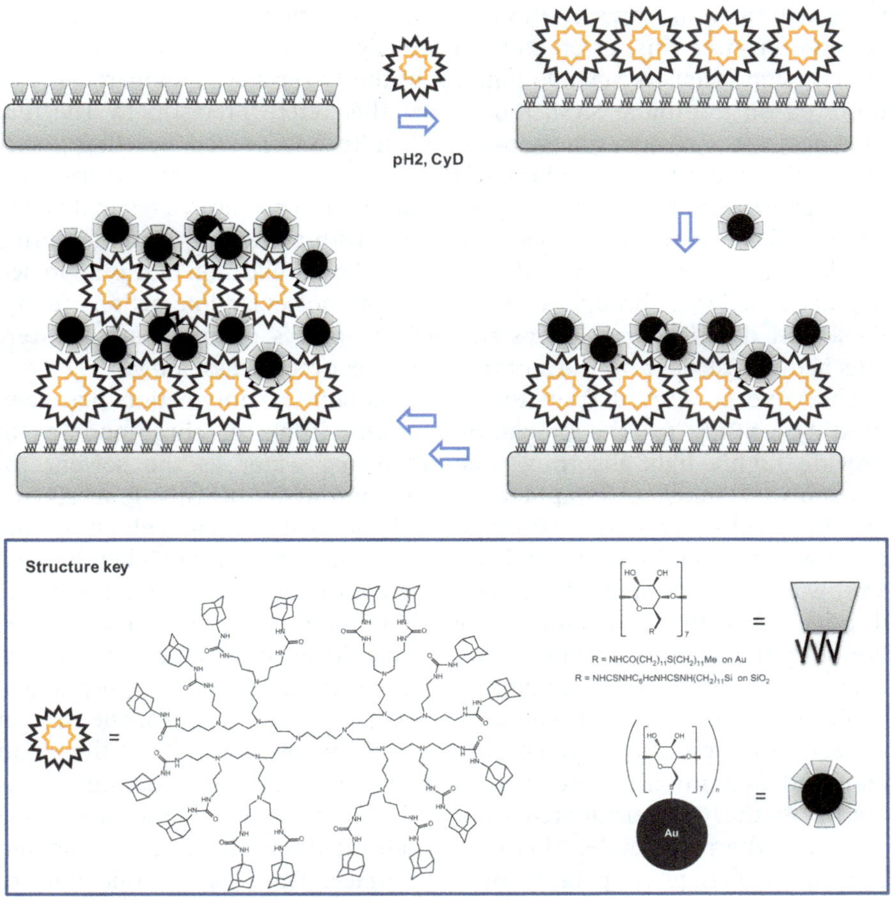

Figure 6.7 The host–guest-based layer-by-layer preparation of alternating adamantyl-coated dendrimers and cyclodextrin-coated gold nanoparticles on SAMs of cyclodextrins on either gold or silicon. The schematic route to this system is shown at the top and the cartoons explained in the box below, the structural key.

the dendrimer has to be solubilised in water, which requires the presence of free cyclodextrin and a low pH. The dendrimer binds to the surface-chemisorbed CyDs as a result of equilibrium of the CyDs in solution and the multivalency at the interface. After rinsing, gold nanoparticles bearing CyDs in the coating groups were adsorbed onto the layer of dendrimers, completing the first cycle of the LbL approach. The layers in the system are not as well-defined as those in the PEM assembly because of the possibility of intermixing of the layers after adsorption. Surface plasmon resonance confirmed *in situ* that the layers had formed in each step, which for a repeat film composition of the complementary components took approximately two minutes, including rinse times. The specificity of the adsorption was probed by using stock solutions of different concentrations, showing that a ten times increment in concentration caused only a one and a half increment in adsorption after rinsing off non-specifically bound material. However, at 58 μM, significant irreversible binding of the CyD-coated nanoparticles was observed on the initial SAM, though not thereafter, interestingly. The film thickness on a silicon–silicon oxide substrate was probed by ellipsometry (estimating the refractive indices of the materials), the results confirming a linear growth and a thickness of approximately 2 nm per layer, a result that was confirmed by scratching a groove in the film with an AFM tip and measuring the depth topographically with the same tool. The application of this approach has been explored through a variety of supramolecular interactions (either those just described or electrostatic), where lattices of microparticles were attached very strongly to the surfaces during convective assembly.[77]

The concept derived from this CyD–adamantane interaction has been used beautifully for the preparation of thin films of chromophores to provide highly light-absorbing material using water as the solvent for assembly.[78] Firstly, a tetraphenyl zinc(II) porphyrin bearing four alkyne-terminated chains on one face and four adamantyl-terminated chains on the other was immobilised covalently on a glass surface bearing a chemisorbed SAM with azide-terminated chains, using click chemistry. Then, a perylene diimide dye with a CyD attached on either side of the chromophore was bound to the priming porphyrin layer. In the idealised structure, one of the CyDs was bound to the adamantyl groups of the immobilised porphyrins, while the other CyD ring would be facing the solvent. Subsequently, an aqueous solution of a perylene diimide with adamantyl units at the chain ends and an ammonium group in-between—to ensure water solubility—was bound to the CyD-terminated film by binding of one of the adamantyl groups to the macrocycle. This process afforded an adamantyl-terminated film. This change in polarity of the surface, from hydrophilic CyD to hydrophobic adamantyl, was proven by using contact angle measurements, where an alternating magnitude of the contact angle was measured. The linear growth of the films was also proven using absorption spectroscopy. Exposure times for each step were around ten minutes, showing that this approach is a relatively fast and robust method for assembly provided the requisites for water solubility can be met.

The combination of chemisorption and physisorption has been exploited for structurally simple and significantly accessible bilayers on gold. Alkanethiol monolayers on gold can be exposed to phospholipid vesicles resulting in the formation of a layer of the lipid with the polar head group directed away from the gold surface, and with the alkane from the thiol and the alkyl chains from the lipid in close contact.[79] The driving force for bilayer formation is believed to be the entropic gain from the release of water from the SAM's surface. The bilayers in this state have very low ion permeability. The layers were made as models of biological membranes, with the advantage that their electrical properties could be studied directly on the gold surface. Indeed, introduction of a pore-forming peptide revealed that after incubation on the bilayer, an increase in ionic transport was detected, showing that the layers were fluid enough to be reorganised by the protein, and therefore the system is an interesting one for modelling biological processes. The absence of significant defects in the SAM layer and also its polarity are important in the formation of the bilayers, otherwise the vesicles of lipid can stay intact.[80] The nature of the SAM on the gold does play an important role in the response of the bilayer to enzymes once the bilayer is formed; when aromatic thiols were used as the monolayer and a lipid layer was formed on top, the response to a lipolytic enzyme was significantly greater than for the alkyl SAM equivalent.[81] It is therefore demonstrated that these combined bilayers are promising for a number of reasons, especially related with the effect of bilayer structure on function.

6.5 Ligand to Metal Ion Coordination for Layer-by-layer Film Growth

Growth of metal–organic frameworks (MOFs) on surfaces is of interest for a number of reasons,[82,83] including separation,[84] sensing,[85] storage and release or remediation.[86] These materials, dubbed surface-grown (crystalline) metal–organic frameworks (SURMOFs) can be prepared by liquid-phase techniques, either using chemisorbed SAMs as templating nucleation surfaces[87] or where the layers of a target structure are introduced sequentially by dipping a suitably functionalised surface into solutions of the reagents using a layer-by-layer approach.[88] In the latter case, use of *in situ* monitoring of adsorption by SPR allows a change in source flows over the substrate and effective in-line preparation. When a carboxylate-terminated SAM of mercaptohexadecanoic acid on gold was treated first with copper(II) acetate—which binds to the carboxylate head groups in the monolayer—and then, after rinsing, 1,3,5-benzenetricarboxylic acid, followed by rinsing and repetition of these steps, an ordered film was formed, which gave an out-of-plane X-ray diffraction pattern that was consistent with the proposed framework structure. The flat morphology of the film is completely different to the result of crystallising a mixture of the components onto a SAM, in which case clear crystallites are formed on the layer that acts as a nucleating agent rather than a true template for oriented growth. The localised and miniaturised growth of surface-based MOF material has even been achieved in a microfluidic device,[89] making the area-specific growth of materials possible.

This method, also referred to as liquid-phase epitaxy (LPE), is analogous to the growth of layers on inorganic materials, although it should be pointed out that while beyond the first few layers of material, epitaxy is clearly taking place, the initial template layer in the SAM, and the layers in close proximity to it, may have a structure that deviates from the ideal one that is generated on top. However, the thin films of MOF material made in this way are often highly aligned and homogeneous, with a flat outer face and molecular level roughness, unlike film materials prepared in other ways.[90] Also, the method does provide a powerful route for the controlled formation of multi-component thin film materials.

The favouring of an open structure rather than an interpenetrated one is an intrinsic property of surface-bound LbL-grown films, because there should never be excess reagents available to form the interpenetrating network. A powerful demonstration of this epitaxial control from the surface is the suppression of interpenetration in a MOF framework,[91] something that can be problematic in homogeneous growth, and the favouring of an open network structure. The MOF in question comprised a 1,4-benzenedicarboxylate bridge between nodes that contained two zinc ions linked by four carboxylates, with the fifth coordination site of the metal ion bound to a 4,4′-bipyridine pillar between layers of the square lattice (Figure 6.8).

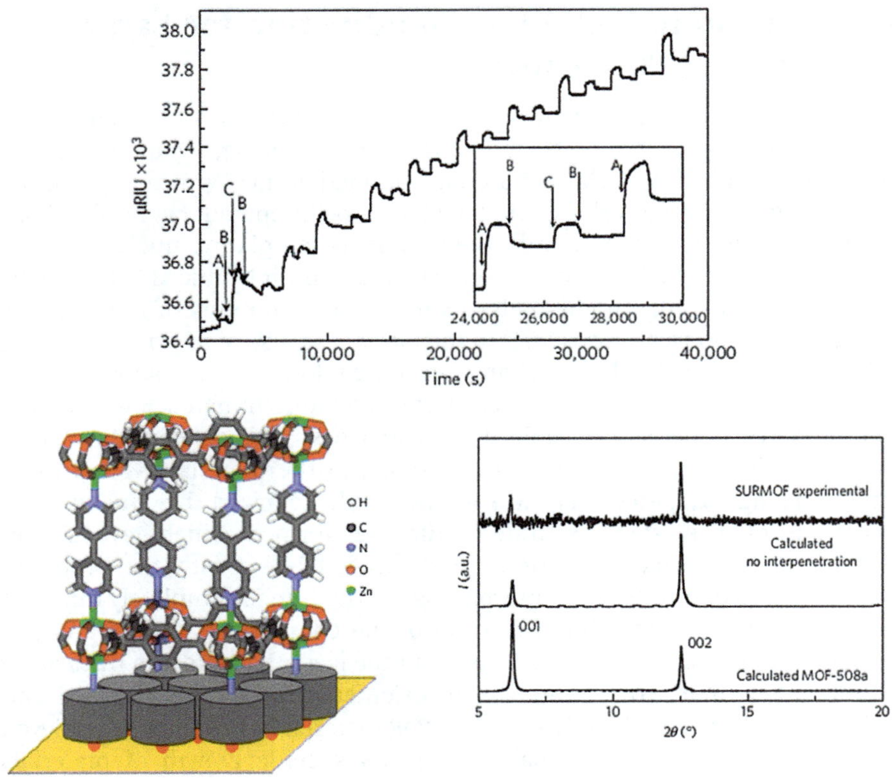

Figure 6.8 The top graph shows the time-dependent SPR signal measured *in situ* while the following sequence of injections was made over a SAM of 4,4-pyridyl-benzenemethanethiolate on gold: zinc acetate (A), ethanol (B) and a mixture of 4,4′-bipyridine and 1,4-benzenedicarboxylic acid. The structure is a model of the framework and the lower right graphs are out-of-plane X-ray diffractograms from top: experimental measurement; middle: calculated for the oriented MOF layer of the structure shown; bottom: calculated for an interpenetrating framework.
Reproduced by permission of Macmillan Publishers Ltd., *Nat. Mater.*, Copyright 2009, ref. 91.

The connecting SAM between the gold surface and the first zinc ion to be adsorbed was made from 4,4-pyridyl-benzenemethanethiol. This terminal group facing away from the surface was used to resemble the bipyridine pillars. The sequence of deposition on this SAM was zinc(II) acetate, followed by washing, followed by exposure to a mixture of 4,4′-bipyridine and 1,4-benzenedicarboxylic acid in ethanol, and the adsorption was followed by SPR measurements in real time (Figure 6.8). An important and interesting point concerning the assembly is that the acid does not apparently remain bound to the free pyridyl moieties that point away from the surface when the second step is carried out. In principle, this might interfere with the assembly process, but the binding between organic acid and base seems to be too weak under the deposition conditions. This observation is important, because it means that trying to form a purely organic system using a similar protocol would be doomed to failure. The strong binding of zinc(II) to the pyridyl ligands and the carboxylates is necessary for the irreversible binding at each layer growth step. X-Ray diffraction proved that the film was crystalline (Figure 6.8), and that the bipyridine pillars were directed perpendicularly away from the surface. AFM was used to scratch away material and then measure the thickness of the film, and showed that each deposition step corresponded to half a unit cell in the structure.

The deposition sequence can be exploited to control the orientation of porous coordination polymer film growth,[92] where the templating SAM on the substrate can determine the direction of framework formation. For a surface framework comprising basic and carboxylate spacers, the surface functionality of the SAM on the subject, base or carboxylate, favours a continuation of that moiety away from the surface because in the structure being formed, they are orthogonal to one another. In Figure 6.9, the principle is shown in cartoon form, where the growth directions corresponding to pillars and struts can be seen. The key to the orthogonality is the selectivity in the assembly that is provided by the metal ion, which in the present case forms a paddlewheel complex.

When a pyridyl-terminated SAM on gold was used as the directing layer, the dicarboxylate metal coordination plane propagated parallel to the initial interface. This orientation is shown in Figure 6.10 with the exact chemical structure (assuming the interface to be parallel to the lines in this script!). A quartz crystal microbalance was used to monitor the growth of the LbL growth for each route. On the other hand, when the carboxylate SAM is used as the growth-directing layer, the step involving coordination of the metal centre to the basic ligand requires a growth of the framework parallel to the interface. Subsequent addition of dicarboxylic acid results in addition perpendicular to the initial interface, and therefore the orientation of the crystal is perpendicular to the case where a pyridyl-terminated SAM was used as the template. This exquisite control over orientation through the supramolecular chemistry during film growth is potentially of interest for the orientation of specific channels in the MOF towards the overlying medium.

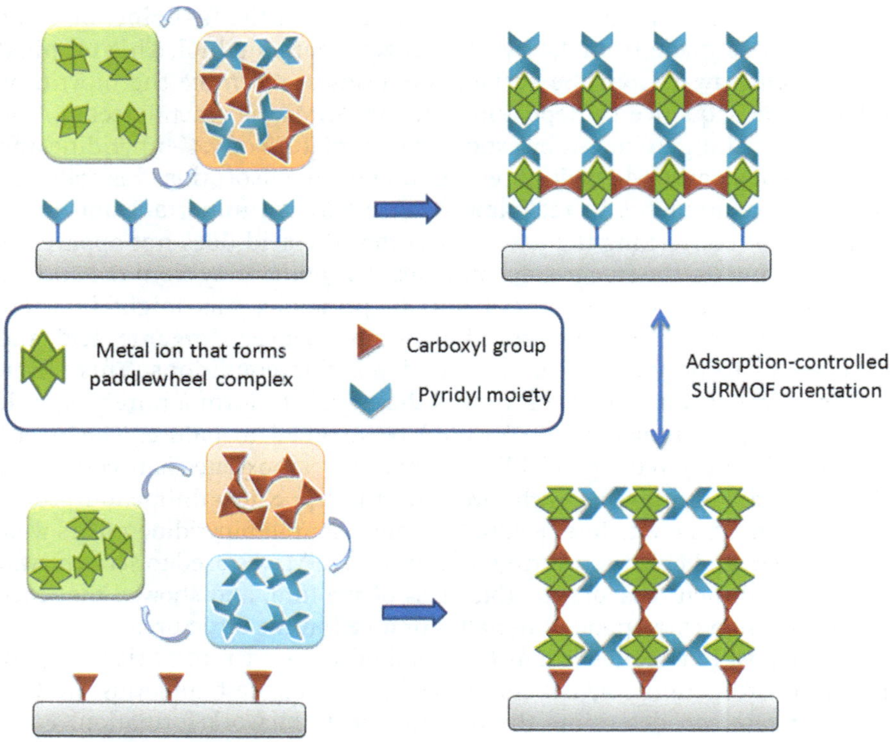

Figure 6.9 The surface-defined directional growth of a SURMOF from substrates functionalised with either basic or carboxylate groups. The principle relies on the unidirectional growth of the base-metal complex and the sheet formation by the metal–carboxylate complex.

The same strategy has been used for the preparation of chiral SURMOFs, where the enantiomers of camphoric acid were used separately as the dicarboxylate struts between zinc(II) ions with DABCO as the linker between the paddlewheel nodes.[93] The orientations of the frameworks were again controlled using either carboxylate or pyridyl-terminated SAMs on gold, and the order was monitored using X-ray diffraction that revealed the directions of the crystal faces with respect to the substrate interface. A quartz crystal microbalance was used not only to monitor the growth of the layers, but also to test their enantioselectivity for binding of small molecules. The response time was of the order of hundreds of minutes. The enantiomers (2*R*,5*R*)-2,5-hexanediol (*R*-HDO) and (2*S*,5*S*)-2,5-hexanediol were distinguished, although their separation could not be evaluated.

The coordination of pyridyl groups to palladium chloride has proved particularly effective for the preparation of a wide variety of multilayers where functional units are embellished with the coordinating group.[94] For example, the surface of glass or silicon oxide on silicon can be functionalised with an extended pyridyl-pyridinium conjugated ligand, then reacted with $PdCl_2(PhCN)$ or K_2PdCl_4 and then with the core ditopic ligand 1,4-bis[2-(4-pyridyl)ethenyl]benzene using solution-based methods to provide

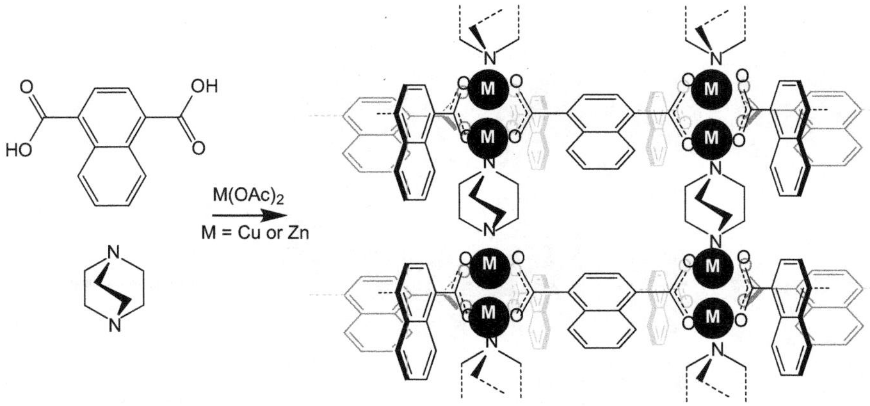

Figure 6.10 A representation of the structure of a MOF whose orientation on a substrate can be defined by the surface functionality.

multilayers (Figure 6.11).[95] The detail in the deposition chemistry from solution in this paper is particularly useful, and the cleaning of the functionalised surface with the initial monolayer with critical CO_2 noteworthy.[96] Optical absorption spectroscopy proved to be most effective in determining the degree of functionalisation, while XPS showed the chemisorption was effective, and provided a film thickness of approximately 1.2 nm. The thickness would be 2 nm assuming a totally upright orientation. This monolayer was used as a template for the growth of linear oligomers by treatment with K_2PdCl_4 in deionised water then (after rinsing), the ditopic ligand in THF (both steps at room temperature), and so on. The linear growth of the layers was proved by absorption spectroscopy where a linear increase of absorption with sequence number was observed. The absorption maximum indicated J-type aggregation (stepped molecules), which is consistent with the angled nature of the linear structures. Importantly, it is the initial templating monolayer (with an angle of approximately 110° to the normal to the interface caused by the methylene bond directions) that determines the structure and physiochemical properties of the multilayer. The (in principle) linear structures thus grow at an angle of about 20° with respect to the interface plane.

This approach has been employed for a number of organic components. Notably, a variety of tetralactam macrocycles have been sequentially deposited on a gold template layer combining the pyridyl-palladium coordination motif as well as iron or nickel-terpyridine, generating layers of distinct chemical nature in a controlled way from solution.[97] Absorption spectroscopy proved particularly appropriate in this case because the metal-to-ligand charge transfer bands provided information as to the relative content of each coordination mode.

The growth of the multilayers using this coordination chemistry-based LbL approach is not always well behaved at the outset. When trigonal or tetrahedral metal complexes with outside-facing pyridyl ligands were used in

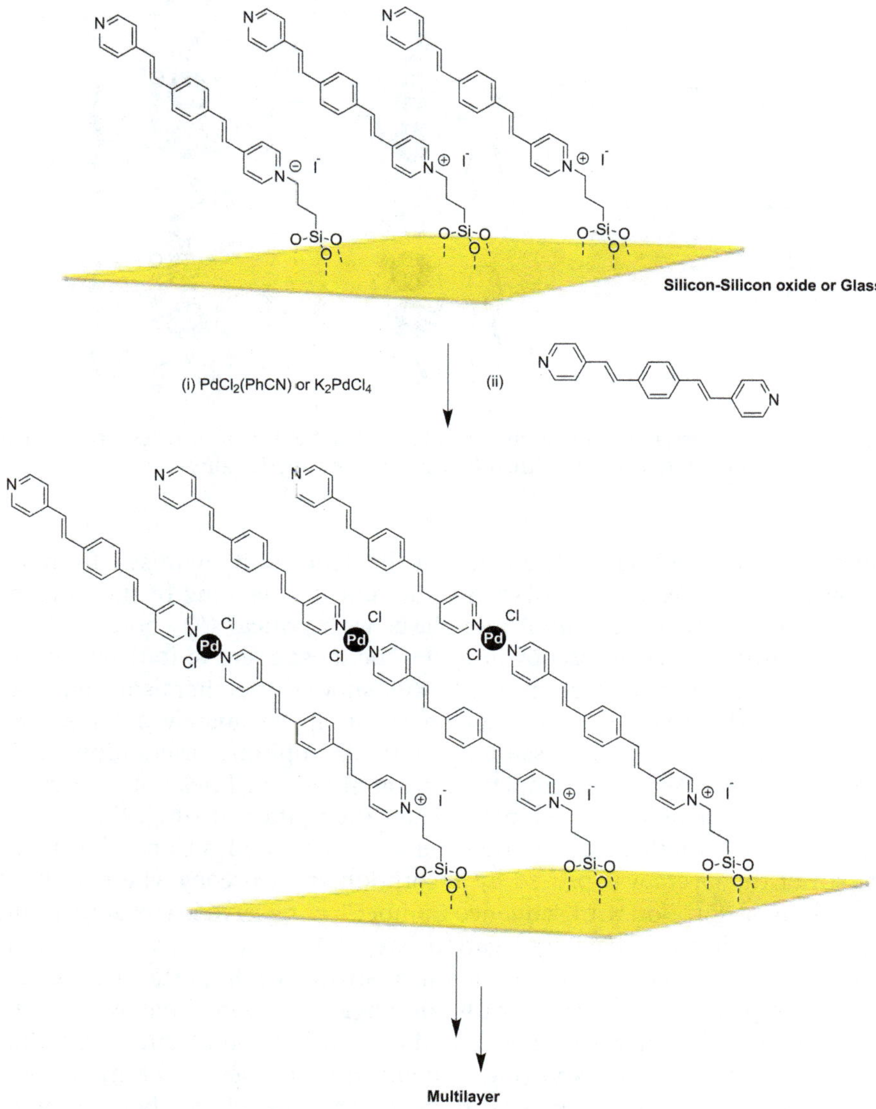

Figure 6.11 The layer-by-layer growth of coordination oligomers based on pyridyl coordination to palladium(II).

the process, initially, absorption spectroscopy and ellipsometry indicated that an exponential growth of the film thickness was observed.[98] On the other hand, organic compounds with similar topology and topography revealed linear growth. It was suggested that the metal complexes form porous structures that are capable of including unreacted palladium salt, which becomes available in the next step resulting in an excess of the reagent. The organic molecules are believed to form more dense networks in the palladium-linked oligomers. Linear growth of the coordination-based pyridyl building blocks was achieved by shortening the time the layer was exposed to palladium salt, not allowing it to enter into the layer. It is a stark reminder of the sensitivity of self-assembly of monolayers to subtleties in adsorbate structure.

The permeability of the multilayers prepared by the LbL method exploiting coordination of palladium(II) chloride by pyridyl ligands can be probed electrochemically. The films made using pyridyl-based trivalent ligands of different character (purely organic or based on ruthenium(II) tris-bipyridine) and with different thickness were probed by cyclic voltammetry using redox-active molecules with different sizes.[99] Recall that penetration through a layer would give rise to an intense and reversible wave. Size-selective permeation was indeed observed, for films either 5 or 20 nm thick (the thicker films were more compact). It was confirmed that the films that displayed exponential growth were the more permeable multilayers. The purely organic pyridyl ligand produced the least permeable films, showing weak and poorly reversible cyclic voltammetry responses for all the probes used, and especially the larger ones.

The preparation of LbL films of self-assembled coordination polymer multilayers has also been performed in a process christened "accelerated self-assembly procedure (ASAP)" that involves the evaporation of small volumes of dilute solutions on the substrate of choice.[100] This evaporation process—that was performed with natural convection conditions—leaves the substrate's surface covered with surplus ligand that must be removed by thorough washing (that is also performed under the more standard solution growth methods) to afford the clean additional layer in the film. It is claimed that, using this protocol, a full organic ligand layer can be made in around one minute. The ASAP approach was demonstrated for the formation of multilayers made up of mercaptoundecanoic acid–Cu(II) and bishydroxamate–Zr(IV) coordination polymers. The layers prepared with overnight adsorption from solution and by ASAP were compared (by ellipsometry, contact-angle measurements and FTIR) and led the authors to conclude that periodic multilayer growth was achieved for both methods, although the fast adsorption achieved with ASAP can result in a different structure to those prepared by more thermodynamic assembly. Fast ligand binding kinetics were observed for ASAP, and were ascribed to larger local ligand concentration during evaporation, as might be expected as the solvent reduces in volume, but especially at the moving liquid front at the three phase contact point. It was claimed in this work that the formation of coordination multilayers by the sequential deposition of metal ions and, in a separate step, ligands, to generate the repeat unit or monomer in the coordination polymer that is grown of the surface can be a slow process. The authors implied that often the binding is left to take place overnight to go to completion, rendering multilayer preparation exceedingly slow and, in many cases, impractical. Clearly this statement is not general, as QCM measurements show that in many cases, binding takes place in five to ten minutes, but the evaporation of adsorpt solutions may be an interesting route if binding from solution is sluggish. In both methods, extensive rinsing is necessary to ensure the absence of any excess reagents at each step for precise LbL assembly. However, it is clear that having a dry layer incorporating the adsorpt favours a high local concentration that could lead to more complete layer formation and, on the other hand, it is formed under more kinetic conditions than the purely solution-based method, and therefore the approach is an interesting one for the preparation of this kind of multilayer.

While SURMOFs are undoubtedly novel materials with unique characteristics, it has been shown that caution is necessary when hoping for continuous growth and properties that will be similar to bulk materials. In the case of the SURMOF that was prepared from copper(II) acetate and 1,3,5-benzenetricarboxylate and grown on a 16-mercaptohexadecanoic acid SAM on gold or hydroxylated silicon dioxide, a change in the growth mode was seen above forty layers of the material, for both templating surfaces.[101] Below forty layers, the LbL mechanism is followed, but above that, octahedral crystals can be seen by SEM, and the XRD patterns show a clear change in the orientation of the crystal lattice at the surface. The growth of the layers was witnessed employing a quartz crystal microbalance with dissipation monitoring, which shows both amount of bound material and the thickness of the film. The linear growth up to forty layers is followed by a large increase in the uptake of the material from the solutions, particularly the copper salt. Curiously, in the initial layers, it seems that more ligand and less copper than would be expected are required for the subsequent growth of the MOF, indicating that the interface region between the substrate and the strict LbL region is a unique kind of material and emphasising the unique nature of the interface regions. This result is interesting in the context of AFM observations of the same material grown on a different surface. The growth of four different island types was observed, which were assigned to differently-oriented crystallites.[102] The AFM images show that within a cycle, the crystallites can grow by 5–10 layers, the degree depending on orientation of the crystallites and the morphology of the gold substrate, and in separate work the same technique showed that temperature plays an important part in the process.[103] In other SURMOF structures, this temperature dependence was shown to be a result of the ligand–metal bond formation activation barrier, and the preferred growth direction could be controlled through variation of this parameter.[104] The CO_2 uptake in the films on silicon dioxide or gold was measured over time, revealing a degradation over eight days, with the uptake falling by 70%. The authors suggested that protective hydrophobic groups on the outer surface of the SURMOF could be used to minimise water incorporation to avoid the degradation of this thin film material.

6.6 Composite Film Preparation

The incorporation of multiple components into films is achieved with relative ease using the layer-by-layer method, giving systems that can be used for encapsulation and release of components. Even the preparation of conducting fabrics has been accomplished.

The combination of chemisorbed self-assembled monolayers and the ionic-covalent strategy to multilayer formation has been used to self-assemble inorganic films, providing a versatile preparative palette for films.[105] An example is shown in Figure 6.12, where a monolayer comprising protonated 2-mercaptoethylamine chemisorbed on gold was used (along with chemisorbed analogues on silicon oxide on silicon in parallel) as the substrate for adsorption of sheets of anionic α-zirconium phosphate as its tetrabutyl ammonium salt. Replacement of the ammonium counter-ions with tetrameric zirconium hydroxide cations that over-compensate the charge in the α-zirconium phosphate afforded a cationic layer that bound one "end" of a small molecule bis(phosphonic acid) (like $H_2O_3P(CH_2)_{16}PO_3H_2$) by adsorption from a 10 mM solution in ethanol. Treatment with tetrameric zirconium hydroxide cations anew generated a polycationic surface that could be employed for fresh deposition of a sheet of α-zirconium phosphate. The disposition steps were followed with ellipsometry—whereby a linear plot of thickness against adsorption step was obtained—as well as RAIR spectroscopy, which showed significant symmetric and asymmetric CH_2 stretching bands (2855 and 2927 cm^{-1}) indicating a tilt to the layer normal and the absence of any P–OH stretch, showing that the compound was fully ionised. X-Ray diffraction and AFM measurements indicated that the organic layer was thinner than expected for a normal orientation to the surface plane, a fact explained by the relatively low cation charge density over the layer. The AFM measurements showed clearly the formation of sheet-like flaky objects a few microns in width on the surface oriented at random, with straight edges but occasional kinks, along with an as yet unexplained grainy texture. Nonetheless, the route opened routes to the self-assembly of films through ionic strategies, notably for lamellar assemblies incorporating porphyrins whose aim is artificial photosynthesis.[106]

The incorporation of dyes with a strong dipole moment into LbL films is potentially interesting because of the non-linear optical properties that the chromophores can display when suitably aligned. A particularly interesting case at the crossroads between atomic layer deposition and LbL preparation is the construction of covalent multilayers by the formation of zirconium phosphate–phosphonate bonds between polar diazo dyes that are interesting for their non-linear optical properties.[107] The multilayers—that were well oriented and were very stable—were constructed on a hydroxylated silicon surface using aminopropyltrimethoxysilane in the first instance,[108] followed by phosphorylation with $POCl_3$–collidine, reaction with $ZrOCl_2$, and reaction with the phosphonic acid group of the dye(4-{4-[*N,N*-bis(2-hydroxyethyl)amino]phenylazo} phenylphosphonic acid). Repetition of the phosphorylation, and so on (with appropriate rinsing protocols in-between) led to multilayer growth. Polycationic layers can also be used to reverse interfacial dipole interactions in the polyelectrolyte superlattices, which promises to have unique effects on the properties of these materials.[109]

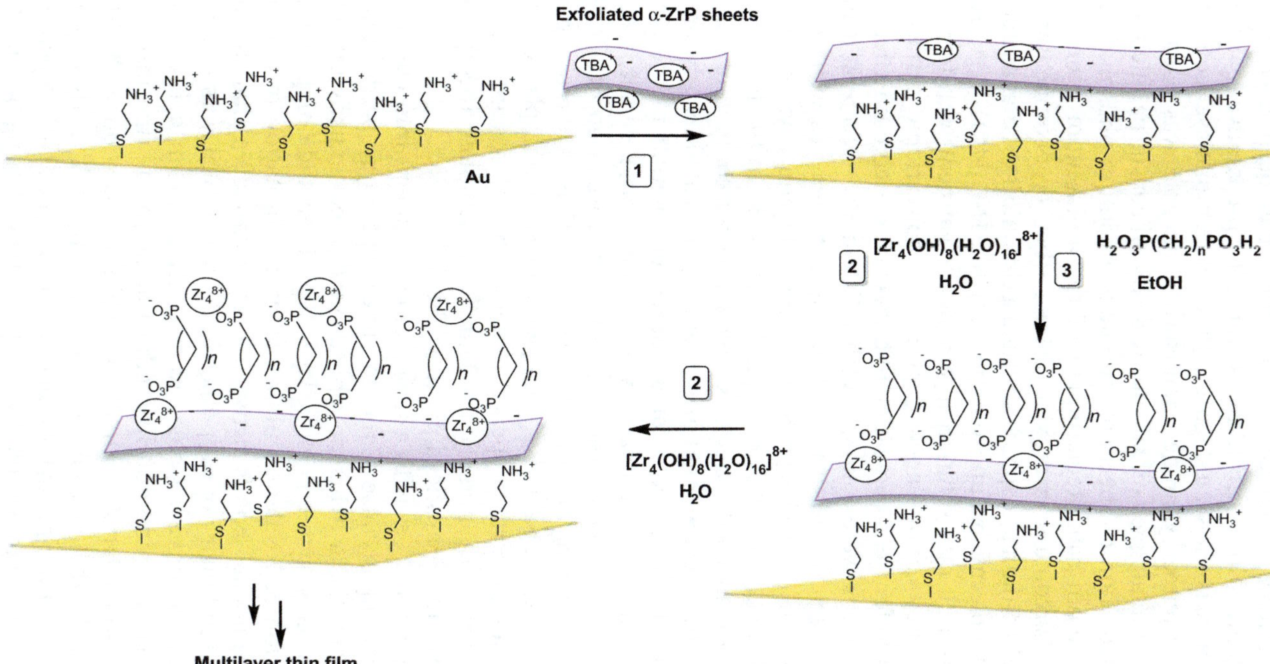

Figure 6.12 The construction of a composite layer built from a chemisorbed self-assembled monolayer on gold, followed by adsorption of α-zirconium phosphate (α-ZrP) in step 1, then adsorption of tetrameric zirconium hydroxide cations (step 2) followed by a bis(phosphonic acid) (step 3) to give the phosphonic acid terminated surface. Repeating step 2 afforded a cationic surface that could again be treated with the sheet-like polyanion, and so on, leading to multilayer films.

SAMs on gold have been employed as substrates for PEMs and reflection absorption infrared spectroscopy (RAIRS) and sum frequency generation (SFG) spectroscopy showed that upon adsorption of the cationic polymer onto a SAM of mercaptoundecanoic acid (MUA), the surface-anchored chain straightened (Figure 6.13), as shown by a reduction in the number of gauche defects in the chain.[110] This feature indicated to the authors that there was at least a partial penetration of the electrostatically-bound polymer into the chemisorbed SAM. The use of per-deuterated MUA in the templating SAM allowed quantification of the relative contributions from the polyelectrolytes and the SAM to the SFG and RAIRS spectra. The spectra—accompanied by absorbed mass information from a quartz crystal microbalance—allowed observation of subsequent LbL growth of cationic and anionic polymers, which interpenetrate as shown by a decrease in the orientation of the cationic polymer's methylene spectroscopic signals.

The incorporation from solution of the lipophilic molecule curcumin (1,7-bis-(4-hydroxy-3-methoxyphenyl)-1,6-heptadiene-2,5-dione) into a polyelectrolyte multilayer (PEM) on quartz made from poly(diallyldimethylammonium chloride) and poly(4-styrene sulfonate, sodium salt) was achieved by adsorption onto a pre-made film of the polymers.[111] The incorporation of the molecule was witnessed by eye and was quantified using absorption spectroscopy. The uptake of the curcumin from mixtures of water and ethanol was in principle governed by a partitioning mechanism whereby the polymer acts as a hydrophobic host for the neutral, largely lipophilic, guest, and uptake was higher for higher water content in the solution (for pure ethanol, no curcumin adsorbed to the multilayer). The amount of curcumin in the film depended on the number of layers in the PEM. The fact that the surface charge of the PEM film did not play a role in the uptake was proven by performing experiments with multilayers terminated either with cationic or anionic polymers, and from the results it appeared that the determining polymer layer in the uptake was the poly(diallyldimethylammonium) part. While release experiments were not performed, it was foreseen that this kind of system could be useful to this end because of the possibility for the film to swell with changing pH of the solvent medium.

The excess charges present in some of the PEMs can be used to advantage for the preparation of composite multilayers. Silver(I) can be incorporated into multilayers comprising poly(diallyldimethylammonium chloride) and poly(styrene sulfonate), prepared in the presence of sodium chloride, through rapid ion exchange of remnant alkali cation in 10 mM silver nitrate solution.[112] This latter process takes only five minutes. The exchange was followed using XPS spectroscopy that shows clearly the presence of silver—and absence of sodium—after the treatment. The XPS also indicated that the cations were associated with the sulfonate groups of the poly(styrene) derivative, as in the spectrum peaks corresponding to small counter-ions were not observed. The contact angle of the PEM with silver incorporated was identical to the film prior to treatment, showing that the ions were included. Reaction of the silver(I) with sodium

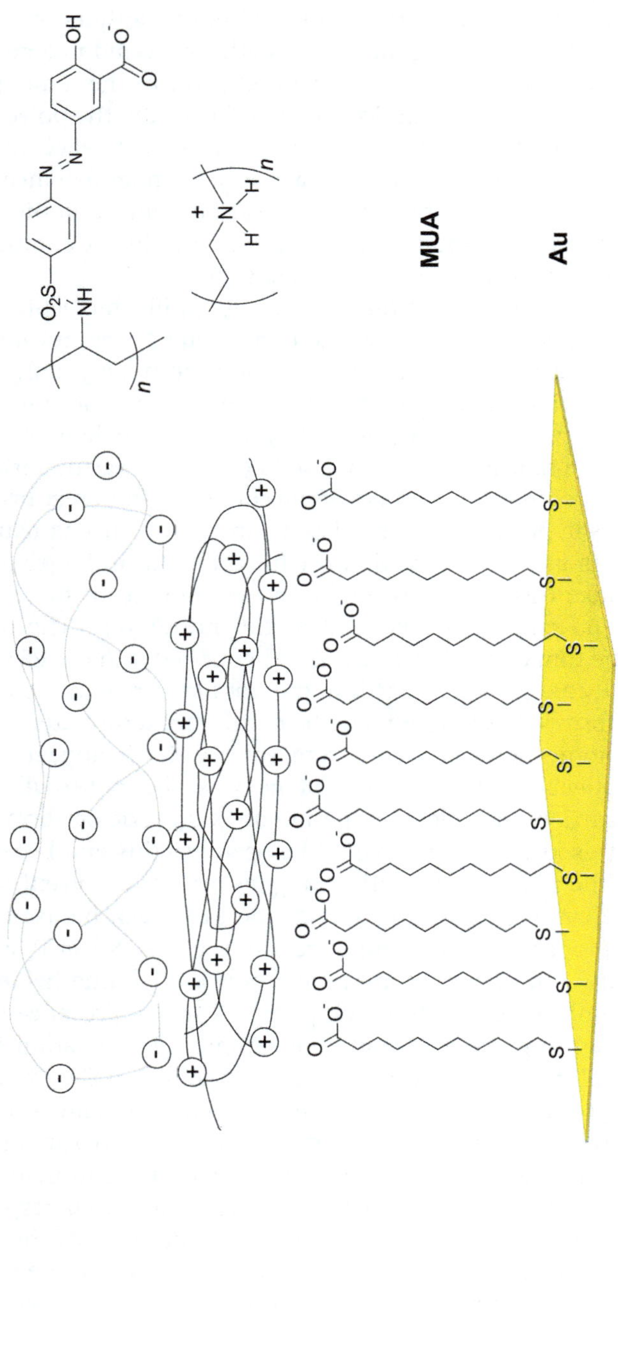

Figure 6.13 A PEM comprising a functional polyanion grown on a gold SAM template.

borohydride resulted in the formation of silver nanoparticles within the PEM. The construction of layers at different concentrations permits the preparation of sodium ion rich layers and others where the anionic and cationic polymers are more densely packed. This effect facilitates the preparation of films with an interior devoid of silver nanoparticles and the area nearer the surface loaded with them (Figure 6.14). The route is similar in a way to an imprinted polymer, but in which reversible supramolecular interactions between ions are used. The route could be general to any ionic chemical, in principle, and therefore is a very interesting supramolecular route to nanocomposite thin films.

The preparation of photovoltaic devices through "electrostatic self-assembly" of anionic and cationic derivatives of electron donors and acceptors,[113] and a supramolecular complex of a C_{60} fullerene sandwiched by two tricationic calix[3]arene derivatives (Figure 6.15) was employed for devices that generated photocurrent.[114] The thin film was grown from indium tin oxide (ITO) in an analogous way to ones grown on gold,[115] where the surface was modified to include sulfonate groups that bound the sandwich complex of the fullerene and then an anionic polymer incorporating porphyrin side groups was adsorbed electrostatically on top from an aqueous solution to give the bilayer film (Figure 6.15). The coated electrode was characterised by absorption spectroscopy, which confirmed the presence of all the chromophores that were expected. The ITO surface is generally too rough to permit accurate characterisation by many surface science techniques. The photocurrent generated by the film was measured by immersing it in an ascorbic acid solution (which acted as an electron sacrifice), and the passage of electrons from porphyrin to fullerene to ITO was demonstrated to

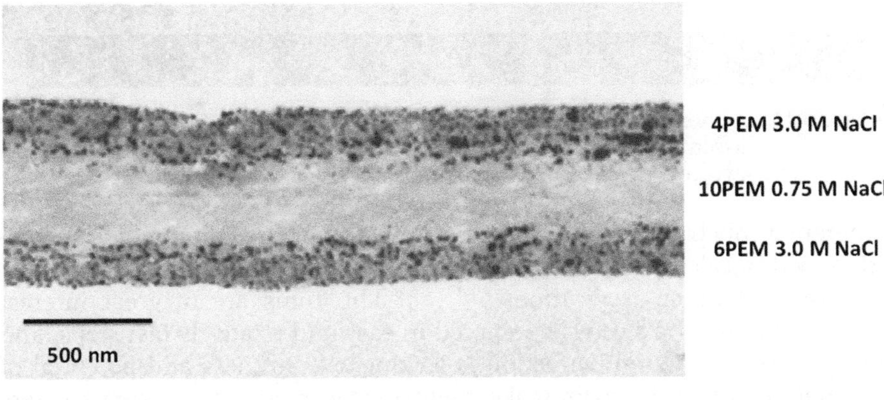

Figure 6.14 A TEM image in cross section of a PEM grown under different salt concentration conditions, subject to cation exchange with silver(I) and then reduced to silver nanoparticles. The areas where the particles are indicate the original locations where sodium ions were incorporated in the composite.
Reprinted with permission from ref. 112. Copyright 2009 American Chemical Society.

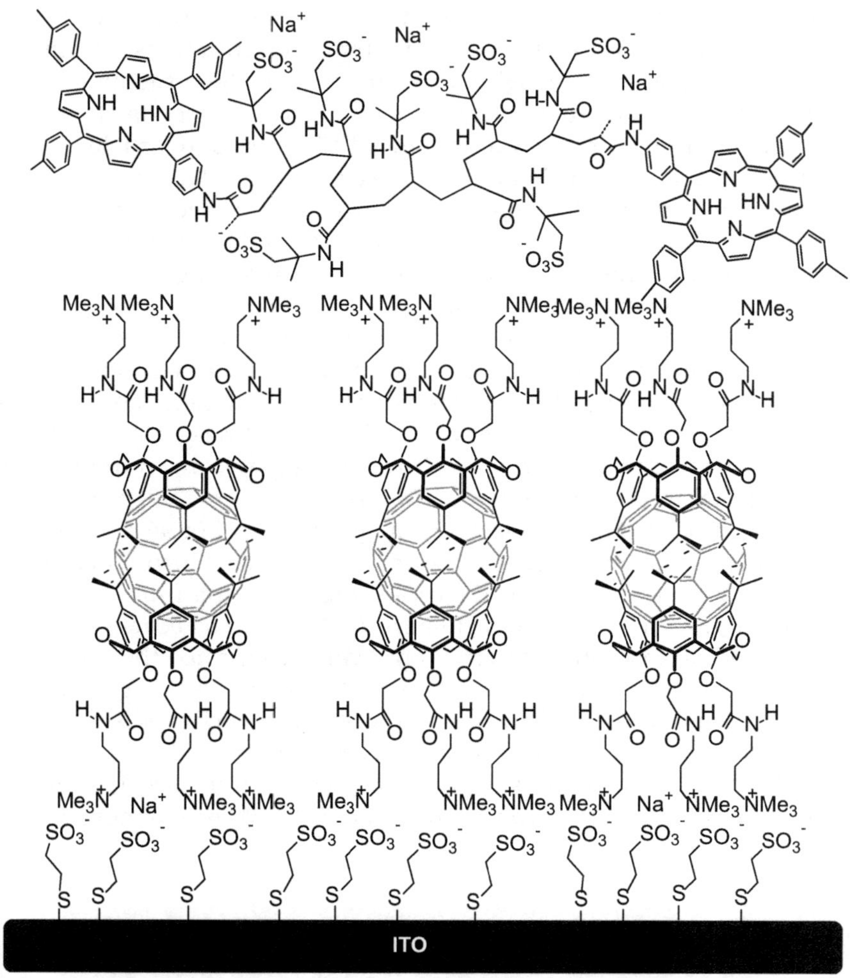

Figure 6.15 A chemical structure of a bilayer film constructed through electrostatic assembly and including the supramolecular sandwich complex between C_{60} fullerene and two calix[3]arenes.

generate a much higher photocurrent than any of the components in isolation, with a calculated efficiency of approximately 20% in the best case.

These and other applications[116,117] of LbL films are of great interest because the materials can be prepared in easy and relatively fast steps, and although the structural perfection is lacking, the softness and potential of the materials to self-repair make them very attractive for systems where lightness and flexibility are desirable.

References

1. J. Borges and J. F. Mano, *Chem. Rev.*, 2014, **110**, 8883.
2. K. Ariga, J. P. Hill and Q. Ji, *Phys. Chem. Chem. Phys.*, 2007, **9**, 2319.

3. S. M. George, *Chem. Rev.*, 2010, **110**, 111.
4. K. Gregorczyk and M. Kneza, *Prog. Mater. Sci.*, 2016, **75**, 1.
5. V. Chechik, R. M. Crooks and C. J. M. Stirling, *Adv. Mater.*, 2000, **12**, 1161.
6. X. Arys, A. M. Jonas, A. Laschewsky and R. Legas, *Supramolecular Polyelectrolyte Assemblies*, ed. A. Ciferri, Marcel Dekker, New York, 1999.
7. A. S. Michaels, *Ind. Eng. Chem.*, 1965, **57**, 32.
8. G. Decher and J. D. Hong, *Makromol. Chem., Macromol. Symp.*, 1991, **46**, 321.
9. G. Decher, J. D. Hong and J. Scmitt, *Thin Solid Films*, 1992, **210**, 831.
10. D. R. Bain, M. C. Cafe, I. D. Robb and P. A. Williams, *J. Colloid Interface Sci.*, 1982, **88**, 467.
11. A. Takahashi, M. Kawaguchi, K. Hayashi and T. Kato, Polymer Adsorption and Dispersion Stability, *ACS Symp. Ser.*, 1984, **240**, 39.
12. H. Ahrens, H. Baltes, J. Schmitt, H. Möhwald and C. A. Helm, *Macromolecules*, 2001, **34**, 4504.
13. C. B. Bucur, Z. Sui and J. B. Schlenoff, *J. Am. Chem. Soc.*, 2006, **128**, 13690.
14. N. Fujii, K. Fujimoto, T. Michinobu, M. Akada, J. P. Hill, S. Shiratori, K. Ariga and K. Shigehara, *Macromolecules*, 2010, **43**, 3947.
15. G. Decher, *Science*, 1997, **277**, 1232.
16. N. Laugel, C. Betscha, M. Winterhalter, J. C. Voegel, P. Schaaf and V. Ball, *J. Phys. Chem. B*, 2006, **110**, 19443.
17. M. Salomaki, I. A. Vinokurov and J. Kankare, *Langmuir*, 2005, **21**, 11232.
18. C. Picart, P. Lavalle, P. Hubert, F. J. G. Cuisinier, G. Decher, P. Schaaf and J. C. Voegel, *Langmuir*, 2001, **17**, 7414.
19. N. Hoda and R. G. Larson, *J. Phys. Chem. B*, 2009, **113**, 4232.
20. A. Izquierdo, S. S. Ono, J. C. Voegel, P. Schaaf and G. Decher, *Langmuir*, 2005, **21**, 7558.
21. F. Xiang, T. M. Givens and J. C. Grunlan, *Ind. Eng. Chem. Res.*, 2015, **54**, 5254.
22. C. Sung, K. Hearn, D. K. Reid, A. Vidyasagar and J. L. Lutkenhaus, *Langmuir*, 2013, **29**, 8907.
23. S. Owusu-Nkwantabisah, M. Gammana and C. P. Tripp, *Langmuir*, 2014, **30**, 11696.
24. C. Porcel, G. Decher, B. Senger, J.-C. Voegel and P. Schaaf, *Langmuir*, 2007, **23**, 1898.
25. N. Cini, T. Tulun, G. Decher and V. Ball, *J. Am. Chem. Soc.*, 2010, **132**, 8264.
26. N. Acar and T. Tulun, *Eur. Polym. J.*, 2001, **37**, 651.
27. G. Ladam, P. Schaad, J. C. Voegel, P. Schaaf, G. Decher and F. Cuisinier, *Langmuir*, 2000, **16**, 1249.
28. P. Fischer and A. Laschewsky, *Macromolecules*, 2000, **33**, 1100.
29. A. P. R. Johnston, E. S. Read and F. Caruso, *Nano Lett.*, 2005, **5**, 953.
30. J. A. Jaber and J. B. Schlenoff, *J. Am. Chem. Soc.*, 2006, **128**, 2940.

31. R. A. Ghostine, M. Z. Markarian and J. B. Schlenoff, *J. Am. Chem. Soc.*, 2013, **135**, 7636.
32. H. M. Fares, Y. E. Ghoussoub, R. L. Surmaitis and J. B. Schlenoff, *Langmuir*, 2015, **31**, 5787.
33. H. Dautzenberg, *Macromolecules*, 1997, **30**, 7810.
34. H. W. Jomaa and J. B. Schlenoff, *Macromolecules*, 2005, **38**, 8473.
35. X. Zan, B. Peng, D. A. Hoagland and Z. Su, *Polym. Chem.*, 2011, **2**, 2581.
36. S. T. Dubas and J. B. Schlenoff, *Langmuir*, 2001, **17**, 7725.
37. M. Salomäki, P. Tervasmäki, S. Areva and J. Kankare, *Langmuir*, 2004, **20**, 3679.
38. M. Salomäki, T. Laiho and J. Kankare, *Macromolecules*, 2004, **37**, 9585.
39. M. Salomäki and J. Kankare, *Macromolecules*, 2008, **41**, 4423.
40. W. J. Dressick, K. J. Wahl, N. D. Bassim, R. M. Stroud and D. Y. Petrovykh, *Langmuir*, 2012, **28**, 15831.
41. S. A. Sukhishvilli, E. Kharlampiva and V. Izumrudov, *Macromolecules*, 2006, **39**, 8873.
42. N. G. Hoogeveen, M. A. C. Stuart, G. J. Fleer and M. R. Böhmer, *Langmuir*, 1996, **12**, 3675.
43. L. Kolarik, D. N. Furlong, H. Joy, C. Struijk and R. Rowe, *Langmuir*, 1999, **15**, 8265.
44. B. Schoeler, G. Kumaraswamy and F. Caruso, *Macromolecules*, 2002, **35**, 889.
45. A. Acevedo-Fani, L. Salvia-Trujillo, R. Soliva-Fortuny and O. Martín-Belloso, *Biomacromolecules*, 2015, **16**, 2895.
46. B. Schoeler, S. Sharpe, T. A. Hatton and F. Caruso, *Langmuir*, 2004, **20**, 2730.
47. H. Krass, G. Papastavrou and D. G. Kurth, *Chem. Mater.*, 2003, **15**, 196.
48. R. R. Netz and J.-F. Joanny, *Macromolecules*, 1999, **32**, 9013.
49. J. Schmitt, T. Grünewald, G. Decher, P. S. Pershan, K. Kjaer and M. Lösche, *Macromolecules*, 1993, **26**, 7058.
50. K. Büscher, K. Graf, H. Ahrens and C. A. Helm, *Langmuir*, 2002, **18**, 3585.
51. G. Decher and J. Schmitt, *Prog. Colloid Polym. Sci.*, 1992, **89**, 160.
52. C. Y. Cho and N. S. Zacharia, *Langmuir*, 2012, **28**, 841.
53. C. Cho, J.-W. Jeon, J. Lutkenhaus and Zacharia, *ACS Appl. Mater. Interfaces*, 2013, **5**, 4930.
54. Q. Li, J. F. Quinn and F. Caruso, *Adv. Mater.*, 2005, **17**, 2058.
55. J.-W. Jeon, J. O'Neal, L. Shao and J. L. Lutkenhaus, *ACS Appl. Mater. Interfaces*, 2013, **5**, 10127.
56. M. Cui, W. S. Ng, X. Wang, P. Darmawan and P. See Lee, *Adv. Funct. Mater.*, 2015, **25**, 401.
57. J.-W. Jeon, S. Ra Kwon, F. Li and J. L. Lutkenhaus, *ACS Appl. Mater. Interfaces*, 2015, 7, 24150.
58. S. Ra Kwon, J.-W. Jeon and J. L. Lutkenhaus, *RSC Adv.*, 2015, **5**, 14994.
59. S. Y. Kim, J. Hong, R. Kavian, S. W. Lee, M. N. Hyder, Y. Shao-Horn and P. T. Hammond, *Energy Environ. Sci.*, 2013, **6**, 888.

60. C. Cho, F. Xiang, K. L. Wallace and J. C. Grunlan, *Macromolecules*, 2015, **48**, 5723.
61. A. Mulder, J. Huskens and D. N. Reinhoudt, *Org. Biomol. Chem.*, 2004, **2**, 3409.
62. Y. Shimazaki, M. Mitsuishi, S. Ito and M. Yamamoto, *Langmuir*, 1997, **13**, 1385.
63. Y. Shimazaki, M. Mitsuishi, S. Ito and M. Yamamoto, *Langmuir*, 1998, **14**, 2768.
64. Y. Shimazaki, M. Mitsuishi, S. Ito and M. Yamamoto, *Macromolecules*, 1999, **32**, 8220.
65. Y. Shimazaki, S. Ito and N. Tsutsumi, *Langmuir*, 2000, **16**, 9478.
66. Y. Shimazaki, R. Nakamura, S. Ito and M. Yamamoto, *Langmuir*, 2001, **17**, 953.
67. L. Wang, Z. Wang, X. Zhang, J. Shen, L. Chi and H. Fuchs, *Macromol. Rapid Commun.*, 1997, **18**, 509.
68. L. Wang, Y. Fu, Z. Wang, Y. Fan and X. Zhang, *Langmuir*, 1999, **15**, 1360.
69. L. Wang, S. Cui, Z. Wang, X. Zhang, M. Jiang, L. Chi and H. Fuchs, *Langmuir*, 2000, **16**, 10490.
70. Y. Fu, H. Chen, D. Qiu, Z. Wang and X. Zhang, *Langmuir*, 2002, **18**, 4989.
71. H. Zhang, Y. Fu, D. Wang, L. Wang, Z. Wang and X. Zhang, *Langmuir*, 2003, **19**, 8497.
72. H. Zhang, Z. Wang, Y. Zhang and X. Zhang, *Langmuir*, 2004, **20**, 9366.
73. O. Crespo-Biel, B. Dordi, D. N. Reinhoudt and J. Huskens, *J. Am. Chem. Soc.*, 2005, **127**, 7594.
74. J. Huskens, M. A. Deij and D. N. Reinhoudt, *Angew. Chem., Int. Ed.*, 2002, **41**, 4467.
75. T. Auletta, B. Dordi, A. Mulder, A. Sartori, S. Onclin, C. M. Bruinink, M. Péter, C. A. Nijhuis, H. Beijleveld, H. Schönherr, G. J. Vancso, A. Casnati, R. Ungaro, B. J. Ravoo, J. Huskens and D. N. Reinhoudt, *Angew. Chem., Int. Ed.*, 2004, **43**, 369.
76. C. A. Nijhuis, J. Huskens and D. N. Reinhoudt, *J. Am. Chem. Soc.*, 2004, **126**, 12266.
77. X. Y. Ling, L. Malaquin, D. N. Reinhoudt, H. Wolf and J. Huskens, *Langmuir*, 2007, **23**, 9990.
78. M. Zhu, G. H. Aryal, N. Zhang, H. Zhang, X. Su, R. Schmehl, X. Liu, J. Hu, J. Wei and J. Jayawickramarajah, *Langmuir*, 2015, **31**, 578.
79. A. L. Plant, *Langmuir*, 1993, **9**, 2764.
80. V. I. Silin, H. Wieder, J. T. Woodward, G. Valincius, A. Offenhausser and A. L. Plant, *J. Am. Chem. Soc.*, 2002, **124**, 14676.
81. C. Li, M. Wang, M. Ferguson and W. Zhan, *Langmuir*, 2015, **31**, 5228.
82. P. Falcaro, R. Ricco, C. M. Doherty, K. Liang, A. J. Hill and M. J. Styles, *Chem. Soc. Rev.*, 2014, **43**, 5513.
83. O. Shekhah, J. Liu, R. A. Fischer and C. Wöll, *Chem. Soc. Rev.*, 2011, **40**, 1081.
84. L. Heink, M. T. S. Wannapaiboon, R. A. Fischer and C. Wöll, *Microporous Mesoporous Mater.*, 2015, **216**, 200.

85. L. E. Kreno, K. Leong, O. K. Farha, M. Allendorf, R. P. Van Duyne and J. T. Hupp, *Chem. Rev.*, 2012, **112**, 1105.
86. A. Bétard and R. A. Fischer, *Chem. Rev.*, 2012, **112**, 1055.
87. E. Biemmi, C. Scherb and T. Bein, *J. Am. Chem. Soc.*, 2007, **129**, 8054.
88. O. Shekhah, H. Wang, S. Kowarik, F. Schreiber, M. Paulus, M. Tolan, C. Sternemann, F. Evers, D. Zacher, R. A. Fischer and C. Wöll, *J. Am. Chem. Soc.*, 2007, **129**, 15118.
89. D. Witters, S. Vermeir, R. Puers, B. F. Sels, D. E. De Vos, J. Lammertyn and R. Ameloot, *Chem. Mater.*, 2013, **25**, 1021.
90. D. Zacher, O. Shekhah, C. Wöll and R. A. Fischer, *Chem. Soc. Rev.*, 2009, **38**, 1418.
91. O. Shekhah, H. Wang, M. Paradinas, C. Ocal, B. Schüpbach, A. Terfort, D. Zacher, R. A. Fischer and C. Wöll, *Nat. Mater.*, 2009, **8**, 481.
92. D. Zacher, K. Yusenko, A. Bétard, S. Henke, M. Molon, T. Ladnorg, O. Shekhah, B. Schüpbach, T. de los Arcos, M. Krasnopolski, M. Meilikhov, J. Winter, A. Terfort, C. Wöll and R. A. Fischer, *Chem. – Eur. J.*, 2011, **17**, 1448.
93. B. Liu, O. Shekhah, H. K. Arslan, J. Liu, C. Wöll and R. A. Fischer, *Angew. Chem., Int. Ed.*, 2012, **51**, 807.
94. G. de Ruiter, M. Lahav and M. E. van der Boom, *Acc. Chem. Res.*, 2014, **47**, 3407.
95. M. Altman, A. D. Shukla, T. Zubkov, G. Evmenenko, P. Dutta and M. E. van der Boom, *J. Am. Chem. Soc.*, 2006, **128**, 7374.
96. M. E. van der Boom, G. Evmenenko, C. Yu, P. Dutta and T. J. Marks, *Langmuir*, 2003, **19**, 10531.
97. S. Richter, C. H.-H. Traulsen, T. Heinrich, J. Poppenberg, C. Leppich, M. Holzweber, W. E. S. Unger and C. A. Schalley, *J. Phys. Chem. C*, 2013, **117**, 18980.
98. J. Choudhury, R. Kaminker, L. Motiei, G. de Ruiter, M. Morozov, F. Lupo, A. Gulino and M. E. van der Boom, *J. Am. Chem. Soc.*, 2010, **132**, 9295.
99. L. Motiei, R. Kaminker, M. Sassi and M. E. van der Boom, *J. Am. Chem. Soc.*, 2011, **133**, 14264.
100. M. Greenstein, R. B. Ishay, B. M. Maoz, H. Leader, A. Vaskevich and I. Rubinstein, *Langmuir*, 2010, **26**, 7277.
101. N. Nijem, K. Fürsich, S. T. Kelly, C. Swain, S. R. Leone and M. K. Gilles, *Cryst. Growth Des.*, 2015, **15**, 2948.
102. A. Summerfield, I. Cebula, M. Schröder and P. H. Beton, *J. Phys. Chem. C*, 2015, **119**, 23544.
103. M. L. Ohnsorg, C. K. Beaudoin and M. E. Anderson, *Langmuir*, 2015, **31**, 6114.
104. J.-L. Zhuang, M. Kind, C. M. Grytz, F. Farr, M. Diefenbach, S. Tussupbayev, M. C. Holthausen and A. Terfort, *J. Am. Chem. Soc.*, 2015, **137**, 8237.
105. M. Fang, D. M. Kaschak, A. C. Sutorik and T. E. Mallouk, *J. Am. Chem. Soc.*, 1997, **119**, 12184.

106. D. M. Kaschak, S. A. Johnson, C. C. Waraksa, J. Pogue and T. E. Mallouk, *Coord. Chem. Rev.*, 1999, **185–186**, 403.
107. H. E. Katz, G. Scheller, T. M. Putvinsk, M. L. Schilling, W. L. Wilson and C. E. D. Chidsey, *Science*, 1991, **254**, 1485.
108. H. E. Katz, M. L. Schilling, C. E. D. Chidsey, T. M. Putvinski and R. S. Hutton, *Chem. Mater.*, 1991, **3**, 699.
109. J. L. Casson, D. W. McBranch, J. M. Robinson, H.-L. Wang, J. B. Roberts, P. A. Chiarelli and M. S. Johal, *J. Phys. Chem. B*, 2000, **104**, 11996.
110. P. J. N. Kett, M. T. L. Casford, A. Y. Yang, T. J. Lane, M. S. Johal and P. B. Davies, *J. Phys. Chem. B*, 2009, **113**, 1559.
111. P. Kittitheeranun, N. Sanchavanakit, W. Sajomsang and S. T. Dubas, *Langmuir*, 2010, **26**, 6869.
112. X. Zan and Z. Su, *Langmuir*, 2009, **25**, 12355.
113. M. Durstock, B. Taylor, R. J. Spry, L. Chiang, S. Reulbach, K. Heitfeld and J. W. Baur, *Synth. Met.*, 2001, **116**, 373.
114. A. Ikeda, T. Hatano, S. Shinkai, T. Akiyama and S. Yamada, *J. Am. Chem. Soc.*, 2001, **123**, 4855.
115. T. Hatano, A. Ikeda, T. Akiyama, S. Yamada, M. Sano, Y. Kanekiyo and S. Shinkai, *J. Chem. Soc., Perkin Trans. 2*, 2000, 909.
116. M. N. Hyder, S. W. Lee, F. Ç. Cebeci, D. J. Schmidt, Y. Shao-Horn and P. T. Hammond, *ACS Nano*, 2011, **5**, 8552.
117. F. Tancini, D. Genovese, M. Montalti, L. Cristofolini, L. Nasi, L. Prodi and E. Dalcanale, *J. Am. Chem. Soc.*, 2010, **132**, 4781.

CHAPTER 7

Supramolecular Chemistry in Thin Film Formation Upon Deposition From Vapour or Solution

7.1 Introduction

The preparation of thin films of molecular materials is of interest principally because it provides a means of obtaining a system that can be more easily manipulated and incorporated into a device than single crystals that are generally fragile and difficult to handle. Thin films of molecular systems can be prepared on surfaces using a wide range of techniques,[1–6] but broadly these fall into two contrasting routes: from solution or by vapour deposition. The solution methods that are presented here refer to self-assembly that takes place during evaporation, in an all-in-one step, rather than the step-wise routes used by the LB and related techniques or LbL approaches to the preparation of films. Like these methods, the solution-based approaches to the deposition of films of materials have no restrictions in terms of the molecular weights of the materials being employed; they can be used for polymers, high molecular weight oligomers, biological materials and electrolytes with the only restriction being the evaporation of the solvent used to deposit the material. On the other hand, the level of precision of growth of layers of molecular systems by vacuum deposition is huge, and the complexity of the system upon film formation is reduced because the chemical components contributing to the final supramolecular organisation are reduced to the minimum. Both media, vacuum and solvent, require the

Monographs in Supramolecular Chemistry No. 19
Supramolecular Chemistry at Surfaces
By David B. Amabilino

Published by the Royal Society of Chemistry, www.rsc.org

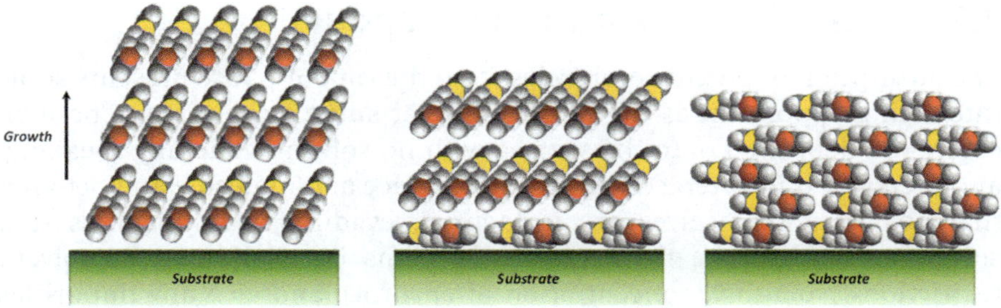

Figure 7.1 Idealised growth models for molecular thin films. In the first case (left), the molecules interact strongly with one another ensuring a quasi-perpendicular arrangement to the interface, while if the adsorbate interacts strongly with the substrate, a parallel arrangement is favoured that can lead to continued growth in this orientation in the case of a favourable commensurate structure (right) or to a change in orientation if the intrinsic packing of the molecule does not match the initial monolayer.

adsorption of material at an interface and there are some common features that must be taken into account.

The first layer of compound between the bulk adsorbent and the material in the film can be critical in determining the structure of the film and in coupling the two materials. When a molecule wets a surface poorly, the first layer may be aligned perpendicularly to the plane of the interface. In this situation, where molecule–molecule interactions dominate, a continued growth of this orientation could take place (Figure 7.1). On the other hand, it was shown that the physisorption of molecular materials to substrates often gives rise to the formation of layers where the molecules are oriented in the plane of the interface, and this situation could lead to either a continued deposition of the molecule in a parallel orientation or to the growth of molecules in a different orientation in the next layers. Which of these growth modes dominate will depend on whether a commensurate packing can be achieved or not, that lead to continued growth or change in orientation, respectively. These growth mechanisms at the molecular level have to be coupled with the morphological considerations concerning the growth modes—Frank–van der Merwe, Stranski–Krastanov and Volmer–Weber—that were highlighted in the introduction to this monograph. The formation of the films involves non-equilibrium phenomena, so equilibrium or near-equilibrium energy considerations alone cannot describe properly all growth effects seen.

The modification of surfaces with intermediate layers can modify the surface energy that the adsorpt experiences when it grows, and the functionality of the surface has not only an effect on the orientation but also the sizes of the domains that are formed and the connectivity between them, which is so important for the properties of the materials.[7] The following sections outline the growth of thin films of materials with an emphasis on those where supramolecular effects have been shown to play an important role in the assembly.

7.2 Thin Film Growth From Vapour

The adsorption of organic molecules from the vapour phase presents some interesting opportunities for supramolecular surface chemistry. The molecules are deposited onto the surface with no solvent, generally speaking, and therefore only interactions between surface and molecule and between the molecules themselves are influential, evading the difficulties that solvent can bring: the thickness of the films is not limited by solvent restrictions (volatility, concentration of adsorpt), and also any impurities that might arise from solution processing are avoided. In addition, the temperature of the adsorbent can be varied in some instruments, allowing control of adsorbate transport after deposition and the possibility of thermal annealing. Molecular beam epitaxy (MBE), or organic molecular beam deposition (OMBD), is a method where samples are sublimed onto a substrate; it has been developed for the extremely precise deposition of fractional monolayer coverages of materials.[8,9] The deposition of organic compounds on organic surfaces to give well-ordered films presents considerable challenges, because of the many kinds of crystal lattice that molecular systems can form in terms of size and symmetry: it is not trivial to identify commensurate structures between disparate molecular materials. However, in the examples that follow, it shall become clear that these difficulties can be overcome by control of the growth parameters and substrate choice. A variety of organic systems have been deposited onto substrates of different kinds and their growth behaviour described in detail, among them phthalocyanines, acenes and perylene derivatives, and this behaviour has been reviewed thoroughly.[10]

The rod-like molecule *para*-sexiphenyl serves as an example of the organisation of rigid organic molecules deposited onto surfaces under vacuum and how temperature can be used to favour thermodynamically more favourable structures.[11] When the compound was deposited onto a cleaved potassium chloride surface at 20 °C, a lying down phase was observed, where the long molecular axis was parallel to the plane of the interface. Here, the molecules grow parallel to the interface because the first layers are kinetically trapped in this position despite the weakness of the interaction between the disparate materials. In contrast, deposition at 150 °C led to the formation of an ordered film with the long axis quasi-perpendicular to the interface. In contrast, deposition onto indium tin oxide led to the formation of a randomly oriented material, showing the incredible importance that the organic–inorganic interface can have on the material being condensed under these conditions. A correlation between the structures and the electroluminescence properties of the materials was put forward. The change from in-plane and standing-up orientation on the surface is not limited to substrate temperature, but can be induced by adsorbate chemical structure. Sexithiophene derivatives generally lie flat on a silver (111) surface, but the compound with two decyl chains at its extremities lies flat for two layers and then packs at a high angle to the interface plane.[12]

Weak interactions between the organic adsorbent and growing supramolecular structure of the adsorbate can give rise to epitaxial systems that are incommensurate; quaterthiophene growth on single crystals of potassium hydrogen phthalate (KHP) is a beautiful example.[13] The phthalate ion has its benzene rings pointing perpendicularly to the plane of the surface, presenting the hydrogen atoms attached to them to any adsorbate, and therefore is an ordered organic surface. The quaterthiophene has a crystal structure that does not permit a matching of the periodicities, nonetheless preferred orientations are observed in both molecular beam epitaxy (MBE) and hot wall epitaxy (HWE). In the former technique, the sample was deposited under UHV onto the KHP crystal at room temperature, while for HWE the wall temperature was 120 °C during adsorption.

The deposition of organic materials onto self-assembled monolayers can give materials that have an electronic utility, such as field effect transistors. Pentacene is a benchmark material in this sense, as it has a mobility similar to that of silicon and is one of the organic semiconductors with the most studied properties in these devices.[14,15] In these systems, the active material—the pentacene—is deposited onto a dielectric material, usually silicon dioxide on silicon.[16] The films' molecular orientation depends on the balance between the pentacene–substrate and pentacene–pentacene interactions.[17–19] On substrates that interact poorly with the molecule, such as silicon dioxide, the vertical orientation is preferred because the crystal face that is in contact with the adsorbent has the lowest energy,[20] with the molecules in the crystal packed closely with close interactions between the aromatic planes. The oxide can be covered with a SAM to improve deposition of the pentacene, but the order in the monolayer has a very important influence on the morphology of the material and therefore on its performance.[21] This effect was proven for monolayers prepared by the reaction of octadecyltrichlorosilane with a highly doped p-Si wafer with a thermally grown oxide layer (300 nm thick), which acts as the gate system. The SAM formation was carried out at 4 and 65 °C to generate ordered and disordered SAMs, respectively. The thickness of the ordered film was 24 Å, while that of the disordered SAM was 16 Å, and the disorder was also reflected in the surface energy determined from the contact angle formed by hexadecane on the surfaces. Infrared spectra also confirmed the differences in the order of the methylene chains in the chemisorbed chains (Figure 7.2). The methylene stretching bands are far sharper for the ordered material and display little variation with temperature, which is an important observation because the pentacene was adsorbed onto these monolayers at a similar range of temperatures. The disordered SAM showed broader peaks with a greater variation (broadening) with increasing temperature, because of the space available for conformational reorganisation in the layer.

The monolayers—both ordered and disordered—were used as adsorbents for pentacene that was deposited at a range of temperatures to generate films. The layers were monitored at different stages of growth by AFM (Figure 7.3). Deposition of pentacene on the ordered SAM quickly generates

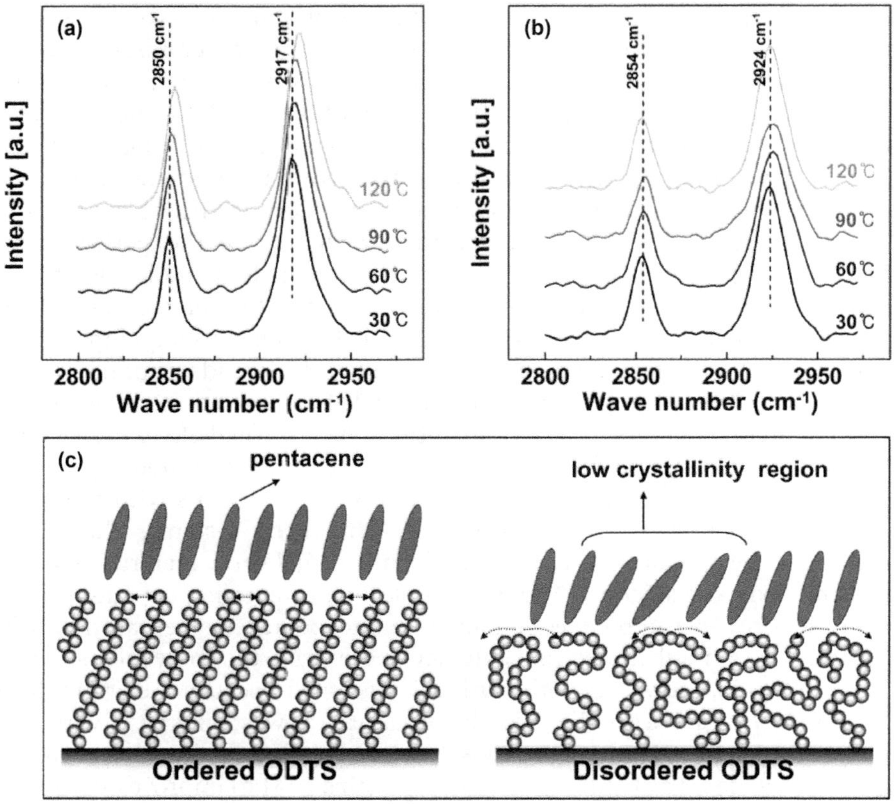

Figure 7.2 Variable temperature infrared external reflection spectra of monolayers of octadecyltrichlorosilane (ODTS) on thermally-grown silicon dioxide on silicon for (a) ordered (prepared at 4 °C) and (b) disordered (prepared at 65 °C) configurations. The spectra show the asymmetric and symmetric methylene stretching modes. Below in (c) are cartoon representations of these monolayers with a film of pentacene (the dark ellipses) grown on them.
Reprinted with permission from ref. 21. Copyright 2008 American Chemical Society.

numerous two-dimensional islands whose heights were approximately 3.2–5.0 nm, around 2–3 pentacene layers. Increasing the amount of pentacene made islands grow and coalesce. This connection of crystalline grains took place at a nominal thickness of 30 Å, and above 100 Å, mainly vertical growth was seen, following a Stranski–Krastanov mechanism. Raising the substrate temperature gave a larger grain size (because of greater transport on the surface) but similar effects.

When the disordered SAM was used as a substrate, conglomerated island-type structures grew in the early stages of adsorption (Figure 7.3(b)). As more material was introduced, similar objects grew at previously empty areas, with virtually constant grain size and no vertical growth. The height of the islands

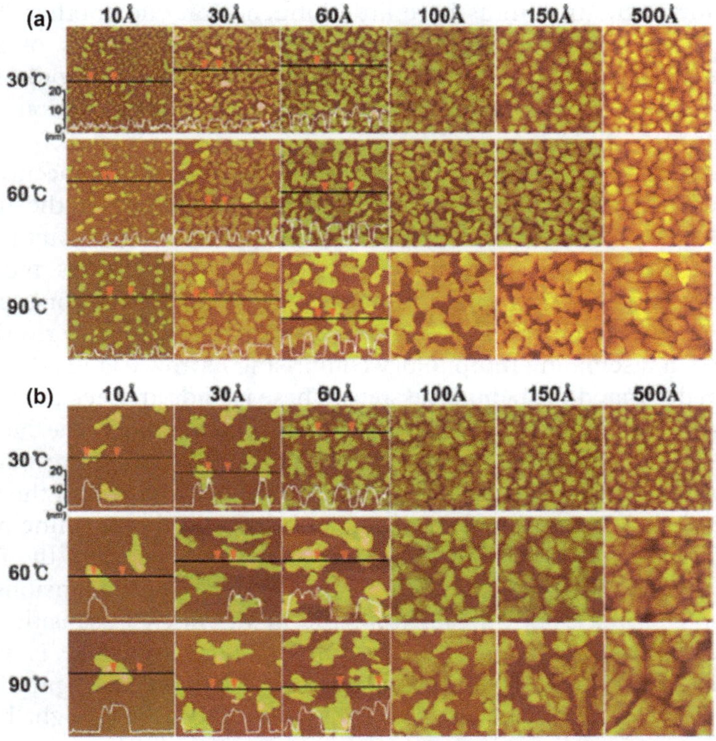

Figure 7.3 Pentacene growing on ordered (a) and disordered (b) octadecyltrichlorosilane monolayers at different deposition temperatures as observed by AFM. Height profiles are shown in white and correspond to the black lines in the images. All images show an area 2 μm×2 μm.
Reprinted with permission from ref. 21. Copyright 2008 American Chemical Society.

(about 15 nm) indicated a columnar growth mode for this combination of substrate and adsorbate,[22] and this mechanism remained across the range of temperatures assayed. Therefore, the uneven film seems to inhibit lateral growth in the plane of the interface. Overall, the pentacene layers formed on highly-ordered SAMs have higher crystallinity and greater interconnectivity of pentacene crystallites, giving better field-effect mobility than the layers deposited on the disordered SAM. These differences are accentuated at higher deposition temperatures, probably because of the high in-plane growth of the pentacene on the ordered layer because of its low energy coinciding with the pentacene plane as well as the increased disorder in the SAMs of lower density with increasing temperature.

The growth of ordered layers of a molecular material on another have been explored by sequential evaporation of the components. This process is sometimes referred to as heteroepitaxy or weak epitaxy,[23] but the processes behind the nomenclature are essentially the same: a molecular material is deposited onto an unlike molecular material such that the crystalline phases developing in the thin films have little influence on one another. There is a very subtle interface between the materials. This area is of growing importance, because these heterostructures are of interest in optoelectronic devices whose performance depends upon the organisation of the materials.[24]

The assembly of perfluoro-copper phthalocyanine on pentacene reveals some remarkable effects in the initial stages of growth of the disk-like molecule.[25] Upon evaporation under UHV, the pentacene grew on the silicon dioxide/silicon surface in the "standing-up" mode that was mentioned previously, and AFM showed the presence of steps in the film corresponding to the length of the molecule. When the phthalocyanine derivative was deposited, a fascinating morphology comprising narrow islands (50–100 nm in width) decorated the pentacene steps. These islands are seen as the bright features in the topographical AFM image in Figure 7.4(a). The heights of these features were between 90 and 170 Å. These pillars correspond to a lying-down mode of the phthalocyanine that stacks, as shown in the cartoon in Figure 7.4(c). It was not clear if this was all the phthalocyanine material though, and the heights of the two molecular components in the film are similar, making distinguishing them on the basis of dimensions alone perilous. The phase-shift signal from the AFM, which is sensitive to the dissipative processes from the tip–sample interactions and is material dependent,[26] reveals the position of both the standing-up configuration that is clear in topography but also the lying-down phase that might be more expected, that is indicated on the profile in the figure. There was no clear correlation between the pentacene terrace size or with the coverage of the phthalocyanine with the distribution of the two packing types, indicating that diffusion was not responsible alone for the effect. It was inferred that the steps in the pentacene steps contain an energetic profile that favours the formation of the lying-down phase. The standing-up configuration (with the phthalocyanine plane perpendicular to the interface) is the one favoured on weakly interacting surfaces.[27–29]

The growth of the film was also measured *in situ* at a synchrotron where X-ray diffraction showed the development of the layers of the "standing-up" structure of perfluoro-copper phthalocyanine, giving rise to the (001) and (002) Bragg peaks in Figure 7.5. A clear Bragg peak was also seen at a q_z value of 2.0011 Å^{-1}, corresponding to an intermolecular distance of 3.14 Å. This feature is consistent with the observations made by AFM of a parallel arrangement of the phthalocyanine derivative with the interface. The fact that the mosaicity of this signal was very small (0.017°) and similar to that of the pentacene (0.015°) implied a good transfer of molecular order. Similar measurements on a thicker film grown at 65 °C confirmed that this arrangement was not a kinetic artefact, and was caused by specific features of the pentacene layer in which a microscopic energetic feature of the step nucleates the formation of columns perpendicular to the interface. The growth of C_{60} on pentacene layers shows similarly spectacular growth characteristics at the edges of the layers,[30] though in this case, diffusion

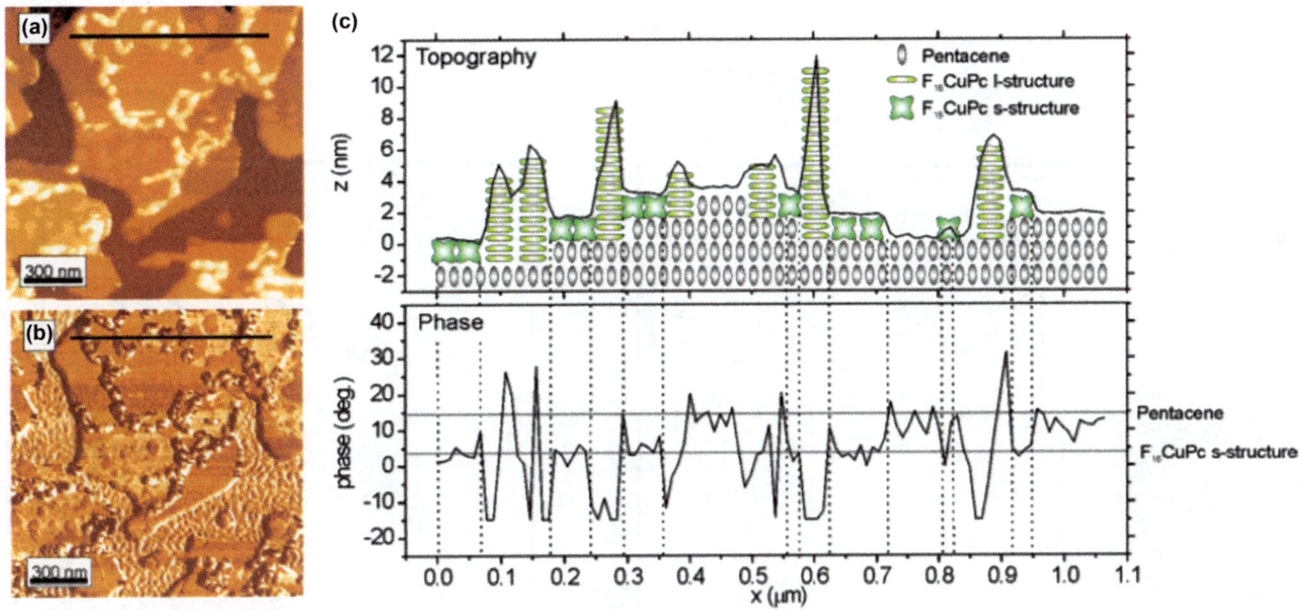

Figure 7.4 The AFM images show (a) the topography of a partial layer of perfluoro-copper phthalocyanine grown on a layer of pentacene and (b) the phase-shift image recorded simultaneously in the same area. Profiles of (c) topographic (top) and phase-signal (bottom) of marked areas in images (a) and (b), and a representation of the molecular packing. The open ellipses correspond to pentacene and the green ellipses and squares to perfluoro-copper phthalocyanine.
Reprinted with permission from ref. 25. Copyright 2006 American Chemical Society.

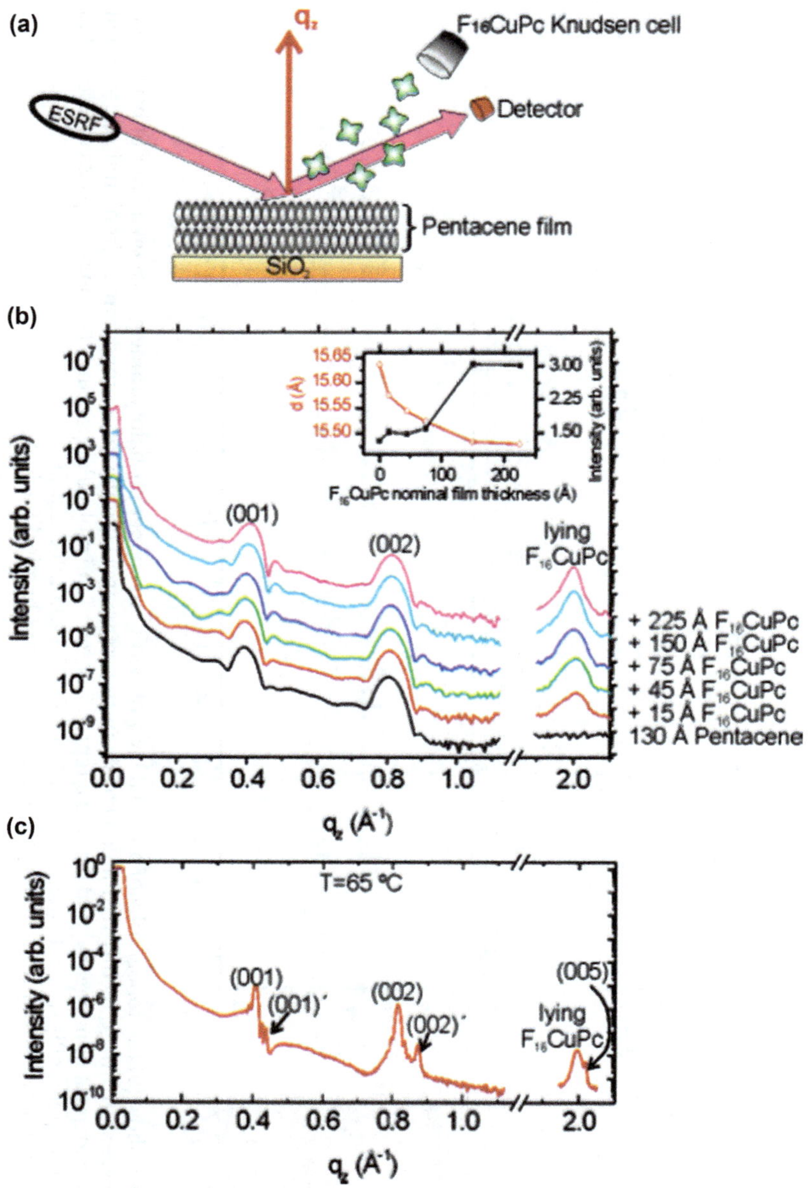

Figure 7.5 A representation (a) of the *in situ* X-ray diffraction experiment of the heterostructures formed by the deposition of copper phthalocyanine grown on a layer of pentacene on SiO_2/Si and (b) the specular X-ray diffraction data of the growth of the film at room temperature (nominally for growth on 130 Å of pentacene). The inset represents the change in the spacing (open symbols) and integrated intensity (filled symbols) obtained from fits to the (002) Bragg reflection. (c) Specular X-ray diffraction data of a thicker heterostructure grown at 65 °C (nominally 100 Å of phthalocyanine on 752 Å of pentacene). The prime symbol indicates the Bragg peaks arising from the pentacene bulk phase.
Reprinted with permission from ref. 25. Copyright 2006 American Chemical Society.

control led to the formation of one- and two-dimensional islands, emphasising the special characteristics that these regions have on these organic crystals. Remarkably for the fullerene, these structures were even stable at room temperature, an incredible feat given the relatively weak interactions at play at the interface. These kinds of interfaces with step edges as recognition sites are a fascinating area for potentially exploring supramolecular chemistry at interfaces.

This kind of assembly is very similar to the "disks on rods" growth of zinc phthalocyanine on a layer of the p-type organic semiconductor 2,5-bis(4-biphenylyl) bithiophene (that are the flat round and linear molecular objects, respectively) that in turn was grown on the native silicon oxide on silicon layer. The rod-like component forms smooth films following a Frank–van der Merwe growth pattern with the molecules standing up quasi-perpendicular to the surface and with a crystallographic plane of this organic ultrathin crystal in the plane of the substrate.[31] When zinc phthalocyanine (a p-type semiconductor also) was grown on the film formed by the 2,5-bis(4-biphenylyl) bithiophene, AFM showed the presence of lines of the disk-like molecule along the crystallographic directions of the rod, which has grooves on its upper surface that presumably guide the formation of the assembly of the phthalocyanine.[32] This result was similar for p-sexiphenyl as the rod-like component, although hole transparent and blocking transport (probed with photoemission spectroscopy and field-effect transistors) were superior for the bithiophene derivative.

The growth of polar layers thanks to hydrogen bonding was elegantly shown in order to prepare films that present non-linear optical behaviour thanks to their orientation.[33] The basis for the preparation of the films is the incorporation of both hydrogen bond accepting and donating units in the same rigid molecule, so that it cannot fold back on itself generating a non-polar structure. In bulk materials, it is habitual for this kind of material to crystallise in centrosymmetric space groups, with chains of molecules packed in a head-to-tail fashion. Indeed, when the hydrazo derivatives in Figure 7.6 bearing a pyridyl function at one extreme and a pyridyl unit at the other were evaporated onto inorganic hydroxylated surfaces, no polar order was observed. However, the preparation of a SAM followed by covalent attachment of the derivative in the first step of film growth led to a thin film with polar order, as shown in the figure. The initial chemisorbed SAM was formed using *p*-(iodomethyl)phenyldiiodochloro-silane, leaving a benzyl halide functionality directed away from the interface. This group was reacted with the pyridyl group of the hydrazo derivative, forming a pyridinium group at the interface. Continued evaporation of the same molecule led to hydrogen-bonded chains growing off the surface in a single direction. The opposite orientation that would give a non-polar film is not thermodynamically favoured. The films are very smooth; they had r.m.s. roughness of 0.37 nm for a 400 nm film and 8.68 nm for a 1470 nm film, as shown by AFM. The polar order was proven by second harmonic generation spectroscopy, that showed that the films have very high second-order optical susceptibility

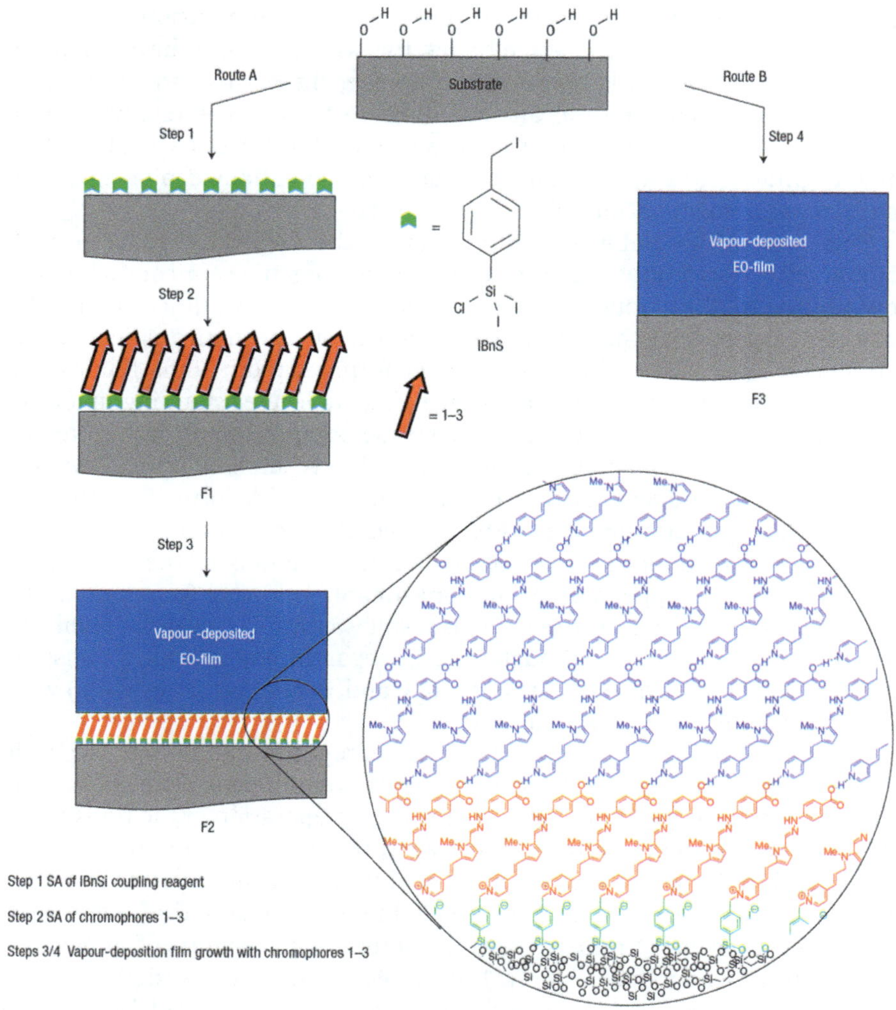

Figure 7.6 Routes to films of polar molecules bearing both hydrogen bond accepting and donating components. The substrate was float glass or single-crystal silicon with a native oxide coating (approximately 1.5 nm thick).
Reproduced by permission of Macmillan Publishers Ltd., *Nat. Mater.*, ref. 33, copyright 2004.

(approximately 100 pm V^{-1}), making them of interest for non-linear optics. The route has advantages over others in that there is no need for electrical poling to generate polar order and it does not rely on the formation of bulk crystals. This phenomenon of polar assembly of rigid molecules incorporating hydrogen bond accepting and donating units is valid for other polar molecules.[34]

Clearly, the material used for the preparation of thin films under evaporation must be stable to the sublimation process; this restriction limits molecular weight and certain groups that might be unstable because of thermochemical reactions. High vacuum is usually required, and even ultrahigh vacuum (UHV) cannot avoid the decomposition of some materials—such as oligomeric and polymeric materials—that inevitably have to be solution processed. The development of electrospray techniques is interesting but as yet not general and in any case may be somewhat limited to the volume and area that can be used. For this reason, many applications require the use of solvent, which will be discussed in the next section. Before that though, the preparation of polymeric materials on surfaces by polymerisation of vapour phase monomers must be highlighted as a method for the growth of interesting materials as well as for the development of strategies to prepare composite films under vacuum and UHV conditions. The chemical vapour deposition polymerisation of *para*-cyclophanes has received particular attention.[35,36] The deposition of these compounds in vacuum followed by pyrolysis leads to the formation of the polymers. For example, 4-(3,4-dibromomaleimide)[2.2]paracyclophane can be sublimed onto a substrate (including gold, silicon and glass) and then pyrolysed to produce a poly(phenylethane) (or poly(*para*-xylene, also known as parylene) terminated with dibromomaleimide moieties that can be used for the immobilisation of thiol-functionalised proteins.[37]

7.3 Thin Film Growth From Solution

Deposition of molecules from solution is an attractive processing method for supramolecular chemists, because no expensive equipment is required and it takes advantage of familiar behaviour of purifying and processing molecular materials in liquids that are the daily bread of researchers active in this area of research. It is, though, a tricky business! As mentioned in Chapter 1, an evaporating liquid may well have local variations in surface tension that cause Marangoni flow, and flow can also be brought about by capillary forces: the wetting of the adsorbent is a critical step in the formation of a film. Apart from any specific supramolecular effect, these factors can lead to pattern formation in the system that is far from equilibrium because of the rapidly increasing concentration, diffusion and flow effects.[38] Nonetheless, supramolecular effects in the assembly of materials upon evaporation undoubtedly play a role in thin film structure, as shall be demonstrated with the examples in this section.

The deposition of calix-4-resorcarene derivatives onto a templating layer of a SAM of a calix-4-resorcarene chemisorbed to gold showed the spontaneous growth of thin films with a bilayer-type structure.[39] The monolayer was exposed to a hexane solution of the alkyl derivative, leading to films that gave infrared spectra consistent with the formation of a hydrogen-bonded layer of the amphiphile on the functionalised metal surface followed by the formation of subsequent strata of the physisorbed molecule by the interpenetration of alkyl chains and then repetition of the formation of the hydrogen-bonded interface (Figure 7.7). The outer surface of the film was

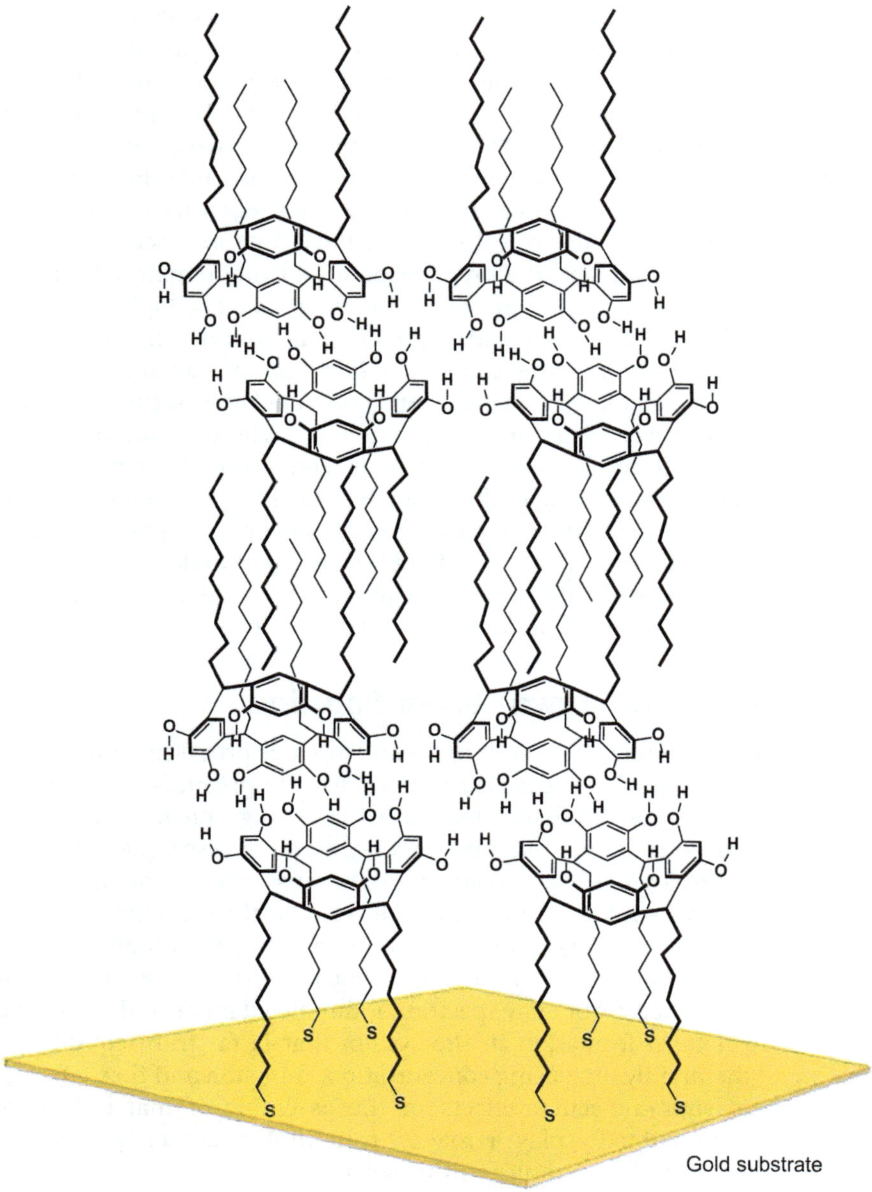

Figure 7.7 A representation of the multilayer films formed by an amphiphilic calix-4-resorcarene derivative on a chemisorbed SAM of a related molecule on gold.

shown (by contact angle measurements) to be hydrophobic, indicating that the last layer to be deposited had its alkyl chains facing the solvent. This principle has been used for the preparation of hydro- and oleophobic films.[40]

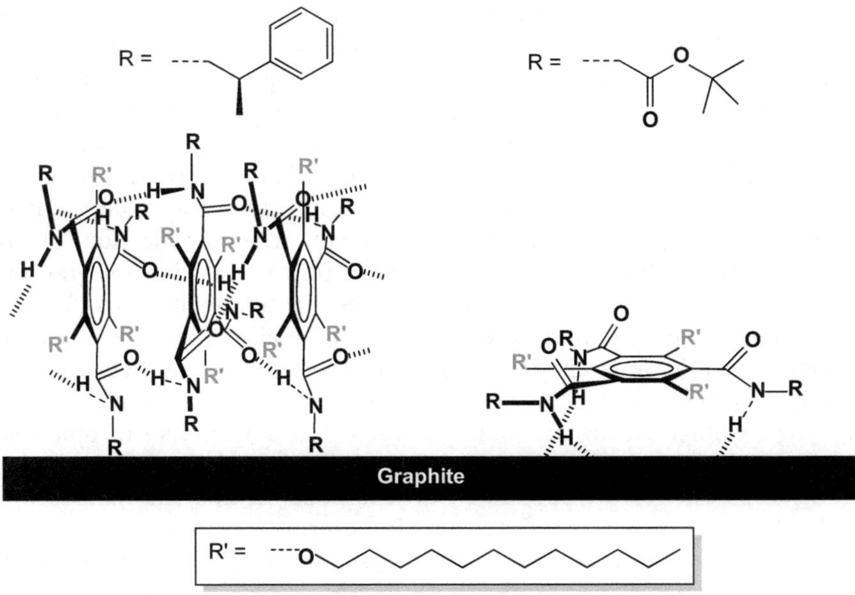

Figure 7.8 The helical supramolecular columns formed by a chiral benzene tricarboxamide that forms polar one-dimensional assemblies on HOPG parallel to the surface both at submonolayer coverage and in thin films, and a derivative with a more bulky side-chain that stacks face-on to the same substrate.

Polar order was also observed in the formation of stacks of small organic molecules that self-assemble through hydrogen bonds. A family of benzene 1,3,5-trialkoxy-2,4,6-carboxamides (Figure 7.8) form helical stacks where all three secondary amide groups form hydrogen bonds with neighbouring molecules in the supramolecular column.[41] This situation exists in the solid state for simple 1,3,5-tricarboxamides,[42] and is particularly true for the cases where the other positions of the benzene ring are substituted, forcing a high angle between the aromatic plane and the amide bond vectors. While the two compounds in Figure 7.8 both assemble in columns, on graphite the bulky achiral side-chained compound prefers a face-on orientation while the chiral derivative forms columns whose axes lie parallel to the graphite plane. AFM and electrostatic force microscopy (EFM) were recorded simultaneously to measure the topography and polarity of both monolayer and multilayer films of the compounds. At sub-monolayer coverage on graphite, after spin casting from dichloromethane, the chiral compound forms long fibres—up to six microns in length—where each component strand is approximately 2 nm in diameter, consistent with a single molecule. EFM showed no net dipole. Apparently, the intermolecular interactions dominate over those between the molecule and the surface. On the other hand, when a more bulky group is attached to the amide, the molecule–surface interaction, through N–H/π interactions as well as alkyl chain–graphite interactions, dominate over the intermolecular forces that lead to column formation. A monolayer is formed where the carbonyl moieties are directed with their dipole away from the surface, as indicated by the negative charge detected by EFM. This physisorbed monolayer can act as a template for the growth of subsequent layers, as shown by the observation of negative charge even when three layers were clearly stacked on top of one another. This observation shows that in the absence of graphite substrate, the molecule itself can stack, giving ordered directional growth. The substitution of the secondary amide thus provides selectivity to film orientation on graphite. This family of molecules also show hierarchical assembly in films.[43] Chiral side-chains emanating from the amide functional group give a film of material five micrometres thick that orders spontaneously between ITO-coated glass slides, with the columns organised parallel to the interfaces.[44] Applying a voltage during the cooling of the material from its isotropic state gave an optically isotropic film, and it was shown that the columns were organised parallel to the applied field. It is believed that the macroscopic dipole of the self-assembled fibres aids in ordering the material.

Good polar and/or halogenated solvents, such as 1,2-dichlorobenzene and the like, are often used for forming films of π-functional molecules. They can give very good results in terms of homogeneity of the film and supramolecular organisation. However, these kinds of solvents are not ideal from a processing perspective because of the difficulty in handling them (toxicity and high boiling point) as well as from the viewpoint of green chemistry[45] because they are environmentally undesirable. A supramolecular solution to the dissolution of hydrophobic molecules in water for the formation of films is by their complexation in water soluble hosts. When complexes are formed[46] between the polysaccharide amylose (made up of α-D-glucose segments) with a molecular weight of approximately 4.5 kDa and the hemicyanine dye 4-(4(dimethylamino)styryl)-1-docosylpyridinium bromide, dissolved in water and cast onto glass, spontaneous order perpendicular to the film surface was inferred.[47] Second harmonic generation measurements indicated a clear alignment of the chromophore, implying that the molecule was oriented at a right angle to the interfaces between the complex film—that was up to six microns thick—and air and the glass substrate over the entire sample. The inclusion of the dye in the polysaccharide helical tube led to an increase of the thermal and optical stability of the chromophore, presumably because it is shielded from chemical attack by the host.[48]

The constitution and effect of alkyl chain length both have an important influence in the formation of aggregates of aromatic systems substituted with two secondary amide groups bearing the aliphatic moiety.[49] The tetrathiafulvalene derivatives shown in Figure 7.9 are configured in a *trans* fashion around the central double bond in the molecules. The difference between them is the length of the alkyl chains that are attached to the nitrogen atom in the secondary amide unit, the number of carbon atoms being 12, 15 or 18 at both ends. All of the compounds form fibres, consistent with the idea that hydrogen bonding—probably aided and abetted by π-stacking

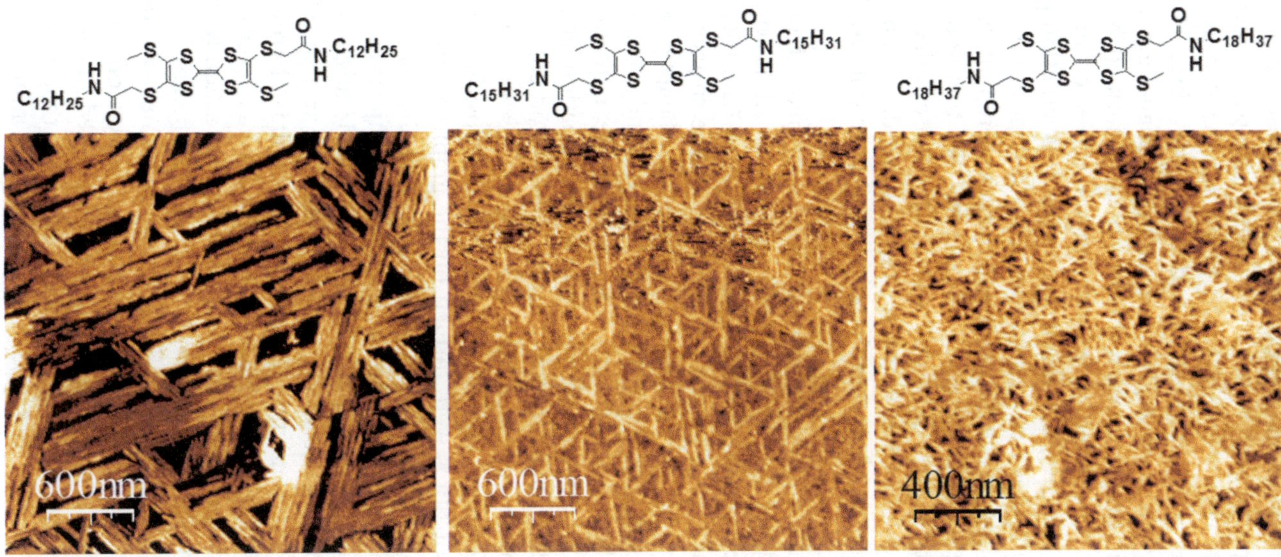

Figure 7.9 AFM images of the fibres formed upon evaporation of chloroform solutions of the tetrathiafulvalene derivatives shown on HOPG.

and van der Waals interactions between the chains—are aiding in the formation of the supramolecular objects. For the compound bearing two dodecyl chains, it is clear that grainy well-oriented fibres are formed that follow the orientation of the graphite adsorbent. These axes cannot be seen in the AFM images presented, but the triangular shapes formed in some of the voids indicate clearly an epitaxial-type growth dictated by the graphite surface. A similar situation is also evident for the fibres formed by the molecule bearing the pentadecyl chains: triangular shapes are formed by the fibres, which this time are much finer than those formed by the dodecyl analogue. On the other hand, very short and non-oriented fibres are formed by the compound bearing the octadecyl chains. This situation points to the formation of the aggregates in solution prior to deposition and therefore the film is not epitaxial. The fibres formed by these compounds get narrower as the alkyl chain length increases (see Figure 7.9), therefore long chains seem to limit lateral growth in this system. For the octadecyl compound, the width and height of the fibres are far more similar than those of the analogues with shorter chains, again pointing to a solution precipitation for this compound.

Deposition of the equivalent compounds bearing alkyl chains but where the central bond has the *cis* configuration leads to films with very short range order. As can be seen in the graphs in Figure 7.10, their height and width are extremely small, and they basically wet the surface, as shown in their physisorbed layers that were discussed in Chapter 4. Therefore, epitaxial growth is not possible for the *cis* isomers under these conditions. The surface makes a significant difference to the outcome of surface assembly as neither isomer type nor chain length variation lead to fibres on mica, where smooth films are formed. The solvent and concentration play very important parts in determining the size and order of the structures formed. When toluene was used as the solvent for deposition, only very short ordered regions of fibres were formed on the graphite surface. When the concentration was reduced for the *trans* isomer that showed fibre formation in solution, long well-formed fibres and rectangular objects like nanocrystals appeared that were clearly epitaxial because of their alignment with the graphite axes. Therefore, the concentration must be at a specific value for each compound independently in order to show epitaxial growth, aligning with the graphite templating adsorbent. Similar effects of constitution, alkyl chain length and adsorbent were also observed for a family of porphyrin derivatives,[50] suggesting the generality of the phenomenon.[51]

When one of the *cis* compounds in the former study was deposited on graphite from octanoic acid, a remarkable formation of disk and spiral nanostructures took place when the solvent evaporated in certain places of the surface (Figure 7.11).[52] AFM images showed that monolayers and multilayers with straight fibres formed domains on much of the surface, similar to the physisorbed layers at the liquid–solid interface (see Chapter 4), but sometimes disk-like aggregates with space in the middle grew into hexagonal close-packed superstructures.

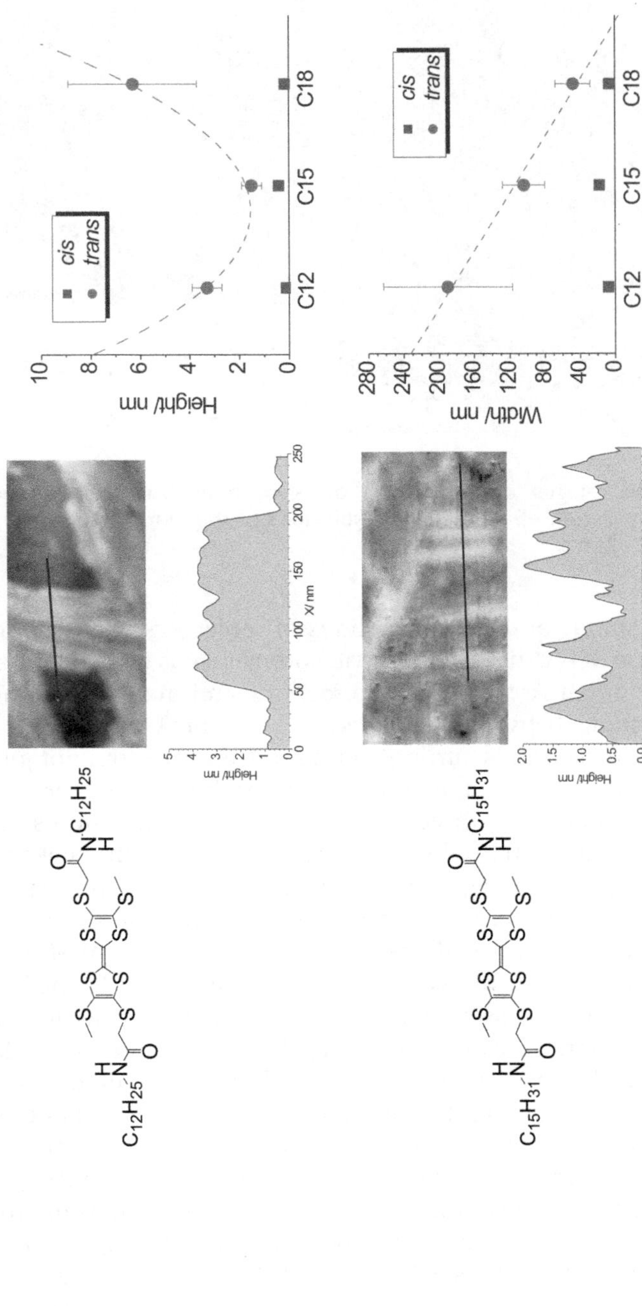

Figure 7.10 Close-ups of the AFM images of the fibres formed upon evaporation of chloroform solutions of the tetrathiafulvalene derivatives shown on HOPG and graphs showing the effect of alkyl chain on the width and height of the nanostructures.

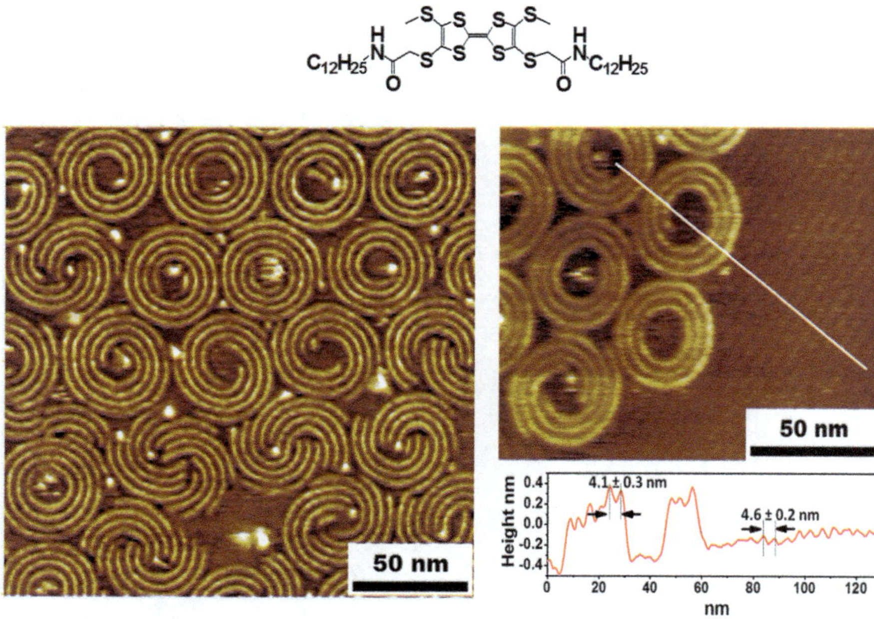

Figure 7.11 AFM images of spiral and disk-like nanostructures formed after evaporation of octanoic acid solutions of the tetrathiafulvalene derivative shown.

In both the spirals and the disks, lines of molecules curl around each other; in the case of the disks, concentric homogeneous curved layers form. The spirals are either Archimedean spirals (a spiral made up of one strip with equal spacing between the lines at each turn) or somewhat like Fermat's spiral (two strands curling around to/from a centre, not joined in the structures but they do in the figure that gives the name). More complex nanostructures comprising three or more strands giving spirals are also formed. There is no chirality beyond each spiral, as both clockwise and anticlockwise ones were imaged. These areas do not seem to be templated by the surface, as they were seen on "bare" graphite and also in areas where monolayers were formed previously (as seen in the AFM image on the right of Figure 7.11). It was suggested that the molecule formed strands in solution and then assembled into the circular structures before adsorbing to the interface. Even thicker films show both fibre-like and discoid objects. The distance between the fibres in the objects is about 4.1 nm, suggesting a different molecular geometry to the lamellar-type monolayers. Also, the very different phase contrast that they display in the AFM images indicates a different orientation with respect to the plane of the surface. The specificity of this assembly route to solvent would indicate that in addition to intermolecular hydrogen bonds bringing the molecules together, solvent interactions with the amide and/or the alkyl chains lead to ring formation with the long axis of the aromatic residues quasi-perpendicular to the surface and the chains filling the space between them. The example shows the remarkable structures that can be formed using a combination of interactions at a surface and also the complexity regarding interactions with solvent. Even today, explaining this kind of assembly at the molecular level is a true challenge, and this example awaits interpretation through detailed molecular modelling.

Evaporation of solvent from solutions of simple surfactants can lead to selective adsorption in certain areas when the adsorbent has two chemically different faces exposed. The deposition of stearic acid on $SrTiO_3$ substrates leads to selective adsorption of the acid-terminated surfactant onto the TiO_2 terraces of the surface, the SrO areas remaining surfactant free.[53] The adsorption of the molecule—presumably through attachment of the carboxylate group to the inorganic surface—improves the frictional properties of the interface and changes the surface contact potential, as shown by extensive AFM experiments.

The formation of films on a surface where adsorbate and adsorbent interact weakly can be used in post-deposition annealing for the formation of ordered structures thanks to the intrinsic interactions between the molecules on the surface. A perylene diimide derivative when cast on glass or mica by spin coating and drop casting, respectively, can be annealed thermally or with solvent annealing to produce "nanobelts".[54] The effects of thermal and solvent annealing are shown for the case of the film of the molecule on mica in Figure 7.12. It is clear that thermal annealing and treatment with different solvents produce nanostructures of different dimensions, and this fact demonstrates the great flexibility and opportunity that these methods give when processing materials on surfaces. When glass was used as the surface, the annealing treatment with chloroform vapour actually yielded extremely long fibres tens of nanometres in height and width. The formation of these intricate and varied structures was ascribed to dewetting of the molecules in the film on mica and glass—crystalline and amorphous, respectively—that give similar results because of the poor interaction between the surface and molecule in both cases. Both mica and glass have high hydrophilicity compared with the hydrophobic nature of the adsorbate that must diffuse and presumably undergoes Ostwald ripening under the annealing treatments. The fact that the amorphous or crystalline nature of the substrates has little role in the outcome is because of the lack of a specific directional interaction between them and the molecule that forms the fibres.

The processing of polymers from solution is of increasing importance because of the great strides being made in the preparation of materials for molecular electronics. Deposition onto silicon–silicon oxide substrates by spin casting from solution affords thin films whose properties can be optimised through thermal annealing of the sample. Figure 7.13 shows an example of a thermal annealing process that is often used to optimise polymer structures where mobility of the chains is limited and thermal energy is required to help the material to reorganise.[55] This particular case shows an

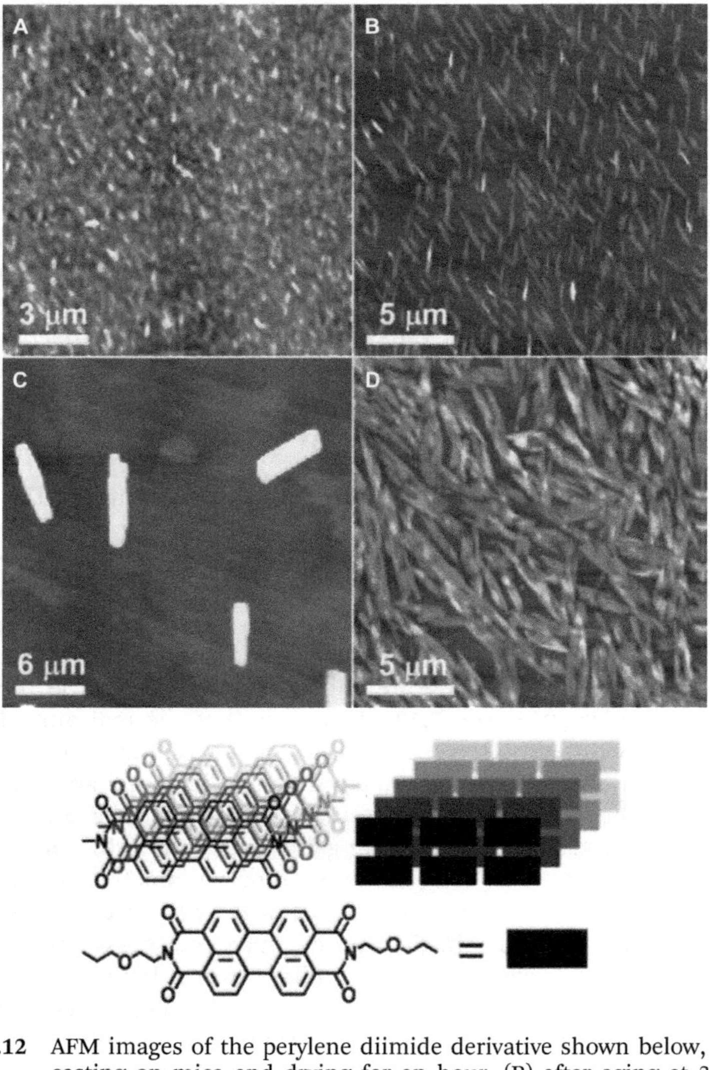

Figure 7.12 AFM images of the perylene diimide derivative shown below, (A) after casting on mica and drying for an hour, (B) after aging at 358 K for 56 hours, (C) after chloroform annealing for 24 hours and (D) after annealing in methanol vapour for 24 hours. The cartoon shows the proposed π stacking in the aggregates.
Reprinted from ref. 54 with permission from the Royal Society of Chemistry.

alternating polymer of a diketopyrrolopyrrole derivative with oligothiophene that forms ordered films by simple evaporation of solvent (chloroform), without the need for solvent annealing, and with an orientation that is perpendicular to the interface with the substrate (hydrophobitised silicon dioxide on silicon).[56] The diffraction studies on the thin films indicated a π–π stacking distance of approximately 3.75 Å (peak at $2\theta = 23.7°$), consistent with a good packing in the film. This π–π stacking is more efficient than

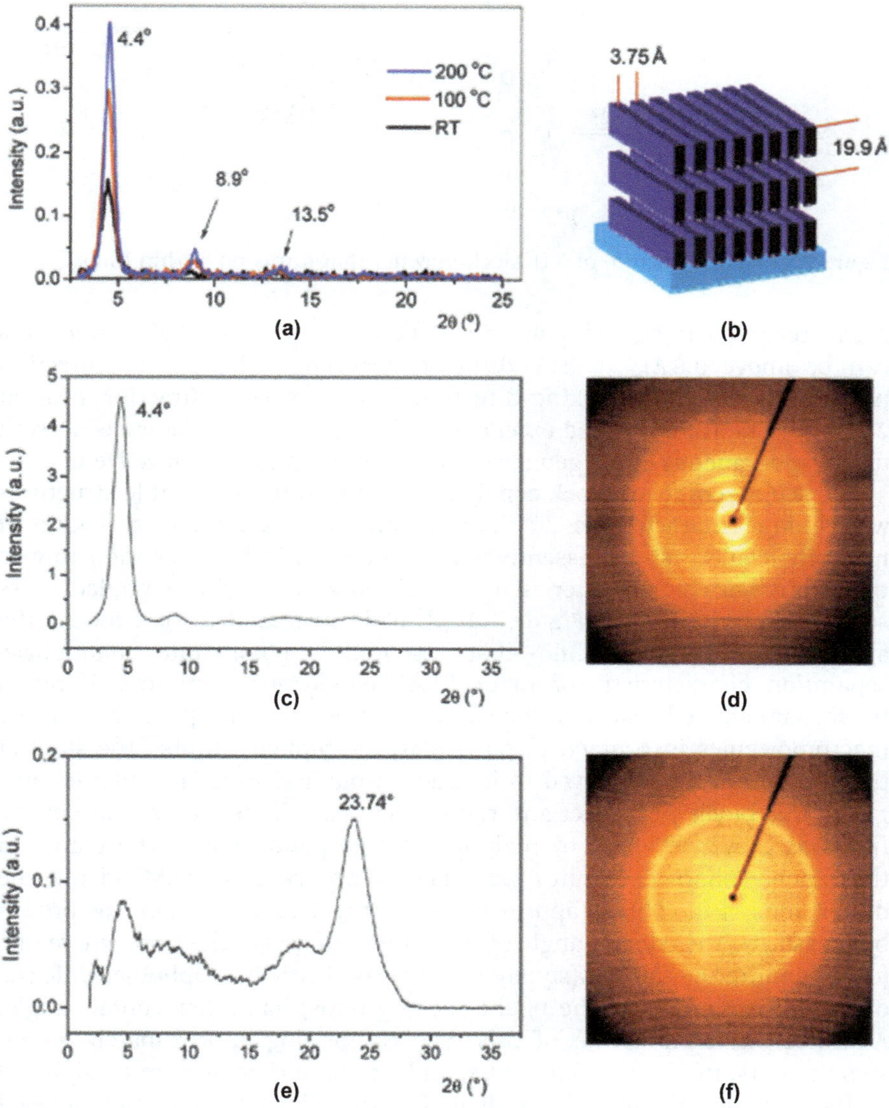

Figure 7.13 X-Ray diffraction (XRD) patterns of a thin film of an alternating polymer of a diketopyrrolopyrrole derivative with oligothiophene on Si–SiO_2 substrates at the annealing temperatures indicated (a). A cartoon showing the polymer's lamellar structure in the thin film is shown in (b). Images (c) and (d) show the one-dimensional XRD 2D XRD patterns with the beam parallel to the film, and (e) and (f) perpendicular to the film, showing the high degree of alignment. Reproduced from ref. 55 with permission from the Royal Society of Chemistry.

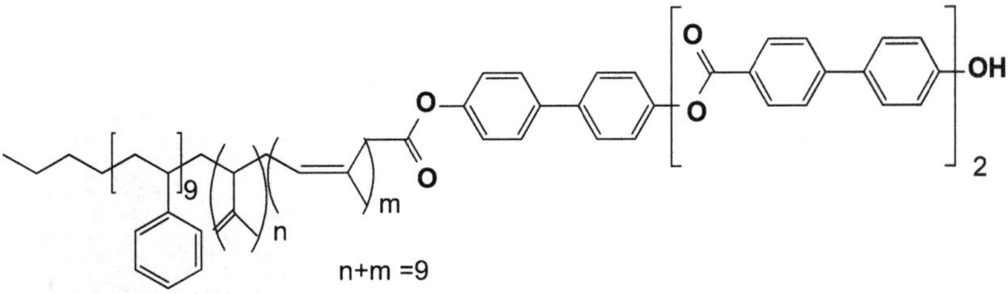

Figure 7.14 The structure of a triblock polymer that forms polar thin films.

many related conjugated polymers (where the distance between chains can be above 3.8 Å).[57,58] The distance between chains in the direction perpendicular to the stacking direction is seen in the diffraction peak at $2\theta = 4.4°$. The films showed excellent performance as transistors as a result of the spontaneous order generation upon evaporation of the solvent.

The self-assembly of block copolymers can lead to remarkable structures with unique characteristics.[59,60] A quite amazing example of this kind of macromolecule that self-assembles upon deposition from solution to give thin films with polar order is a triblock copolymer with a graded cross section along its length (Figure 7.14).[61] This structural feature means the aggregates of the molecule—that essentially demonstrates nanophase separation characteristic of many block copolymers—are forced into a mushroom shape because of the splaying necessary to keep all areas of the macromolecules in contact. In particular, the biphenyl units—the stalk of the mushroom—are believed to interact through edge-to-face interactions, giving this block a compact and rigid structure. A lamellar structure of the film was proven by TEM by making a cut perpendicular to the plane. In the plane, spherical features were clearly observed by TEM, giving the dimensions of the object (approximately 5 nm). The polar order was proven by measuring the contact angle of the different faces of the film, where one (corresponding to the poly(styrene) head) was clearly hydrophobic while the other (corresponding to the phenol head groups) had a low contact angle. A triblock macromolecule of this type comprising approximately eleven styrene units, fifteen isoprene units, and a precise rigid section of biphenyl and stilbene terminated with a nitrile functionality gave films that showed piezoelectric activity.[62]

The complexation of functional small molecules into block copolymer architectures through supramolecular chemistry can give thin film materials with interesting structures and properties.[63–65] When polystyrene-*block*-poly(4-vinylpyridine) (PS-*b*-P4VP) was mixed with hydroxymethyl-ferrocene in chloroform, and the thin films were formed by evaporation of solvent upon spin coating followed by solvent annealing in THF vapours, a variety of morphologies were observed depending on the proportion of the components.[66] Investigation of the films (that were 40–50 nm thick) by AFM showed that the composites contained spherical blocks for a low proportion of ferrocene derivative, through a cylindrical phase where the rods

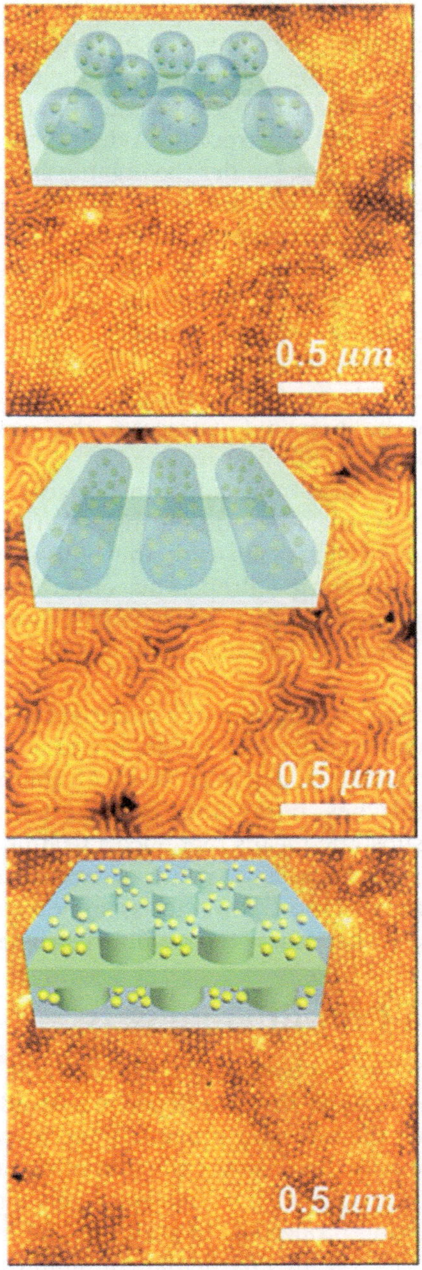

Figure 7.15 AFM images of thin films containing mixtures of polystyrene-block-poly(4-vinylpyridine) (PS-*b*-P4VP) and hydroxymethyl-ferrocene with increasing proportion of the latter from top to bottom. The films were prepared by spin-coating chloroform solutions onto SiO_2–Si followed by THF solvent vapour annealing. The cartoons represent the phase separation, with the darker regions the pyridyl segment and the yellow dots the small molecule component.

incorporated the pyridyl-ferrocene complex, up to a cylindrical phase with (excess) free ferrocene in the polystyrene phase (Figure 7.15). The presence of hydrogen bonding between the pyridyl group in the more polar region of the block copolymer and the hydroxyl hydrogen atom in the ferrocene derivative was proven through infrared spectroscopy. When the material was included in a field effect transistor device under the organic active layer, it was shown that effective charge-trapping was achieved with the nanocomposite, which behaves as an electret (the electrostatic equivalent of a permanent magnet). The ability of the material to trap charge was easily modulated by changing the proportion of ferrocene derivative in the material. The memory performance of the device was manipulated through the morphological transition accompanying change in the polymer's block proportions or through the loading with ferrocene derivative. The ease with which these combinations could be prepared vindicates the use of supramolecular approaches to this kind of thin film material.

The self-assembly of large molecules on surfaces can combine effects intrinsic to the process, meaning the intrinsic thermodynamic preference for a particular kind of aggregate, with the bulk processes of dewetting associated with an evaporating drop. This effect has been shown for a tris-porphyrin derivative based on a benzene tricarboxamide core that self-assembles through hydrogen bonding,[67] and was shown in the introductory chapter of this book.

We have seen that the organisation of a material being deposited as a thin film can depend greatly on the concentration and solvent used for the deposition process. The production of thin films with control over the size of the aggregates and the thickness of the films can be extremely challenging, and the dependence on substrate means that once a technique is optimised for a given substrate then this knowledge is not necessarily transferable to other supports. A way around this limitation is to assemble a structure in solution to form gel-like materials where this pre-assembled structure is deposited.

7.4 Thin Film Growth From Gels

Gels[68] are interesting for the preparation of thin films because their structure is already formed before they are deposited, avoiding the potential effects of the surface on the assembly process.[69] The preparation of inorganic films by sol–gel chemistry that involves the formation of covalent bonds has received significant attention,[70] but molecular gels also lead to materials that can have interesting properties. In addition, the microstructure in the thin film material formed from a particular compound as a gel can be quite distinct from that formed from a bulk gel.[71]

A donor–acceptor material that can be employed for charge transport and charge generation used a perylene bisimide gel network as an acceptor conductive system self-assembled in an amorphous p-donor conductive polymer matrix comprising a derivative of tetraphenyl benzidine.[72] An interpenetrating mesh of fibres with many electron donor–electron acceptor contacts was made employing this gel-embedded polymer matrix route. Smooth films could be made in a single step, by drying a mixture of the two components on a surface, or by carrying out sequential depositions. In devices used for photovoltaic current generation, smooth films with high optical quality were achieved and the performance of the films prepared from the mixture of components provided the greater efficiency, surely a result of the high interface area between the donor and acceptor moieties. Films prepared with a non-gelator derivative exhibited very low efficiency in the devices, pointing to a potential advantage of the gel route for the processing of these materials.

A number of tetrathiafulvalene derivatives have been used as gelators that can conduct electrons.[73–75] A representative example is shown in Figure 7.16. This three-fold symmetric molecule assembles through stacking of the bipyridine units (that are rigidified by an intramolecular hydrogen bond) and forms gels in halogenated solvents, the films of which can be doped employing iodine vapours.[76] The existence of the mixed valence state was proven by near-infrared absorption spectroscopy, where a broad band corresponding to charge transfer was evident. The electrical conductivity in the film was evident from AFM studies, where the passage of current through the material was registered at the same time as the topography. The micrograph revealed a complex mesh comprising thick and thin conductive nanofibres. The larger fibres are the main charge carriers in the material. The advantage of this kind of material is that in principle any surface could be coated with the gel and made conducting, with no need for pre-coating or surface-dependent assembly.

It is not necessary to have the gel state to be able to form a conducting thin film. The gelator shown in Figure 7.17 forms disordered fibres in the gel state (with hexane) and when deposited on either glass or graphite and doped with iodine vapours becomes a bulk semiconductor.[77] The material also undergoes a phase change upon annealing the film to give a more conducting state.[78] In order to produce smoother and thinner films for a detailed study of the nanostructure of the conducting fibres, the material was deposited onto graphite below the critical gel concentration, leading to a mesh of straight fibres (Figure 7.17) that proved to be conductive when doped. When this material was scanned with a conducting AFM tip, the presence of a smooth network showing conductivity was clear, although upon application of high potential, fibres were removed from the material by the tip.[79] This effect is probably a general one for thin films of this kind of material, although it needs further research. It is potentially important because of the interfaces that exist between metals and soft organic matter in many devices.

The preparation of thin films on a substrate is useful when large areas of material are required for the particular application that one has in mind, be it optical, electronic and so on as exemplified by recent examples.[80,81] However, the localisation of the growth of materials on surfaces can require specific patterning methods and particular deposition techniques, which are covered in the Chapter 9.

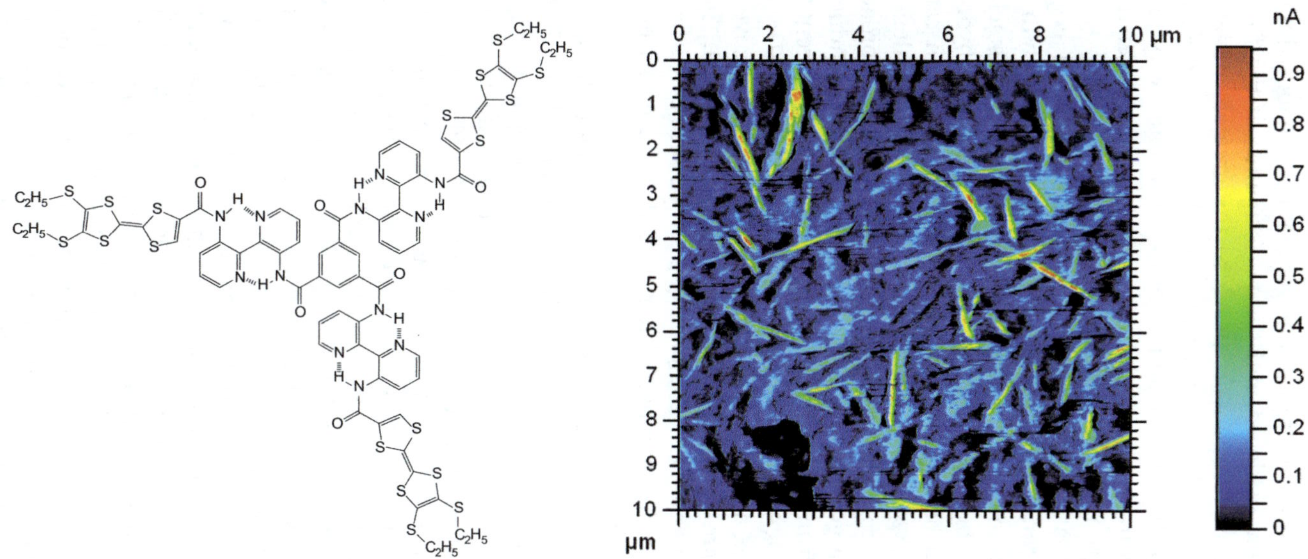

Figure 7.16 A three-fold symmetric gelator that can be prepared in thin film form and doped so that it becomes a bulk semiconductor. The image is a conducting AFM image showing the routes of highest current.

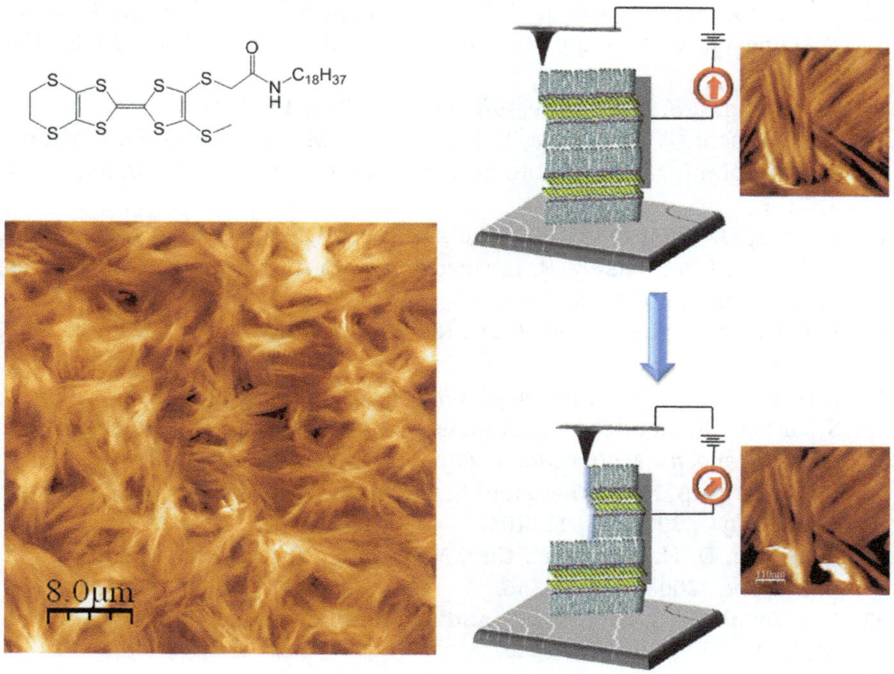

Figure 7.17 A molecular gelator that forms thin films with very well defined fibres, as shown in the large scale AFM image. The material once doped with iodine was probed by conducting AFM that showed that the lamellae of the material are removed under potential control, as indicated in the cartoons and the conducting AFM images that show partial removal of a piece of material by the scanning tip.

References

1. P. W. Bohn and D. J. Walls, *Mikrochim. Acta*, 1991, **1**, 3.
2. F. Schreiber, *Prog. Surf. Sci.*, 2000, **65**, 151.
3. D. B. Mitzi, *Chem. Mater.*, 2001, **13**, 3283.
4. V. Palermo, S. Morelli, C. Simpson, K. Müllen and P. Samori, *J. Mater. Chem.*, 2006, **16**, 266.
5. E. Gomar-Nadal, J. Puigmartí-Luis and D. B. Amabilino, *Chem. Soc. Rev.*, 2008, **37**, 490.
6. K. Sakakibara, J. P. Hill, Jonathan and K. Ariga, *Small*, 2011, 7, 1288.
7. J. Yang, D. Yan and T. S. Jones, *Chem. Rev.*, 2015, **115**, 5570.
8. A. Koma, *Prog. Cryst. Growth Charact. Mater.*, 1995, **30**, 129.
9. S. R. Forrest, *Chem. Rev.*, 1997, **97**, 1793.
10. F. Schreiber, *Phys. Status Solidi*, 2004, **201**, 1037.
11. H. Yanagi, S. Okamoto and T. Mikami, *Synth. Met.*, 1997, **91**, 91.
12. S. Duhma, Q. Xin, N. Koch, N. Ueno and S. Kera, *Org. Electron.*, 2011, **12**, 903.

13. M. Campione, A. Sassella, M. Moret, A. Papagni, S. Trabattoni, R. Resel, O. Lengyel, V. Marcon and G. Raos, *J. Am. Chem. Soc.*, 2006, **128**, 13378.
14. A. Babel and S. A. Jenekhe, *Adv. Mater.*, 2002, **14**, 371.
15. S. Kobayashi, T. Nishikawa, T. Takenobu, S. Mori, T. Shimoda, T. Mitani, H. Shimotani, N. Yoshimoto, S. Ogawa and Y. Iwasa, *Nat. Mater.*, 2004, **3**, 317.
16. R. Ruiz, D. Choudhary, B. Nickel, T. Toccoli, K.-C. Chang, A. C. Mayer, P. Clancy, J. M. Blakely, R. L. Headrick, S. Iannotta and G. G. Malliaras, *Chem. Mater.*, 2004, **16**, 4497.
17. F.-J. Meyer Zu Heringdorf, M. C. Reuter and R. M. Tromp, *Nature*, 2001, **412**, 517.
18. J. H. Kang and X. Y. Zhu, *Appl. Phys. Lett.*, 2003, **82**, 3248.
19. S. Verlaak, S. Steudel, P. Heremans, D. Janssen and S. M. Deleuze, *Phys. Rev. B: Condens. Matter Mater. Phys.*, 2003, **68**, 195409.
20. J. E. Northrup, M. L. Tiago and S. G. Louie, *Phys. Rev. B: Condens.Matter Mater. Phys.*, 2002, **66**, 121404.
21. H. S. Lee, D. H. Kim, J. H. Cho, M. Hwang, Y. Jang and K. Cho, *J. Am. Chem. Soc.*, 2008, **130**, 10556.
22. J. A. Venables, G. D. T. Spiller and M. Hanbücken, *Rep. Prog. Phys.*, 1984, **47**, 399.
23. J. Yang and D. Yan, *Chem. Soc. Rev.*, 2009, **38**, 2634.
24. A. Hinderhofer and F. Schreiber, *ChemPhysChem*, 2012, **13**, 628.
25. D. G. de Oteyza, E. Barrena, J. O. Ossó, S. Sellner and H. Dosch, *Chem. Mater.*, 2006, **18**, 4212.
26. R. Garcia and R. Perez, *Surf. Sci. Rep.*, 2002, **47**, 197.
27. J. O. Ossó, F. Schreiber, V. Kruppa, H. Dosch, M. Garriga, M. I. Alonso and F. Cerdeira, *Adv. Funct. Mater.*, 2002, **12**, 455.
28. J. O. Ossó, F. Schreiber, M. I. Alonso, M. Garriga, E. Barrena and H. Dosch, *Org. Electron.*, 2004, **5**, 135.
29. Y. Oh, S. Pyo, M. H. Yi and S.-K. Kwon, *Org. Electron.*, 2006, **7**, 77.
30. T. Breuer and G. Witte, *ACS Appl. Mater. Interfaces*, 2013, **5**, 9740.
31. L. Huang, C. Liu, B. Yu, J. Zhang, Y. Geng and D. Yan, *J. Phys. Chem. B*, 2010, **114**, 4821.
32. F. Zhu, M. Grobosch, U. Treske, L. Huang, W. Chen, J. Yang, D. Yan and M. Knupfer, *ACS Appl. Mater. Interfaces*, 2011, **3**, 2195.
33. A. Facchetti, E. Annoni, L. Beverina, M. Morone, P. Zhu, T. J. Marks and G. A. Pagani, *Nat. Mater.*, 2004, **3**, 910.
34. A. N. Rashid and P. Gunter, *Org. Electron.*, 2004, **5**, 147.
35. X. Deng and J. Lahann, *J. Appl. Polym. Sci.*, 2014, **131**, 40315.
36. H. Y. Chen and J. Lahann, *Langmuir*, 2011, **27**, 34.
37. A. Ross, H. Durmaz, K. Cheng, X. Deng, Y. Liu, J. Oh, Z. Chen and J. Lahann, *Langmuir*, 2015, **31**, 5123.
38. M. C. Cross and P. C. Hohenberg, *Rev. Mod. Phys.*, 1993, **65**, 851.
39. F. Davis and C. J. M. Stirling, *J. Am. Chem. Soc.*, 1995, **117**, 10385.
40. F. Davis, E. Frary and C. J. M. Stirling, *Langmuir*, 2004, **20**, 9075.

41. T.-Q. Nguyen, M. L. Bushey, L. E. Brus and C. Nuckolls, *J. Am. Chem. Soc.*, 2002, **124**, 15051.
42. M. P. Lightfoot, F. S. Mair, R. G. Pritchard and J. E. Warren, *Chem. Commun.*, 1999, 1945.
43. M. L. Bushey, A. Hwang, P. W. Stephens and C. Nuckolls, *J. Am. Chem. Soc.*, 2001, **123**, 8157.
44. M. L. Bushey, A. Hwang, P. W. Stephens and C. Nuckolls, *Angew. Chem., Int. Ed.*, 2002, **41**, 2828.
45. P. Anastas, N. Eghbali and Nicolas, *Chem. Soc. Rev.*, 2010, **39**, 301.
46. O.-K. Kim and L.-S. Choi, *Langmuir*, 1994, **10**, 2842.
47. O.-K. Kim, L.-S. Choi, H.-Y. Zhang, X.-H. He and Y.-H. Shih, *J. Am. Chem. Soc.*, 1996, **118**, 12220.
48. O.-K. Kim, L.-S. Choi, H.-Y. Zhang, X.-H. He and Y.-H. Shih, *Thin Solid Films*, 1998, **327–329**, 172.
49. J. Puigmartí-Luis, A. Minoia, A. Pérez del Pino, G. Ujaque, C. Rovira, A. Lledós, R. Lazzaroni and D. B. Amabilino, *Chem. – Eur. J.*, 2006, **12**, 9161.
50. P. Iavicoli, M. Linares, A. Pérez del Pino, R. Lazzaroni and D. B. Amabilino, *Superlattices Microstruct.*, 2008, **44**, 556.
51. P. Iavicoli, M. Simón-Sorbed and D. B. Amabilino, *New J. Chem.*, 2009, **33**, 358.
52. B. Li, J. Puigmartí-Luis, A. M. Jonas, D. B. Amabilino and S. De Feyter, *Chem. Commun.*, 2014, **50**, 13216.
53. M. Paradinas, L. Garzón, F. Sánchez, R. Bachelet, D. B. Amabilino, J. Fontcuberta and C. Ocal, *Phys. Chem. Chem. Phys.*, 2010, **12**, 4452.
54. A. Datar, R. Oitker and L. Zang, *Chem. Commun.*, 2006, **42**, 1649.
55. Y. Li, P. Sonar, L. Murphy and W. Hong, *Energy Environ. Sci.*, 2013, **6**, 1684.
56. Y. Li, P. Sonar, S. P. Singh, M. S. Soh, M. van Meurs and J. Tan, *J. Am. Chem. Soc.*, 2011, **133**, 2198.
57. R. J. Kline, M. D. McGehee, E. N. Kadnikova, J. Liu, J. M. J. Frechet and M. F. Toney, *Macromolecules*, 2005, **38**, 3312.
58. H. Pan, Y. Li, Y. Wu, P. Liu, B. S. Ong, S. Zhu and G. Xu, *J. Am. Chem. Soc.*, 2007, **129**, 4112.
59. P. Alexandridis and B. Lindman, *Amphiphilic Block Copolymers: Self-Assembly and Applications*, Elsevier, Amsterdam, 2000.
60. Y. Maia and A. Eisenberg, *Chem. Soc. Rev.*, 2012, **41**, 5969.
61. S. I. Stupp, V. LeBonheur, K. Walker, L. S. Li, K. E. Huggins, M. Keser and A. Amstutz, *Science*, 1997, **276**, 384.
62. M. U. Pralle, K. Urayama, G. N. Tew, D. Neher, G. Wegner and S. I. Stupp, *Angew. Chem., Int. Ed.*, 2000, **39**, 1486.
63. W. van Zoelen, T. Asumaa, J. Ruokolainen, O. Ikkala and G. ten Brinke, *Macromolecules*, 2008, **41**, 3199.
64. B. J. Rancatore, C. E. Mauldin, S.-H. Tung, C. Wang, A. Hexemer, J. Strzalka, J. M. J. Frechet and T. Xu, *ACS Nano*, 2010, **4**, 2721.
65. B. K. Kuila, M. S. Rama and M. Stamm, *Adv. Mater.*, 2011, **23**, 1797.

66. H.-Y. Chi, H.-W. Hsu, S.-H. Tung and C.-L. Liu, *ACS Appl. Mater. Interfaces*, 2015, 7, 5663.
67. R. van Hameren, P. Schon, A. M. van Buul, J. Hoogboom, S. V. Lazarenko, J. W. Gerritsen, H. Engelkamp, P. C. M. Christianen, H. A. Heus, J. C. Maan, T. Rasing, S. Speller, A. E. Rowan, J. A. A. W. Elemans and R. J. M. Nolte, *Science*, 2006, **314**, 1433.
68. *Functional Molecular Gels*, ed. B. Escuder and J. F. Miravet, RSC Publishing, 2014.
69. D. B. Amabilino and J. Puigmartí-Luis, *Soft Matter*, 2010, **6**, 1605.
70. H. Schmidt, *Struct. Bonding*, 1992, 77, 119.
71. Y. Liu, W.-J. Zhao, J.-L. Li and R.-Y. Wang, *Phys.Chem.Chem.Phys.*, 2015, **17**, 8258.
72. A. Wicklein, S. Ghosh, M. Sommer, F. Würthner and M. Thelakkat, *ACS Nano*, 2009, **3**, 1107.
73. M. Jørgensen, K. Bechgaard, T. Bjørnholm, P. Sommer-Larsen, L. G. Hansen and K. Schaumburg, *J. Org. Chem.*, 1994, **59**, 5877.
74. T. Kitahara, M. Shirakawa, S-i. Kawano, U. Beginn, N. Fujita and S. Shinkai, *J. Am. Chem. Soc.*, 2005, **127**, 14980.
75. D. Canevet, Á. Pérez del Pino, D. B. Amabilino and M. Sallé, *J. Mater. Chem.*, 2011, **21**, 1428.
76. I. Danila, F. Riobé, J. Puigmartí-Luis, A. Pérez del Pino, J. D. Wallis, D. B. Amabilino and N. Avarvari, *J. Mater. Chem.*, 2009, **19**, 4495.
77. J. Puigmartí-Luis, V. Laukhin, A. P. del Pino, J. Vidal-Gancedo, C. Rovira and D. B. Amabilino, *Angew. Chem., Int. Ed.*, 2007, **46**, 238.
78. J. Puigmartí-Luis, E. E. Laukhina, V. N. Laukhin, Á. Pérez del Pino, N. Mestres, J. Vidal-Gancedo, C. Rovira and D. B. Amabilino, *Adv. Funct. Mater*, 2009, **19**, 934.
79. C. Munuera, J. Puigmartí-Luis, M. Paradinas, L. Garzón, D. B. Amabilino and C. Ocal, *Small*, 2009, **5**, 214.
80. K. Kinjo, T. Hirao, S.-i. Kihara, Y. Katsumoto and T. Haino, *Angew. Chem. Int. Ed.*, 2015, **54**, 14830.
81. C.-H. Hsu, X.-H. Dong, Z. Lin, B. Ni, P. Lu, Z. Jiang, D. Tian, A.-C. Shi, E. L. Thomas and S. Z. D. Cheng, *ACS Nano*, 2016, **10**, 919.

CHAPTER 8

Supramolecular Chemistry for the Formation of Dynamic Systems at Interfaces

8.1 Introduction

The movement of molecules on surfaces is seen as important in the areas of nanoscience and technology,[1–3] because the motions created by the changes in monolayers can give rise to chemical or physical effects thanks to supramolecular phenomena.[4] The general approach has been to apply "machines" developed in solution[5] and adsorb them at an interface.[6] For example, several types of mechanically interlocked systems and their supramolecular analogues have been incorporated into interfaces.[7] Figure 8.1 shows a variety of these simple machine types that will be shown here. It is important to note that the surface forms a vital part of these molecules. For example, in the case of the rotors, the surface is the stator, connected to the axle that in turn is connected to the rotor. In the altitudinal rotaxanes, the surface actually behaves as a stopper, preventing the ring(s) coming off the thread at the connected end. These structures are quite different to "simply" immobilised molecular machines, in which a solution-type system is anchored through a long tether to a surface: in this situation, in all likelihood, the system acts as it would "free" in solution. The solvation would be similar away from the surface, although tethered, the molecule would have a large number of the degrees of freedom available to it in the non-adsorbed state, and there would be little influence of neighbouring molecules. Contrast this situation to the cases in Figure 8.1. The components are very close to the surface, any area free of molecular motion component would be solvated very differently to the bulk solvent which

Monographs in Supramolecular Chemistry No. 19
Supramolecular Chemistry at Surfaces
By David B. Amabilino

Published by the Royal Society of Chemistry, www.rsc.org

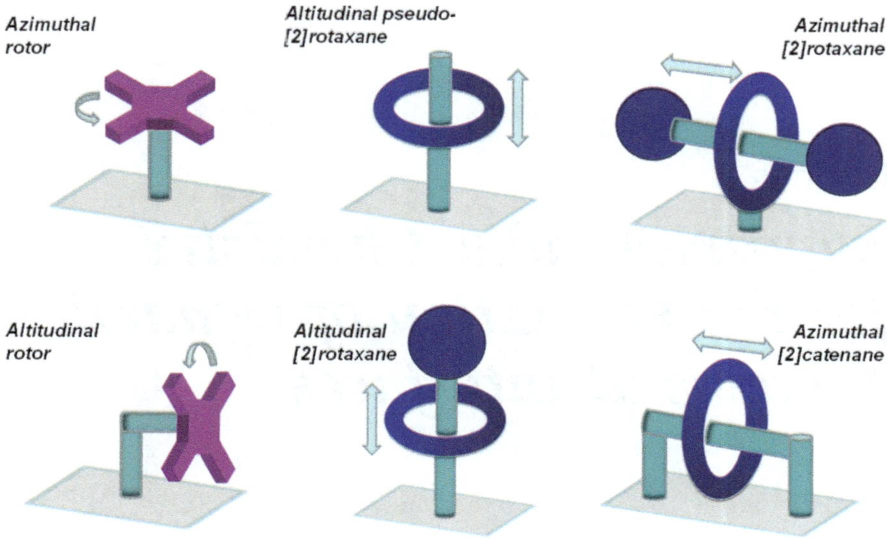

Figure 8.1 Molecular-based systems that are capable of dynamic motion between components at interfaces, where the surface forms an integral part of the system, except for the case of the azimuthal [2]rotaxane where the link in this example is through the side of the macrocyclic component. One could imagine that this cyclic component were a hoop, as in the case of the azimuthal [2]catenane, where the dumbbell component would be trapped by a cycle incorporating the surface, as if it were stapled to the surface.

might itself be temporarily trapped. The degrees of freedom are very limited, and surely influenced by interactions between the surface and the adsorbate. In the layer, the components could be surrounded by others, which could interfere with any motion very significantly. Thus, dynamic systems at interfaces offer unique situations in terms of their mobility. They also offer unique opportunities because they can be studied with the battery of surface science techniques, and can be reversibly addressed in a robust environment, for the systems are more often than not chemisorbed.

Aside from the mechanically-entrapped molecules on surfaces, isomerisation can also, in principle, give rise to long-range effects from surface reactions. In addition, host–guest systems—that have been forwarded as potential delivery systems—have also been demonstrated.[8] Self-assembled monolayers can be switched with light and the supramolecular effects that the isomerisation processes have can be significant.[9,10] The nature of the surface is highly important, as excited state quenching of photo-excited species near a metal would perhaps be expected. The use of applied potential to move or switch molecules through an electric field[11] or by formal oxidation or reduction is a facile route to switching on a surface. Indeed, much of the work in this area is being driven by the capacity of scanning tunnelling microscopy to measure molecular dynamics.[12] In the following sections, the assembly of these systems will be presented, and the micro- and macroscopic motions that they undergo will be discussed.

8.2 Self-assembly Approaches to Surface-based Dynamic Systems

The most used and in a way most trivial way to attach molecules that could be dynamic and responsive, or both, to a surface is by using moieties on the pre-assembled molecule that can react with the adsorbent to give chemisorbed layers.

The first example of a surface-attached catenane is actually a nice example of how the surface can actually form part of a ring.[13] The cyclophane cyclobis(paraquat-*para*-phenylene) when mixed with a bisthiol derived from hydroquinone in acetonitrile solution forms a complex (Figure 8.2) where the π-electron rich and deficient units interact with one another in a pseudorotaxane configuration. This complex was confirmed by absorption spectroscopy that shows the charge transfer band typical of these complexes at 460 nm. Despite the relatively low binding constant (about 250 M^{-1} at room temperature), when an excess of the cyclophane was used and the solution was exposed to gold electrodes, electrochemistry indicated that a monolayer was formed. The proof that catenation incorporating the gold surface as a part of the cycle was achieved came by control experiments, including replacing a monothiol for the dithiol. In this case, no reduction waves corresponding to the surface-bound cyclophane were observed. Calculations of coverage (taking into account the current flowing through the monolayer) showed that coverage was low, probably because of the low binding constant. Nonetheless, the approach is an elegant supramolecular way to close an interlocked and potentially dynamic structure to the surface. In these days, before STM, no imaging of the complex was achieved, and the system awaits reinvestigation to see if this kind of complex could be a feasible route to surface-based machines.

An approach somewhat similar in concept is the formation of copper(I)-templated catenanes[14] and rotaxanes[15] on the surface of gold (Figure 8.3). Here, for the catenane, the disulfide bridge was incorporated into one of the rings in a pre-formed [2]catenate with copper(I) at the templating centre, while a [3]rotaxane was employed to form a monolayer of [2]rotaxanes wherein the gold surface acts as one of the stoppers. For these rotaxanes, it is possible to change the coordination sphere of the copper ion from two phenanthroline ligands for copper(I) to one phenanthroline and one terpyridine ligand for copper(II), a fact proven by careful electrochemical experiments in solution, although on the surface, very different behaviour is observed. In the case of the catenane, although switching behaviour of the copper between terpyridine and phenanthroline sites was observed in solution when oxidising copper(I) to copper(II), this process did not take place

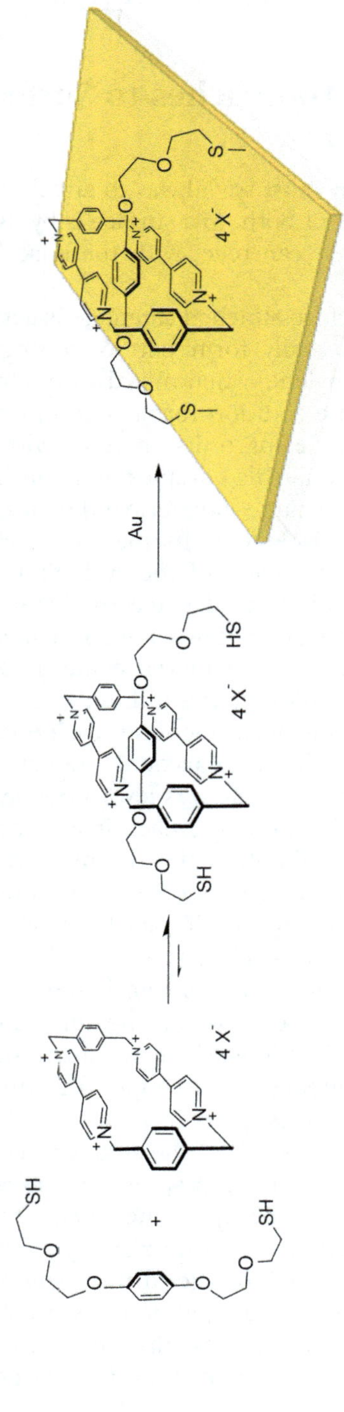

Figure 8.2 Self-assembly of a surface-incorporating [2]catenane.

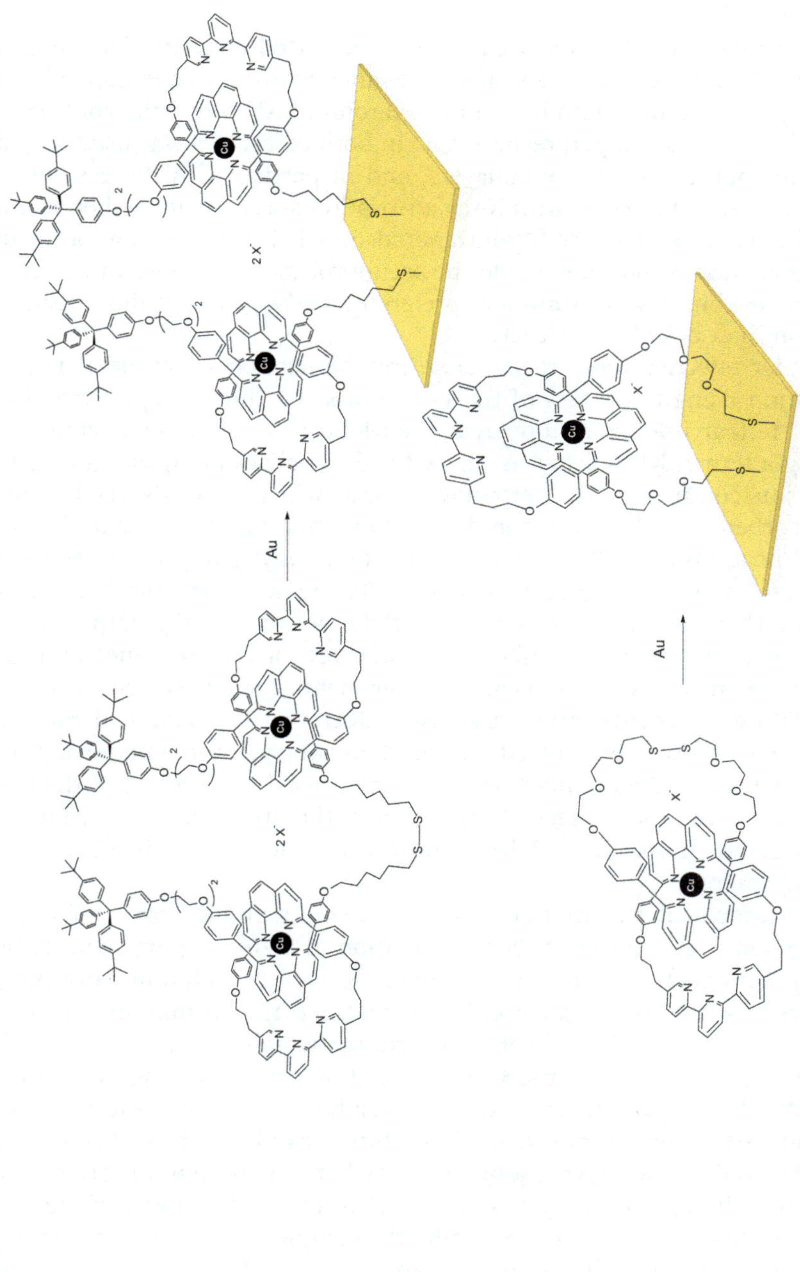

Figure 8.3 Formation of gold surface-anchored rotaxane (top) and catenane (bottom) from disulfides.

on the surface of gold in the surface-catenated self-assembled monolayer.[16] It is likely that the interaction of the catenane's rings with the surface and neighbouring molecules impedes the dynamics of the linked rings. It could be that assembling these machine-like components in a matrix and studies by STM may shed light on what could be happening in these systems. For now, only STM studies showing a smooth monolayer of the catenane have been obtained.

Conceptually similar approaches have been used elegantly for the preparation of chloride anion template catenanes with ferrocene appended to one ring[17] and ammonium ion templated rotaxanes[18] where the gold acts as part of a ring or a stopper, respectively. In both cases, XPS was used to good effect to characterise the monolayers, and in particular in the case of the rotaxane—in conjunction with X-ray adsorption spectroscopy—showed that only the use of a disulfide [3]rotaxane adsorpt led to the formation of the surface-anchored rotaxane, while the pseudorotaxane route was ineffective, and the low binding constant and preferential adsorption of the thread led to a non-interlocked monolayer.

Another effective route to rotaxane immobilisation is by metal ion coordination using the tenets of the LbL approach. Thus, a [2]rotaxane with a rigid tetraarylmethane-stoppered phenylene rod and a tetralactam ring bearing a terpyridyl moiety was reacted with a palladium(II) salt, giving the metal ion complex of the terpyridyl ligand, which was adsorbed onto a chemisorbed monolayer anchored on gold with a thiophenylate and with a pyridyl head group.[19] The orientation of the pyridyl group in the rigid-rod SAM was important, and angle-resolved NEXAFS was particularly useful in showing that the long axis of the adsorbate was practically perpendicular to the plane of the gold surface. The adhesion of the rotaxanes (and the component macrocycle in separate experiments) was observed clearly by TOF-SIMs experiments. Angle-resolved NEXAFS on the adsorbed rotaxane layers showed a well-ordered structure. Interestingly, when a poorly-ordered SAM was used as the template, the rotaxane caused an ordering of the layer in contact with the surface. It is possible that the efficient packing of the rotaxane in the second layer away from the surface instigates this rearrangement.

A supramolecular azimuthal rotor has been assembled on a gold (111) surface taking advantage of the coordination of a zinc(II) porphyrin derivative by a pyridyl thiol derivative (Figure 8.4).[20] In solution, absorption spectroscopy proved that the porphyrin and the pyridyl compound formed a complex. It was also shown that the coordination aided solubilisation of the porphyrin, an important observation regarding the processing and surface attachment of apparently intractable solids: here, the supramolecular complex aids the formation of layers. This complex can be chemisorbed directly onto the gold surface, giving a monolayer where the porphyrins are evident in the STM images of the system. It was clear that the majority of the layer (and presumably part of the layer under the complex) was dissociated pyridyl thiol. The bright disks did not present any distinctive features corresponding

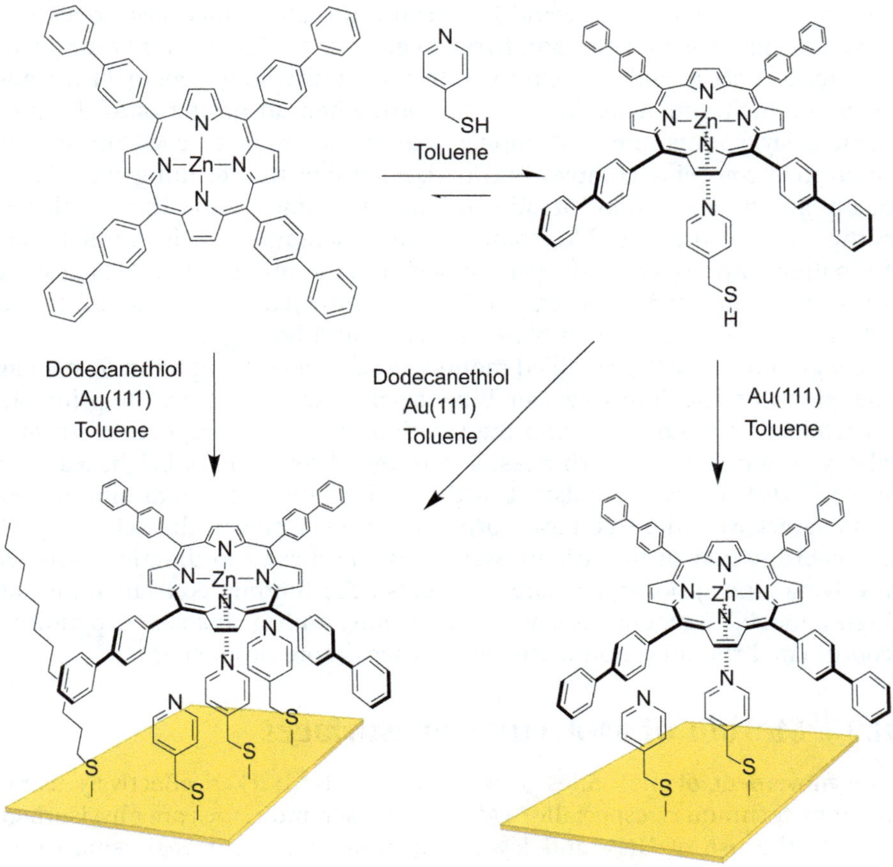

Figure 8.4 Formation of gold surface-anchored rotors through diverse routes.

to static biphenyl units (that were attached to the porphyrin core as paddles) and therefore motion of the complex was inferred. In addition, different contrasts of the disks were observed, implying different adsorption geometries of the complex. An alternative route to this monolayer is by the chemisorption of the pyridyl compound to the gold followed by complex formation from the porphyrin in solution (Figure 8.4). However, this route did not work because the arrangement of the thiolate on the surface apparently did not point toward the solution, thus precluding coordination to the porphyrin. The situation was redressed by making a mixed monolayer with dodecanethiol that changes the packing of the coadsorbate: STM showed that this mixed monolayer incorporated pyridyl groups directed perpendicular to the plane of the interface. The porphyrin adsorbed to this mixed monolayer from solution, giving few but homogeneous coordinated molecules that were observed by STM. The coverage of the surface by the putative rotor is lower when adsorbing the porphyrin from solution onto the already chemisorbed pyridyl adsorbate (even accounting for the character of the mixed layer) and this could be because the pre-formed layer does not have enough space for the porphyrin to adsorb, as if there were a negative allosteric effect in action.[21] On the other hand, the homogeneity of the adsorption environment is far more uniform when adsorbing onto the preformed layer. This effect is emphasised by the result of coadsorption of preformed rotor-like complex and dodecanethiol: the resulting monolayer again presents a variety of adsorption environments for the porphyrin compound, so the rotor-like complex can apparently exist in a tilted conformation with respect to the perpendicular to the surface plane. The results show the importance of equilibria in components and the synthetic adsorption approach to monolayer structure and homogeneity.

On graphite, a self-assembled monolayer of a zinc(II) porphyrin formed at the phenyloctane interface can bind axial ligands that are in principle switchable.[22] Introduction into the solution of an azo compound with one phenyl and one 4-pyridyl ring resulted in the observation of bright features in the STM images consistent with axial ligation. In addition, mixed monolayers with the free base porphyrin adsorpt proved the selectivity of the coordination in the ternary system. While these coordination systems involving zinc(II) porphyrins are of interest, the binding constants are not high enough in general to have robust stems, and in this sense, pyridine-coordinated rhodium porphyrins are perhaps more promising.[23]

8.3 Motion of Molecules on Surfaces

The movement of molecules on surfaces can be tracked effectively using imaging techniques, especially STM.[24] Even when molecules are physisorbed weakly, the use of UHV and low-temperature imaging reveal remarkable effects with sub-molecular resolution. A number of single-component systems have been described comprising an open network—with the network plane formed by the adsorbate parallel to the adsorbent plane—with a fortuitously-incorporated mobile guest in the void. This guest molecule that also adsorbs coplanar to the surface often has a relatively high degree of mobility. The porphyrin network shown in Figure 8.5 forms on a copper (111) surface in UHV and the same adsorbate fills in some of the voids in the supramolecular surface structure.[25] At low temperatures (around 77 K), STM images show that the molecule in the space in the network is immobile, with the molecule in a well-defined position presumably with the nitrile group located between the bulky tertiary butyl groups in the host network. Three equivalent relative orientations of network and included molecule are observed, coinciding with the three-fold symmetry of the surface. Warming to 112 K gives indications of "shivering" of the tertiary butyl containing groups, at 115 K the porphyrin starts to appear "fuzzy" (indicating motion) and at 150 K an averaged STM image of the included porphyrin is observed corresponding to the molecule being partially present in its three possible equivalent positions during the timescale of imaging (148 seconds per image). No hopping between cavities was observed. It should be noted that

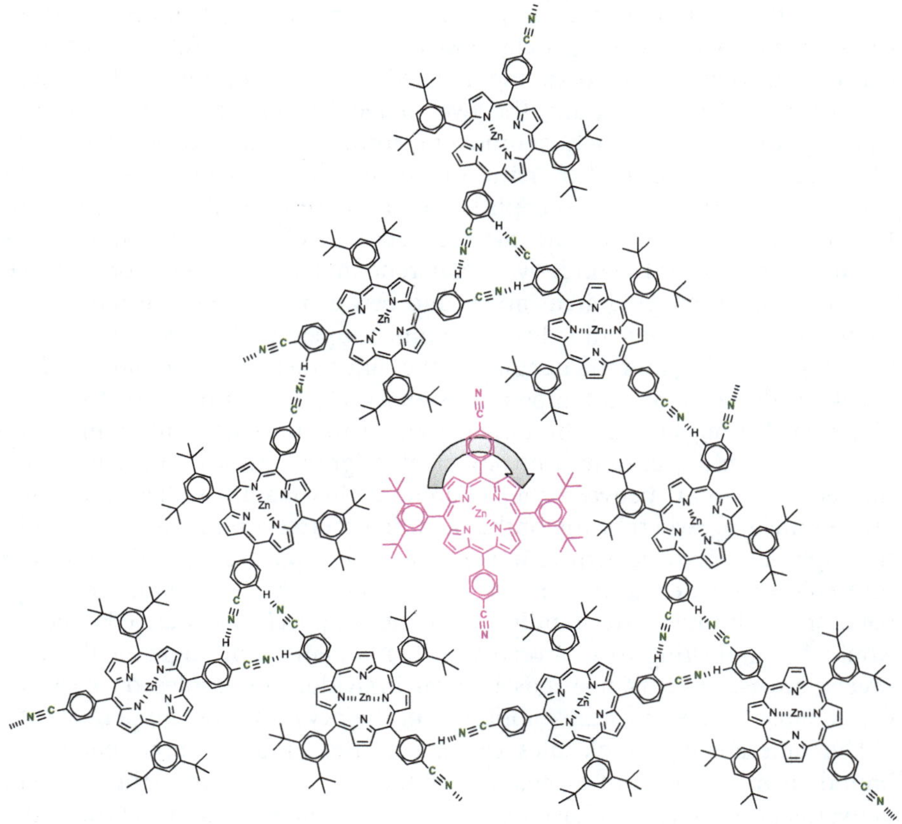

Figure 8.5 A partial two-dimensional open network of a porphyrin on the (111) surface of copper (not shown) where a molecule (coloured fuchsia) of the network-former is incorporated into the void and acts as a multipositional rotor. Here, it is shown in one of its three possible orientations, with the nitrile groups located between the bulky tertiary butyl moieties.

the network remains immobile thanks to the non-covalent interactions holding it together (bonds between the nitrile and aromatic C–H groups). Switching of the included molecules at low temperature was achieved using voltage pulses from the STM tip, although control of position and direction of rotation were clearly not possible. Nonetheless, the example shows how suitably designed combinations of supramolecular networks and surfaces can result in dynamic systems that might one day be addressable in a controlled manner. A conceptually similar situation was observed for *para*-sexiphenyl-dicarbonitrile deposited onto the (111) face of silver under UHV and coordinated to cobalt atoms, where the hexagonal network incorporated the rod-like molecules in some of the voids.[26] There is actually selectivity for non-covalent trimers within the cavities. In this case, chiral orientations in the host network were observed, and these were conserved up to 70 Kelvin, when occasional inversions of chirality started to set in.

A beautiful example of an in-vacuum constructed supramolecular rotor system is that where a hexagonally close packed layer of C_{60} fullerene on gold (111) has a copper phthalocyanine molecule adsorbed on top.[27] The system was prepared by evaporating the two molecules onto the metal surface, whereby domains of each component form, but also ones where the phthalocyanine is located over the fullerene two-dimensional lattice. STM showed that below 80 K the phthalocyanine is located as indicated in Figure 8.6, where one of the pyrrole subunits of the phthalocyanine is located directly over the underlying fullerene molecule, with the other three like positions occupying areas above and between the spherical molecules. Under these conditions, the phthalocyanines appear in the STM images with their expected size, and the imaging of the fullerene lattice gives an excellent check for dimensions. At temperatures above 150 K, ring-like structures are observed, far bigger than the dimensions of the phthalocyanine and with uneven tunnelling current with six clear brighter areas (relatively higher tunnelling current) between the fullerenes. This feature indicates a time-arrived imaging, whereby the phthalocyanine rotates around as indicated in the superposition of structures in Figure 8.6. The appearance of this rotating molecule varies depending on the bias applied. The independence of appearance on tunnel current indicates that it is a purely bias-derived effect. It would be interesting to re-investigate this system in the light of the very precise measurements now possible on individual molecules on surfaces where noise is used to study motion in phthalocyanines on copper (111).[28]

The fullerene-type molecules can act as the mobile component when bound to monolayers of aromatic compounds. In solution, trimesic acid networks (see Chapter 4) have been used to complex C_{60} and the tip of the STM was employed to visualise the complexes as well as to induce motion of the "ball" at room temperature.[29] The voids in the surface network (filled with solvent in the solution) have a diameter of approximately 1.1 nm, the same size as the van der Waals radius of C_{60}. When an open network of the acid with the chicken wire topology was physisorbed on graphite from heptanoic acid (other non-carboxylic acid solvents did not give monolayers) and a solution of C_{60} in the same solvent was added, STM showed the presence of bright protrusions in the trimesic acid hexagonal voids. These were assigned to the fullerene: no internal structure could be observed because of the supposed motion of the spherical molecule in the "socket". However, at low tunnelling current (approximately 70 pA), the layers were perfectly stable and the fullerenes stayed put. When a slightly high tunnelling reference current (approximately 150 pA) was used, the fullerenes could be moved from one open cell to an adjacent one. The modest increase in tunnelling current required to perform this movement implied to the authors that there is a low energetic barrier to fullerene motion, which is probably caused by "pushing" the molecule with the STM tip.

Under UHV conditions, an open network of porphyrin derivatives with void diameters of approximately 1.2 nm and an inter-void distance of 3.3 nm was formed upon deposition of the molecule onto silver (111).[30] The spaces

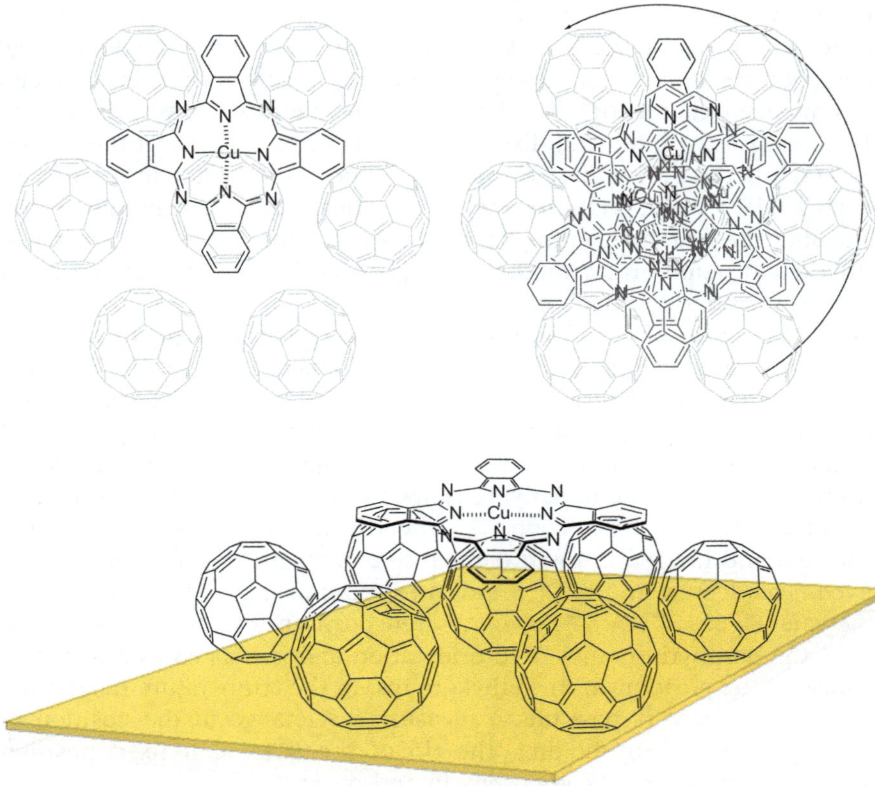

Figure 8.6 The energy minimum superstructure of copper phthalocyanine on a hexagonal close packed structure of C_{60} (top left), a view showing the superposition of all six minima (top right) that lead to imaging of the molecule as a dynamic structure larger than the actual phthalocyanine dimensions, and a side-on view of the system that operates on a gold (111) surface (bottom).

are suitable for filling by fullerenes with either sixty or seventy carbon atoms.[31] At low coverages, STM showed that both fullerenes occupy totally separated voids, and at 298 K are partially mobile between the voids in the surface network. Over a period of approximately half an hour, around ten per cent of the fullerenes were observed to jump from one void to a neighbouring one, one per cent moved to the next nearest void and the remainder of the molecules remained stationary. The C_{70} had a relatively greater mobility (lower energy barrier to movement) attributed to its greater size and therefore poorer fit to the void space compared with C_{60}. An intriguing effect took place when the coverage was increased. The larger fullerene started to form chains and islands in adjacent voids and a corresponding lowering of mobility was witnessed using STM, consistent with the formation of well-ordered superstructures with some kind of long-range non-covalent interaction. On the other hand, the C_{60} showed an increase of mobility at slightly higher coverages (up to 0.3), followed by a dip in mobility and clustering of molecules in nearby voids at still higher coverage (0.5). The long-range interaction responsible for these phenomena is not absolutely clear, though it seems likely that a slight local distortion of the host network is responsible, because the inter-void distance is too large to be a direct interaction between fullerenes, and therefore porphyrin-mediated interactions are likely to play a role. It is important to note that identical results were obtained after annealing the system, pointing to a true equilibrium state and not metastable out of equilibrium motion. The possible involvement of the silver adsorbent cannot be ruled out.

The adsorption of four-fold symmetric zinc-octaethylporphyrin into the spaces in a hexagonal molecular network on copper (111) provides a system with hindered thermally-activated rotation.[32] The network of spaces with exposed surfaces was prepared under UHV by thermal dehydrogenation of 4,9-diaminoperylenequinone-3,10-diimine, that gives a hexagonal network—commensurate with the underlying copper surface—with a 16.3 Å vertex-to-vertex diameter of exposed copper.[33] When the zinc-octaethylporphyrin was deposited and imaged at room temperature, only circular objects were observed. This feature in the STM indicates rotation that is fast on the STM timescale (Figure 8.7). At 77 Kelvin, discrete positions can be detected, while at 5 K the porphyrin is in a fixed orientation and is imaged as a rectangle because of the positions of the ethyl groups. In the temperature range from 77 to 84 Kelvin, it is possible to measure the change in the position of the porphyrin ring by holding the tip of the STM at a fixed position offset from the centre of the porphyrin and measuring the variation in the tunnelling current. This measurement permitted an estimation of the magnitude of the activation barrier between the local energy minima upon rotation (Figure 8.7), that was approximately 16 kJ mol^{-1} (0.17 eV per molecule). In addition, this motion could be modified by moving the tip of the STM close to the molecule. The approach is a fascinating one, in that in principle an array of identical rotors is generated within the voids of the network.

The motion of molecules on surfaces can be directional because of the dissymmetry of the particular surface, by directional diffusion,[34,35] where molecular shape plays an important role.[36] A specific interaction between surface and molecule can actually lead to a walking-type motion along a surface, as is the case for a horseshoe-shaped molecule with two imidazolyl groups that bind copper atoms on the (110) surface of the metal (Figure 8.8).[37] Theory showed that the molecule "stretches" to bridge between one carbon atom and the fourth one along, then contracts to bridge one carbon atom and the third one, with the methylene spacer flexing and the central phenyl ring weakly adsorbed on an adjacent row. The directional movement was observed by STM under UHV conditions. When the bisimidazolyl compound was deposited on copper (110) alone, streaks were observed, but little could be said of its directionality. However, when deposited on a surface where organometallic chemisorbed porphyrin rows had been pre-adsorbed,[38] the picture

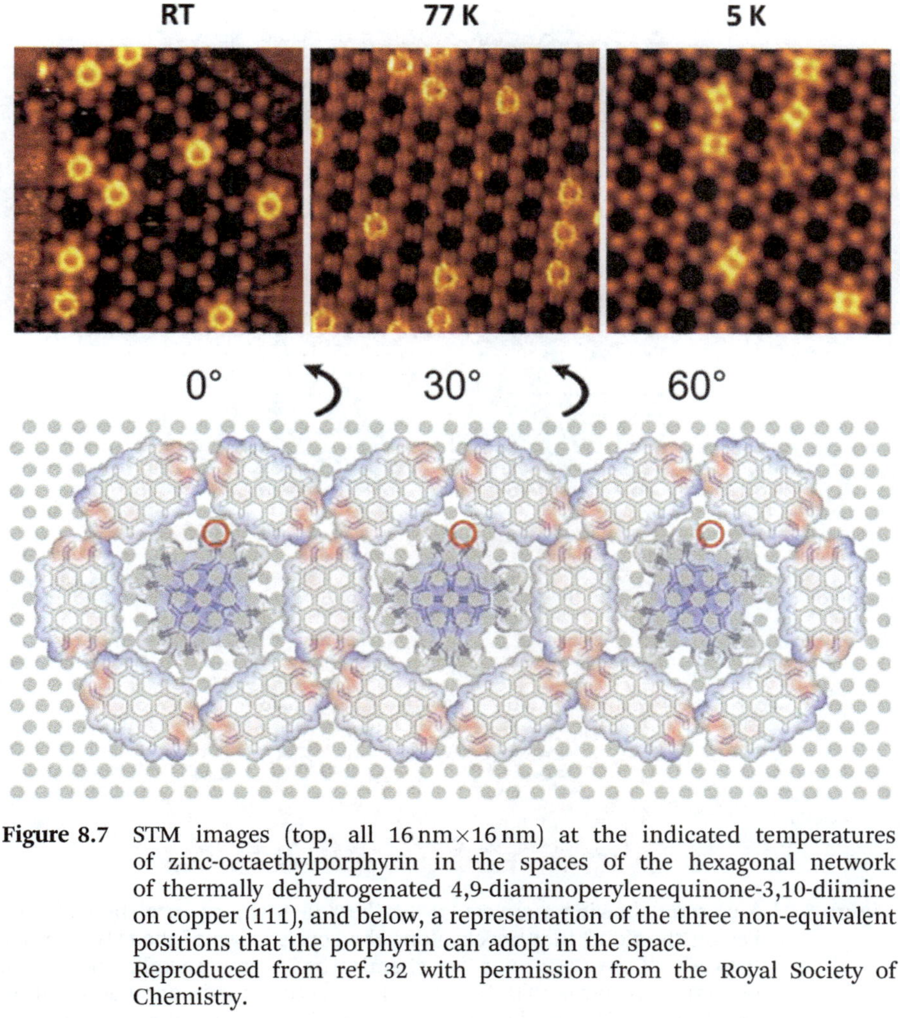

Figure 8.7 STM images (top, all 16 nm×16 nm) at the indicated temperatures of zinc-octaethylporphyrin in the spaces of the hexagonal network of thermally dehydrogenated 4,9-diaminoperylenequinone-3,10-diimine on copper (111), and below, a representation of the three non-equivalent positions that the porphyrin can adopt in the space.
Reproduced from ref. 32 with permission from the Royal Society of Chemistry.

became much clearer. The molecule clearly prefers motion along the [1$\bar{1}$0] direction. When a molecule was prevented from doing so because it was trapped between porphyrin rows, the motion in the perpendicular direction on the surface was (relatively) extremely slow. This system is useful because the direction of motion of the walker is perpendicular to the growth direction of the organometallic chemisorbed porphyrin rows. There are even random positions (for the fences are formed as spontaneously) where there is a small gap in-between the porphyrin rows that acts as a "turnstile", meaning that diffusing molecules can pass through, but are usually stopped there briefly on the STM timescale. An important feature of this system is that it operates at room temperature (albeit in UHV) and therefore acts as inspiration for the preparation of walking molecules that might one day emulate biology's walking macromolecules that carry cargos in and out of cells.[39,40]

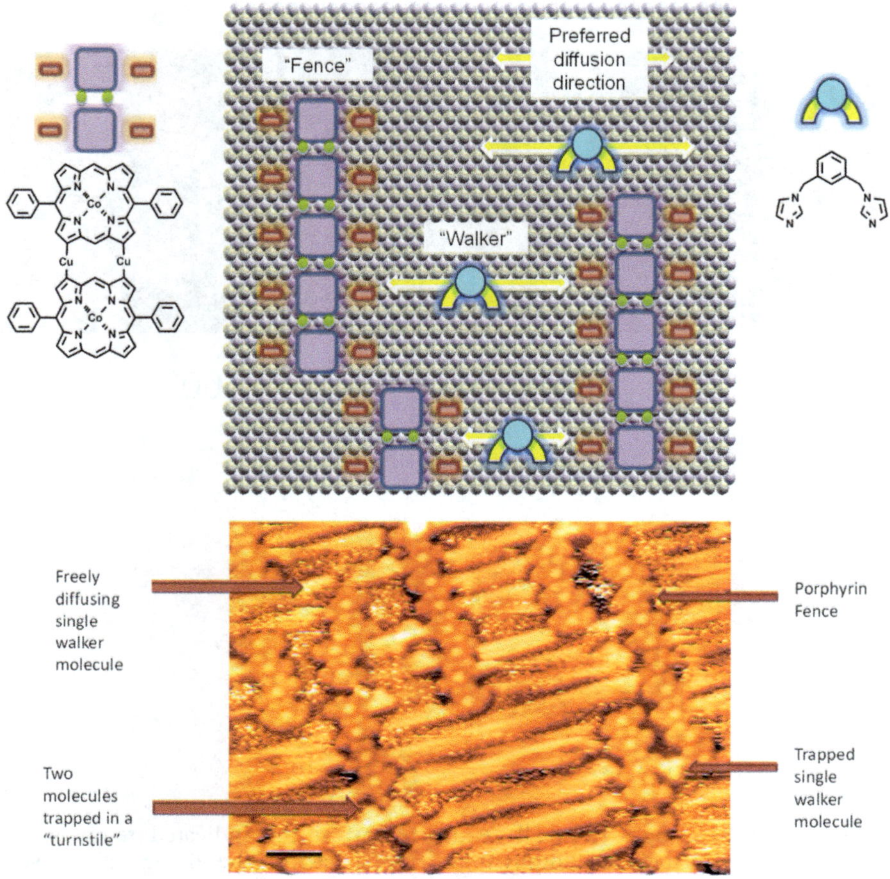

Figure 8.8 A cartoon showing a molecular "walker" (actual structure shown in the key on the right) that travels along the copper rows of the (110) surface between "fences" formed by chemisorbed organometallic porphyrin oligomers. Below is an STM image showing the experimental observation of diffusion of the walker, with trap sites for the molecule, and showing clearly the porphyrin fences. The scale bar is 2 nm.

Molecular motion in continuous layers of material can also be observed using STM, as demonstrated for the chiral switching in monolayers of bisuccinate on Cu (110).[41] As seen in the chapter on chemisorbed systems, this adsorbate is bound to the rows of copper through its two carboxylate groups. On occasion, the STM images of a high coverage layer of this compound showed abnormally bright contrast (Figure 8.9), and close inspection of the images showed that these incidences corresponded to changes in the twist angle displayed by the achiral molecule. It is an interesting case perhaps to be pursued in terms of a domino effect on the surface through chiral flipping.

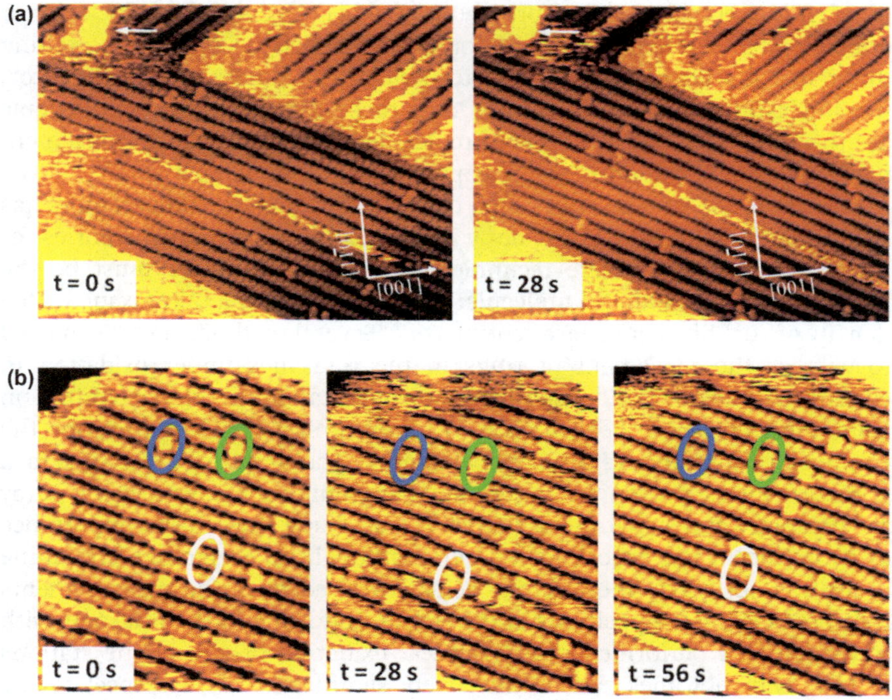

Figure 8.9 STM images showing the reorientation of succinate on copper (110), corresponding to a flipping of chirality.
Reproduced from ref. 41 with permission from the Royal Society of Chemistry.

8.4 Responsive (Sub)Monolayer Systems

Single molecules adsorbed onto metal surfaces have shown absolutely remarkable phenomena. A "four-wheeled molecule"—where each "wheel" was a sterically encumbered alkene unidirectional rotor—was adsorbed onto the (111) surface of copper and shown to "drive" upon an electrical impulse.[42] The electric field causes the rotors to isomerise and make a paddle-like motion, whereby flipping over results in a "forward" movement, defined by the chirality of the "wheels". Ten excitation steps with the tip resulted in a 6 nm trajectory in one direction of the surface. It was recognised that direct excitation of all rotors simultaneously is difficult, but nevertheless STM does show changes in contrast of the molecule after motion that do imply changes in the rotors' state. An important point is that the chirality of "left" and "right" wheels must be opposite in order for motion to take place. Key control experiments involved excitation of the molecule with opposite handed front and back wheels, where no directional motion was observed (as was the case for the homochiral molecule). Here, the chirality of the molecule drives the directional motion. In a "piano stool" complex based on a ruthenium core and chemisorbed onto gold (111), controlled clockwise or anticlockwise motion of a top rotor were induced by selectively exciting different subunits of the rotary motor (which had a de-symmetrised rotor) with inelastic tunnelling electrons.[43] Both these adsorbed molecular systems are the state of the art in terms of molecular design coupled to movement using electrons from an STM tip as power.

The formation of rotaxanes based on π electron donors and acceptors has led to remarkable feats, particularly in the field of molecular memories for electronics.[44] The ultimate technological achievement here must be the fabrication of a 160 kilobit molecular memory based on a [2]rotaxane with a density of 10^{11} bits per square centimetre.[45] Several methods have been used to witness the movement of rings in the rotaxanes under oxidation in condensed phases. Condensed Langmuir and Langmuir–Blodgett films on solid substrates have been proven to be dynamic using XPS to probe the composition of the layers when switched.[46] The shuttling of a ring in a [2]rotaxane was proven in condensed Langmuir monolayers using X-ray reflectometry at a synchrotron source.[47] That study showed the importance of hydration in the order in the layers, although no long-range crystalline order was observed. The electron density derived from the measurements indicated clearly the positions of the stations and ring in the rotaxane, with measurements performed after chemical oxidation of one of the stations *in situ*. The rotaxanes can be tilted very significantly with respect to the normal to the water's surface, while the component dumbbells do not show this effect.[48] The hydrophilic nature of the tetracationic cyclophane component and associated counter-ions is likely to cause this situation. A particularly interesting study with relevance to the function of the rotaxanes in devices was that of the infrared spectroscopic characterisation of the switches upon metal ion deposition onto the transferred Langmuir–Blodgett layers.[49] The RAIR spectra of the monolayers transferred at different pressures in the Langmuir trough showed that when the layer was well compressed, only the part of the molecule very close to the metal contact had its structure changed significantly, leaving the functional part well-ordered as it was in the Langmuir layer. In contrast, if transfer was performed on a less dense layer, the rotaxane core structure was severely affected; the IR spectra underwent very significant modifications. This structural change was evidenced in the devices made using the monolayer that had been transferred at low pressure, where poor switching and reversibility were evident. The study shows the importance in the organisation of layers for the performance of devices, especially when "hard" materials are evaporated onto them to prepare them.

A [2]rotaxane chemisorbed from solution onto the (111) surface of gold—and in the process becoming a catenane where the surface is part of the ring—at a coverage of between five and six molecules every 1000 square Ångstroms showed electrochemical movement of the ring on the anchored thread.[50] The rotaxane, drawn in Figure 8.10, deposits with its axle parallel to the gold surface and has a length (if fully extended) of approximately 7.5 nm. As deposited, the molecules show fuzzy protrusions from the gold surface all

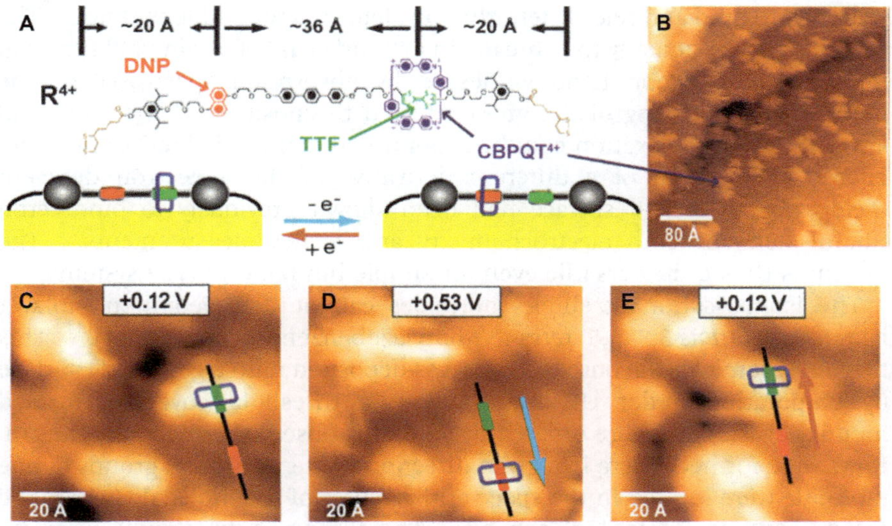

Figure 8.10 (A) Representation of the surface-anchored rotaxane on gold, above the chemical structure and below a cartoon showing the switching process taking place upon oxidation and reduction of the tetrathiafulvalene unit. (B) shows a relatively large-scale STM image with the isolated molecules. (C)–(E) show the reversible movement visualised using ECSTM where the tetracationic cyclophane component moves to and fro during oxidation and reduction of the most electron rich unit in the thread.
Reprinted with permission from ref. 50. Copyright 2010 American Chemical Society.

over the atomically flat terraces (Figure 8.10) and are clearly independent from one another when imaged by STM. In the ground state, the molecule exists in solution with the ring component (cyclobis(paraquat-*para*-phenylene)) circling the tetrathiafulvalene component (in green in Figure 8.10(A)) of the thread, as this unit is the more electron rich. With electrochemical scanning tunnelling microscopy (ECSTM) with a bias voltage of 120 mV, the molecule is in its ground state and images were registered to show this position. Subsequently, a potential of 530 mV was applied, which oxidises the TTF component forming the cation radical. In this situation, the ring preferably resides over the 1,5-dioxynaphthalene component. Reversible changes in the ECSTM images, by oxidising and then reducing, were demonstrated (Figure 8.10(C)–(E)), although many motions did not follow the same forward and reverse trajectories. It was proposed that the thread component could undergo a conformational change during the redox processes, a situation that might be remedied using a more rigid axle according to the authors.

The immobilisation of a cyclobis(paraquat-*para*-phenylene) derivative onto a chemisorbed siloxane layer on an ITO surface was achieved using click chemistry, and in a simpler way, from the chemical perspective, was used to bind and release tetrathiafulvalene in a reversible manner.[51] The ITO surface was actually the "bread" in a "sandwich" of a thin gold film that presented a plasmon band witnessed by absorption spectroscopy. The binding and releasing of the guest affected the position of the band and showed how complexation could be monitored using this kind of plasmon effect.[52] Adsorption of a different derivative of the π-electron deficient cyclophane to a gold surface had been shown previously to bind tetrathiafulvalene by both electrochemistry and RAIRS measurements.[53] The system is thus quite versatile even for simple bind and release systems.

The isomerisation of single molecules can be achieved in monolayer islands for *trans*-3,3′,5,5′-tetra-*tert*-butylazobenzene (Figure 8.11).[54] This thermodynamically-favoured isomer was deposited onto a gold (111) surface by evaporation under UHV, and above a threshold coverage, formed monolayer islands where ordered packing was observed by STM at 5 Kelvin. Rows were formed where in a given direction the tertiary butyl groups of one molecule were located in the space at the centre of the molecule (where the diazo group is located) of the next. The tertiary butyl groups were incorporated so as to "decouple" the π system of the molecule from the surface. This has the effect of making mobility fast at higher temperatures, but allows switching of the molecule from the *trans* to the *cis* isomer with voltage pulses from the STM tip. In an island comprising approximately 400 molecules, nine voltage pulses were sufficient to switch 43 of them. The *cis* isomer is clear in the STM images as an area much higher than the rest of the monolayer is observed. This protrusion was interpreted as the di-*tert*-butylphenyl group pointing away from the surface, as this conformation cannot lie flat on the gold. The pulsing did not affect the number of molecules in the island, so the supramolecular host surrounding the switched molecules is maintained. The exact position of the switching cannot be controlled, but a threshold voltage does exist, below which switching does not take place. This fact, and the tip height dependence of the switching, indicate that it is the electric field that is responsible for the isomerisation process. A similar effect has been witnessed in the ring closing and opening of diarylethylene molecules physisorbed onto a silver (111) face.[55] To close the open form, an electric field-assisted lifting was proposed, that should be followed by disrotatory ring closure. Returning to azobenzenes, the isomerisation has also been observed (on bismuth (111)) with resonant X-ray illumination.[56] In this case, at 110 K, approximately 26% of the adsorbed molecules undergo isomerisation according to XPS and NEXAFS data that show the reorientation of one of the phenyl rings. In the same system, the photoisomerisation is believed to proceed *via* substrate-mediated electron transfer, because of a partial overlap of the LUMO of the molecules with the unoccupied p-bands of the substrate.[57] So, although the direct photoexcitation is quenched, the indirect mechanism is effective on this surface whose energy levels match the adsorbate for this process well.

Deposition of a six-armed molecule based on hexaarylbenzene onto a copper (111) surface generates islands of molecules that at their edges show

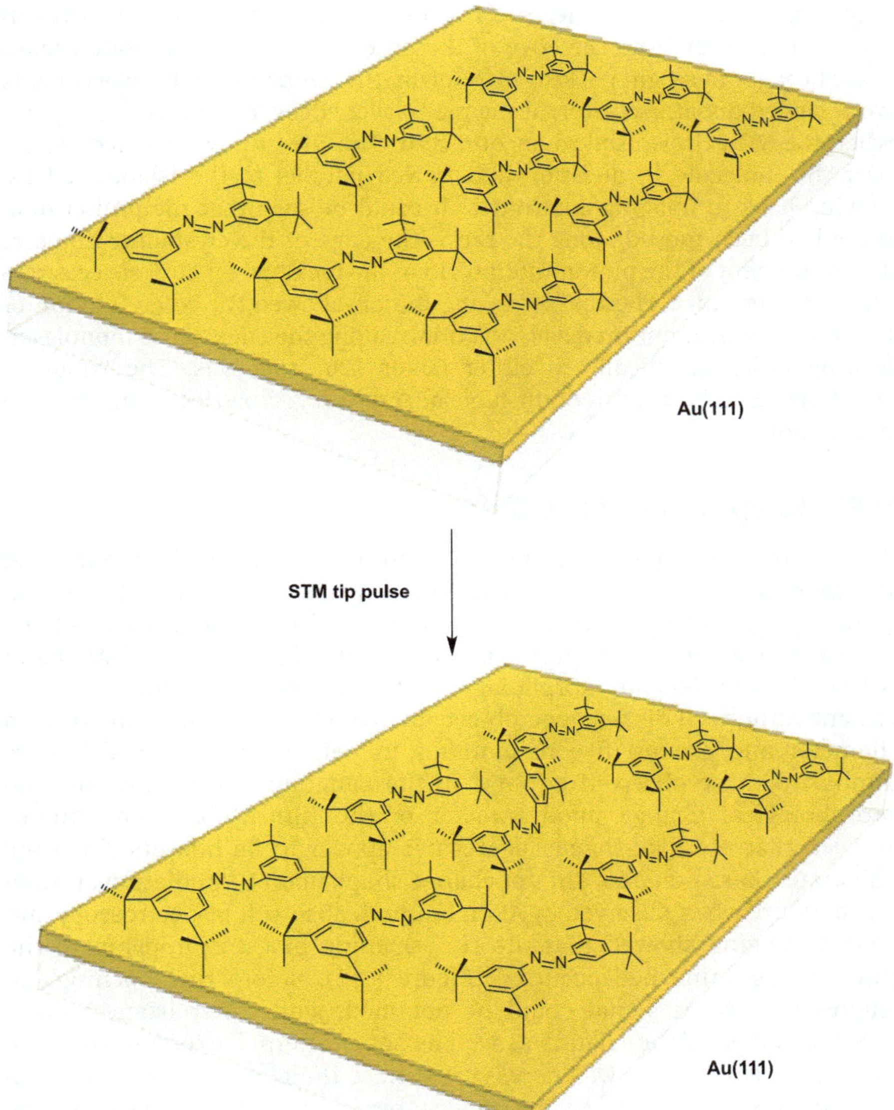

Figure 8.11 A representation of the tip-induced isomerisation of an azobenzene derivative that takes place in monolayer islands and is observed with STM at 5 K under UHV.

evidence of a rack-and-pinion type motion.[58] The compound—hexa-*t*-butyl-pyrimidopentaphenylbenzene—was carefully made so as to incorporate a "marker" (the pyrimido moiety) that provides a different STM contrast to the five other phenyl substituents. The molecule has space between its "arms" for neighbouring molecules to lock into, since the aromatic rings are twisted with respect to the benzene core because of the steric repulsion between them. Thus, away from the edge of the ordered domains, no motion takes place because of a tight packing at 7 Kelvin, the temperature the experiments were carried out at. However, at the monolayer edges, motion can take place when the STM tip is used to encourage it! The STM tip was located closely over the molecule in question and moved parallel to the islands' edges. Moving back to imaging conditions, it could be seen that the pinion molecule had been moved along the rack—the edge of the crystalline domain. The movement of the pinion motion along the rack had an 80% success rate after 150 attempts. The marker in the pinion allowed the rotary motion of the pinion to be monitored as it was drawn along the edge of the monolayer, leading quite specifically to either 60 or 120° rotations. The "double" rotations could be a result of unstable adsorption geometries along the edge of the monolayer.

8.5 Responsive Thin Films

The incorporation of molecular switches into thin films can have effects on the morphology of the layers. The photochromism exhibited by amphiphilic spiropyran derivatives has significant effects on the shape of the films made by the vertical dipping Langmuir–Blodgett method.[59] AFM and SNOM showed that the morphology of the transferred LB film of the surfactant comprised disk-shaped objects between ten and twenty microns in diameter and around five nanometres in height. These domains were similar to those observed by BAM in the Langmuir layer, indicating no morphological change upon transfer of the film. Subsequent studies showed that the disk-shaped aggregates appear to be bilayers.[60] On the substrate, these domains did not change shape upon UV radiation at room temperature when observed by AFM, but both absorption spectroscopy and SNOM imaging showed that the ring opening of the spiropyran to the photomerocyanine had occurred (Figure 8.12). Before ring opening, no fluorescent SNOM signal could be obtained, while after isomerisation, the disk shape of the islands as well as some internal structure could be discerned. However, it was necessary to raise the temperature of the film during isomerisation until the metastable photomerocyanine could reorganise structurally to give the J-aggregate form of this state. A dramatic rearrangement of the film takes place: AFM images reveal the formation of cone-shaped structures that grow up 40 nm in height from the original disk (from which several cones grow as can be seen in the AFM images) with a base width of approximately 600 nm. Absorption spectroscopy witnessed the formation of the J-aggregate clearly.

While the morphological changes in the previous example were apparently not reversible, an amphiphilic diazo compound that was used to make bilayer LB films did show noteworthy reversible morphological changes upon irradiation to interchange *cis* and *trans* isomers.[61] The Langmuir films comprised the diazo compound—a sodium sulfonate salt that was soluble in

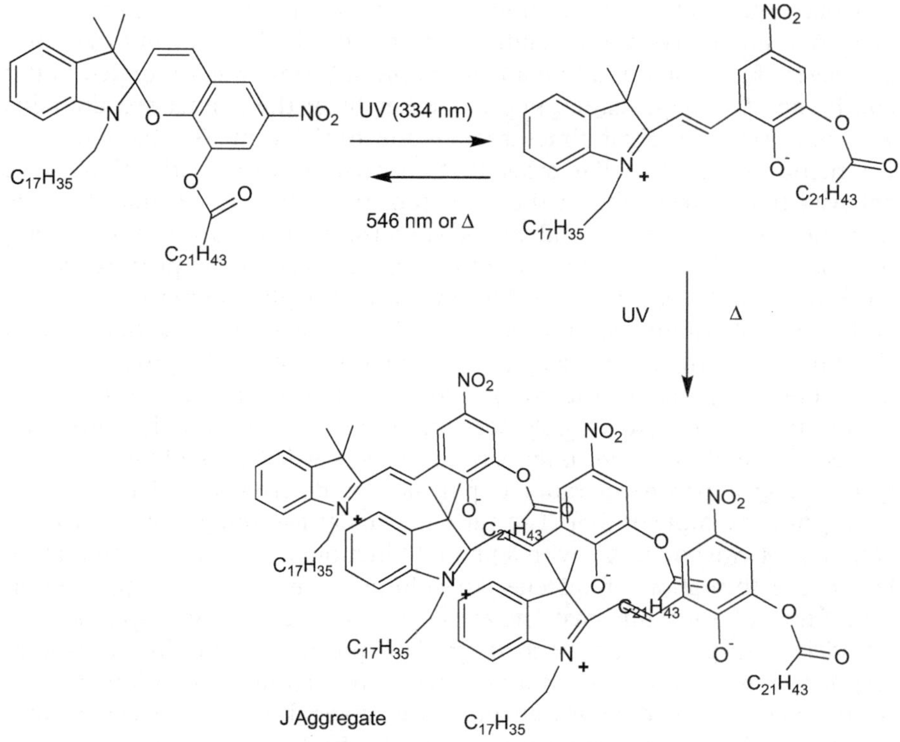

Figure 8.12 The photoisomerisation of an amphiphilic spiropyran derivative and the aggregation of the ring-opened photomerocyanine to give a J-aggregate.

water—and a water-soluble polycation (poly(diallyldimethylammonium chloride)). The Langmuir layers were transferred onto solid substrates using the vertical dipping method. In the initial downward strokes, the effective transfer ratio was *ca.* 0.1 before the fourth cycle, when it became 0.5, making the authors presume a Z-type layer up to the third cycle and Y-type thereafter. The ion exchange of each component was found to be complete, proof coming from XPS measurements. The Z-type layer was used for the isomerisation studies, with initial UV-visible absorption and IR studies indicating that reversible photoisomerisation took place in these films. For the LB films on mica, AFM revealed the morphological change going from a quite flat (roughness less than 1 nm) sponge-type open cell morphology to a multilayer-type structure upon formation of the *cis* isomer, with a height difference of approximately 5 nm between hills and valleys. The process was completely reversible. The mechanism for the formation of these hills was proposed to be bending of the layers upon formation of the *cis* isomer that cannot fit into a lamellar structure as well as the *trans* isomer, although the height difference observed and the fact that it is an ionic complex make this supramolecular rearrangement intriguing.

Azobenzene photoisomerisation has been used for patterning of morphology of a liquid crystalline dendrimer in a thin film.[62] A 90 nm thick spin-cast layer of the compound on a hydrophilic substrate was irradiated under continuous UV light. Astoundingly, the height of the isomerised domains was approximately eight times that of the initial film. GIXRD showed a destruction of the lamellar order that characterised the initial films. It is believed that a dewetting of the film from the substrate occurs. The phenomenon is thickness dependent, as when the films are 100 nm or thicker, only occasional protrusions were observed. The size of the protrusions increased as the thickness of the film increased for photopatterned systems, and the protrusions became thicker with increasing laser power, and therefore mass migration also plays an important role in the process, which is an interesting one for the bottom-up formation of surface features.

The area of responsive polymer thin films is huge and has been reviewed,[63] therefore, here, only certain aspects will be highlighted. Many systems rely on phase separation of polymers, and perhaps block copolymers are a prime example of that. The supramolecular assembly of poly(styrene)-*block*-poly(4-vinyl pyridine) with a phenol-functionalised diazo chromophore led to the formation of micelles in solution that were also apparent in thin films of the spin cast materials, whose morphology—spherical or cylindrical—depended on the loading of the polymer.[64] The thin films were responsive to irradiation, in that they demonstrated photoinduced birefringence that could be written and erased reversibly. The block copolymer micelle provided an environment where this effect was far more prominent than in other materials, such as the homogeneous material made by mixing poly(4-vinyl pyridine) and the dye.

8.6 Surface-based Responsive Systems Showing Macroscopic Effects

The wetting of monolayers by a liquid is perhaps the most obvious macroscopic demonstration of changes in supramolecular chemistry at interfaces. An azimuthal molecular rotor with a tripodal base to ensure its well defined orientation and a light-activated rotor with a perfluoroalkane chain attached to it forms a SAM that shows very different contact angles depending on the position of the halocarbon section (Figure 8.13).[65] The spacer between the tripodal base and the rotor is to ensure rotation of the sterically-encumbered alkene without further impediment from the base. The thermal isomerisation of the rotor is slow at room temperature, but sequential irradiation at determined wavelengths drives the rotor controllably around. Depending on the isomer present in the rotor component, *cis* or *trans*, the perfluoro chain is exposed to the water or "tucked away" towards the metal surface. Thus the *trans* isomer in the SAM gives a high contact angle because of the hydrophobic nature of the organofluorine segment while the *cis* isomer presents a lower contact angle with a drop of water. Irradiation of either deposited

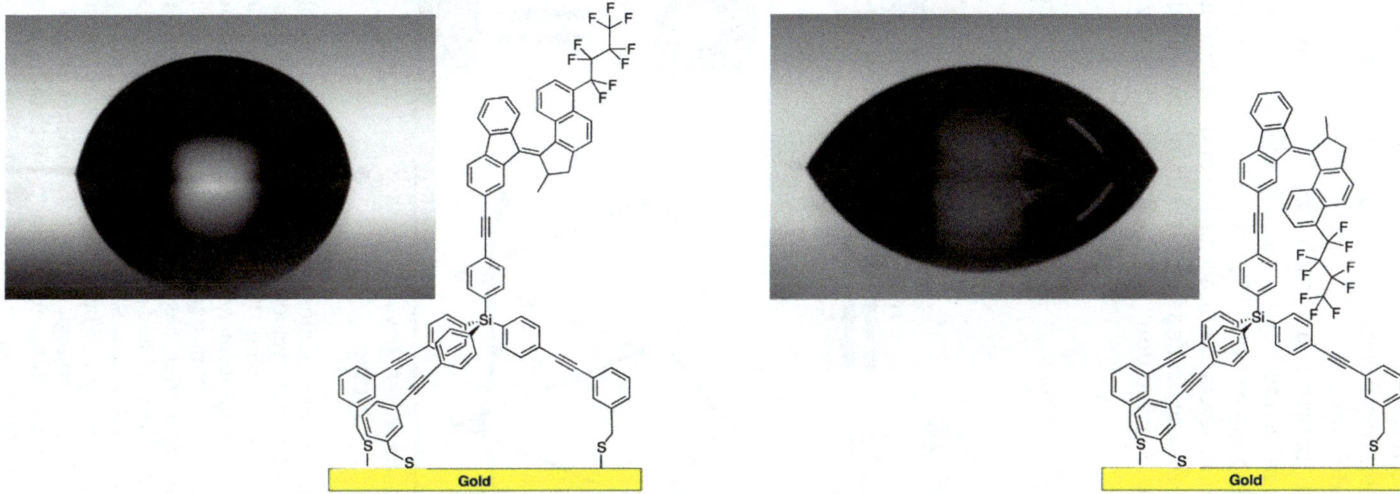

Figure 8.13 A chemisorbed molecular altitudinal rotor that displays different wetting by water depending on the stage of rotation. On the left is the high contact angle resulting from the *trans* isomer (pictured) and on the right is the *cis* isomer where the water wets the surface more efficiently.
Reprinted with permission from ref. 65. Copyright 2014 American Chemical Society.

isomer resulted in a change of the water contact angle of approximately 16°. The difficulty in making these extremely elegant and complex systems work is in achieving photostationary states that show a significant bias towards one of the isomers of the rotor.

A complex formed between an azobenzene derivative and an α-cyclodextrin forms a surface-mounted molecular shuttle that shows reversible changes in its wettability by water (Figure 8.14).[66] The threaded complex was formed from the macrocyclic oligosaccharide and a thread-like molecule incorporating the azobenzene unit and a decyl chain terminated with a thiol group: the former acts as a "station" for the shuttle while the thiol group

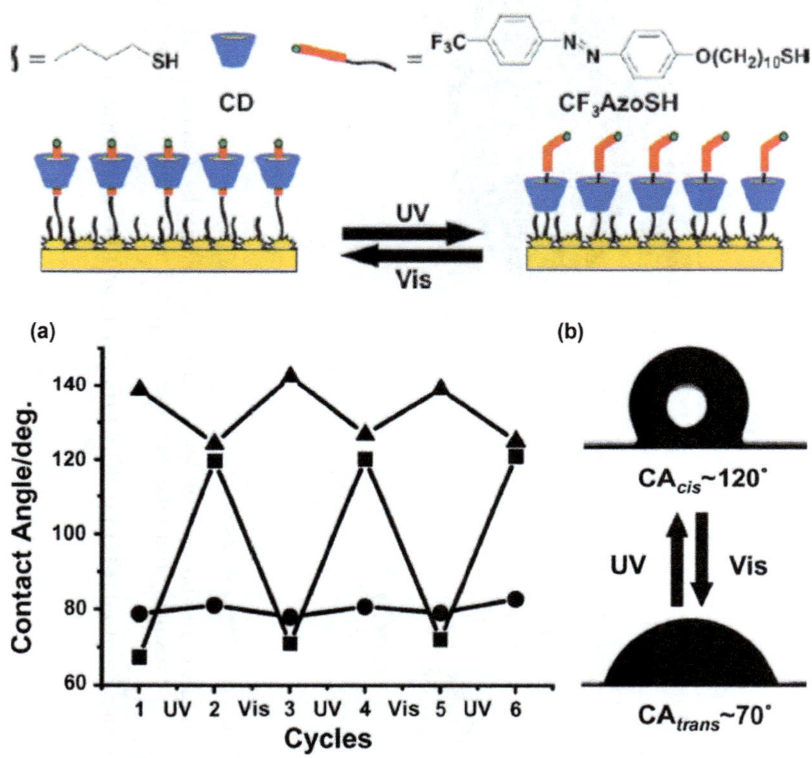

Figure 8.14 The components of a photoswitchable system that displays a change in wettability (CD is α cyclodextrin) and a representation of the mixed SAM that is switched with UV light to form the *cis* isomer and visible light to give the *trans* isomer, with movement of the thread component. (a) shows the water contact angle of films before and after switching to the *cis* azobenzene (that has the higher contact angle) for the mixed monolayer of alkyl thiol and pseudorotaxane on a rough surface (filled squares), the same mixed SAM on a flat gold surface (filled circles) and a mixed monolayer of butanethiol and CF_3AzoSH (no CD present) on rough gold (filled triangles). (b) shows a water droplet on the mixed layers of the shuttle on the rough gold surface in the two states. Reproduced from ref. 66 with permission from the Royal Society of Chemistry.

allows attachment to a gold surface. The fact that it is saturated is also important, because conductive tethers can reduce the efficiency of the photoisomerisation process.[67] The trifluoromethyl group prevents dethreading of the complex by electrostatic interactions.[68,69] Three types of SAM were produced: a complete monolayer of the threaded complex on a rough gold surface, a mixed monolayer of butanethiol and the threaded complex on a smooth gold surface and the same mixed layer on a smooth gold surface. The mixed monolayers were prepared by first treating the gold to a solution of butanethiol for a short immersion time followed by the adsorption of the threaded complex in the spaces in the loose layer. The contact angle of the mixed monolayer prior to irradiation was approximately 70°, while after twenty minutes' exposure to UV light (365 nm), the angle was about 120°, a change of 50°. The origin of this remarkable change was the motion of the cyclodextrin ring toward the gold adsorbent because of the kink introduced into the azobenzene unit. Before irradiation, the polar groups of the threaded ring are exposed to the water, providing a relatively low contact angle, while after switching, the cyclodextrin is more sheltered by the apolar *cis* azobenzene moiety (Figure 8.14). The process is reversed by exposure to visible light. When a complete monolayer of the threaded complex was treated in a similar fashion, only very slight changes in the water contact angle were observed. This modest change was ascribed to the difficulty of the threaded component to undergo motion because of the surrounding complexes that must either prevent photoisomerisation or discourage the movement of the ring. On the other hand, if a smooth gold surface is used, the contact angle changes are again modest because the enhanced contact angle effect on the rough gold.[70] A mixed monolayer of the azobenzene thread molecule and butanethiol on rough gold exhibits an opposite and more modest change in contact angle, from 140° to 125° in the *trans* and *cis* states, respectively, as a result of the change in the dipole moment of the switched moiety.[71–73] Changes in the SAM structure can also be monitored electrochemically using a redox active probe in the electrolyte solution. When the thread was switched to the *cis* state, the ferricyanide ion that was used to measure the film permeability could not penetrate the film as well as in the *trans* state, because of the movement of the thread toward the surface and a closing of the voids that are present in the initial state. While the changes in this system can be repeated, over repeated cycles, the response is lost slightly (perhaps because of dethreading) and the switch from a more hydrophobic to hydrophilic surface is relatively slow, leaving room for further research into this remarkable system.

A landmark result in the macroscopic expression of microscopic motion was the use of a palindromic [3]rotaxane as a molecular muscle,[74] both by sequential chemical[75] and electrical[76] oxidation and reduction. The system is shown in Figure 8.15: two of the π-electron deficient cyclobis(paraquat-*para*-phenylene) cyclophane derivatives act as rings on a C_2 symmetrical dumbbell with two tetrathiafulvalene derivatives nearest the stoppers and two naphthalene moieties near the centre. The palindromic [3]rotaxane was

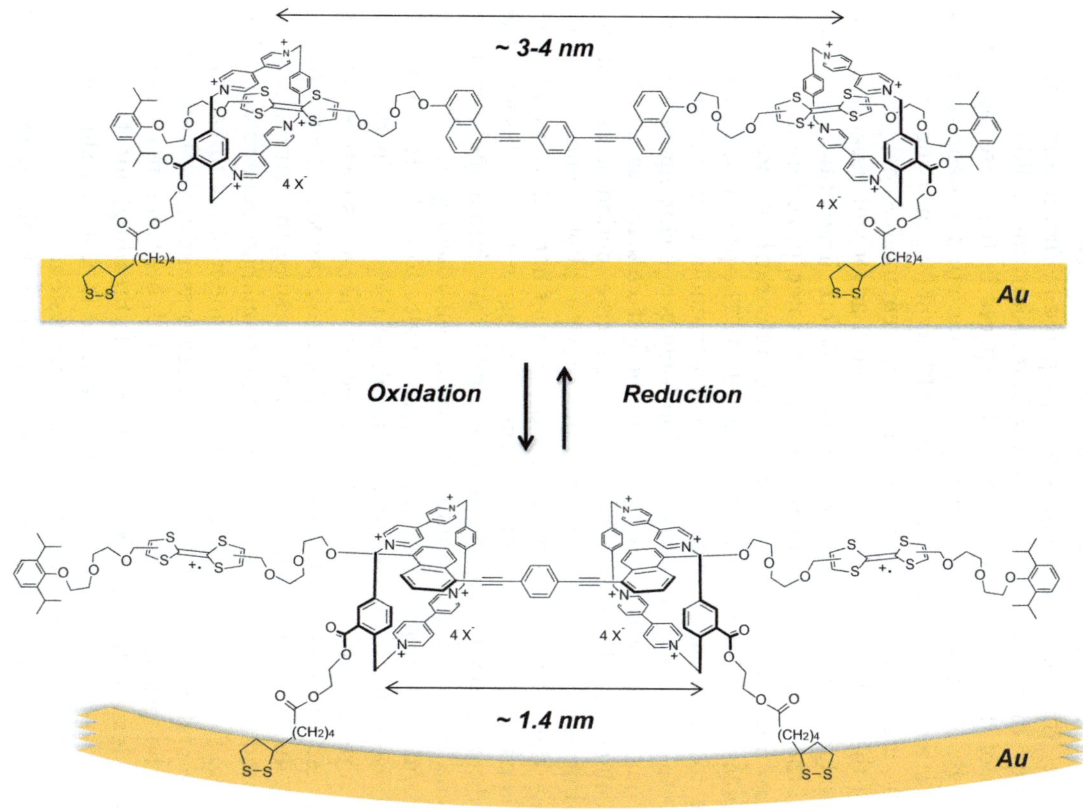

Figure 8.15 A chemisorbed [3]rotaxane that acts as a molecular muscle when chemisorbed to a gold surface. The drawing at the top shows the ground state of the muscle with the rings encircling the tetrathiafulvalene units that when oxidised have poor binding to the rings that move to the naphthalene moieties thus generating the contracted form at the bottom. The process is reversible, chemically or electrochemically.

self-assembled by "clipping" the cyclophane around the dumbbell template. Also, notice the very rigid middle section of the dumbbell, preventing conformational collapse. In the ground state, the cyclophanes encircle the tetrathiafulvalene units. The pre-formed rotaxane was anchored to the gold surface of cantilevers through a lipoic acid functionality where the two sulfur atoms interact with the metal adsorbent. In this ground state, the two π-electron deficient cyclophanes are favourably located over the most π-electron rich site and the average distance between the rings is between about three and four nanometres. Oxidation of the tetrathiafulvalene results in the cation radical of this unit that binds much more weakly to the cyclophanes (partially because of like charge repulsion presumably), which now prefer the naphthalene residues as their host unit. This movement brings the rings much closer together, approximately 1.4 nm. The effect of a whole monolayer of these compounds on a gold cantilever is a bending of the solid, with a deflection of 35 nm for the solution-based chemical redox processes. This experimental value agreed well with that calculated assuming six billion molecules on the $500 \times 100 \times 1$ μm surface and a force per molecule of between 21 and 14 pN for different conformations on the surface. The calculated force per molecule is 10.2 pN. For the electrochemically controlled mechanical actuation, different cantilevers gave a deflection of approximately 550 nm that indicated a force per molecule of approximately 650 pN. Both the electrochemical and chemical processes showed fatigue and "creep", which was assigned to a reorganisation of the molecules. This change in the monolayer under cycled movement could conceivably be caused by the mobility of the gold surface atoms, as has been demonstrated in other self-assembled monolayers that can reconstruct under certain conditions. It is certainly of interest to exploit these remarkable feats on surfaces where the surface atoms are immobile.

Using the cantilever of an AFM, measurement of the mechanical work performed by a rotaxane—in which hydrogen bonds are found between the string and ring components—has been achieved.[77] The synthetic molecular shuttle has a biased Brownian motion, in a space less than 5 nm (the length of the molecule). The AFM cantilever was used to apply a mechanical load to the ring (that was harnessed to the tip), pulling it away from its thermodynamically preferred location on the axle, and then allowing it to wiggle back, for many individual molecule pulling–relaxing cycles. The ring found its way back to the most favourable site against an external load of 30 pN. Calculation of the free-energy change taking place led the authors to conclude that these synthetic rotaxanes can afford directional forces of similar strength to ones found in natural molecular machines. In these experiments, the surfaces play the role of an immobilising superstructure, but it is perhaps a fascinating goal to be able to measure the force exerted by molecules on the surface as a function of their supramolecular surroundings. To achieve this goal, it would be necessary to have reference points nearby, and for this reason, a patterning of the surface would be required. We now turn to this subject.

References

1. V. Balzani, A. Credi and M. Venturi, *ChemPhysChem*, 2008, **9**, 202.
2. B. H. Northrop, A. B. Braunschweig, P. M. Mendes, W. R. Dichtel and J. F. Stoddart, in *Handbook of Nanoscience Engineering and Technology*, ed. W. A. Goddard III, D. W. Brenner, S. E. Lyshevski and G. J. Iafrate, CRC Press, 2nd edn, 2007, vol. 11, p. 1.
3. E. Coronado, P. Gavina and S. Tatay, *Chem. Soc. Rev.*, 2009, **38**, 1674.
4. D. Li, W. F. Paxton, R. H. Baughman, T. J. Huang, J. F. Stoddart and P. S. Weiss, *MRS Bull.*, 2009, **34**, 671.
5. E. R. Kay, D. A. Leigh and F. Zerbetto, *Angew. Chem., Int. Ed.*, 2007, **46**, 72.
6. K. Ariga, T. Mori and J. P. Hill, *Soft Matter*, 2012, **8**, 15.
7. J. J. Davis, G. A. Orlowski, H. Rahman and P. D. Beer, *Chem. Commun.*, 2010, **46**, 54.
8. Y.-W. Yang, Y.-L. Sun and N. Song, *Acc. Chem. Res.*, 2014, **47**, 1950.
9. N. Katsonis, M. Lubomska, M. M. Pollard, B. L. Feringa and P. Rudolf, *Prog. Surf. Sci.*, 2007, **82**, 407.
10. Y. B. Zheng, B. K. Pathem, J. N. Hohman, J. C. Thomas, M. Kim and P. S. Weiss, *Adv. Mater.*, 2013, **25**, 302.
11. F. Moresco, in *Advances in Atom and Single Molecule Machines*, ed. C. Joachim and G. Rapenne, Springer, Switzerland, 2015, pp. 165–186.
12. S. Uemura, R. Tanoue, N. Yilmaz, A. Ohira and M. Kunitake, *Materials*, 2010, **3**, 4253.
13. T. Lu, L. Zhang, G. W. Gokel and A. E. Kaifer, *J. Am. Chem. Soc.*, 1993, **115**, 2542.
14. L. Raehm, C. Hamann, J.-M. Kern and J.-P. Sauvage, *Org. Lett.*, 2000, **2**, 1991.
15. N. Weber, C. Hamann, J.-M. Kern and J.-P. Sauvage, *Inorg. Chem.*, 2003, **42**, 6780.
16. L. Raehm, J.-M. Kern, J.-P. Sauvage, C. Hamann, S. Palacin and J.-P. Bourgoin, *Chem. – Eur. J.*, 2002, **8**, 2153.
17. N. H. Evans, H. Rahman, A. V. Leontiev, N. D. Greenham, G. A. Orlowski, Q. Zeng, R. M. J. Jacobs, C. J. Serpell, N. L. Kilah, J. J. Davis and P. D. Beer, *Chem. Sci.*, 2012, **3**, 1080.
18. A. L. Vance, T. M. Willey, T. van Buuren, A. J. Nelson, C. Bostedt, G. A. Fox and L. J. Terminello, *Nano Lett.*, 2003, **3**, 81.
19. S. Richter, J. Poppenberg, C. H.-H. Traulsen, E. Darlatt, A. Sokolowski, D. Sattler, W. E. S. Unger and C. A. Schalley, *J. Am. Chem. Soc.*, 2012, **134**, 16289.
20. J. Puigmartí-Luis, W. J. Saletra, A. González, D. B. Amabilino and L. Pérez-García, *Chem. Commun.*, 2014, **50**, 82.
21. J. Puigmartí-Luis, W. J. Saletra, A. González, L. Pérez-García and D. B. Amabilino, in *Advances in Atom and Single Molecule Machines*, ed. C. Joachim and G. Rapenne, Springer, Switzerland, 2015, pp. 127–141.
22. J. Otsuki, E. Seki, T. Taguchi, M. Asakawa and K. Miyake, *Chem. Lett.*, 2007, **36**, 740.

23. T. Ikeda, M. Asakawa, M. Goto, K. Miyake, T. Ishida and T. Shimizu, *Langmuir*, 2004, **20**, 5454.
24. D. Lensen and J. A. A. W. Elemans, *Soft Matter*, 2012, **8**, 9053.
25. N. Wintjes, D. Bonifazi, F. Cheng, A. Kiebele, M. Stöhr, T. Jung, H. Spillmann and F. Diederich, *Angew. Chem., Int. Ed.*, 2007, **46**, 4089.
26. D. Kühne, F. Klappenberger, W. Krenner, S. Klyatskaya, M. Ruben and J. V. Barth, *Proc. Natl. Acad. Sci. U. S. A.*, 2010, **107**, 21332.
27. M. Stöhr, T. Wagner, M. Gabriel, B. Weyers and R. Möller, *Phys. Rev. B: Condens. Matter Mater. Phys.*, 2001, **65**, 033404.
28. J. Schaffert, M. C. Cottin, A. Sonntag, H. Karacuban, C. A. Bobisch, N. Lorente, J.-P. Gauyacq and R. Möller, *Nat. Mater.*, 2013, **12**, 223.
29. S. J. H. Griessl, M. Lackinger, F. Jamitzky, T. Markert, M. Hietschold and W. M. Heckl, *J. Phys. Chem. B*, 2004, **108**, 11556.
30. H. Spillmann, A. Kiebele, M. Stöhr, T. A. Jung, D. Bonifazi, F. Cheng and F. Diederich, *Adv. Mater.*, 2006, **18**, 275.
31. A. Kiebele, D. Bonifazi, F. Cheng, M. Stöhr, F. Diederich, T. Jung and H. Spillmann, *ChemPhysChem*, 2006, **7**, 1462.
32. M. Wahl, M. Stöhr, H. Spillmann, T. A. Jung and L. H. Gade, *Chem. Commun.*, 2007, 1349.
33. M. Stöhr, M. Wahl, C. H. Galka, T. Riehm, T. A. Jung and L. H. Gade, *Angew. Chem., Int. Ed.*, 2005, **44**, 7394.
34. S. C. Jensen, A. Shank, R. J. Madix and C. M. Friend, *ACS Nano*, 2012, **6**, 2925.
35. X. Bouju, F. Cherioux, S. Coget, G. Rapenne and F. Palmino, *Nanoscale*, 2013, **5**, 7005.
36. D. M. Huang and P. Harrowell, *J. Phys. Chem. C*, 2011, **115**, 9526.
37. S. Haq, B. Wit, H. Sang, A. Floris, Y. Wang, J. Wang, L. Pérez-García, L. Kantorovitch, D. B. Amabilino and R. Raval, *Angew. Chem., Int. Ed.*, 2015, **54**, 7101.
38. S. Haq, F. Hanke, M. S. Dyer, M. Persson, P. Iavicoli, D. B. Amabilino and R. Raval, *J. Am. Chem. Soc.*, 2011, **133**, 12031.
39. D. J. G. Bakewell and D. V. Nicolau, *Aust. J. Chem.*, 2007, **60**, 314.
40. S. S. Patel and I. Donmez, *J. Biol. Chem.*, 2006, **281**, 18265.
41. N. Liu, G. R. Darling and R. Raval, *Chem. Commun.*, 2011, **47**, 11324.
42. T. Kudernac, N. Ruangsupapichat, M. Parschau, B. Maciá, N. Katsonis, S. R. Harutyunyan, K.-H. Ernst and B. L. Feringa, *Nature*, 2011, **479**, 208.
43. U. G. E. Perera, F. Ample, H. Kersell, Y. Zhang, G. Vives, J. Echeverria, M. Grisolia, G. Rapenne, C. Joachim and S.-W. Hla, *Nat. Nanotechnol.*, 2012, **8**, 46.
44. C. P. Collier, E. W. Wong, M. Belohradský, F. M. Raymo, J. F. Stoddart, P. J. Kuekes, R. S. Williams and J. R. Heath, *Science*, 1999, **285**, 391.
45. J. E. Green, J. W. Choi, A. Boukai, Y. Bunimovich, E. Johnston-Halperin, E. DeIonno, Y. Luo, B. A. Sheriff, K. Xu, Y. S. Shin, H.-R. Tseng, J. F. Stoddart and J. R. Heath, *Nature*, 2006, **445**, 414.

46. T. J. Huang, H.-R. Tseng, L. Sha, W. Lu, B. Brough, A. H. Flood, B.-D. Yu, P. C. Celestre, J. P. Chang, J. F. Stoddart and C.-M. Ho, *Nano Lett.*, 2004, **4**, 2065.
47. K. Norgaard, B. W. Laursen, S. Nygaard, K. Kjaer, H.-R. Tseng, A. H. Flood, J. F. Stoddart and T. Bjornholm, *Angew. Chem., Int. Ed.*, 2005, **44**, 7035.
48. K. Nørgaard, J. O. Jeppesen, B. W. Laursen, J. B. Simonsen, M. J. Weygand, K. Kjaer, J. F. Stoddart and T. Bjørnholm, *J. Phys. Chem. B*, 2005, **109**, 1063.
49. E. DeIonno, H.-R. Tseng, D. D. Harvey, J. F. Stoddart and J. R. Heath, *J. Phys. Chem. B*, 2006, **110**, 7609.
50. T. Ye, A. S. Kumar, S. Saha, T. Takami, T. J. Huang, J. F. Stoddart and P. S. Weiss, *ACS Nano*, 2010, **4**, 3697.
51. F. Barka-Bouaifel, J. Niedziólka-Jönsson, X. Castel, O. Saison, A. Akjouj, Y. Pennec, B. Djafari-Rouhani, P. Woisel, L. Sambe Lyskawa, G. Cooke, N. Bezzi, R. Boukherroub and S. Szunerits, *J. Mater. Chem.*, 2011, **21**, 3006.
52. S. Szunerits and R. Boukherroub, *Chem. Commun.*, 2012, **48**, 8999.
53. B. C. Bunker, D. L. Huber, J. G. Kushmerick, T. Dunbar, M. Kelly, C. Matzke, J. Cao, J. O. Jeppesen, J. Perkins, A. H. Flood and J. F. Stoddart, *Langmuir*, 2007, **23**, 31.
54. M. Alemani, M. V. Peters, S. Hecht, K.-H. Rieder, F. Moresco and L. Grill, *J. Am. Chem. Soc.*, 2006, **128**, 14446.
55. J. Wirth, N. Hatter, R. Drost, T. R. Umbach, S. Barja, M. Zastrow, K. Rück-Braun, J. I. Pascual, P. Saalfrank and K. J. Franke, *J. Phys. Chem. C*, 2015, **119**, 4874.
56. A. Krüger, M. Bernien, C. F. Hermanns and W. Kuch, *J. Phys. Chem. C*, 2014, **118**, 12916.
57. C. Bronner, B. Priewisch, K. Rück-Braun and P. Tegeder, *J. Phys. Chem. C*, 2013, **117**, 27031.
58. F. Chiaravalloti, L. Gross, K.-H. Reider, S. M. Stojkovic, A. Gourdon, C. Joachim and F. Moresco, *Nat. Mater.*, 2007, **6**, 30.
59. H. Tachibana, Y. Yamanaka and M. Matsumoto, *J. Phys. Chem. B*, 2001, **105**, 10282.
60. H. Tachibana, Y. Yamanaka and M. Matsumoto, *J. Mater. Chem.*, 2002, **12**, 938.
61. M. Matsumoto, D. Miyazaki, M. Tanaka, R. Azumi, E. Manda, Y. Kondo, N. Yoshino and H. Tachibana, *J. Am. Chem. Soc.*, 1998, **120**, 1479.
62. W. Li, T. Dohi, M. Hara, S. Nagano, O. Haba, K. Yonetake and T. Seki, *Macromolecules*, 2012, **45**, 6618.
63. M. A. Cohen-Stuart, W. T. S. Huck, J. Genzer, M. Müller, C. Ober, M. Stamm, G. B. Sukhorukov, I. Szleifer, V. V. Tsukruk, M. Urban, F. Winnik, S. Zauscher, I. Luzinov and S. Minko, *Nat. Mater.*, 2010, **9**, 101.
64. L. M. Saiz, P. A. Oyanguren, M. J. Galante and I. A. Zucchi, *Nanotechnology*, 2014, **25**, 065601.

65. K.-Y. Chen, O. Ivashenko, G. T. Carroll, J. Robertus, J. C. M. Kistemaker, G. London, W. R. Browne, P. Rudolf and B. L. Feringa, *J. Am. Chem. Soc.*, 2014, **136**, 3219.
66. P. Wan, Y. Jiang, Y. Wang, Z. Wang and X. Zhang, *Chem. Commun.*, 2008, 5710.
67. B. K. Pathem, Y. B. Zheng, J. L. Payton, T.-B. Song, B.-C. Yu, J. M. Tour, Y. Yang, L. Jensen and P. S. Weiss, *J. Phys. Chem. Lett.*, 2012, **3**, 2388.
68. Y. Kawaguchi and A. Harada, *J. Am. Chem. Soc.*, 2000, **122**, 3797.
69. E. J. Mahan and J. A. Dennis, *Org. Lett.*, 2006, **8**, 5085.
70. Y. Jiang, Z. Wang, X. Yu, F. Shi, H. Xu, X. Zhang, M. Smet and W. Dehaen, *Langmuir*, 2005, **21**, 1986.
71. L. M. Siewierski, W. J. Brittain, S. Petrash and M. D. Foster, *Langmuir*, 1996, **12**, 5838.
72. W. Jiang, G. Wang, Y. He, X. Wang, Y. An, Y. Song and L. Jiang, *Chem. Commun.*, 2005, 3550.
73. H. Lim, J. Han, D. Kwak, M. Jin and K. Cho, *J. Am. Chem. Soc.*, 2006, **128**, 14458.
74. C. J. Bruns and J. F. Stoddart, *Acc. Chem. Res.*, 2014, **47**, 2186.
75. Y. Liu, A. H. Flood, P. A. Bonvallet, S. A. Vignon, B. H. Northrop, H.-R. Tseng, J. O. Jeppesen, T. J. Huang, B. Brough, M. Baller, S. Magonov, S. D. Solares, W. A. Goddard, C.-M. Ho and J. F. Stoddart, *J. Am. Chem. Soc.*, 2005, **127**, 9745.
76. B. K. Juluri, A. S. Kumar, Y. Liu, T. Ye, Y.-W. Yang, A. H. Flood, L. Fang, J. F. Stoddart, P. S. Weiss and T. J. Huang, *ACS Nano*, 2009, **3**, 291.
77. P. Lussis, T. Svaldo-Lanero, A. Bertocco, C.-A. Fustin, D. A. Leigh and A.-S. Duwez, *Nat. Nanotechnol.*, 2011, **6**, 553.

CHAPTER 9

Patterning of Surfaces for Supramolecular Chemistry and Template Effects

9.1 Introduction

The growth of layers of material at specific locations on a surface, to give either a locally-modified area on a bare surface or a template where subsequent layers could be grown, is of interest for two principal reasons for scientists:

(i) Patterning and area-selective growth gives relative information on the outcome of self-assembly compared with a reference surface *in situ*. Very often, uniform surface modification is very hard to "see" by any analytical technique because of the lack of any internal standard to judge the thickness of the film, its fluorescence, its polarity, and so on.

(ii) To provide local and sequential changes that can endow the surface with directionality to its properties as a result of the change in functionality. This could be to affect wettability, conductivity, optical properties, or other characteristics. Thus, unique surface behaviour is achieved.

The way these (usually micrometre- but possibly nanometre-scale) patterns are made can be divided into two main generic routes: photolithography and related techniques, and soft lithography.[1] The former has been used on large scales in the microelectronics industry, and is largely responsible for the dramatic advances present generations have seen in

Monographs in Supramolecular Chemistry No. 19
Supramolecular Chemistry at Surfaces
By David B. Amabilino

Published by the Royal Society of Chemistry, www.rsc.org

computing capacity. It is a truly top-down technique where chemical treatments are used to remove material from a surface to generate a pattern. Soft lithographies (for there are many distinct variants that have been reviewed thoroughly[2]) are generally associated with the patterned deposition of materials; therefore, material is usually built up on a substrate, and as such, one could consider that it is in a way a bottom-up approach (though it is usually considered top-down as a lithography and the preparation of the stamps involves top-down methods). Both are inspired by antique lithography—the name comes from the Ancient Greek for stone and to write, *litho* and *graphein*—where a drawing was made in limestone with some kind of lipophilic substance (grease or wax, *etc.*) and the exposed mineral was etched with acid, as in the classical etching that was established in the Middle Ages. This etched region was exposed to water, then oily ink was applied and directed to the greasy areas. This pattern—the same as the original drawing—could then be printed on another surface, such as paper. The parallels with many modern day microscopic versions based on different chemistry are palpable, and its basis is the separation of hydrophilic and hydrophobic materials. The process played its part in making printing and therefore distribution of knowledge hugely available. The opportunities to extend the concept to supramolecular interactions in general are clear. Yet the lithography that has perhaps revolutionised civilisation more recently is based on photochemical reactions, photolithography, that produced microelectronics leading to the information revolution we have witnessed in recent decades. Beyond these classically-derived lithographic techniques that rely on blocks of material, there are also direct patterning techniques, such as deposition of material through masks, using irradiation like an electron beam to alter near-surface material, or the writing on surfaces with a tool such as the tip of an AFM.[3]

Photolithography generally consists of using monomers or polymers (termed "resists") to cover a surface uniformly and irradiating certain areas selectively through a photomask, whereby only certain areas are affected photochemically by exposure to light (be it visible, ultraviolet, or even higher energy, Figure 9.1). If the polymer reacts where the light is incident at the interface, it can be removed with washing, and if the monomer is polymerised by light, it is the non-irradiated area that is rinsed off.[4–6] What remain are the unreacted resist material in the case that a polymer is attacked, or the polymerised material in the case that a monomer is polymerised, and the bare (original) substrate material. This substrate material can subsequently be etched chemically. The etch should not affect the resist. So, this is a hard core top-down method where material is blasted with light, brushed aside and eaten away with aggressive solution! But the technique has truly remarkable resolution: extreme ultraviolet light has already been used to generate features only 10 nm wide.

These hard surface patterns prepared by photolithography, not dissimilar to the wooden blocks humans once used for printing, can be used as

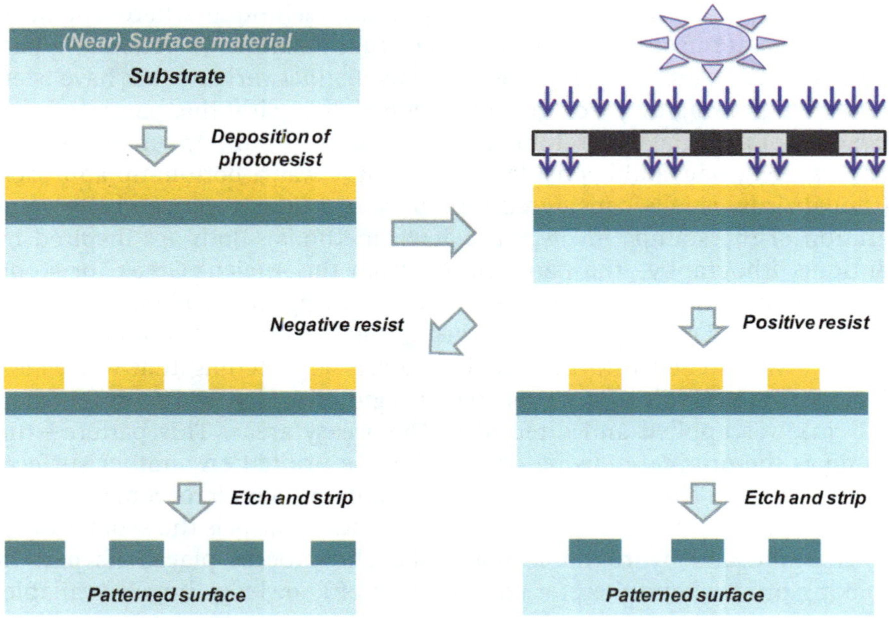

Figure 9.1 A cartoon showing possible photolithographic processes for patterning surfaces.

moulds and for embossing. In this sense, they are in turn the basis for soft lithography. When a polymerisable material is introduced onto a patterned surface, and then allowed to react before being peeled off, a negative pattern results that can be used for printing, for example, as we shall see for the case of microcontact printing.

Patterns are useful for the creation of defined areas of materials on surfaces for a number of reasons. The pattern can act as a template[7] in two main ways (Figure 9.2): either (i) as a passive template, where the newly adsorbed material fills the spaces around a pattern on the substrate or (ii) where the pattern acts as a positive template and material is deposited on it because there is a thermodynamic driving force. An alternative way of looking at surface template effects is to look at the strength of the interaction between adsorbates themselves and between the adsorbate and the adsorbent.[8] According to this vision, that applies to uniform surfaces as well, self-assembly is dominated by adsorbate–adsorbate interactions, with increasing influence of the substrate there is template-assisted assembly, and when the adsorbent forms strong bonds with the adsorbate the resulting pattern is template controlled.

In order for these template reactions, positive or negatively blocking, to be able to function effectively, the routes to developing the different areas of the surface must function independently of one another, in other words they should be orthogonal pathways. These orthogonal interactions can be

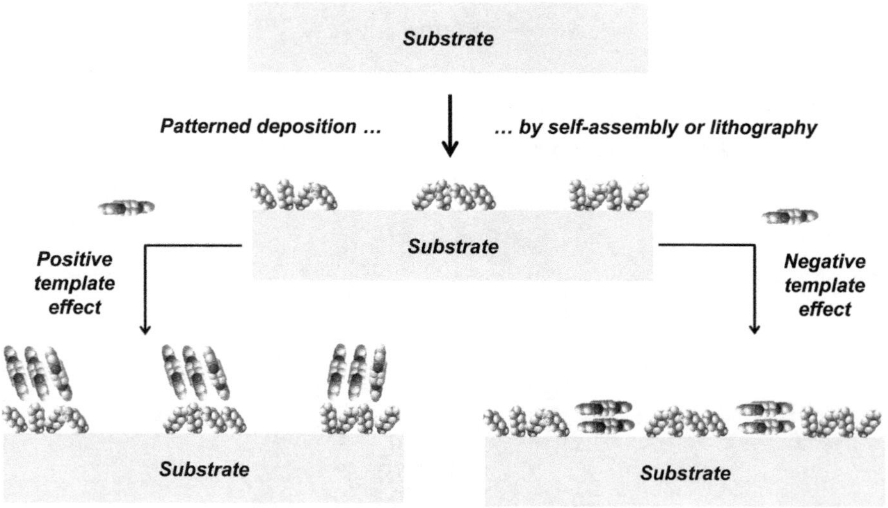

Figure 9.2 Patterning and the roles of patterned layers in templating molecular adsorption through positive or negative/passive templating.

designed by exploiting supramolecular motifs that operate in parallel and in a non-interfering way with each other.[9]

The main alternatives to preparing patterns will now be presented, focussing on those that can be used afterwards for some supramolecular effect. That will draw the initial discussion to systems in which very strong bonds exist between adsorbate and adsorbent, although non-covalent approaches can be used for spontaneous pattern formation, which is treated separately, and could be of interest for the development of systems that self-organise to give functional materials.[10]

9.2 Patterning Self-assembled Monolayers

Monolayers formed by siloxanes on silicon can be patterned using deep ultraviolet ($\lambda < \sim 250$ nm) irradiation.[11] Organosilane SAMs are removed through a photocleavage of the adsorbate, specifically at the Si–C bond,[12] leaving a hydrophobic region that can be filled with an alternative SAM-forming material. The patterned exposure results in the intact initial SAM in the areas that are masked from the UV irradiation, and reactive hydrophilic regions in the exposed ones. This patterning technique has been used for patterning different SAM termini for the area-selective adhesion and growth of cells[13] as well as for the electroless deposition of metals in the polar irradiated regions[14] that can then be used for multiple chemical or physical devices.[15]

A number of techniques exist for patterning SAMs of thiols on gold, and related monolayer systems.[16] The most widely used and versatile method for

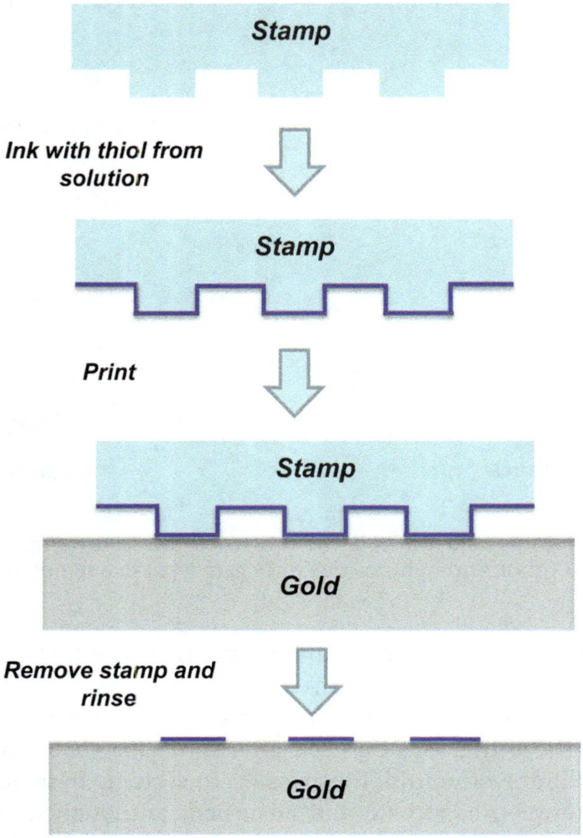

Figure 9.3 The microcontact printing protocol that is frequently used for the patterning of gold surfaces.

patterning chemisorbed layers on gold is microcontact printing (μCP).[17] The procedure is very much like a conventional ink stamp that we may have played with as a child or used in everyday life. In this technique and its derivatives, a soft stamp, most often made of poly(dimethylsiloxane) (PDMS), is inked with a solution of the adsorpt that is then transferred onto the substrate by pressing the stamp down (Figure 9.3). Releasing the stamp leaves the surface inked with the adsorpt. The substrate is then rinsed to remove any excess material. In the case of thiols on gold, the thiol is usually applied to the stamp in ethanol, and the dry stamp is then pressed against the gold surface. The thiols are left attached to the gold, in principle through the same covalent bonds that are responsible for the formation of SAMs. However, printed patterns should not generally be considered to have the same structure as SAMs.[18] They are not thermodynamic minima by any means, as free diffusion is not available to the molecules in what is in principle a solventless process,[19] although for small molecules diffusion does take place and results in a slight broadening of the printed patterns.[20] Controlling the ink transfer and stamping for long periods of time improves the quality of certain patterned monolayers.[21]

The preparation of the stamps is actually facilitated by photolithography that is used to make the masters for the stamps. Silicon wafers are usually patterned using this now classical top-down technique. The surfaces of the patterned chips can be made hydrophobic—to facilitate release of the PDMS once cured—and then the precursors of the polymer are added and reacted to form the elastomer. Once curing has been carried out, the stamp can be removed from the wafer. The use of PDMS as the stamp material is not essential, other polymers have been employed,[22] although the silicone-based elastomer is the best understood and widely used. Microcontact printing is relatively easy and does not require expensive equipment. It has been used for a wide variety of chemistry to pattern surfaces.[23] The stamps are often only used once, as material can be absorbed or the stamps damaged during the processes.[24] The PDMS that is used habitually is limited in terms of the solvents that can be used: very hydrophobic organic solvents can swell the polymer stamp that leads to distortion of the features.

It is clear that once a pattern has been formed on a surface by μCP, it is possible to fill the remainder of the surface with another reagent using self-assembly by chemisorption, or physisorption if a robust film is not required. This situation corresponds to the negative template effect described in Figure 9.2. If the printed pattern contains the supramolecular unit of interest, an inert layer—typically an alkyl thiol—is self-assembled around it. An inverse route can also be performed if a labile self-assembled monolayer is formed first, in a process termed "microdisplacement printing".[25] Thus, SAMs of 1-adamantanethiolate on gold when stamped with 11-mercaptoundecanoic acid are replaced in the areas where the pressing is felt. This process improves the precision of the more usual μCP because of the mobility on the substrate of the ink molecules during and after the stamping.

Stamps with shapes such as the ones used in μCP can also be employed to guide the deposition of materials on surfaces by spatially controlled dewetting.[26,27] Recalling that dewetting is a process whereby a thin film spontaneously separates into a group of separate entities, here, the flow of a solution into a stamp or the deposition of liquid from a stamp to a surface leads to the formation of organised conglomerates of material on a surface of choice. The method can also be used for deposition onto pre-patterned surfaces, made either by μCP or other methods.[28]

Scanning probe-derived lithographies provide unique ways to perform local patterning on a surface,[29–32] through the four basic routes shown in Figure 9.4. These approaches could be aimed at very local patterning of a material to probe nanometre-scale modifications in terms of structure and properties or could include the massively parallel writing with AFM tips using thermal treatment for data storage.[33]

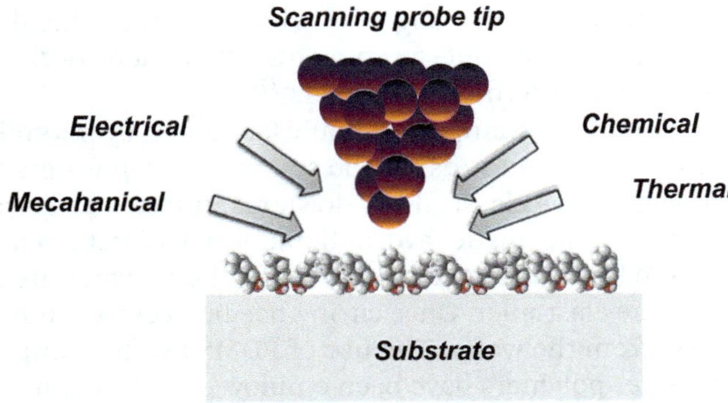

Figure 9.4 Frequently used ways to pattern surfaces using scanning probe lithographies.

Nanoshaving and nanografting use the tip of a scanning probe microscope to remove part of a self-assembled monolayer and fill it with another, either in two steps or simultaneously (Figure 9.5). This was proven using a scanning tunnelling microscope to remove material[34] and using an AFM tip to perform "shaving" of a decylthiol monolayer in the presence of octadecylthiol.[35] In this way, features as small as ten by fifty nanometres were prepared, that appeared as protrusions above the surrounding decylthiol monolayer because of the longer alkyl chain of the "grafted" octadecylthiol.

In an ambient atmosphere in most places on our planet where laboratories are located, water is present at interfaces, at least between polar surfaces and air. When a current is passed between an STM tip and a silicon surface, anodic oxidation takes place generating silica, with the water acting as an electrolyte.[36] An AFM tip can also be used to carry out the same process,[37] and features sizes down to around 10 nm. These polar areas can be modified chemically, or they can be used as anchoring points for molecules or solvents. For example, water droplets around these oxidised areas act as a template during the dewetting of passivated gold nanoparticle solutions, giving gold nanoparticle rings.[38]

"Dip-pen" nanolithography was developed using the tip of an AFM to write lines or dots using thiols as ink onto a gold surface.[39] The adsorpt molecules undergo capillary transport from the tip to the adsorbent surface. The method has been developed to exploit a huge variety of adsorpts and designs for machinery to perform the deposition.[40] The technique is an attractive alternative to microcontact printing, since the resolution of the procedure can be higher and the possibility for generating multiple structures on the same surface exists. In particular, the study of biomolecules at interfaces using the technique to apply tiny quantities of material is promising.[41] In supramolecular science, the technique awaits exploitation, largely.

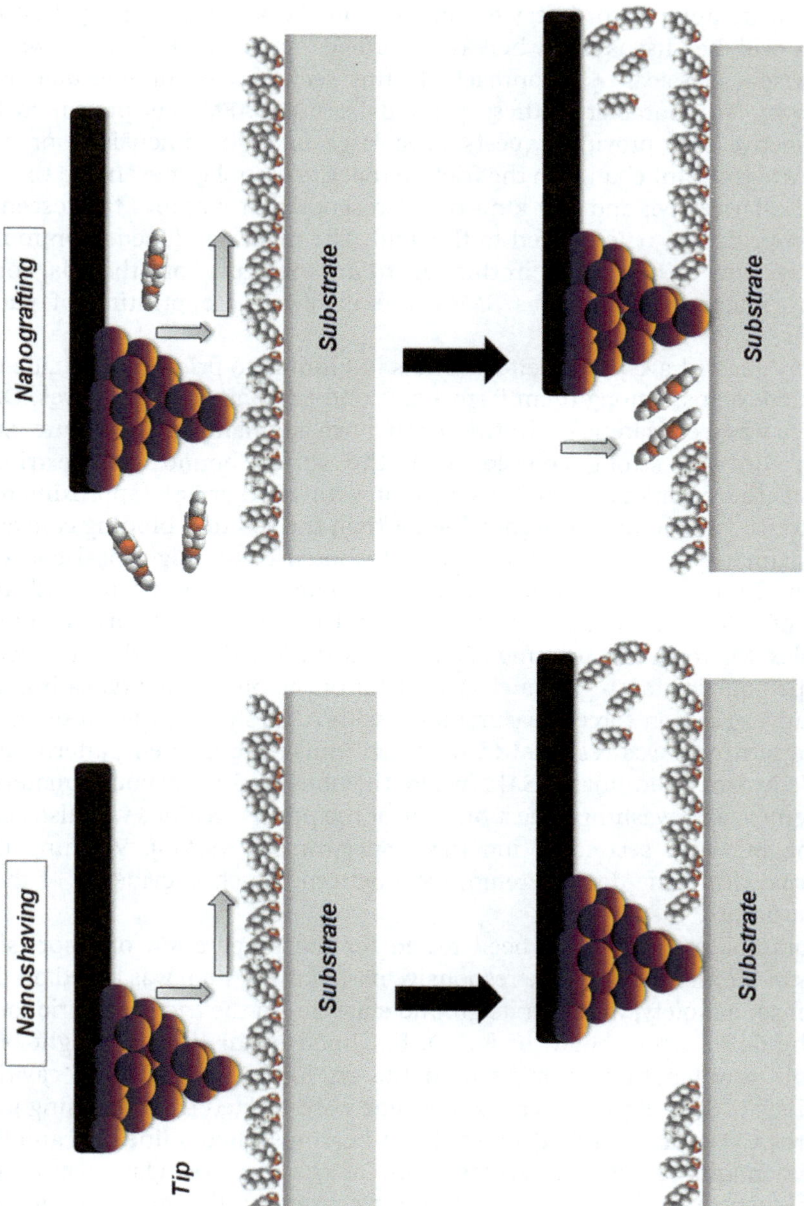

Figure 9.5 Using a scanning probe tip to perform nanoshaving and nanografting.

9.3 Patterned Deposition of Guests on Substrates

Patterning allows the molecular recognition between solution-borne guests to interface-confined guests to be visualised provided a suitable analytical tool can be applied to detect the change on the surface. Here, patterned guests will be discussed, whereas patterned surface-bound hosts would comprise a classical μCP approach, in this section a supramolecular μCP is shown. To characterise these printed features, AFM has proven to be an effective tool provided guests have large enough dimensions or can cause a significant change in the friction that the monolayer exerts on the tip of the microscope, and any kind of microscopy that employs fluorescence can, in principle, be exploited to this end. The most developed supramolecular systems to be studied in this regard are molecular printboards, comprising macrocyclic hosts as SAMs and involving the printing of guest species.[42]

Many guests have proven effectiveness in binding to printboards made up of cyclodextrins, among them ferrocene[43] and adamantyl[44] derivatives. One of them was a calixarene derivative bearing two adamantyl units (Figure 9.6), which forms a strong complex with the surface-bound cyclodextrin.[45] In fact, the calculated binding constant with the surface (approximately $K = 10^9$–10^{10} M^{-1}) was apparently higher than the solution binding constant (approximately $K = 10^7$ M^{-1}), and was attributed to the high local concentration of the host, though presumably entropic effects associated with the water of solvation are increasingly important in this kind of surface-bound complex. Microcontact printing of a cyclodextrin-functionalised surface with a stripe pattern stamp gave a clear transfer of the bis-adamantyl calixarene derivative. Friction force measurements with AFM showed the presence of the pattern that was resistant to washing, unlike the printed pattern on a simple hydroxyl-terminated SAM, where the non-specifically bound material was removed by washing. Guest binding at the printed regions was also seen in imaging with secondary ion mass spectrometry (SIMS). Washing the patterns with brine did not remove the pattern, which is evidence of their robust nature.

Supramolecular μCP has been tested for the preparation of responsive arrays of biomolecules.[46] The responsive part of the system was based on the azobenzene moiety, which in its ground state lies in the *trans* isomeric form that binds to a printboard of β-CyD, but upon irradiation with light of a suitable wavelength is converted to the *cis* form that exits the cavities (Figure 9.7). Two derivatives of azobenzene were printed together using μCP onto the CyD-coated surface, one of them bearing a rhodamine unit and the other a mannose residue. Visualisation of the layer on glass using epifluorescence microscopy showed clearly the pattern of the mixture of guests. After irradiation with UV light, no significant fluorescence could be detected where there had been stripes, showing that the majority of the guests had been expelled from the interface. The mannose moiety in one of the guests is known to bind the carbohydrate-specific lectin concanavalin A (ConA).

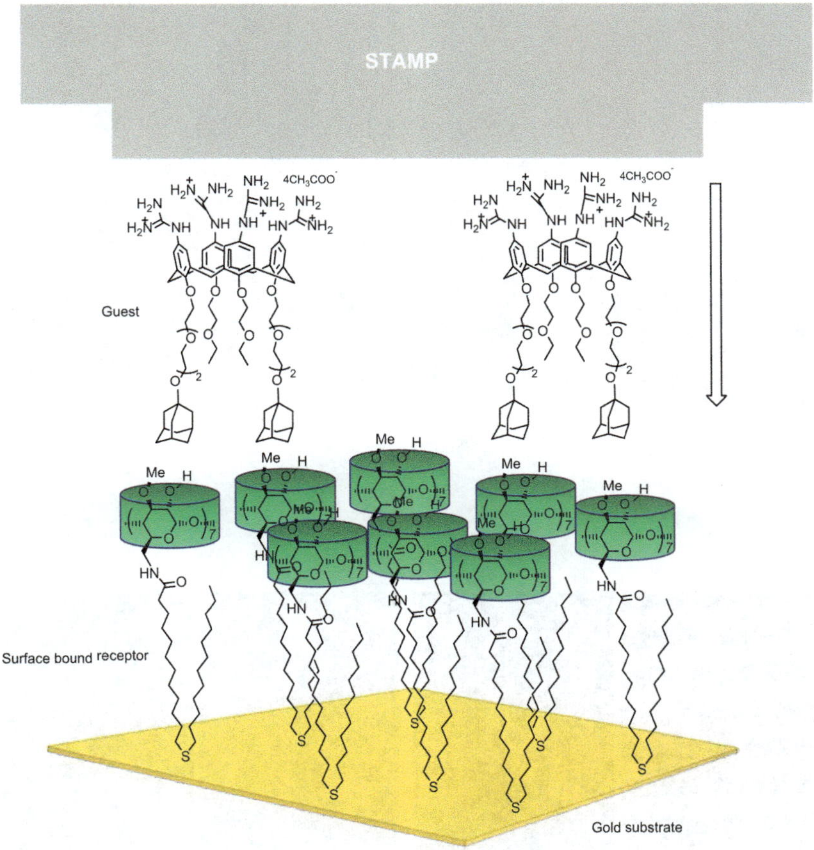

Figure 9.6 A "molecular printboard" of cyclodextrin hosts for adamantyl-functionalised guests that are patterned on the printboard by microcontact printing, and are robust to washing of the surface.

When the mannose-azobenzene compound was printed onto the printboard and fluorescein-conjugated ConA was exposed to the interface, binding was effective, as seen in the fluorescence images of the sample. Remarkably, when this sample was irradiated with UV light, the fluorescein-conjugated ConA was expelled from the surface, because of the *trans–cis* isomerisation of the mannose-azobenzene unit and its exit from the CyD cavity. The effect is seen clearly by monitoring the fluorescence intensity from the interface. It is believed that this approach can lead to the analysis of proteins through surface binding and removal protocols.

The detection of certain molecules from solution by reaction with a patterned monolayer is a useful way to determine if chemisorption has taken place, and this will surely be combined with supramolecular routes. Pyrylium monolayers have proven effective to this end.[47] The monolayer of these materials was made by preparing an azide-terminated SAM[48] which

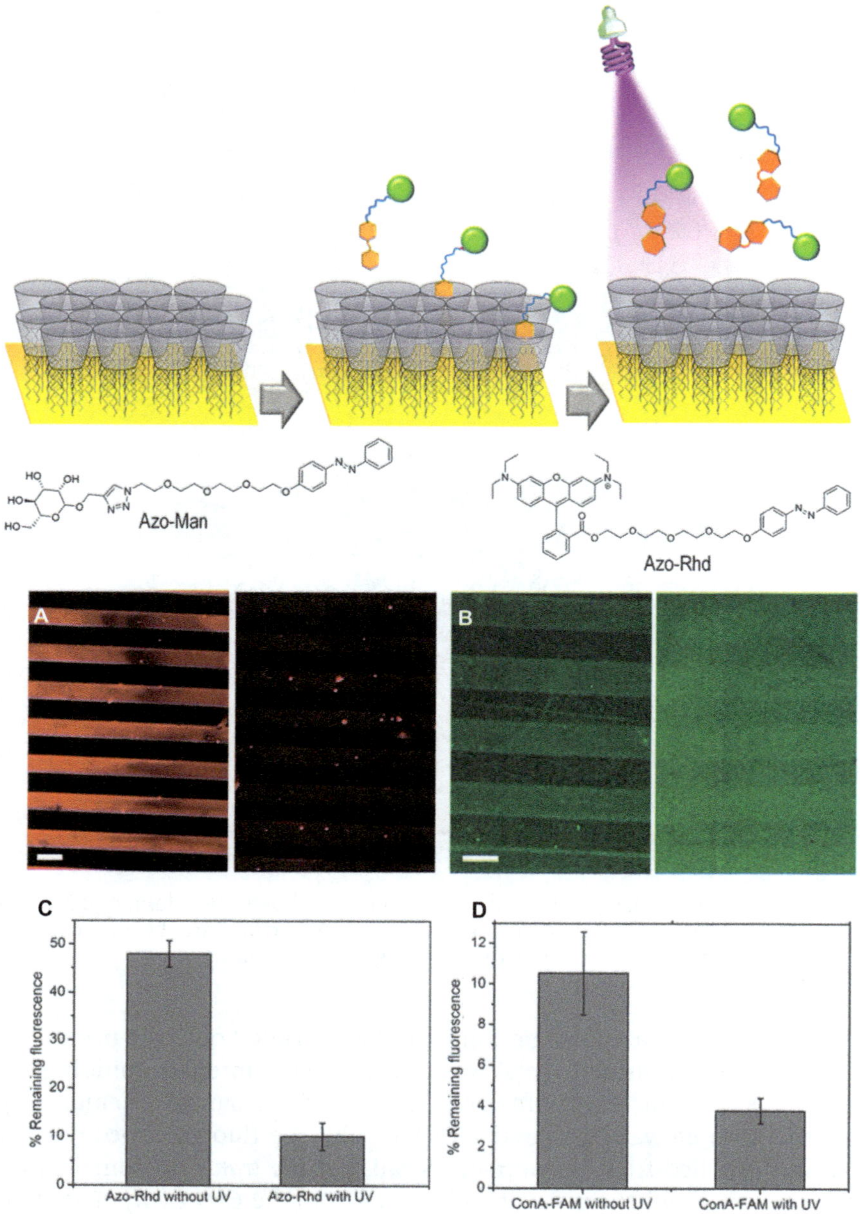

Figure 9.7 A photoresponsive supramolecular microcontact printed pattern comprising cyclodextrin hosts and azobenzene derivatives as ink that can be removed by photoisomerisation. (A) Azo-Rhd immobilised on CyD SAM as seen by fluorescence microscopy before (left) and after irradiation. (B) A fluorescence microscopy image of fluorescein-labelled ConA bound to a layer of Azo-Man in the CyD SAM before (left) and after (right) irradiation with UV light. (C) and (D) show the fluorescence intensity profiles of the images above them. Scale bars represent 100 mm. Reprinted from ref. 46 with permission from the Royal Society of Chemistry.

was reacted with an alkyne-derived pyrylium ion upon printing the reagent on the surface (Figure 9.8). This kind of click reaction was proven previously through application of the concept of heterogeneous catalysis through μCP.[49] The reaction was proven by measuring a fluorescence microscope image of the surface after rigorous solvent and ultrasound treatment. The reaction of the pyrylium ion with amines to generate a pyridinium ion renders the layer non-fluorescent so that covalent attachment can be measured optically. In addition, the attachment of fluorescent proteins was carried out by solution treatment of the patterned surface, as demonstrated by the reaction of the pyrylium ions with the lysine residues of yellow fluorescent protein. Apart from reaction of a patterned surface with amine ligands, the reactivity of the pyrylium layers with amines was also shown with dip-pen-nanolithography (DPN). A bespoke atomic force fluorescence microscope[50] was used to deposit amine derivatives on the surface and image *in situ* the result of the patterning. These chemical and microscopic tools are sure to have an impact in the study of supramolecular chemistry at interfaces. The interactions of the pyrylium ions with anions have been demonstrated in homogeneous media.[51] The toolkit for patterning *via* combinations of covalent and non-covalent interactions is increasing all the time.

The degree to which covalent and supramolecular μCP techniques can be integrated is demonstrated by the orthogonal covalent and non-covalent modification of a glass surface patterned with CyD.[52] The components of this sophisticated system are shown in Figure 9.9. The SAM used as the basis for the system is an alkyne-substituted coumarin as part of a siloxane-based SAM on glass: this unit acts as a reporter group for the effectiveness of the printing, because when the alkyne is converted to the triazole after a click chemistry reaction, its fluorescence intensity is enhanced dramatically. The patterning was checked by binding with the bis-adamantyl fluorophores from solution, thereby proving the location, binding ability and orientation of the immobilised CyD. Printing and solution steps can then be used to pattern the surface. Three sequences are shown in Figure 9.9(A) shows the supramolecular printing of a fluorescein bis-adamantyl compound into the CyD area, (B) azide-labelled Oregon-green 'click' μCP onto the uncovered coumarin alkyne area, and (C) the simultaneous cross-printing of a mixture of rhodamine bis-adamantyl compound and the azide-functionalised Oregon green. Figure 9.10 shows fluorescence micrographs of the experimental realisation of this method. The use of the profluorogenic group to observe the linking of components onto the SAM surface gives an *in situ* control of the covalent chemistry taking place, while the supramolecular μCP is seen by direct visualisation of the dyes. Dip-pen nanolithography is also a viable route to patterning of this kind of supramolecular assembly.[53] This work paves the way for extremely complex supramolecular patterning of surfaces using different deposition techniques and orthogonal pathways perhaps exploiting different supramolecular synthons.

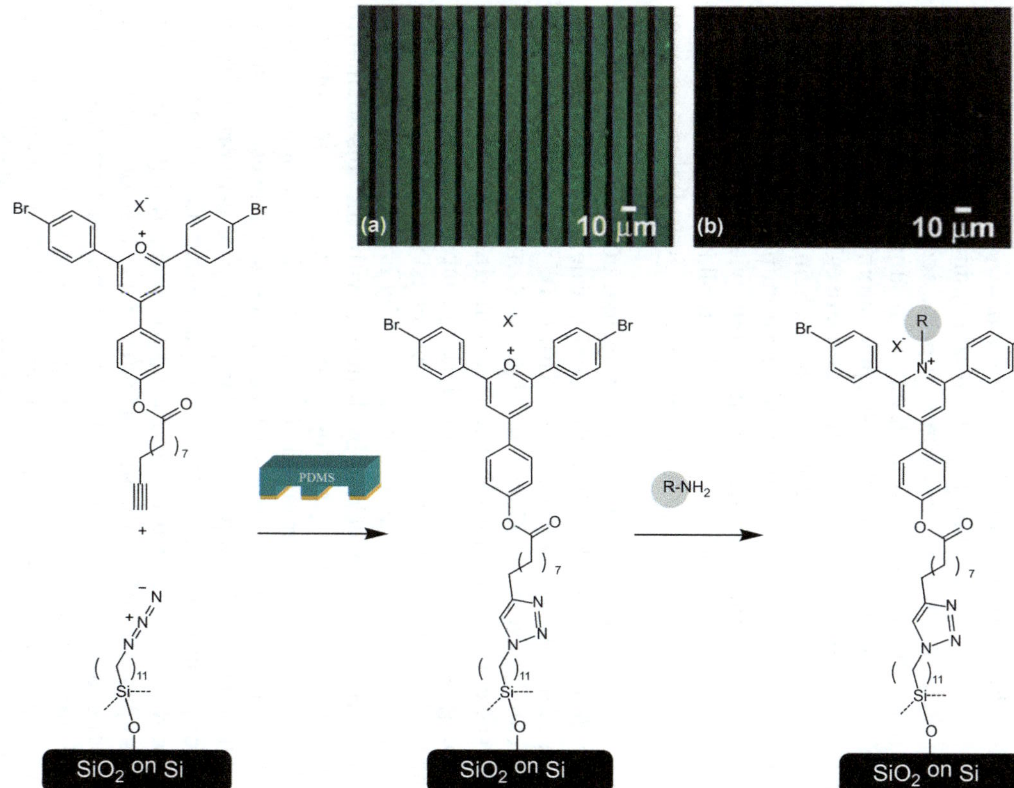

Figure 9.8 The attachment of a pyrylium moiety to a chemisorbed SAM bearing an azide function through stamping with a PDMS stamp to give a patterned surface, which is visualised by fluorescence microscopy in image (a), followed by reaction with an amine that renders the pattern dark by fluorescence microscopy (image (b)).
Reproduced from ref. 47 with permission from the Royal Society of Chemistry.

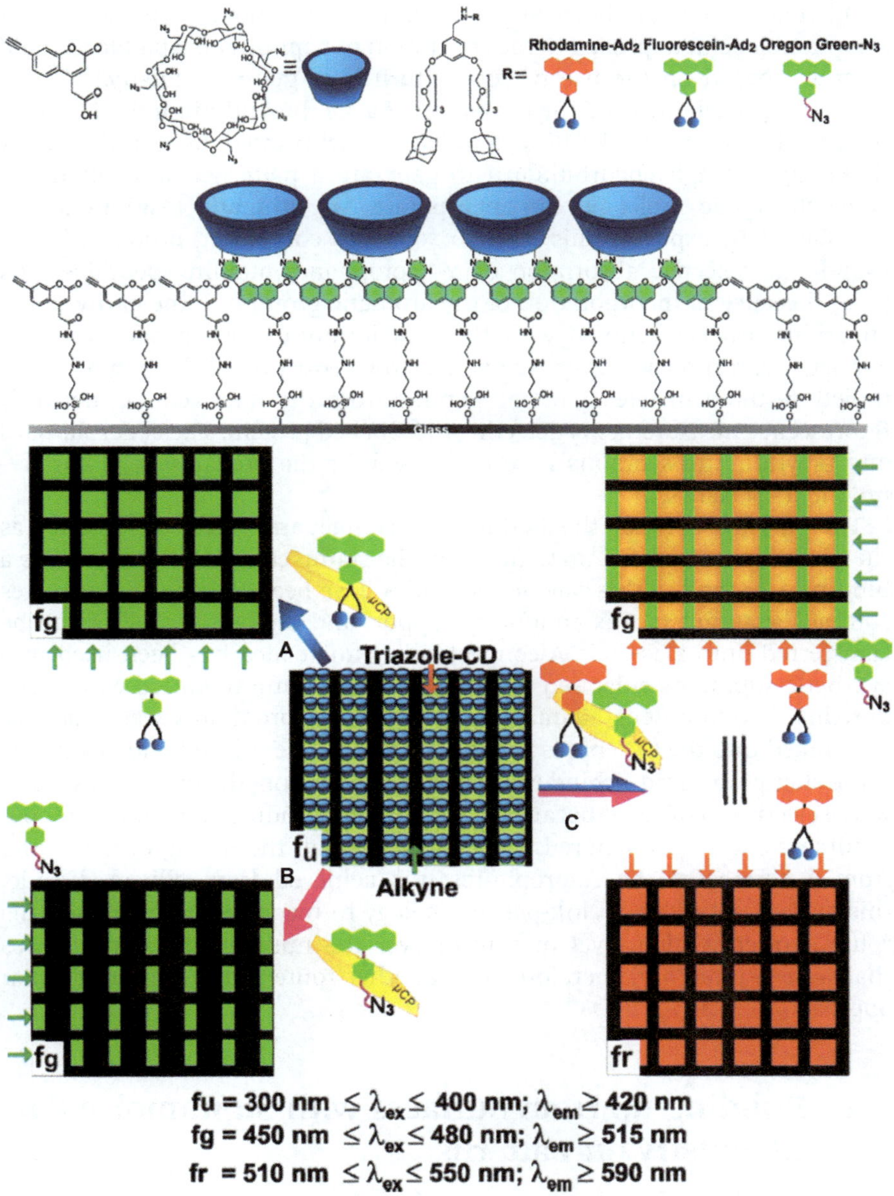

Figure 9.9 The components of an orthogonal assembly route for the preparation of multifunctional surfaces through click chemistry by stamping and host–guest interactions between adamantyl groups and CyDs. Below are shown the printing steps that lead to the theoretical patterns on the surface, starting at the centre from a striped clicked CyD pattern on the alkyne-terminated coumarin SAM on glass and performing patterning with the three dye adsorpts.
Reprinted with permission from ref. 52. Copyright 2010 American Chemical Society.

Other synthons have been employed for other functions, as is the case for a supramolecular approach to the separation of peptides on a patterned gold surface[54] based on the use of cucurbituril host–guest chemistry.[55,56] The elegant approach involved the formation of a 4,4′-bipyridinium thiol-derived patterned SAM by back-filling a microcontact-printed SAM followed by threading through cucurbit[8]uril to generate a patterned layer of hosts, since the cyclic molecule can incorporate two aromatic residues at the interface.[57] By exposing this layer to solutions containing proteins, it was found that selective absorption of *N*-tryptophan-containing peptides took place, because of incorporation of the aromatic group into the cavity of the cucurbit[8]uril co-occupied with the bipyridinium unit. In addition, this surface was responsive, because one electron reduction of the bipyridinium moiety resulted in release of the peptide from the interface, as shown by fluorescence measurements detecting the labelled protein. The cucurbit[8]uril and bipyridinium synthons have been used for the preparation of surface-mounted rotaxanes.[58]

The interface between the SAM and the contact area has been regarded as the place for recognition up to now, but the stamp can actually be used as a binding area, and in this way biomolecules have been fished from mixtures so that the stamp acts as an affinity support and the bound species can be transferred onto glass.[59,60] A conceptually similar idea has been used in a synthetic supramolecular sense, using a PDMS stamp modified with cyclodextrins.[61] Monovalent adamantyl-substituted fluorophores were successfully inked into the CyD pores of the stamp and were printed onto a surface. If the stamp was used to bind molecules from solution, di-adamantyl species were selectively collected because of their higher binding constant, and non-binding solutes were ignored. On the other hand, the inking of the stamp from a surface of the fluorophores was achieved from silicon dioxide, this surface acting as an ink pad in analogy to macroscopic printing. This route showed a high level of control over the number of ink molecules that were printed and therefore adds another route to the supramolecular approach to μCP.

9.4 Building up from Surfaces with Supramolecular Chemistry on Patterns

Patterning of a surface allows construction of supramolecular architectures perpendicular to the plane in defined places of the surface through the use of specific supramolecular synthons. This process can either take place in solution, as in the LbL methods, or during evaporation of a solution, and these quite different situations will be discussed in turn.

Regarding the application of patterning to LbL methods, the use of multivalent complexes between cyclodextrins and adamantyl units has proven effective in patterning,[62] but perhaps the more widely used materials are metal–organic frameworks[63] for which many patterning routes

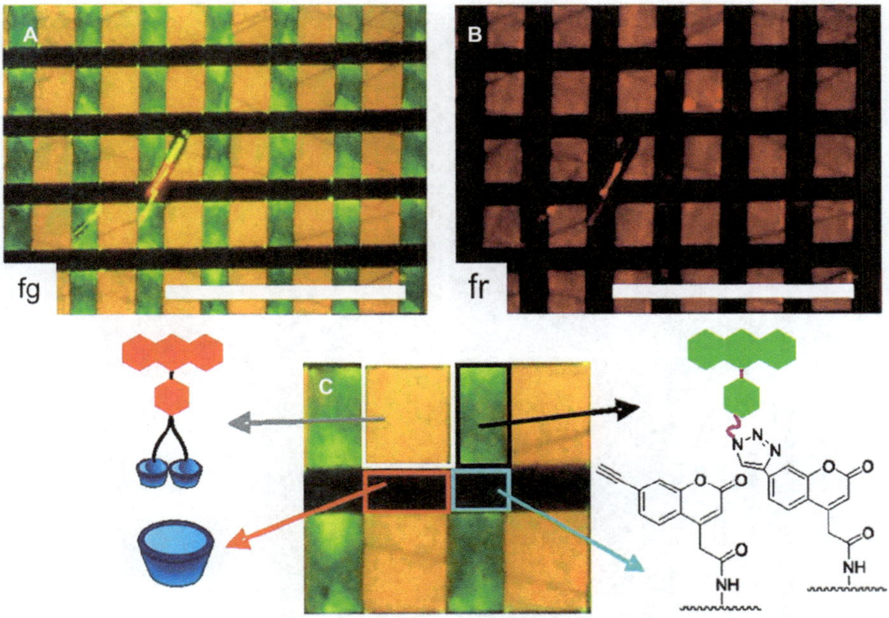

Figure 9.10 The result of the one-step functionalisation of the surface shown in Figure 9.9 with orthogonally-patterned coumarin-CyD surfaces by printing a mixture of Oregon green-N3 and rhodamine-Ad2 as viewed with fluorescence microscopy. The images are excited with (A) blue light (green filter, $\lambda_{em} \geq 515$ nm), (B) green light (red filter, $\lambda_{em} \geq 590$ nm), and (C) the four different areas with indication of the functionality. Scale bars correspond to 55 µm.

have been employed.[64] Doubts concerning the variation of thickness in SURMOF can be allayed by patterning because of the presence of a surround that is inert to the growth process. The oriented growth of a SURMOF comprising dimeric copper(II) paddlewheel complexes and benzenetricarboxylate bridging ligands was prepared on a patterned surface of thiolate squares (3×3 µm) from 16-mercapto-hexadecanoic acid in a surround of hexadecanethiol by sequential dipping of the substrate in solutions of copper acetate and benzenetricarboxylic acid.[65] The AFM images (topography and friction,[66] Figure 9.11) of the patterned surface show that although the substrate has a uniform height (the thiols have very similar length), the friction presented by each area is completely different. For a patterned surface with a nominal thickness of forty-five layers, the size of these areas is the same (Figure 9.11), and the thickness corresponds to that expected for the crystal of the compound on the surface, assuming one unit cell for a complete cycle. The growth in the average height was linear with the number of cycles up to this value. On the other hand, the roughness of the sample increased substantially with the number of cycles, levelling off

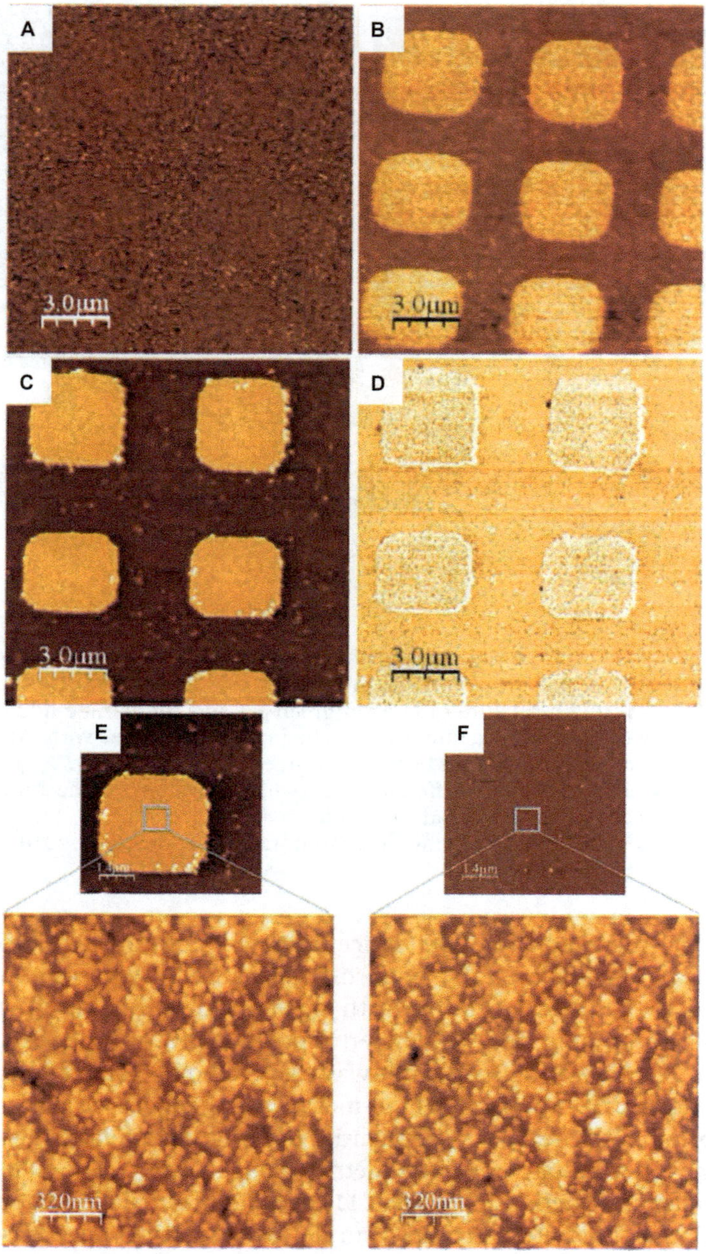

Figure 9.11 AFM images of the patterned monolayer formed from 16-mercapto-hexadecanoic acid (the squares) and hexadecanethiol. (A) Shows topography and (B) friction, (C) and (D) show topography and phase contrast images, respectively, of the SURMOF formed from copper(II) and benzene-tricarboxylate grown in an LbL way on the patterned surface. (E) Shows a close-up of one of the squares and compares it with a SURMOF prepared on a full monolayer (F).
Reproduced from ref. 65 with permission from the Royal Society of Chemistry and the PCCP owners' societies.

at approximately 5 nm after thirty cycles. Nonetheless, the speckled topography in the patterned surface was very similar to that observed in a full monolayer of the material. Therefore, this route to patterned SURMOFs is a viable one for their possible exploration in this environment. These SURMOFs were also grown into wells in monolayers generated by nanografting, with similar results.[67]

MOFs have been nucleated on surfaces as well, rather than being formed in an LbL approach. Using a similar patterning, the carboxylic acid group was shown to nucleate MOFs,[68] albeit to generate a non-homogeneous layer, proving the advantages of the LbL method for the formation of homogeneous patterns of crystallised material. Patterning of the SAM templating layer for framework preparation can also be performed using the microcontact click chemistry approach. A copper-catalysed stamp alkyne to surface azide cycloaddition was employed to pattern silicon wafers with carboxylic acids.[69] This functional group acted as a nucleating area for the growth of MOFs on surfaces in a single solvothermal step.

Patterned surfaces have been used frequently for the study of crystallisation of inorganic and organic materials on surfaces.[70,71] In principle, a pattern of two monolayers with contrasting polarities on the same surface permits induction of mass transport that helps the crystallisation process.[72,73] In principle, the nature of the terminal group on the monolayer surface should influence the growth in some form of template mode. As an example from the author's group, the crystallisation of an organic chiral cyclic phosphate on a patterned monolayer of a thiol derived from the same functionality was used to test the viability of using this approach to resolve enantiomers.[74] It had already been shown that self-assembled monolayers of the compound influenced the crystallisation of the acid from the surface, although the lack of confinement of the process did not allow clear conclusions concerning the diastereoisomeric selectivity of the method.[75] The substrates were prepared by stamping one enantiomer of the chiral thiol-derived phosphate onto polycrystalline gold with a PDMS stamp (Figure 9.12). After rinsing, the bare gold regions were covered by exposure of the whole surface to an ethanol solution of dodecanethiol, affording a surface with hydrophilic dots and a hydrophobic surround. Crystallisation of the enantiomers of a phenyl-derived cyclic phosphate on the chiral dots provided crystals of similar appearance from chloroform. The crystals clearly grew from the surface and up away from it, for both enantiomers of the acid (Figure 9.12). On the other hand, a greater difference was seen when isopropanol was used as the solvent for the crystallisation. For one enantiomer (the opposite one to the monolayer chirality), crystals were formed in the plane of the interface, as can be appreciated in the SEM images of the samples (Figure 9.12). The opposite enantiomer could not be crystallised on the chiral dots, hinting that nucleation was not favoured there. From the SEM images, it can be appreciated that the crystals formed from isopropanol could be associated with growth from the interface between the printed pattern and the surrounding SAM, whereas when chloroform was used, the

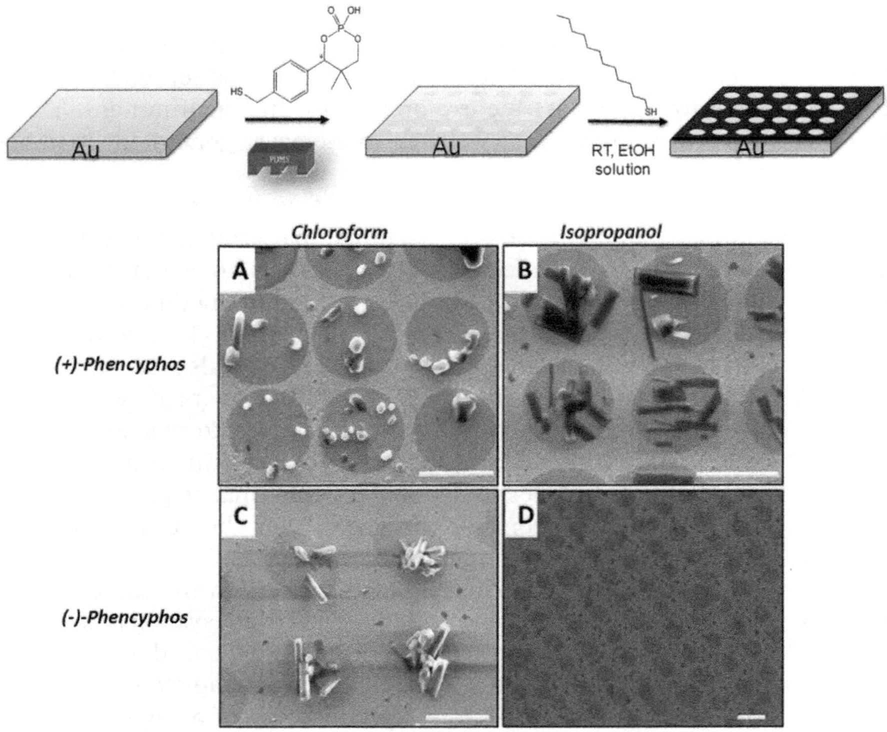

Figure 9.12 The route for the patterning of gold with hydrophilic dots of a chiral organic phosphate functionalised with a thiol group surrounded by a hydrophobic SAM through microcontact printing. The SEM images show the crystals of the enantiomers of the chiral phosphate Phencyphos grown on the dots from different solvents. The scale bar is 5 μm in all cases.

crystals tended to grow within the centre of the patterned dots. In any case, the crystallisation of these compounds from organic solutions is extremely hard, and a deeper understanding of the role of interfaces in the crystallisation of materials, particularly in the absence of water, is a valuable challenge.

An important control experiment in these patterned crystallisation templates is to inverse the pattern in the monolayer. In the present case, this meant printing the alkyl thiol followed by SAM formation with the phosphate derivative on the gold. When this substrate was used for the growth of crystals, it was evident from the microscope images (Figure 9.13) that crystallisation is promoted exclusively by the phosphate-terminated monolayer regions.

The combination of top-down patterning techniques with self-assembly approaches for the formation of complex chemical systems is, in principle, a powerful way to form patterned surfaces with great complexity and

Figure 9.13 A SEM image (with a close-up in the inset) showing crystals of the chiral phosphate Phencyphos grown on an inverse pattern, where the dots comprise a dodecanethiol monolayer. The scale bar is 100 μm in the main image.

potentially function. Sequential employment of passive or negative and positive templates and subsequent removal to permit growth has to be approached carefully so that each step is compatible with the stability limits of the surface pattern, allowing the desired system to be constructed. Template preferences must be high leading to efficient and selective growth.

A beautiful example of this concept is the patterned bottom-up assembly of antibody nanotubes on an array of antigens (Figure 9.14).[76] Self-assembled nanotubes—average diameter 100 nm—formed from bola-amphiphile amide derivatives[77,78] can bind peptides that can subsequently be used for mineralisation,[79] or, as we shall see here, for molecular recognition. Antibodies were covalently attached to the nanotubes and these objects were assembled onto patterned areas of antigens on a gold surface. The patterns of antigens were prepared by nanoshaving a SAM of octadecanethiol with the tip of an AFM, and topographical analysis revealed a trench 10 nm deep (surprising, given the length of the thiol). This trench was filled with antigen, which attached to the gold through thiol groups in the molecule, whose insertion was shown again by AFM, with a protrusion of 10 nm. After incubation with the antibody-coated nanotubes, the height of the protrusions increased to 100 nm, indicating that the antigen had templated the adhesion of the tubes to the interface. This kind of "bio-nanopatterning" of surfaces taking advantage of Nature's molecular toolkit opens many interesting avenues for exploration of nanostructure growth and investigation of interface supramolecular science.[80]

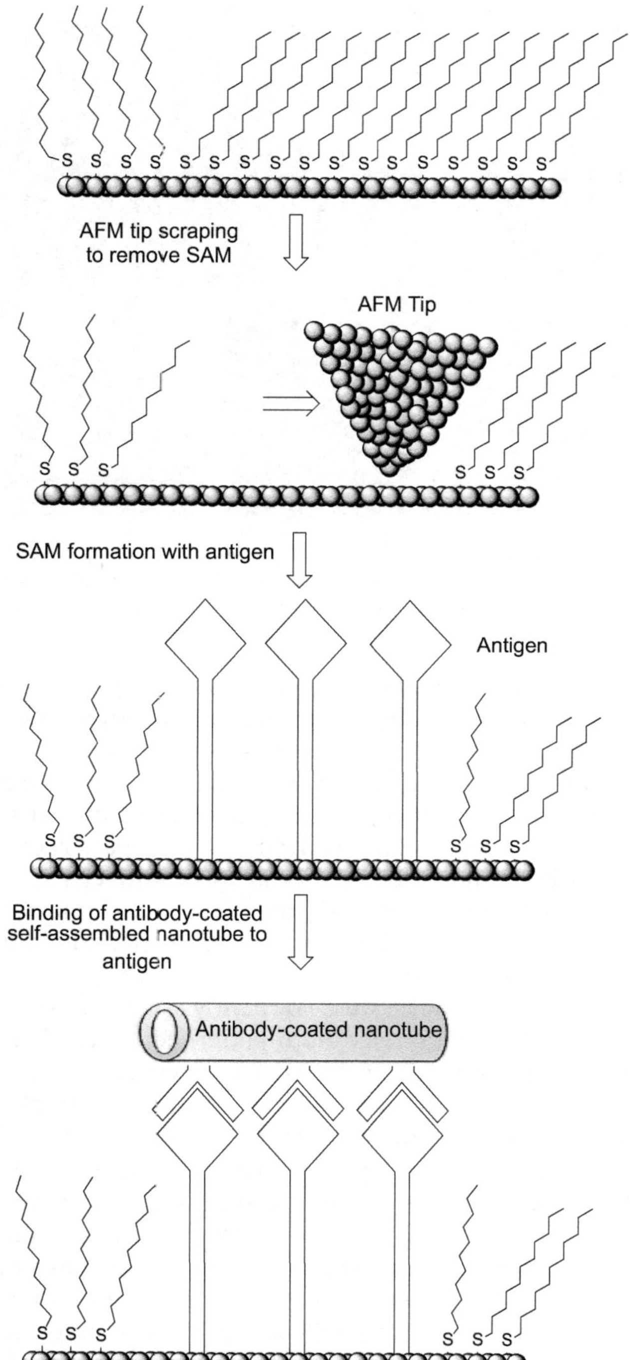

Figure 9.14 The preparation of arrays of nanotubes based on an antigen–antibody interaction after nanoshaving a SAM to generate a template area of antigen that locates the antibody-coated nanotube.

The lithographic techniques for the formation of patterned substrates are not limited to gold surfaces. The preparation of protein patterns through selective adsorption onto SAMs on aluminium oxide have been achieved through line formation with AFM shaving followed by bottom-up assembly.[81] SAMs of *n*-11-(4-azido-benzoylamino)-undecyl-phosphonic acid on the oxide were prepared by exposure of the oxide to an ethanol solution of the adsorbate. The formation of the SAM was confirmed by water contact angle measurements as well as by XPS, especially observing the position of the peak in the P2p region. This SAM was patterned (Figure 9.15) by removing the adsorbate by nanoshaving (400 nN to 1 μN load force) under ambient conditions. The chemical pattern was achieved by adsorbing aminobutylphosphonic acid into the exposed oxide. AFM was used to confirm the removal of the initial SAM and filling by the shorter phosphonic acid. The remaining azide groups in the patterned film were reacted photochemically (UV light) with an amine-terminated PEG (poly(ethylene glycol)) that prevents non-specific protein adsorption. In the final step, protein-coated and fluorescently-labelled polymer nanoparticles were immobilised on patterned SAMs. The selective immobilisation of the polymer nanoparticles thanks to the interaction of the protein coating with the amine functional groups in the trench of the patterned surface was confirmed by confocal microscopy, where the lines of fluorescent particles is evident. This work proves that oxide surfaces can be manipulated to produce nanopatterns similar to those more frequently seen on gold, with the potential advantages that this might bring, such as robustness and transparency.

The patterning of surfaces with responsive molecular systems has proven beneficial for observing the changes in film structure. Patterned layers of a bipyridine derivative chemisorbed onto gold (111) afforded a system that undergoes surface potential switching caused by complexation and release of palladium(II) chloride.[82] Kelvin probe force microscopy (KFM) showed a decrease in signal upon complexation in the monolayers. The signal returned to its original position after treatment with ethylenediamine solution, which complexes the palladium(II), as proven by XPS. The whole process is reversible, and picomolar- sensitive. A molecular reorientation in a patterned monolayer is also seen for rotaxanes on gold.[83] The effect is seen in amide-type rotaxanes that reorganise in the presence of chloride ions, as shown by NEXAFs measurements on the full monolayer. The monolayer has to be well ordered to display the phenomenon, indicating the importance of supramolecular effects between adsorbate molecules.

9.5 Spontaneous Pattern Formation and Directed Self-assembly

The template and lithography-free formation of ordered periodic structures from which patterning could be achieved is a truly appealing prospect.

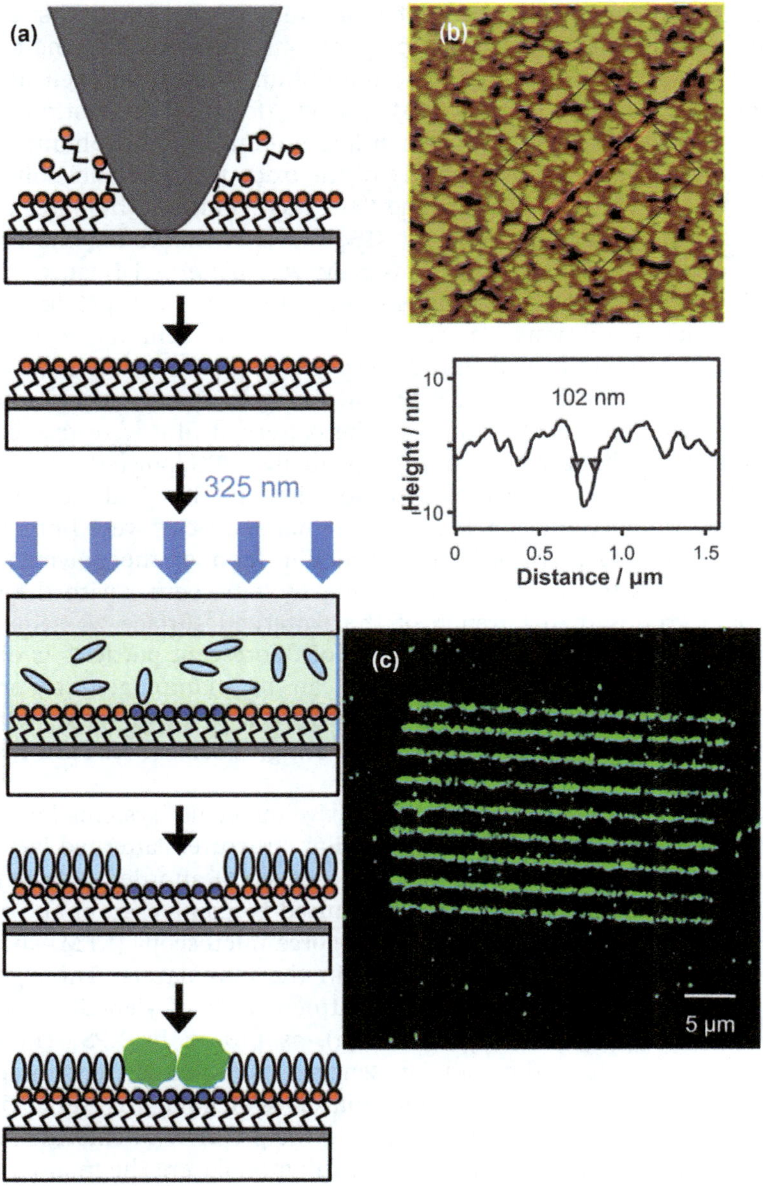

Figure 9.15 The multistep AFM lithography followed by the self-assembly approach to lines of proteins on aluminium oxide. (a) A diagram with cartoons to illustrate the fabrication sequence. Firstly, removal of part of the SAM of an azide (red dot) by nanoshaving allows adsorption of aminobutylphosphonic acid into the bare substrate. Next, PEG-NH_2 (light blue ellipse) is coupled to the surface photochemically and then NeutrAvidin particles (green blob) are adsorbed. (b) Topographical AFM image (3.5×3.5 µm) of a single line at the step before adsorption of NeutrAvidin particles. (c) A confocal fluorescence microscopy image (40×40 µm) of the patterned surface after adsorption of NeutrAvidin.
Reproduced from ref. 81 with permission from the Royal Society of Chemistry.

Convection can play a determining role in the self-assembly processes acting during evaporation of solutions. The self-assembly of zeolite nanoparticles under dynamic convection-assisted conditions leads to porous films with remarkable knotted-rope web and wrinkled honeycomb surface patterns.[84] These phenomena can involve supramolecular effects, and in particular the formation of ordered structures in soft matter is interesting, if not yet mature in terms of its use for studying interfaces. Directed self-assembly usually refers to the generation of periodic structures as a result of the self-organisation of material under some restriction, be it in a channel or a thin film.[85] The phase separation of block copolymers means that the spheres, cylinders and lamellae they form are particularly suited for this purpose.[86] The formation of nanostructures using confined routes for evaporation of solutions containing block copolymer systems is a promising route for the patterned growth of supramolecular systems. A beautiful recent example is the formation of square lattices of domains in a system where hierarchical self-assembly of hydrogen-bonding units was combined with controlled phase separation of diblock copolymers.[87] The polar regions in these structures could be of interest for the generation of patterns, using block copolymer lithography,[88] for exploring supramolecular chemistry at interfaces.

The nature of supramolecular interactions in a thin film can afford surprising effects of patterning upon a mechanical perturbation. The scanning of an AFM tip has been proven sufficient to re-organise thin films of [2]rotaxanes made through hydrogen bond interactions between thread and axle.[89,90] The films of the rotaxane were prepared by spin-coating graphite with an acetone solution of the molecule, followed by a brief thermal annealing (40 °C). These films were stable in air for several months. The repeated scanning of an AFM tip in contact with the surface produced a line of nanometre-scale hillocks, observed as dots in the AFM image recorded at a force below that required to induce the movement of material. Repeated line scans are necessary for the dots to reach their optimal height and spacing, as if the process is a ripening effect (scan rate does not affect the outcome). This phenomenon is seen in Figure 9.16 for sequential scans. The structures can be controlled above the threshold force of 2 nN to sizes down to 40 nm which have a periodicity of 100 nm for the thinnest spin cast films (approximately 3 nm thick). Increased thickness in the film yields a larger dot size. The material that moves to form the dots originates from the scanned lines exclusively, because it is only the molecules touched by the tip of the AFM that receive sufficient mechanical energy to reorganise. The effect was not observed for films of the component of the rotaxane individually or for different types of molecules (including sexithiophene and aluminium tris(8-hydroxyquinoline)). It is believed that the inter-component mobility within the rotaxane molecules in the films facilitates reorganisation that resembles a dewetting transition that could take place in elastic thin films. In the rotaxane films, it is the self-affinity that drives the effect, as it takes place regardless of the

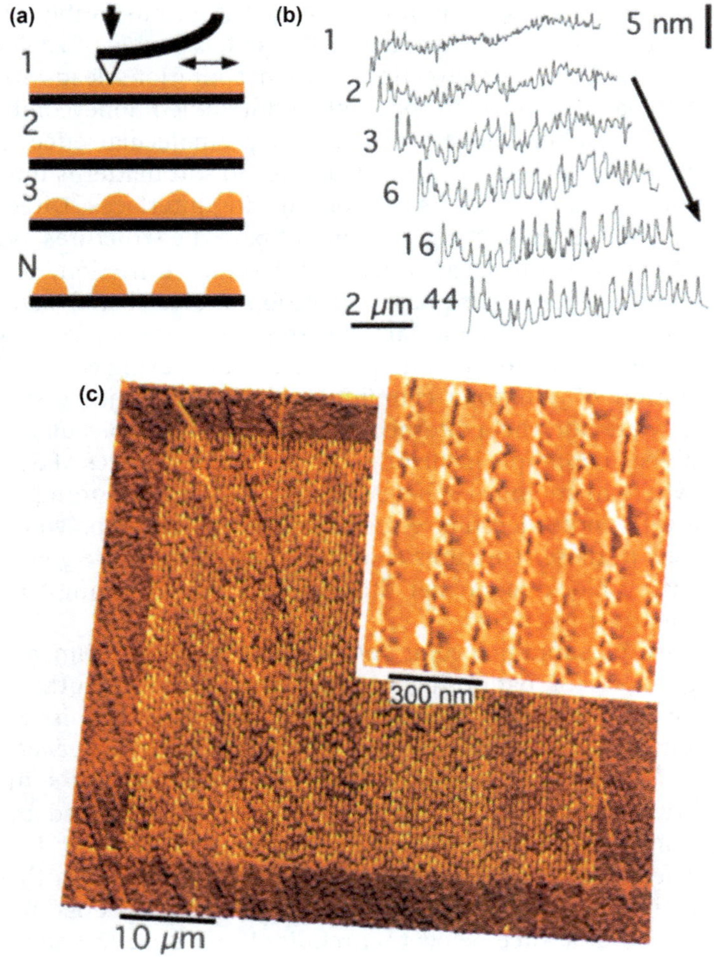

Figure 9.16 Mechanically-induced patterning of rotaxane films with an AFM tip. (a) The process that involves contact of the tip moving along the material, where 1, 2, 3 and N are the number of sweeps, N being the equilibrium state. (b) A succession of topography line scans measured at the threshold force. (c) A square array of individual lines comprising dots of material made with line scans of the AFM over several tens of microns, and in the inset is a close up showing seven lines of dot.
Reproduced with permission of the American Association for the Advancement of Science, ref. 90.

substrate used. The low energy barriers to reorganisation apparently arise from the mechanical bond reorientation.

The spontaneous formation of metal clusters on surfaces of different metals can be used as a basis for the patterned formation of SAMs.[91] When cobalt was deposited onto the (111) surface of a gold crystal, the cobalt

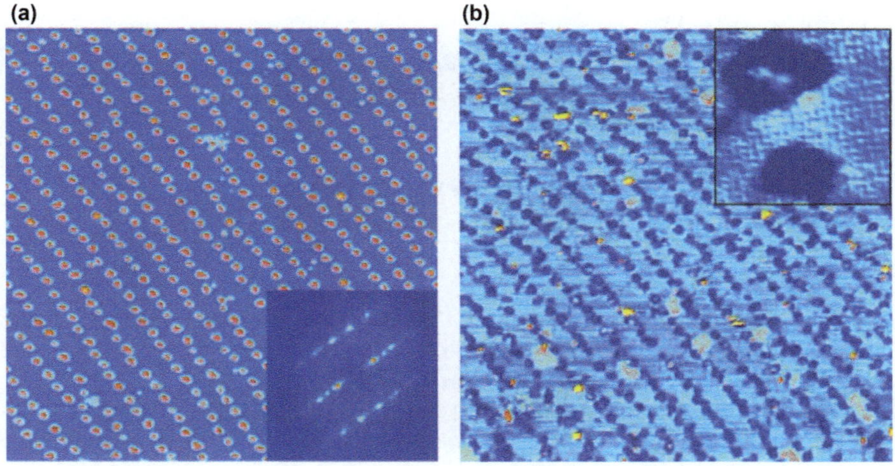

Figure 9.17 STM images (200×200 nm) of (a) cobalt clusters on gold (111) after deposition of 0.2 ML (the inset shows a Fourier transform of the image demonstrating the order in the pattern) and (b) the same surface after oxidation of the cobalt and dodecanethiol SAM formation. (The inset is a close up with molecular resolution, image size 10×10 nm.)
Reproduced with permission from AIP, ref. 91.

formed nanoclusters at the elbows on the surface reconstruction of the gold. These objects are about 3.4 nm in diameter and four Ångstroms in height, and they are spaced from one another by seven nanometres in a line and twelve nanometres between lines. Removal of this substrate from the UHV chamber resulted in formation of cobalt oxide. Dodecanethiol was then evaporated onto the resulting substrate giving a SAM with pits where the cobalt oxide resided, as witnessed clearly in the STM images of the sample (Figure 9.17).

With this patterned surface in hand, following the route shown in Figure 9.18, the metal oxide was removed with aqueous acid leaving bare gold at the bottom of the pits in the SAM of dodecanethiol. Into these spaces, electroactive molecules based on terthiophene and juglone (5-hydroxy-1,4-naphthalenedione) bearing a thiol-terminated chain[92] were introduced by self-assembly. The islands of electroactive components were obvious in the STM images of the monolayer, where they appeared as very bright patches. STS measurements confirmed efficient electron tunnelling through the compounds, and in addition revealed a rectifying effect for the juglone derivative. Cyclic voltammetry showed that for the juglone derivative patterned surface, the kinetic limitation of electron transfer was less than for a full monolayer of the compound, demonstrating an advantage of this nanostructuring in the electrical performance. This combined UHV–air–solution methodology is relatively rare in the preparation of patterned layers, but shows the benefits of combining the strengths of each.

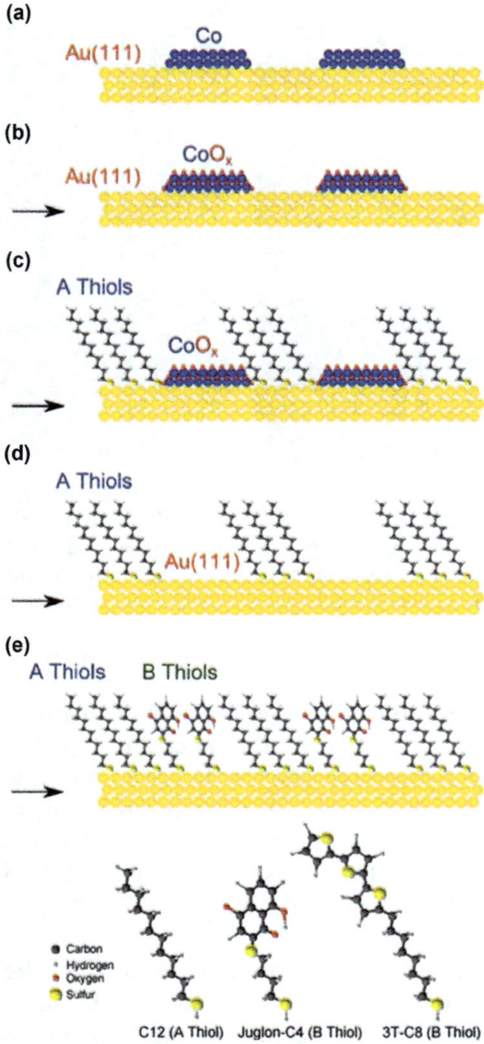

Figure 9.18 The route to a patterned SAM based on spontaneous separation of cobalt islands on gold, oxidation of the cobalt and covering with a dodecanethiol SAM to give a pitted surface. Removal of the oxide exposed gold again so that functional thiols could be introduced to give well-defined sizes of moieties, interesting for organic electronics. Reproduced with permission from AIP, ref. 91.

References

1. R. Smith, P. A. Lewis and P. S. Weiss, *Prog. Surf. Sci.*, 2004, **75**, 1.
2. B. D. Gates, Q. Xu, M. Stewart, D. Ryan, C. G. Willson and G. M. Whitesides, *Chem. Rev.*, 2005, **105**, 1171.
3. R. Garcia, A. W. Knoll and E. Riedo, *Nat. Nanotech.*, 2014, **9**, 577.

4. M. D. Stewart, K. Patterson, M. H. Somervell and C. G. Willson, *J. Phys. Org. Chem.*, 2000, **13**, 767.
5. H. Ito, *J. Polym. Sci., Part A: Polym. Chem.*, 2003, **41**, 3863.
6. C. G. Willson and B. C. Trinque, *J. Photopolym. Sci. Technol.*, 2003, **16**, 621.
7. D. B. Amabilino, in *Supramolecular Chemistry: From Molecules to Nanomaterials*, ed. J. W. Steed and P. A. Gale, Wiley-VCH, Weinheim, 2012, vol. 5, pp. 1967–1992, ISBN: 978-0-470-74640-0.
8. C. Becker and K. Wandelt, *Top. Curr. Chem.*, 2009, **287**, 45.
9. M. D. Yilmaz and J. Huskens, *Soft Matter*, 2012, **8**, 11768.
10. E. Menard, M. A. Meitl, Y. Sun, J.-U. Park, D. J.-L. Shir, Y.-S. Nam, S. Jeon and J. A. Rogers, *Chem. Rev.*, 2007, **107**, 1117.
11. C. S. Dulcey, J. H. Georger Jr., V. Krauthamer, D. A. Stenger, T. L. Fare and J. M. Calvert, *Science*, 1991, **252**, 551.
12. C. S. Dulcey, J. H. Georger Jr., M.-S. Chen, S. W. McElvany, C. E. O'Ferrall, V. I. Benezra and J. M. Calvert, *Langmuir*, 1996, **12**, 1638.
13. J. H. Georger Jr., D. A. Stenger, A. S. Rudolph, J. J. Hickman, C. S. Dulcey and T. L. Fare, *Thin Solid Films*, 1992, **210–211**, 716.
14. J. M. Calvert, J. H. Georger, J. M. Schnur, P. E. Schoen, M. C. Peckerar and P. E. Pehrsson, *Thin Solid Films*, 1992, **210–211**, 359.
15. N. Herzer, S. Hoeppener and U. S. Schubert, *Chem. Commun.*, 2010, **46**, 5634.
16. S. A. Claridge, W.-S. Liao, J. C. Thomas, Y. Zhao, H. H. Cao, S. Cheunkar, A. C. Serino, A. M. Andrews and P. S. Weiss, *Chem. Soc. Rev.*, 2013, **42**, 2725.
17. S. A. Ruiz and C. S. Chen, *Soft Matter*, 2007, **3**, 168.
18. H. Bayat, D. Tranchida, B. Song, W. Walczyk, E. Sperotto and H. Schonherr, *Langmuir*, 2011, **27**, 1353.
19. E. Delamarche, H. Schmid, A. Bietsch, N. B. Larsen, H. Rothuizen, B. Michel and H. Biebuyck, *J. Phys. Chem. B*, 1998, **102**, 3324.
20. G. Gannon, J. A. Larsson, J. C. Greer and D. Thompson, *Langmuir*, 2009, **25**, 242.
21. J. J. Schwartz, J. N. Hohman, E. I. Morin and P. S. Weiss, *ACS Appl. Mater. Interfaces*, 2013, **5**, 10310.
22. T. Kaufmann and B. J. Ravoo, *Polym. Chem.*, 2010, **1**, 371.
23. C. Wendeln and B. J. Ravoo, *Langmuir*, 2012, **28**, 5527.
24. T. E. Balmer, H. Schmid, R. Stutz, E. Delamarche, B. Michel, N. D. Spencer and H. Wolf, *Langmuir*, 2005, **21**, 622.
25. A. A. Dameron, J. R. Hampton, R. K. Smith, T. J. Mullen, S. D. Gillmor and P. S. Weiss, *Nano Lett.*, 2005, **5**, 1834.
26. D. Gentili, G. Foschi, F. Valle, M. Cavallini and F. Biscarini, *Chem. Soc. Rev.*, 2012, **41**, 4430.
27. D. Gentili, F. Valle, C. Albonetti, F. Liscio and M. Cavallini, *Acc. Chem. Res.*, 2014, **47**, 2692.
28. C. P. Martin, M. O. Blunt, E. Pauliac-Vaujour, A. Stannard, P. Moriarty, I. Vancea and U. Thiele, *Phys. Rev. Lett.*, 2007, **99**, 116103.

29. G.-Y. Liu, S. Xu and Y. Qian, *Acc. Chem. Res.*, 2000, **33**, 457.
30. S. Krämer, R. R. Fuierer and C. B. Gorman, *Chem. Rev.*, 2003, **103**, 4367.
31. R. Garcia, R. V. Martinez and J. Martinez, *Chem. Soc. Rev.*, 2006, **35**, 29.
32. B. Pignataro, *J. Mater. Chem.*, 2009, **19**, 3338.
33. U. Dürig, G. Cross, M. Despont, U. Drechsler, W. Häberle, M. I. Lutwyche, H. Rothuizen, R. Stutz, R. Widmer, P. Vettiger, G. K. Binnig, W. P. King and K. E. Goodson, *Tribol. Lett.*, 2000, **9**, 25.
34. C. B. Ross, L. Sun and R. M. Crooks, *Langmuir*, 1993, **9**, 632.
35. S. Xu and G.-y. Liu, *Langmuir*, 1997, **13**, 127.
36. J. Dagata, J. Schneir, H. H. Harary, C. J. Evans, M. T. Postek and J. Bennett, *Appl. Phys. Lett.*, 1990, **56**, 2001.
37. H. C. Day and D. R. Allee, *Appl. Phys. Lett.*, 1993, **62**, 2691.
38. A. Stannard, H. Alhummiany, E. Pauliac-Vaujour, J. S. Sharp, P. Moriarty and U. Thiele, *Langmuir*, 2010, **26**, 13892.
39. R. D. Piner, J. Zhu, F. Xu, S. H. Hong and C. A. Mirkin, *Science*, 1999, **283**, 661.
40. D. S. Ginger, H. Zhang and C. A. Mirkin, *Angew. Chem., Int. Ed*, 2004, **43**, 30.
41. C.-C. Wu, D. N. Reinhoudt, C. Otto, V. Subramanian and A. H. Velders, *Small*, 2011, **7**, 989.
42. M. J. W. Ludden, D. N. Reinhoudt and J. Huskens, *Chem. Soc. Rev.*, 2006, **35**, 1122.
43. L. Yang, A. Gomez-Casado, J. F. Young, H. D. Nguyen, J. Cabanas-Danés, J. Huskens, L. Brunsveld and P. Jonkheijm, *J. Am. Chem. Soc.*, 2012, **134**, 19199.
44. S. Onclin, J. Huskens, B. J. Ravoo and D. N. Reinhoudt, *Small*, 2005, **1**, 852.
45. T. Auletta, B. Dordi, A. Mulder, A. Sartori, S. Onclin, C. M. Bruinink, M. Péter, C. A. Nijhuis, H. Beijleveld, H. Schönherr, G. J. Vancso, A. Casnati, R. Ungaro, B. J. Ravoo, J. Huskens and D. N. Reinhoudt, *Angew. Chem., Int. Ed*, 2004, **43**, 369.
46. J. Voskuhl, S. Sankaran and P. Jonkheijm, *Chem. Commun.*, 2014, **50**, 15144.
47. F. A. Scaramuzzo, A. González-Campo, C.-C. Wu, A. H. Velders, V. Subramaniam, G. Doddi, P. Mencarelli, M. Barteri, P. Jonkheijm and J. Huskens, *Chem. Commun.*, 2010, **46**, 4193.
48. N. Balachander and T. Bein, *Langmuir*, 1990, **6**, 1621.
49. J. M. Spruell, B. A. Sheriff, D. I. Rozkiewicz, W. R. Dichtel, R. D. Rohde, D. N. Reinhoudt, J. F. Stoddart and J. R. Heath, *Angew. Chem., Int. Ed*, 2008, **47**, 9927.
50. R. Kassies, K. O. Van der Werf, A. Lenferink, C. N. Hunter, J. D. Olsen, V. Subramaniam and C. Otto, *J. Microsc.*, 2005, **217**, 109.
51. A. Franconetti, L. Contreras-Bernal, S. Jatunov, M. Gómez-Guillén, M. Angulo, R. Prado-Gotor and F. Cabrera-Escribano, *Phys. Chem. Chem. Phys.*, 2014, **16**, 18442.
52. A. González-Campo, S.-H. Hsu, L. Puig, J. Huskens, D. N. Reinhoudt and A. H. Velders, *J. Am. Chem. Soc.*, 2010, **132**, 11434.

53. C. M. Bruinink, C. A. Nijhuis, M. Péter, B. Dordi, O. Crespo-Biel, T. Auletta, A. Mulder, H. Schçnherr, G. J. Vancso, J. Huskens and D. N. Reinhoudt, *Chem. – Eur. J.*, 2005, **11**, 3988.
54. F. Tian, M. Cziferszky, D. Jiao, K. Wahlström, J. Geng and O. A. Scherman, *Langmuir*, 2011, **27**, 1387.
55. A. González-Campo, M. Brasch, D. A. Uhlenheuer, A. Gómez-Casado, L. Yang, L. Brunsveld, J. Huskens and P. Jonkheijm, *Langmuir*, 2012, **28**, 16364.
56. H. Yang, B. Yuan, X. Zhang and O. A. Sherman, *Acc. Chem. Res.*, 2014, **47**, 2106.
57. F. Tian, N. Cheng, N. Nouvel, J. Geng and O. A. Scherman, *Langmuir*, 2010, **26**, 5323.
58. C. Hu, Y. Lan, F. Tian, K. R. West and O. A. Sherman, *Langmuir*, 2014, **30**, 10926.
59. A. Bernard, D. Fitzli, P. Sonderegger, E. Delamarche, B. Michel, H. R. Bosshard and H. Biebuyck, *Nat. Biotechnol.*, 2001, **19**, 866.
60. J. P. Renault, A. Bernard, D. Juncker, B. Michel, H. R. Bosshard and E. Delamarche, *Angew. Chem., Int. Ed.*, 2002, **41**, 2320.
61. V. B. Sadhu, A. Perl, X. Duan, D. N. Reinhoudt and J. Huskens, *Soft Matter*, 2009, **5**, 1198.
62. X. Y. Ling, I. Y. Phan, D. N. Reinhoudt, G. J. Vancso and J. Huskens, *Int. J. Mol. Sci.*, 2008, **9**, 486.
63. A. Bétard and R. A. Fischer, *Chem. Rev.*, 2012, **112**, 1055.
64. P. Falcaro and M. J. Styles, *Encyclopedia of Inorganic and Bioinorganic Chemistry*, 2011–2014 John Wiley & Sons, Ltd, DOI: 10.1002/9781119951438.eibc2227.
65. C. Munuera, O. Shekhah, H. Wang, C. Wöll and C. Ocal, *Phys. Chem. Chem. Phys.*, 2008, **10**, 7257.
66. C. Munuera, E. Barrena and C. Ocal, *Langmuir*, 2005, **21**, 8270.
67. T. Ladnorg, A. Welle, S. Heißler, C. Wöll and H. Gliemann, *Beilstein J. Nanotechnol.*, 2013, **4**, 638.
68. S. Hermes, F. Schröder, R. Chelmowski, Christof Wöll and R. A. Fischer, *J. Am. Chem. Soc.*, 2005, **127**, 13744.
69. J. J. Gassensmith, P. M. Erne, W. F. Paxton, C. Valente and J. F. Stoddart, *Langmuir*, 2011, **27**, 1341.
70. J. Aizenberg, A. J. Black and G. M. Whitesides, *Nature*, 1999, **398**, 495.
71. A. Singh, I. S. Lee, K. Kim and A. S. Myerson, *CrystEngComm*, 2011, **13**, 24.
72. A. Y. Lee, I. S. Lee, S. S. Dettet, J. Boerner and A. S. Myerson, *J. Am. Chem. Soc.*, 2005, **127**, 14982.
73. K. Kim, I. S. Lee, A. Centrone, T. A. Hatton and A. S. Myerson, *J. Am. Chem. Soc.*, 2009, **131**, 18212.
74. M. Lingenfelder, A. Bejarano-Villafuerte, M. van der Meijden, R. M. Kellogg and D. B. Amabilino, *Chem. – Eur. J.*, 2014, **20**, 10466.
75. A. Bejarano-Villafuerte, M. van der Meijden, M. Lingenfelder, K. Wurst, R. M. Kellogg and D. B. Amabilino, *Chem. – Eur. J.*, 2012, **18**, 15984.

76. N. Nuraje, I. A. Banerjee, R. I. MacCuspie, L. Yu and H. Matsui, *J. Am. Chem. Soc.*, 2004, **126**, 8088.
77. M. Kogiso, S. Ohnishi, K. Yase, M. Masuda and T. Shimizu, *Langmuir*, 1998, **14**, 4978.
78. H. Matsui and B. Gologan, *J. Phys. Chem. B*, 2000, **104**, 3383.
79. L. Yu, I. A. Banerjee and H. Matsui, *J. Am. Chem. Soc.*, 2003, **125**, 14837.
80. P. M. Mendes, C. L. Yeung and J. A. Preece, *Nanoscale Res. Lett.*, 2007, **2**, 373.
81. O. El Zubir, I. Barlow, G. J. Leggett and N. H. Williams, *Nanoscale*, 2013, **5**, 11125.
82. T. Nakamura, E. Koyama, Y. Shimoi, S. Abe, T. Ishida, K. Tsukagoshi, W. Mizutani, H. Tokuhisa, M. Kanesato, I. Nakai, H. Kondoh and T. Ohta, *J. Phys. Chem. B*, 2006, **110**, 9195.
83. T. Heinrich, C. H.-H. Traulsen, M. Holzweber, S. Richter, V. Kunz, S. K. Kastner, S. O. Krabbenborg, J. Huskens, W. E. S. Unger and C. A. Schalley, *J. Am. Chem. Soc.*, 2015, **137**, 4382.
84. H. Wang, Z. Wang, L. Huang, A. Mitra and Y. Yan, *Langmuir*, 2001, **17**, 2572.
85. R. A. Griffiths, A. Williams, C. Oakland, J. Roberts, A. Vijayaraghavan and T. Thomson, *J. Phys. D: Appl. Phys.*, 2013, **46**, 503001.
86. H. Hu, M. Gopinadhan and C. O. Osuji, *Soft Matter*, 2014, **10**, 3867.
87. C. Tang, E. M. Lennon, G. H. Fredrickson, E. J. Kramer and C. J. Hawker, *Science*, 2008, **322**, 429.
88. T. P. Russell and C. J. Hawker, *MRS Bull.*, 2005, **30**, 952.
89. M. Cavallini, F. Biscarini, S. Leon, F. Zerbetto, G. Bottari and D. A. Leigh, *Science*, 2003, **299**, 531.
90. F. Biscarini, M. Cavallini, R. Kshirsagar, G. Bottari, D. A. Leigh, S. Leon and F. Zerbetto, *Proc. Natl. Acad. Sci., U. S. A.*, 2006, **103**, 17650.
91. R. Breitwieser, M. Marsault, V. Repain, J. Lagoute, C. Chacon, Y. Girard, S. Rousset, Z. Qin, N. Battaglini, S. Zrig and P. Lang, *J. Chem. Phys.*, 2013, **139**, 204703.
92. G. March, S. Reisberg, B. Piro, M.-C. Pham, M. Delamar, V. Noel, K. Odenthal, D. B. Hibbert and J. J. Gooding, *J. Electroanal. Chem.*, 2008, **622**, 37.

CHAPTER 10

Supramolecular Surface Systems: Which and Whither?

10.1 Introduction

In air or vacuum, on hard immobile or soft and dynamic surfaces, at the interfaces with liquid or gas, huge advancement has been made in the ability of scientists to explore molecules and their assemblies in layers. It was stated in the first chapter of this book, and demonstrated in the subsequent ones, that the importance of analytical methods in the understanding of surface structure and properties cannot be underestimated. It is a far cry from early observations of adsorption of polar organic molecules to polished metal surfaces,[1] where basically the hydro- or oleo-phobic character was the main observable feature. Now with this barrage of techniques, one can understand the adsorption of molecules to surfaces and probe their growth. Even so, there are still very significant challenges associated with the sometimes hard quantification of surface groups and solutions are still needed to this problem.[2] Also, accurate control of the preparation of surfaces and the precise growth of films on them can be complex and is only made easy through experience. The adsorption of molecules on surfaces is still not a precise science. However, a high degree of control in thin film growth is now in hand, and supramolecular chemistry has a huge part to play. Beyond pure assembly, the beautiful examples of surface supramolecular chemistry presented in this monograph are a solid foundation for the realisation of a number of potential, and a few actual, applications, some of which are presented here.

The hierarchical self-assembly of small molecules is difficult to design in an interface environment (Figure 10.1); there are so many factors to be taken into account—such as the balance of interactions and the effects of mass transport, and so on—that means that there is huge knowledge still to be

Monographs in Supramolecular Chemistry No. 19
Supramolecular Chemistry at Surfaces
By David B. Amabilino

Published by the Royal Society of Chemistry, www.rsc.org

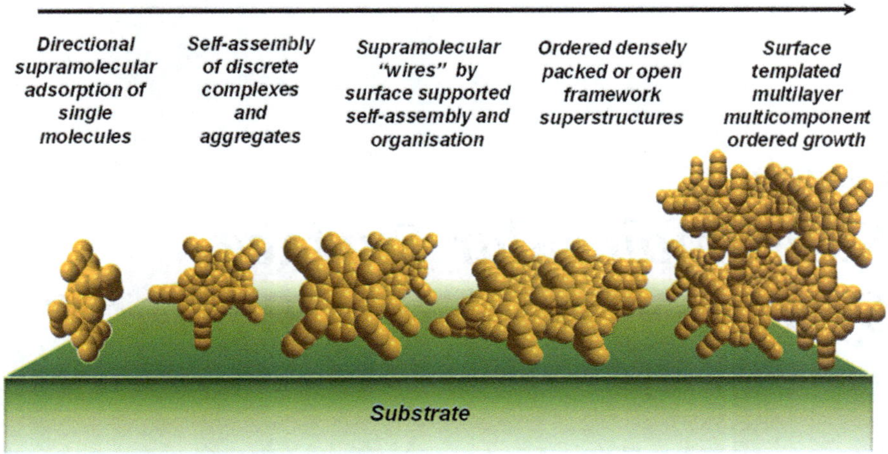

Figure 10.1 A representation of the hierarchy of growth of molecular adsorbates on a substrate through non-covalent interactions.

garnered by performing studies on supramolecular interactions at interfaces. It is presently possible to explain *a posteriori* rearrangements in structure on going from single molecules up to small specific aggregates provided strong interactions are present, but the often weak interactions taking place that determine the formation of densely-packed structures or open frameworks are hard to predict. And in the formation of thin films, whether "standing-up" or in-plane structures will be attained, especially in heterostructures, the full understanding of interface influence on packing is an emerging area where huge advancement can be made. More information and theoretical insight is required before a full understanding of hierarchical growth at interfaces can be attained. In the sections that follow, within this overarching view, challenges and opportunities in different areas are presented.

10.2 Supramolecular Surface Science

Starting from the bottom and looking up, the formation of supramolecular clusters at interfaces is frequent, but its observation is difficult. At the surface of water, the generation of small assemblies is perhaps possible by fluorescence microscopy, but otherwise there are no truly direct methods of observation. The movement of the barrier in a Langmuir trough would eventually coalesce any small self-assembled aggregates before any increment in surface pressure could be observed. When physisorbed systems at solid interfaces are kinetically stable, they can be monitored by scanning probe microscopy, taking advantage of low temperature control. Naturally, cooling during the measurement will increase the kinetic stability under the conditions of the experiment. If the adsorpt chemisorbs to the substrate then at submonolayer coverage, any specific aggregation will be visible using an appropriate technique. There are certain groups of compounds that one might expect to form ordered structures at an interface, through hydrogen bonds or metal coordination,[3] but looking back, why rubrene[4] and why nitronaphthalene[5]? Could we design a compound to do these remarkable hierarchical assembly routes?

In solution, at the liquid–solid interface, it is very rare to observe aggregates of molecules. Their imaging requires very favourable absorption energy, without the formation of complete layers. Large clusters can be observed with relative ease using scanning probe microscopes on cast solutions where the solvent has evaporated from the system and the aggregate is left essentially immobile on the dry surface, but is obtained in a far from equilibrium situation. Therefore, it is interesting to find cases where different environments can be compared. The assembly of clusters of perfluoroalkane–alkane block molecules that takes place at the interface of water can also be observed on other surfaces.[6] Single molecule thick aggregates of eicosylperfluorotetradecane ($F(CF_2)_{14}(CH_2)_2H$) were prepared principally on mica either from direct casting of a solvent solution or by first depositing on water in a Langmuir trough and then performing an LB transfer onto the substrate. The solvents used for deposition were selected carefully, so that they would be good either for the hydrocarbon block (decalin, xylene, chloroform), the fluorinated block (perfluorodecalin, perfluorohexane), or both (hexafluoroxylene). Upon casting from hexafluoroxylene, AFM (Figure 10.2) revealed a hexagonal superstructure of spirals with about 80 nm diameter (with no preference for left- or right-handed structure), with a ribbon width of approximately 35 nm and height

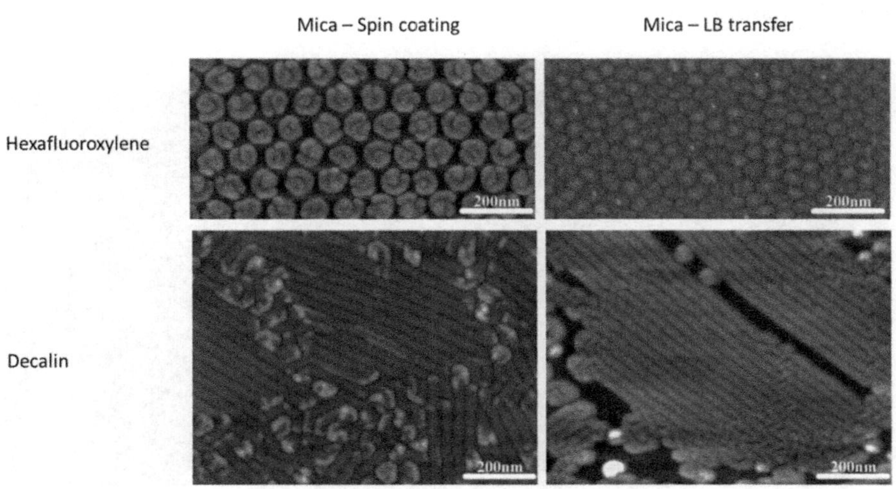

Figure 10.2 AFM images of a spin-coated and LB monolayer film of eicosylperfluorotetradecane on mica when deposited from either hexafluoroxylene or decalin.
Reprinted with permission from ref. 6. Copyright 2005 American Chemical Society.

3 nm, in line with the molecule's length taking into account a slightly bent shape (an angle of 122° between the straight hydrocarbon and fluorocarbon segments explains this non-linear conformation). When the film was made by the LB technique, initially after transfer, a denser array of hexagonal structures was observed that were about three quarters the size of the objects from spin coating. Over time, these structures do relax, however, as a result of the metastable nature of the LB film that has compressed during deposition. So, in this case and from both solvents, spin coating or LB deposition both lead to similar supramolecular structure. Deposition from decalin gives a completely different morphology, with co-parallel ribbon-like objects formed, again under both conditions, although the domains appear somewhat longer and more well defined using the LB method, although the coverage of the surface is more homogeneous using spin casting. Some spiral structures co-exist with these ribbons using both techniques. When fluorinated solvents that dissolve the fluorocarbon segment were used for deposition, ribbon-like structures were also observed. Therefore, selective solvation tends to provide ribbons. Overall, the results demonstrate that for this case, the somewhat simpler spin-coating technique is as effective as the LB method, while being considerably simpler.

Solvent treatment of the films of fluorocarbon–alkane block molecules deposited by spin coating or LB transfer leads to reorganisation. The decalin-deposited ribbons are transformed into toroids by exposure to hexafluoroxylene vapour, and hexafluoroxylene-deposited toroids are partially converted into ribbons upon exposure to perfluorodecalin vapour (Figure 10.3). It is believed that co-adsorption of the solvent vapours and facilitated mass transport by this adsorption leads to this morphological transformation.

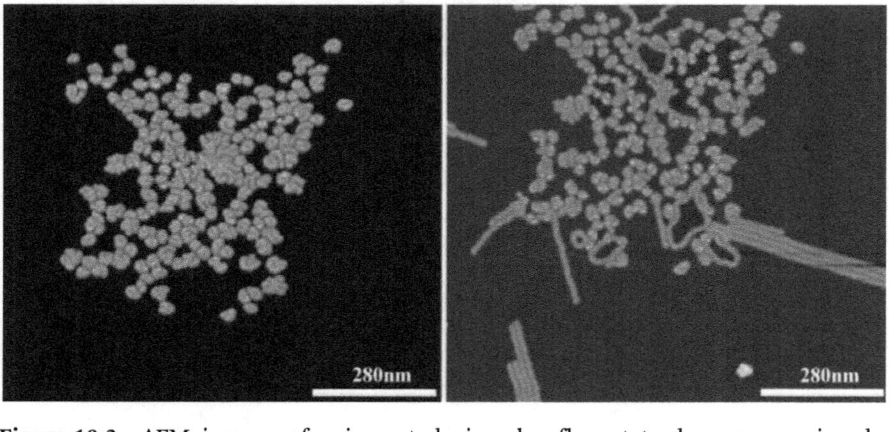

Figure 10.3 AFM images of spin-coated eicosylperfluorotetradecane on mica deposited from hexafluoroxylene (left) and the same sample after treatment with solvent vapour of perfluorodecalin.
Reprinted with permission from ref. 6. Copyright 2005 American Chemical Society.

These kinds of perfluoroalkyl–alkanes can also be deposited onto solids from supercritical carbon dioxide.[7] In this solvent, the viscosity of the depositing medium is very low and the formation of a meniscus during drying can be avoided. In this case, the toroidal shape is again formed on both mica and HOPG for the block molecule $F(CF_2)_{14}(CH_2)_{18}H$, but in the case of $F(CF_2)_{12}(CH_2)_{19}H$, toroids were formed on mica while co-existing toroid and stripe structures were observed on HOPG (Figure 10.4). The toroids had a uniform diameter of approximately 50 nm, and close packing of the phase-separated blocks was proposed to explain this effect. The ribbon-like structures on HOPG, which were only observed under very specific deposition

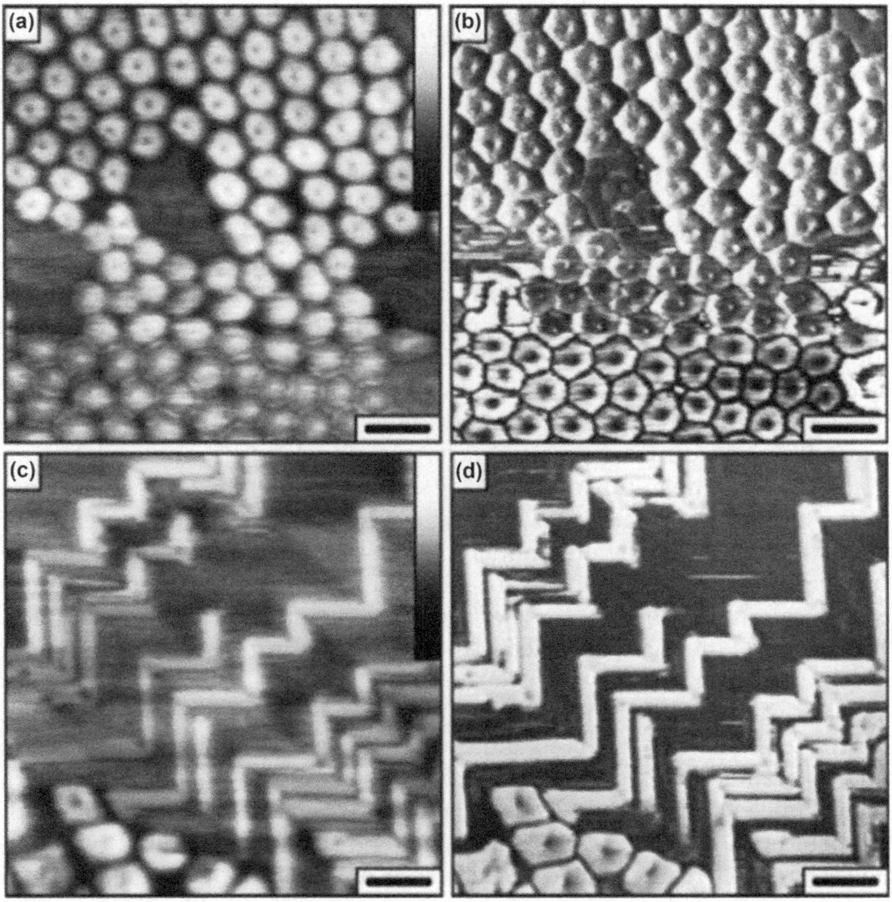

Figure 10.4 AFM images of ($F(CF_2)_{12}(CH_2)_{19}H$) on mica deposited from supercritical CO_2 onto mica ((a) and (b)) and HOPG ((c) and (d)). The scale bar is 75 nm in all cases and topography images are shown in (a) and (c) (height scale 10 nm) and (b) and (d) are phase signals.
Reproduced from ref. 7 with permission from the Royal Society of Chemistry and the PCCP owners' societies.

conditions, were presumably formed because of the tendency of the alkane section to align with the graphite symmetry axes. The use of carbon dioxide as a deposition solvent for these compounds, which are solubilised well, is an attractive route that could be of wider interest, and in fact has already been exploited for the deposition of fluorocarbon polymers on surfaces.[8,9]

After these preliminary considerations, it might be apparent that the use of UHV STM provides both the spatial and temporal resolution to be able to record precisely the shape and dynamics of surface-adhered supramolecular clusters. The constraint of the technique is that because STM measures tunnelling current across a junction, it is restricted to metallic surfaces or metals with a very thin layer of semiconducting or insulating material deposited on top. Often the systems are limited by the sublimation temperature of the adsorpt, so that even under the ultrahigh vacuum used, the sublimation temperature is not reached before decomposition. The use of electrospray inlets—whereby solutions of the adsorpts are introduced into the vacuum chamber and the carrier solvent evaporated during the process—can overcome these restrictions.[10]

The growth of organic materials on inorganic surfaces can be influenced greatly by the temperature at which the deposition is carried out, and domain sizes and multilayer formation can be influenced because of the rates of surface diffusion and growth. However, the initial stages of crystal growth are relatively poorly understood at the molecular level. In general, this is because of the enigmatic nature of nucleation, where a critical nucleus must form before ordered islands can be formed. The critical island size of all 4-sexiphenyl grown on native silicon oxide by OMBE on silicon under UHV was analysed by AFM and by comparing the size and distribution of the islands formed with theoretical models.[11,12] Islands of the molecule grown both at room temperature and at 373 K showed a height of 2.6 nm, very close to the theoretical length of 2.7 nm, and had a corresponding coverage of approximately 12%. Therefore, the molecules are quasi-perpendicular to the surface showing the dominance of molecule–molecule interactions in their condensation and the weakness of their interaction with the oxide surface. The shapes of the islands formed at the two temperatures are quite different; at room temperature, they were disk-like with a limited number of branches (and occasionally with small bilayer areas) while at the higher temperature, far greater branching was seen and the objects were much larger, because of the more efficient diffusion. Analysis of the condensed islands indicated a critical nucleus size of only two molecules. This value seems to apply to both temperatures, although the island shape is different and caution was emphasised. The irregular shape of the islands does not imply crystallinity, and this value is not likely to apply to the formation of crystalline islands. The very initial stages of growth—comprising only a few molecules—could not be observed under these conditions and it was pointed out that both experimental and molecular modelling approaches are likely to give greater insight into the growth of this kind of molecular island on weakly interacting substrates.

In the formation of monolayers by chemisorption on copper, the growth of ordered domains can apparently be arrested by molecules that occupy orientations that do not allow the periodic structures to grow.[13] The chiral molecule shown in Figure 10.5 when deposited on a copper (110) surface adheres through its four carboxyl oxygen atoms (the acidic hydrogen atoms are lost from the molecule and desorb during adsorption), each group attached to a different row. The bent shape of the compound ensures that a C_2 arrangement, which might otherwise be preferred in a more linear molecular structure, is not favoured, and a model of the adsorption (Figure 10.5) shows that an asymmetric orientation results. This feature results in a clearly distinct contrast for the two enantiomers by STM, something not so common for adsorbates on metals. The monolayers of either of the two enantiomers show well-defined homochiral domains over several

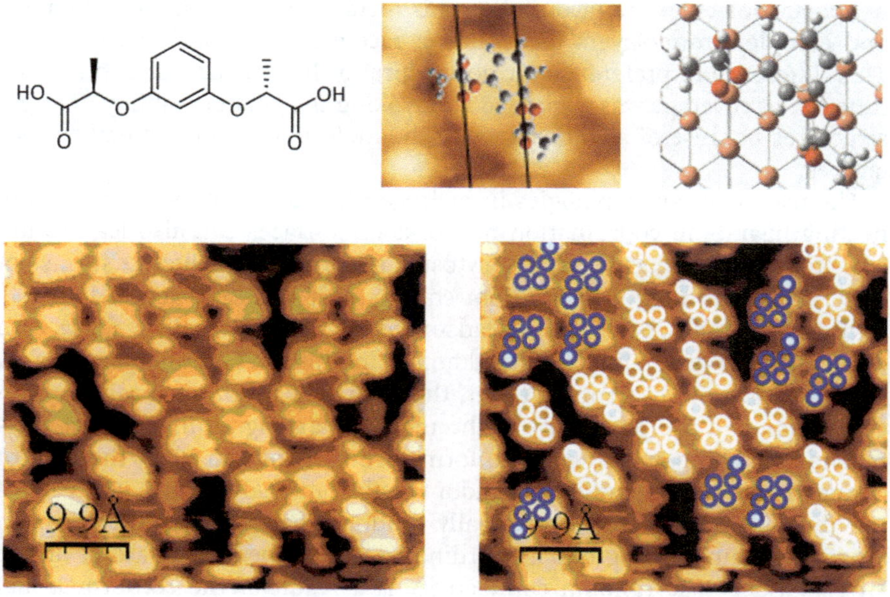

Figure 10.5 A chiral molecule that when adsorbed into a copper (110) surface under UHV conditions gives a clearly asymmetric contrast in the STM images of the compound: the image in the middle at the top shows this feature—that appears as five protrusions—with a model of the molecule and the directions of the copper rows of the surface overlaid. Next to it is a simulation of the chemisorbed carboxylate showing the oxygen atoms attached to neighbouring copper atoms on a row. Note that the identical chirality of the stereogenic centres combined with the "bent core" structure disfavours a C_2 symmetry in the adsorbed molecule. The bottom row shows two identical STM images of the racemic mixture of this molecule adsorbed on the same surface. The image with the overlaid objects comprising five circles indicates the positions of the enantiomers, and shows how mirror image molecules effectively block monolayer domain progression.

tens of nanometres. The chirality of these domains is determined by the stereochemistry of the adsorbate. In contrast, deposition of equal amounts of (*R*,*R*) and (*S*,*S*) molecules leads to a monolayer without long-range order. Very small domains are observed. Close inspection of these layers with STM shows that areas of (*R*,*R*) or (*S*,*S*) molecules are formed, but the domains are very small and are restricted in their growth by the opposite enantiomer. The restricted mass transport in the system—caused by both the chemisorption and the high surface coverage—surely contribute to the lack of large domains, but the apparent inhibition of homochiral domain growth by opposite enantiomers is a recurring feature in the monolayer, as seen in the STM image in Figure 10.5. A similar effect was also observed in monolayers of the racemic modification of an amphiphilic acid physisorbed at a solution–graphite interface, where again small ordered domains were observed in a monolayer largely devoid of periodicity.[14] The growth of heptahelicene on silver's (100) surface shows that while small homochiral clusters of apolar molecules are favoured on this metal crystal, when large domains and complete monolayers form then a racemic compound is favoured.[15] This difference in preference for homochiral or heterochiral structures as a function of coverage was explained by invoking the greater stability of larger structures because of a coincidence of the molecular and the metal surface lattices.

The use of certain topologies in hydrogen bond-forming systems or connecting ligands in coordination networks on surfaces can also lead to unexpected outcomes, as shown for cytosine adsorbed onto the 111 surface of gold[16] or the formation of random strings on a surface when [1,1′,40,1″]-terphenyl-3,3″-dicarbonitrile was adsorbed in UHV onto silver (111) with cobalt as the coordinating atom and node;[17] both form random systems akin to glasses. Devoid of periodic order, the promiscuity of the cytosine hydrogen bonds and the possibility of the molecule arranging in different ways means that complex patterns are formed, but thanks to STM, the type of connectivity at any point in the random chain can be discerned. The case of the dinitrile compound gives equally random strings where the nitrogen atom of the functional group is coordinated to adsorbed cobalt atoms on the silver surface. The random connectivity is a result of the conformational freedom of the ligand, which can exist as a pair of surface enantiomers—because of the prochiral nature of the molecule when the nitriles are located on opposite "sides" of the molecule—as well as a form with a mirror plane through the centre ring. A particularly common motif is the one where a metallocycle incorporating two of the molecules with the mirror plane through the centre ring coordinates two cobalt atoms, with the connectivity beyond this cycle being arbitrary. Cobalt atoms were observed with either three or four nitrile groups coordinated to them. The random incorporation of these units into the coordination network results in the glassy-state-like structures that were observed by STM. These systems provide models for the formation of glassy systems in bulk materials, except here the structures can be monitored at the sub-molecular level.

The organisation of materials in different interfaces can be completely different. Complete monolayers can be formed at virtually every interface imaginable provided the supramolecular interactions between the adsorpts and either continuous phase and themselves favour that. A great deal could be understood about the supramolecular chemistry of interfaces if identical molecules could be tracked across different interfaces. A direct comparison of the assembly of a chiral amphiphile in a Langmuir layer and the SAM physisorbed at the interface between HOPG and 1-phenyloctane showed the propensity for spontaneous resolution of enantiomers,[18] and the structures were compared with the solid state conformation (Figure 10.6). In this instance, the packing on graphite is very similar to the solid state, while in Langmuir films, a racemic compound is favoured when the two enantiomers are deposited. The bent structure in the head group of this compound proved important for its spontaneous resolution, which does not help in the arrangement—perpendicular to the interface—that it has in the Langmuir layer, where other symmetry options are available to favour chiral arrangements,[19] and where chirality can even be amplified.[20]

The attainment of highly ordered and periodic mono- and multi-layered structures at interfaces is a significant challenge, given the generally small sizes of the domains that are formed, but understanding of boundaries in-between these domains and the control of deposition can alleviate some of the issues and even lead to extremely ordered structures. In addition, reorganisation of layers after deposition is a possible challenge to be controlled. In LB layers, stable ordering of the salt of cadmium(II) and arachidic acid [$CH_3(CH_2)_{18}COOH$] has been achieved by stabilising a certain phase of the material using the subphase pH to favour a stable dense phase that obviates rearrangement.[21] The fatty acids in general can exhibit polymorphism both at the interface of water and air[22] as well as in LB layers.[23,24] For these compounds and their salts, changes in supramolecular organisation involve reorganisation of the as-deposited phase to a thermodynamically more stable structure. At pH 7, a 1 : 2 complex between cadmium and the acid is formed, giving charge neutralisation, that leads to short-range hexagonal order in the monolayer and the LB film. In the transferred films with three or more layers, however, there is a structural rearrangement to give long-range crystalline order with herringbone packing.[25,26] This thermodynamically more favourable phase comprises centrosymmetric packing with one carboxylate on either side of a plane formed by the cadmium ions that is parallel to the substrate.[27] The concomitant reduction in unit cell volume causes many defects in the layer. However, at pH greater than 8.5, a monolayer with a relatively dense pseudo-herringbone orthorhombic structure is formed that in LB films cannot rearrange to the other structures even when exposed to water at neutral pH, partially because of the denser lattice that would require expansion of the film and also because of the small energy difference between this structure and the more thermodynamically stable phase. This stabilisation by controlling the packing could be important in other materials where stable and dense layers are desirable.

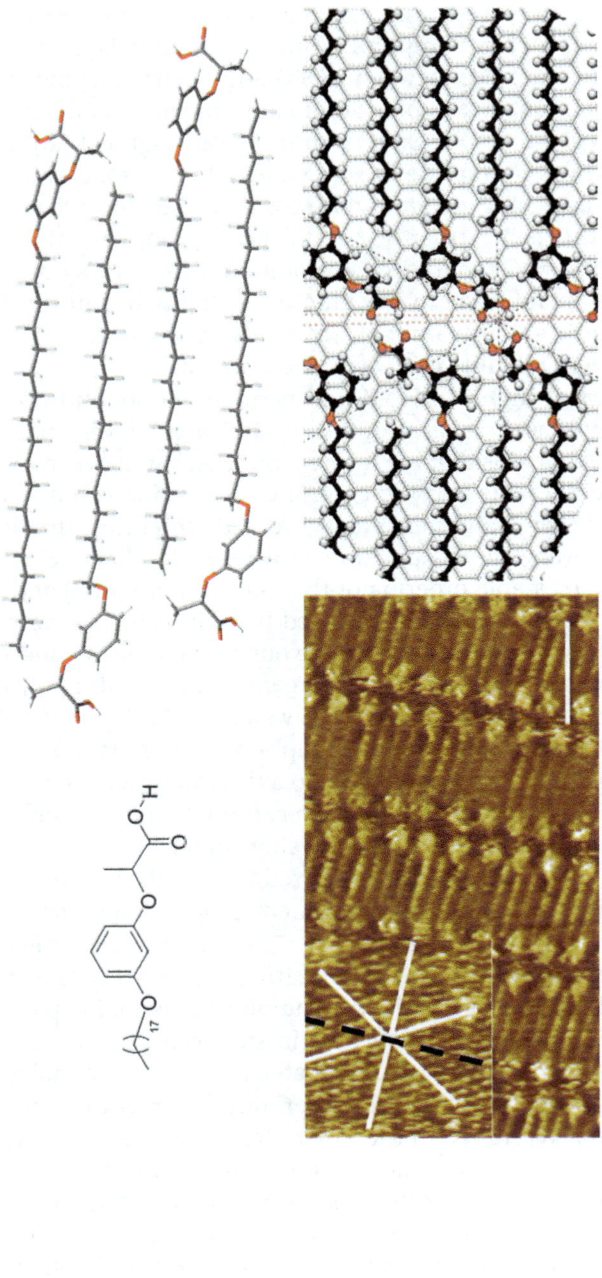

Figure 10.6 A chiral amphiphile, a partial sheet in its crystal structure (DMSO solvate molecules not shown), an STM image (scale bar 3 nm and graphite reference shown top left) from the HOPG-1-phenyloctane interface and a molecular model of the monolayer.

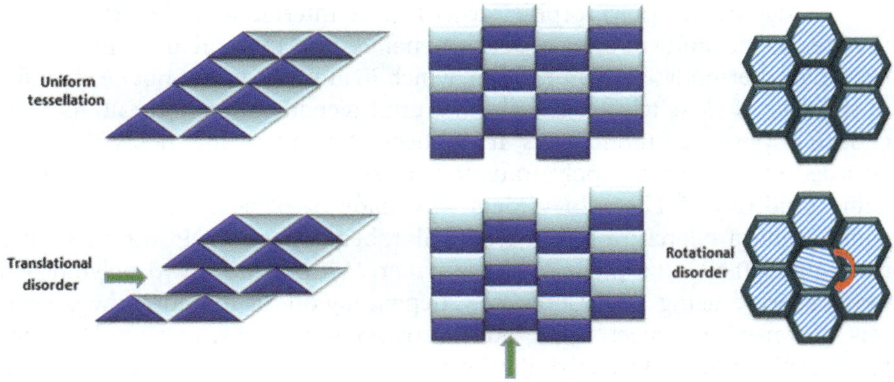

Figure 10.7 Tessellation of triangles, tetragons and hexagons with order and potential defect packing.

The order in films over centimetre-length scales has been achieved taking into account space filling and hexagonal packing of molecular components to generate periodic structures by evaporation in vacuum, spin-coating from solution, and cooling from the isotropic liquid.[28] Thinking of regular tessellated two-dimensional patterns, the triangle, tetragon and hexagon are candidates, but only the latter can pack uniformly if it is C_3 symmetrical. Both triangular and tetragonal patterns can give rise to translational disorder (Figure 10.7). Therefore, a C_3 symmetrical adsorpt was designed. The compounds used to achieve this remarkable feat in packing were triptycene derivatives bearing alkyl substituents on one face of the molecule that can pack into nested patterns, filling space and giving a hexagonal network (Figure 10.7). For example, both the melt and the toluene solution spin-cast film of the symmetrical dodecyl derivative show well-ordered materials. The film on a silicon wafer showed that the long axis of the molecules was perpendicular to the substrate, and time-of-flight low-energy atom scattering showed that the related compound with CF_3 groups at the chain termini had the fluorocarbon located at the interface with air, and presumably the triptycene moiety in contact with the silicon surface. Films prepared by evaporation onto silicon showed similar order, as revealed by two-dimensional grazing incidence X-ray diffraction that proved periodicity in the plane of the interface as well as perpendicular to it. Annealing did sharpen the diffraction peaks, showing that some slight rearrangement is necessary to reach the optimum packing and eradicate defects. AFM showed a flattening of the film, confirming this thermally-aided ordering effect. The octadecyl derivative reveals similar ordering, while the compound bearing only two chains, which in principle might display rotational disorder, actually needs no thermal annealing to reach a highly periodic structure. These compounds could prove useful scaffolds for molecular compounds, through weak molecular epitaxy for example, because they show long-range periodic order with no need for specific interactions with the adsorbate.

The existence of polymorphic structures at interfaces could provide difficulties in the study of certain phenomena. Beyond thermal annealing, to go towards a thermodynamic minimum, which in any case might not be effective in the case of close lying local minima, until recently, there were no tools to choose physisorbed monolayers. It has been shown that the metastable physisorbed monolayer of a polyaromatic hydrocarbon bisimide can be selected using fluid flow.[29] The achievement was made by placing a solution of the compound—a rylene dye—in 1,2,4-trichlorobenzene on HOPG and drawing the liquid off using a piece of tissue paper. Without flow, two polymorphs were observed using STM. Conversely, depending on the direction the solvent was drawn with respect to the adsorbent reference axes, one or the other polymorph could be favoured. Furthermore, the sizes of the domains formed by the dye were far bigger than those formed from a static drop on the graphite. The work is an exciting development, because using this principle combined with the employ of microfluidic devices might lead to the growth of layers that are oriented, in large domains, and with a selected solid form. The possibilities for hierarchical growth are equally exciting.

The use of non-covalent interactions to build specifically and selectively complex architectures on a surface, taking advantage of molecule–surface interactions, is an attractive and significant challenge. In Chapter 7, it was shown how hydrogen bonds can direct growth in a direction quasi-perpendicular to a surface. On the other hand, two-dimensional systems parallel to surfaces have reached a significant stage of development. Theory is increasingly useful for predictive determination of surface structure, particularly with the advent of strategies to probe the supramolecular assembly of large organic molecules on metal surfaces over large areas and showing the importance of surface reconstruction in ordering.[30] To be able to hierarchically assemble specific heteromolecular bi-, tri- or more layers in a specific fashion by physisorption is a true and valuable goal. To this end, the complexation of anions by the newly developed family of hosts termed "cyanostars"[31] is a remarkable development, in that they form bilayers of the functional molecules upon complexation with the charged guests at a solution–graphite interface.[32] Crystal structures of the compounds show that in the three-dimensional bulk solids, the anions are located at the centre of a non-covalent dimer, as illustrated here for a perchlorate complex where the oxygen atoms of the guest form hydrogen bonds with the protons coming from the electron-deficient macrocycle (Figure 10.8). This kind of assembly orthogonal to any interactions between the adsorbates in the plane of the interface is a very novel and promising area.

10.3 Supramolecular Chemistry at Interfaces for Affecting Charge Transport

The molecular materials used most often for charge transport—either in the ground state in doped (oxidised or reduced) materials or excited by light or

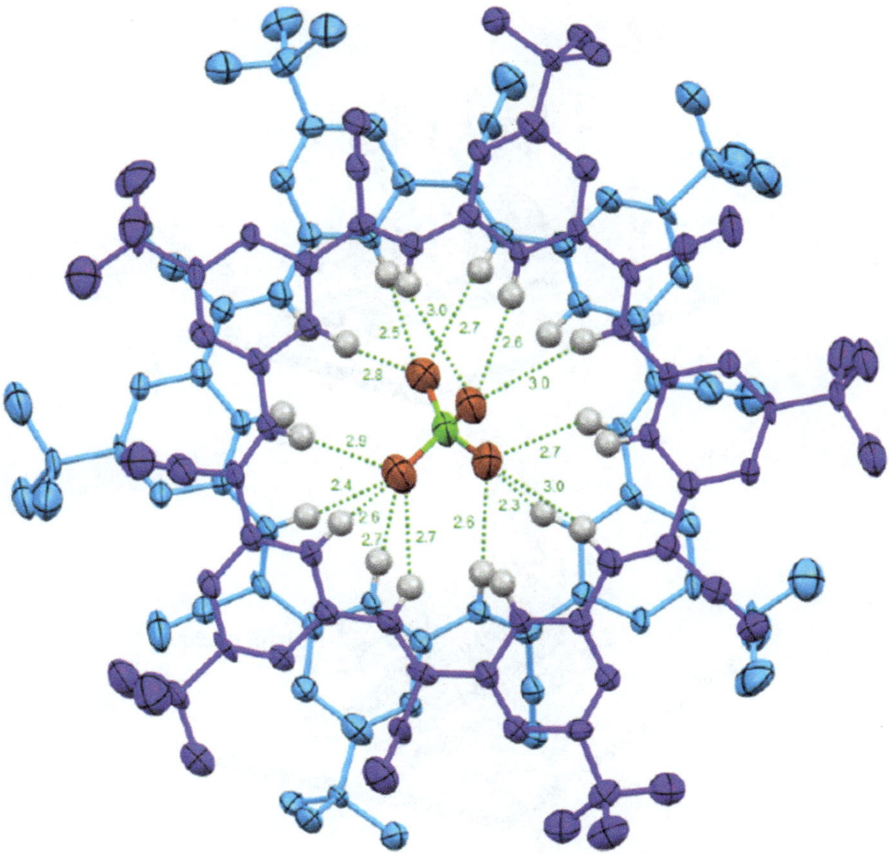

Figure 10.8 The crystalline complex formed between two cyanostar macrocycles and perchlorate, with hydrogen bonding interactions indicated. Reproduced from the supporting information to ref. 32 with permission from the Royal Society of Chemistry.

electric field in neutral systems—are those where relatively large aromatic systems overlap. The packing of these π-functional units in thin molecule-based films[33] is interesting from a supramolecular chemistry perspective. Face and edge-on aromatic interactions determine the structure, and the overlap between orbitals facilitates charge transport: the two are related intimately (Figure 10.9). The arrangement of the stacks of molecules on a surface determines which direction the most favoured charge transfer takes place. For field effect transistors—where an electric field induces charge mobility—the stacks are usually oriented perpendicular to the electrode surface to which the molecules are deposited. In solar cells, the orientation is preferentially parallel (as we shall see later) because charge must enter the electrode. In addition, the sizes of the domains of materials must be controlled because the connections between domains usually introduce a barrier to charge transport.[34] Therefore, a control of stacking and orientation as well as morphology is required.[35]

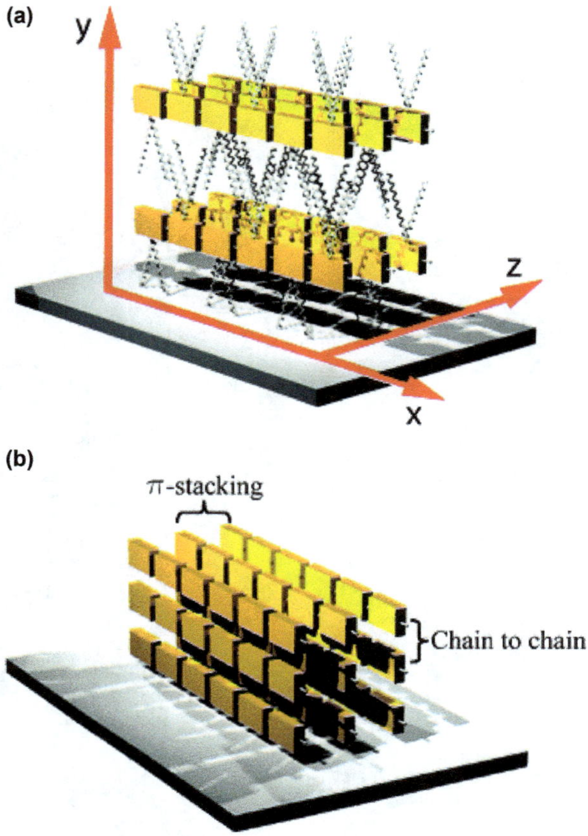

Figure 10.9 Cartoons showing the directions of charge transport in aromatic molecule-based materials ((a) the x direction indicates transport along chains, z direction between stacked chains, and y direction between the edges of chains) and the non-covalent interactions (b) that help the supramolecular organisation and facilitate the electron mobility modes.
Reproduced with permission from the Royal Society of Chemistry, ref. 33.

Control of the orientation with respect to the interface can be achieved by coating electrode surfaces with self-assembled monolayers to affect the wetting of the surface or to provide a specific interaction. In addition, the nature of any SAM on an electrode-bulk active material interface can influence the energy levels of the material and how these energy levels are aligned.[36] So, by choosing adsorbates that are electro-active or have large dipole moments, the charge distribution near the interface can be influenced and the charge transport across it can be modulated.[37–39] On the other hand, changing the length of the adsorbates that form the SAMs on the inorganic substrate in transistors leads to dramatic changes in their efficiency, with those longer than tetradecyltrichlorosilane providing the best electron mobility and on–off ratio for an n-type material based on thiazolothiazole.[40] The origin of this improvement is understood to be the effective shielding of trap sites (localising charge) in the silicon dioxide dielectric by the compact monolayer.

The strength of interaction between the functional units in molecular adsorbates and the surfaces onto which they deposit can be modulated by changing the substitution pattern and bulkiness of the compounds. These factors in turn affect supramolecular structure and electronic interaction. A series of rubrene derivatives with different substituents was prepared such that they would interact weakly with flat transition metal surfaces (the 111 faces of gold and copper) whose work function was slightly higher and close to the effective ionisation potential of the adsorbates.[41] On the surface of gold, the compounds display anomalous coarsening—a surprising coexistence of small and large supramolecular islands—as a result of competing intermolecular attraction from van der Waals interactions and repulsion because of reversible integer charge transfer with the metal. It turned out that by modifying the "legs" attached to the rubrene core, this coarsening—and by inference the charge transfer—could be modulated on the two metals. The coarsening effect is interesting, in that it is almost the opposite of Ostwald ripening. It is fascinating to think what might be done with this charge transfer and the resulting coarsening in nanostructured interfaces in devices, where the interface structure is so important. The work highlights the effect of small structural changes leading to organisational modification, and implies that a further understanding of these effects could lead to novel (nano)structure and function at interfaces.

On the other hand, single molecule wide devices are a fascinating prospect, and here control over orientation, union to interfaces, and even supramolecular connectivity are appealing prospects. In the following sections, we shall discuss these different situations that all require control of supramolecular chemistry at surfaces. Molecule-containing systems capable of acting as memory devices are appealing,[42] and the assembly of the active components on inorganic surfaces is a requirement that still needs to be mastered. The use of molecules that contain both donor and acceptor regions, be they π-electron deficient and rich or hydrogen bond donating and accepting in nature,[43] is a promising way to create polar order at interfaces and one that has not received a great deal of attention. The use of steric bulk around a thiophene moiety in a molecule incorporating a thiazole acceptor has afforded thin films with efficient performance as transistors.[44]

Organic molecules are studied intensely for their application in molecular electronics.[45] In 2013, several authors wrote about their "visions for a molecular future".[46] One of them mentions "supramolecular aggregates", but it seems to me that supramolecular systems have a very great deal to offer for the study of electron transport, and here some examples are discussed that show the remarkable advances and excitement in the area.

The Langmuir–Blodgett technique has been important for the development of molecular electronics[47] and is particularly useful for the assembly of thin films of π-functional polymers for field effect transistors, where a number of alternative fabrication methods are available.[48] The deposition of regioregular poly(3-hexylthiophene) using the LB technique afforded films that were used for field effect transistors that had among the highest charge mobility of any polymeric thin films of this type.[49] In general, these films show more homogeneous orientation—that favours electron transport—when compared with spin cast films, a recent example being a poly(thiophene) substituted with mesogenic groups.[50]

The solar cells where supramolecular chemistry has the greatest potential to play important roles comprise two main groups: the bulk heterojunction and dye-sensitised varieties.[51] The (mainly organic material) bulk heterojunction devices comprise donor and acceptor materials combined so that the donor material is in close proximity to the anode and the acceptor to the cathode. Both materials should be good conductors (of holes/p-type or electrons/n-type), and the two should phase separate giving nanometre-scale domains of each component. Supramolecular chemistry can help to organise the materials at interfaces and aid the orientation and structural perfection of the bulk. Dye-sensitised solar cells use organic molecules and metallo-organic compounds as electron donors (generally) at interfaces with inorganic electron conducting materials (most commonly); when illuminated, the photoexcited states of the adsorbates transfer electrons into the adsorbent. Supramolecular chemistry at these inorganic–organic interfaces can greatly influence the efficiency of the devices derived from the systems.

As the carriers of electrons in many solar cells, the electron acceptor of choice is very frequently a derivative of a fullerene. The development of a wide range of fullerene derivatives immobilised onto metal or oxide surfaces either directly or through supramolecular routes has been described.[52] The search for non-fullerene materials that might provide transport of both electrons and holes, be it for solar cells or organic transistors, has also led to the investigation of oligothiophene derivatives of diketopyrrolopyrroles that organise into very well-defined arrays when cast from chloroform solutions onto HOPG.[53] The supramolecular organisation of the molecules depended on their topology in general, in particular the position of substitution with alkyl chains and the lengths of these groups. STM of the monolayers allowed their direct comparison with diffraction data and is therefore a useful way to study their structural, and eventually electrical, properties.

The hierarchical assembly of all-organic photovoltaic devices based on hydrogen bonding between melamine and barbiturate derivatives, using a surface modified with derivatives of these synthons, was shown to improve the performance of devices compared with derivatives without these recognition features.[54] The reason for this effect is that often donors and acceptors of very different chemical character have a tendency to form separate phases in thin films. In addition, the interface between inorganic electrodes and organic material can introduce a barrier to charge mobility. The thin film formed by a pentathiophene capped with melamine groups (Figure 10.10) and the C_{60} derivative bearing a barbiturate functionality has a

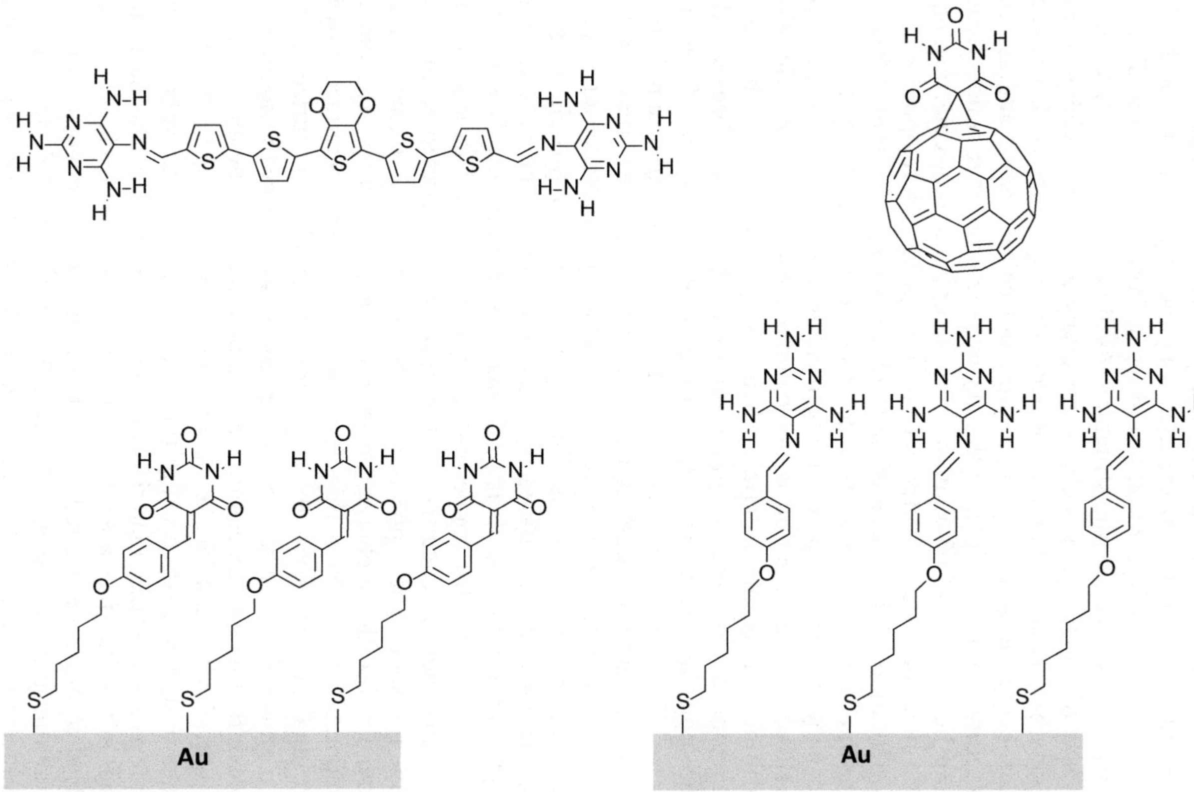

Figure 10.10 Components for the self-assembly of thin films of active layer materials for solar cells, whose organisation is influenced by SAMs.

2.5-fold better photocurrent than non-hydrogen-bonding C_{60}–oligomer systems, and the efficiency was increased further by functionalising a gold surface with a SAM terminated with groups capable of interaction with the active layer system. This work has apparently not been followed up, despite the fact that the concept is quite beautifully conceived. Perhaps the modest improvements in photocurrent are caused by the very small size of the domains of donor and acceptor that would result from hydrogen bonding between them; analysis of the film structure was difficult in this case. A certain domain size of donor and acceptor is necessary for efficient charge separation. Nonetheless, taking the concepts from this paper, one could envisage interfaces being "engineered" in a supramolecular way to address the nanostructure of these devices. In this regard, the proof that amide derivatives have better control of domain morphology, intermolecular arrangement and therefore solar cell efficiency is an important development.[55]

The orientation of the stacks of molecules must be such that it favours transport of charge between the two electrodes. For molecular systems, this would usually involve a parallel arrangement with the surface of the electrodes. However, this situation is not common, especially when the electrodes comprise inorganic materials, when (whichever deposition method is used) the molecules tend to stand up from the surface because the molecule–molecule interaction is greater than that between the π-faces and the surface. This obstacle can be overcome using "structural templating" layers[56] that make the conducting material's affinity for the surface greater. It has been demonstrated that the photovoltaic performance of organic materials can be increased significantly using this approach. Organic photovoltaic (OPV) devices based on heterojunctions of fullerene C_{60} and copper phthalocyanine had their performance improved in this way.[57] On bare indium tin oxide, the copper phthalocyanine lies with its stacking arrangement parallel to the surface, perpendicular to the preferred direction for charge transport. The α polymorph of the macrocycle is formed, which has strong unidirectional charge transport that is favoured by π-stacking of the compound.[58,59] When a thin layer of 3,4,9,10-perylenetetracarboxylic dianhydride is deposited onto the oxide, it lies parallel to the surface and it has been shown that free-base phthalocyanine has its growth modified by this layer so that its stacked columns are perpendicular to the interface.[60–62] The same trick was used for the vacuum deposition of copper phthalocyanine onto indium tin oxide using the 3,4,9,10-perylenetetracarboxylic dianhydride as the interface material (as a proof of concept, as the electronic properties of this interface are not ideal) and columns of the electron donor did indeed form in the desired orientation. Subsequent deposition of C_{60} and an aluminium top electrode led to a functioning OPV device with potentially increased charge collection. A conducting polymer layer turns out to be suitable for this effect, and combined with the deposition of tetracene between the bulk material and the aluminium anode, solar cells incorporating zinc phthalocyanine as the donor material and the same fullerene acceptor have reached impressive efficiency.[63]

The interface between electrodes and active layer materials (those that generate and transport the separate charges) as well as the interfaces between the donor and acceptor materials themselves are extremely important for the charge separation and collection processes in solar cells, and the characterisation of the structure in thin films is challenging. However, the use of high resolution transmission electron microscopy (HRTEM) on a solar cell of the type described (indium tin oxide–copper phthalocyanine–C_{60}–aluminium) showed with remarkable detail the structures that were formed during the growth of the layers, with crystalline regions of the donor oriented in different directions with respect to the interface with the oxide.[64] The images showed precisely the donor's orientation (with or without the templating molecular layer) as well as the sizes of the crystalline domains, which is important for device performance. The improvements in the sample preparation of these systems composed of many materials (in this case using a focused ion beam) and the use of powerful microscopy techniques is bound to afford valuable knowledge that will help in understanding the ways that supramolecular chemistry can influence the growth of materials in these important devices.

The potential of Langmuir–Blodgett films is possibly under-exploited for the fabrication of bulk heterojunction or multilayer active materials for solar cells. The ability to specifically layer either complete layers or nanoscale aggregates by horizontal lift-off could be promising. Certainly they allow the preparation of vectorial systems. It has been shown how the layering of electron acceptors (derivatives of perylene diimide and violanthrone) followed by their combination with fullerene-porphyrin diads and a polythiophene leads to functioning devices.[65] The multilayers were prepared by ensuring a perpendicular orientation of the molecules to the interface thanks to the appropriate positioning of hydrophilic groups.[66,67] The LB transfer method ensures a controlled deposition of each layer and the absorption spectra show that chromophores at different levels do not interact significantly in their ground states. Single LB layers of each component were placed atop one another and the heterobilayers were probed for photoinduced current. The results indicated that a higher interaction between layers of the electron acceptors did not necessarily produce a higher photocurrent. In general, it was shown that increase of the lifetime of charge separation and absorption can be achieved using complex layer architectures and that photocurrent is optimised for a given set of components thanks to the sequential deposition by the LB method.

Singlet fission is an interesting phenomenon that is potentially revolutionary for solar cells. A single photon generates a singlet excited state that decays to two separate triplet excitations by transferring energy to a neighbouring ground state molecule.[68] Thus, two charge carriers can be generated from one photon. It has been shown using molecular dynamics calculations that the polarity of surfaces can control the clustering and packing of tetracene—a molecular material known to show significant singlet fission.[69] It was shown that the proportion of herringbone packing—an arrangement that is believed to favour singlet fission—to stacked arrangement (where π faces overlap) in the supramolecular clusters varies greatly with the polarity of the surface to which the molecule is adsorbed. The nature of the polarity at the interface is also important, as either ordered or disordered surfaces affect the proportion of clustered molecules. While these theoretical results await a thorough experimental explanation, they demonstrate how simulation can raise new insight into microscopic phenomena at interfaces, which could improve the efficiency of materials for applications such as photovoltaic devices.

The adsorption of strongly electron accepting molecules onto ITO reveals the great importance that the atmosphere can have on the electronic properties of the resulting materials.[70] The adsorpts used (Figure 10.11) were tetrafluoro-tetracyanoquino-dimethane (F4TCNQ) and hexaaza-triphenylene-hexacarbonitrile (HATCN), which both increase the work function (Φ) of ITO, and therefore reduce the barriers to hole injection into organic P-type (hole/cation radical) conducting materials from the inorganic oxide. The monolayers were formed by immersing a cleaned and dried ITO surface into organic solutions of either acceptor at a concentration of either 1.2 or 50 mg l^{-1}. The latter formed far more quickly and allowed monitoring of

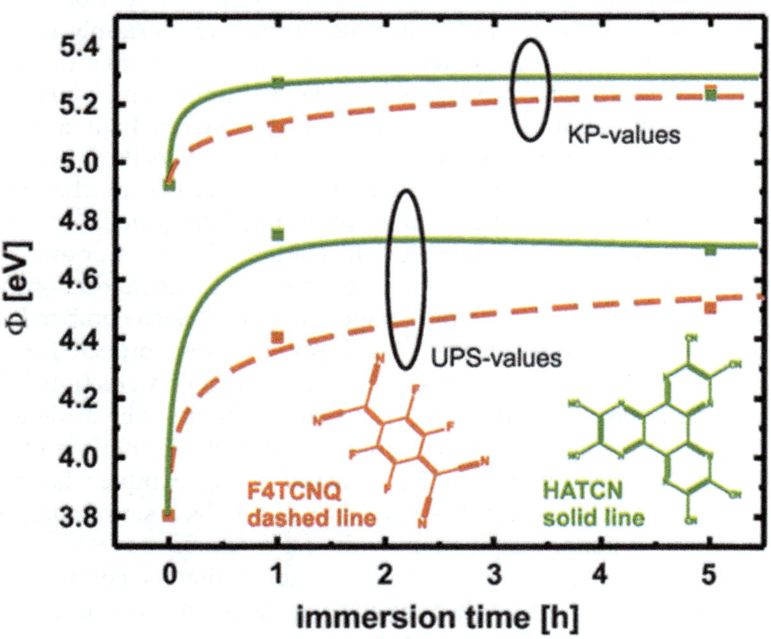

Figure 10.11 High concentration adsorption profiles of the electron accepting molecules F4TCNQ and HATCN onto ITO as monitored by ultraviolet photoelectron spectroscopy (UPS, inert conditions) and Kelvin probe (KP, ambient conditions)—measurements that quantify the work function of the surface (Φ). The lines are to guide the eye. Reproduced with permission from Springer-Verlag, ref. 70.

the monolayer formation (Figure 10.11). It is clear from the adsorption curves that the monolayer of HATCN forms far more quickly than the F4TCNQ monolayer. Ultraviolet photoelectron spectroscopy (UPS) showed that while no formal charge transfer was observed from the ITO to the electron acceptors (that did not appear in their radical anion form), surface dipoles arising from a charge transfer interaction were inferred from peaks in the ITO not seen in the pristine oxide. As can be seen from the figure, adsorption experiments were performed under completely dry conditions (and the samples were measured with UPS) or under ambient conditions (and monitored using Kelvin probe measurements). Especially important for the latter is the presence of atmospheric water. The graph in Figure 10.11 shows that even the pristine ITO has a different work function in air when compared to its vacuum level, and this is caused by the presence of water molecules whose dipoles interact with the surface. However, even under these conditions, the work function is increased further by the adsorption of either of the electron accepting molecules. On the basis of these very precise interface measurements, HATCN was identified as a promising molecule for interface modification because of its electronic properties as well as low desorption rate (arising from its strong interaction and relatively high molecular weight), that would limit diffusion away from the interface in a device.

The layer-by-layer approach to linear coordination complexes is useful for the study of electrical conduction through the resulting molecular "wires". The stepwise coordination of iron(II) to terpyridine ligands generates wires that can be anchored to gold through different templating molecules.[71] Results of studies of conductivity using electrochemical methods showed that the surface-anchoring template moiety modified the rate constant of electron transport, although it did not change the intrinsic transport properties of the molecular wire. Therefore electron conduction kinetics, which could be important in all kinds of optoelectronic devices, can be modulated through the anchoring group between the molecular wire and adsorbent as well as the units within the adsorpt. These results are not only important for surface-based systems, but also for systems anchored between two bulk electrodes. The approach for bottom-up construction and electrochemical analysis has been applied to a novel branched surface-anchored system,[72] shown in Figure 10.12. The complex supramolecular system was prepared by inserting the anchoring terpyridyl thiolate into a monolayer of thiophenol, and constructing the branched macromolecule in a stepwise layer-by-layer way. The objects were visualised using STM, showing the expected dimensions.[73] The electron transport characteristics were measured using an electrochemical method, potential-step chronoamperometry. The response of current over time followed quasi-non-exponential decay (rather than the exponential decay habitual in linear molecular wires, both electrochemically and by STM[74]) especially in the forward electron-transport direction (oxidation, the electron from the ferrocene to the gold electrode). A mechanism of electron transport comprising intramolecular electron hopping between adjacent metal ions was proposed.

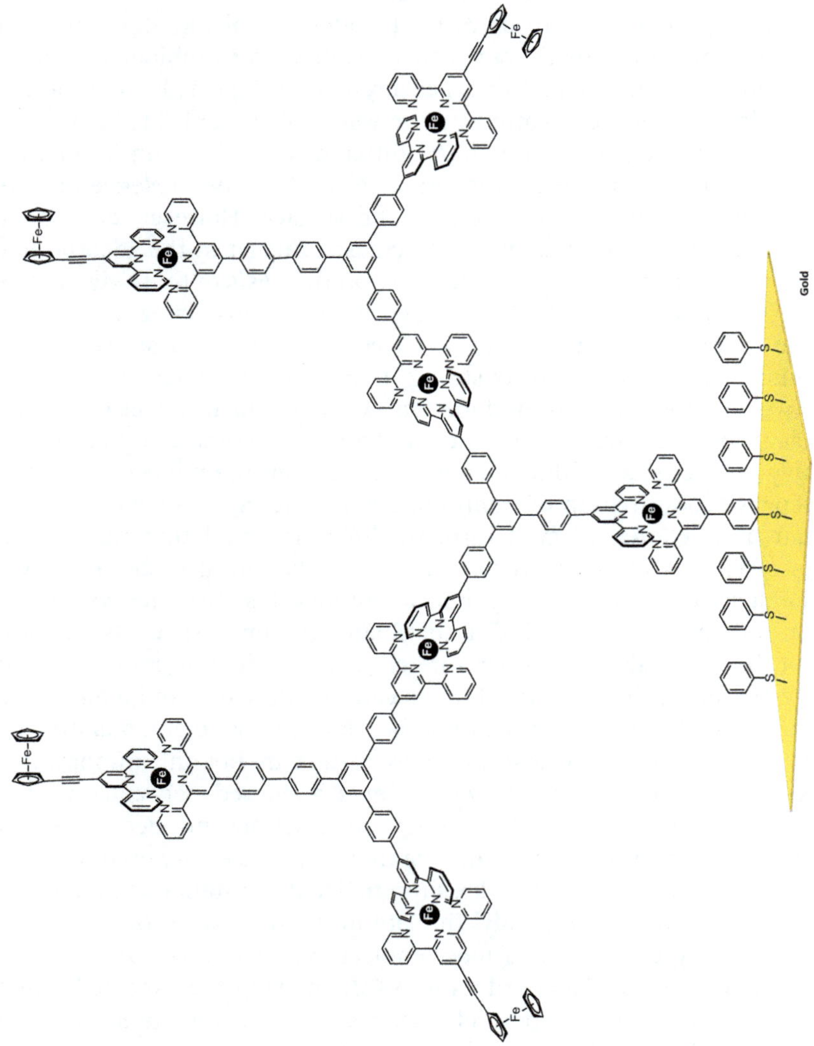

Figure 10.12 A branched surface-based supramolecular system showing unusual electron transport phenomena.

Supramolecular complexes can be formed between surface-based supramolecular systems and additional complementary molecules adsorbed to a scanning probe tip. Cobalt(II) tetraphenylporphyrin bound to a monolayer of 4-aminothiophenol was contacted with a tip to which a fullerene derivative had been chemisorbed.[75] Clear evidence of electronic conduction was observed when the complex between the fullerene and the surface-bound porphyrin was formed (with a conductance G_0 of approximately 6×10^{-5}) and a rectifying effect was observed. Very recently, it was shown how the orientation of a cobalt(II) porphyrin can lead to very high electrical conduction when it is contacted by two thiomethylpyridines attached to a surface and an STM tip (Figure 10.13).[76] This supramolecular junction was prepared *in situ* by introducing a solution of the cobalt porphyrin to a gold surface substituted with the pyridine derivative and then lowering the functionalised STM tip on top.

A conceptually similar work that actually allows the formation of very long supramolecular wires off a surface from a base-physisorbed layer is that of zinc(II) octaethylporphyrin complexed with 4,4′-bipyridine.[77] The growth of the wires is done in a stepwise manner and is monitored by STM *in situ* (Figure 10.14),[78] firstly adsorbing the porphyrin from tetradecane followed by rinsing and adsorption of the bipyridine from the same solvent, and so on. Despite the fact that there is no strong bond between the surface and the porphyrin, remarkably complete layers with epitaxial growth are formed where the porphyrins are linked through the ligand that results in a six coordinate zinc(II) ion (except at the extremities of the wire). The electronic properties of the wires were probed by STM, which showed that the shorter wires had a very low decay of conductance with distance (the attenuation factor β is $0.015\pm0.006\,\text{Å}^{-1}$) and a change in the conduction mechanism occurred at approximately 6.5 nm length. It should be noted that the longest wires were approximately 14 nm from end to end.

The coordination to zinc(II) complexes has also been used in the preparation of quartz balance-type detectors of biomolecules, where a combined assembly using chemisorption, LB film preparation and detection from solution was employed.[79] The combined approach to preparation of films and the study of binding might be useful for future studies on this kind of system where layers are grown off a surface for materials of interest in molecular electronics.

Supramolecular chemistry-based assembly followed by covalent modification can lead to systems where a wire is connected to a contact, as shown for a poly(diacetylene) connected to fullerene.[80] The bottom-up approach first involved simultaneous physisorption on molybdenum disulfide or graphite of a diacetylene molecule bearing a straight alkyl chain at one end and a carboxylic acid terminus at the other, which forms a complete monolayer, and C_{60} all under UHV conditions. The resulting system comprised lamellar structures of the diacetylene and islands of the fullerene, where there is full phase segregation and where the fullerene has a similar packing to its pure state on these adsorbents.

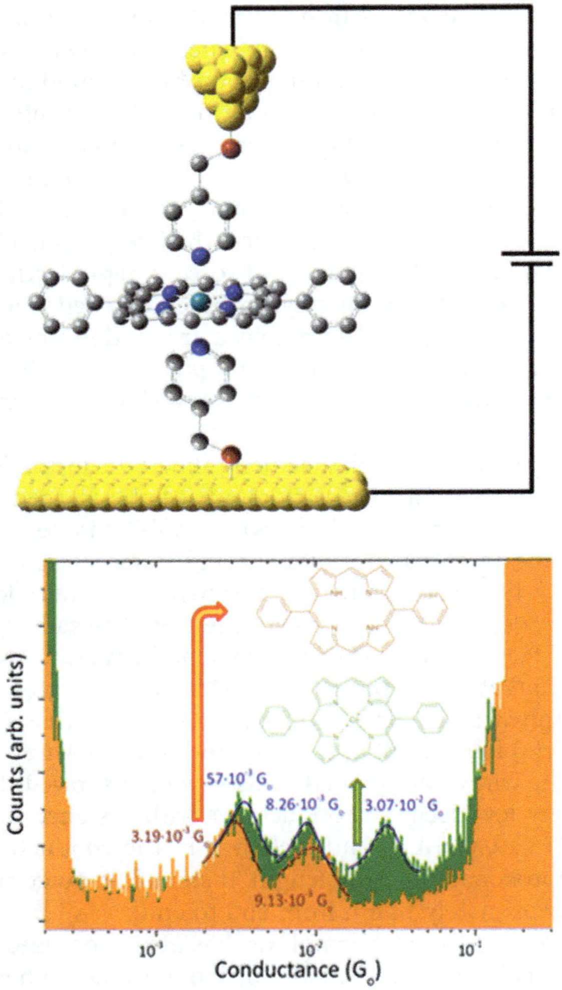

Figure 10.13 A supramolecular electrical junction formed by trapping a cobalt(II) porphyrin in-between a gold tip and a surface functionalised with pyridylmethanethiol (top). Below, the conductance curves of this junction and the one formed by the free-base porphyrin (drawn in orange) are shown.

The monomeric diacetylene was polymerised either by UV light irradiation in the vacuum chamber or using electronic excitation from the STM tip that was also used to visualise the process. The images of the product clearly showed the formation of the poly(diacetylene), and at the ends, fullerenes with bright contrast were observed. Indeed, the fullerene seemed to terminate the polymerisation, probably through cyclopropanation between a terminal carbene of the growing polydiacetylene and the fullerene on the surface. The termini were often contained at islands, but could also occur in the middle of domains of the aliphatic compound.

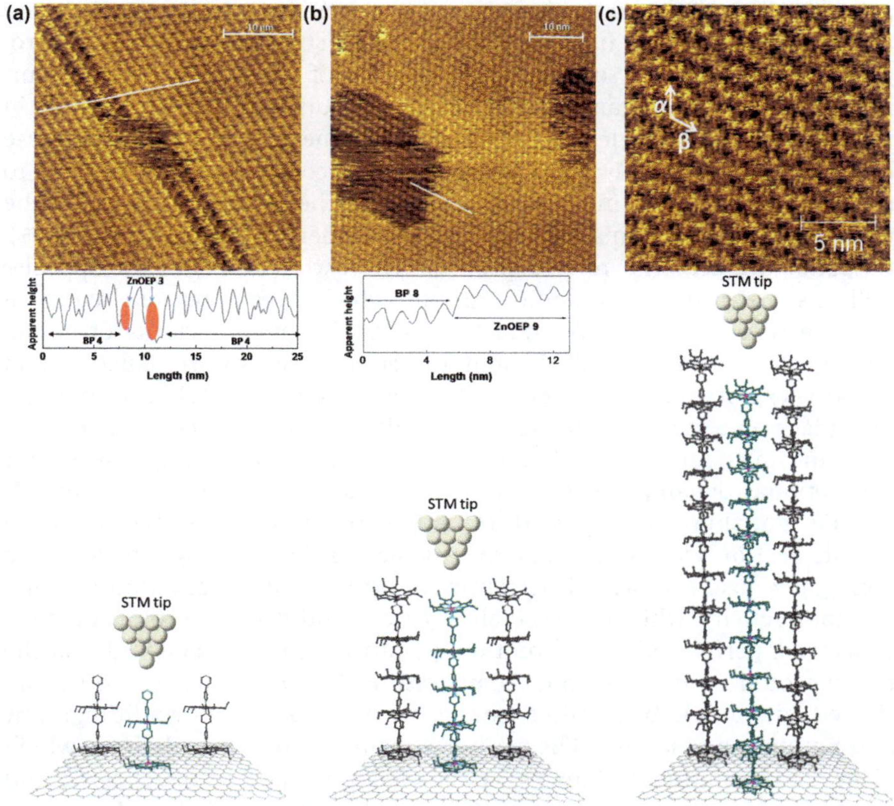

Figure 10.14 A series of physisorbed supramolecular wires comprising zinc(II) octaethylporphyrin and 4,4′-bipyridine grown on the surface of graphite and the STM images and line profiles corresponding to them: (a) is after four adsorptions with a bipyridine terminus (the dark stripes in the STM correspond to missing ligand), (b) corresponds to nine steps of adsorption and (c) to twenty-three steps, the latter both terminated with porphyrin molecules.
Reprinted with permission from ref. 77. Copyright 2014 American Chemical Society.

The result shows that by combining control over self-assembly and covalent bond forming reactions, it is possible to prepare robust systems of relevance for molecular electronics on surfaces.

10.4 Supramolecular Chemistry for Dynamic Surfaces

There is tremendous potential for understanding motion at surfaces by employing known dynamic systems from solution chemistry to interfaces. Some of these were presented in Chapter 8, but there is a treasure trove awaiting swashbuckling scientists! Intuitively, perhaps, electrochemically and chemically induced motions hold great promise.[81] In this sense, there are some systems that have been described in solution that await interrogation on surfaces. In solution, the reduction of a copper(II) complex surrounded by a chiral ligand incorporating two quinoline moieties results in an inversion of the optical activity because of the reverse twisting of these chromophores.[82] The five-coordinate copper(II) complex upon reduction to copper(I) has a change in coordination sphere: the carboxylate group of the chiral side-arm is replaced with the thioether moiety (Figure 10.15). Oxidation induces the reverse process. The two states have an opposite helical sense of the two quinoline groups, which gives opposite circular dichroism response. The structural foundation[83,84] and versatility of homologues have been proven,[85,86] and therefore it seems a robust system awaiting exploration at an interface. It is interesting to speculate what might take place on a surface either as a molecular or supramolecular system.

Quantum chemistry calculations have identified an exciting prospect for a functional dynamic system: "an all-electric single molecule motor."[87] The design, shown in Figure 10.16, is ingenious in that it incorporates a mobile section with a dipole perpendicular to the rotor axis as well as a conjugated axle that would permit monitoring of the orientation through the conductivity, while the oscillating gate would drive the rotor. Calculations were performed at different temperatures, and it is believed that the device would function at room temperature. The device has yet to be fabricated, but should be possible using state-of-the-art top-down lithography and bottom-up assembly. The work has inspired other calculations, which claim conformational changes in molecules subject to a voltage[88] and thermally-driven "molecular fuses" have been suggested, based on chromium arenes.[89]

A far simpler experimental system has been claimed as a single molecule rotor on copper (111).[90] The adsorbate butyl methyl sulfide binds to the surface through its sulfur atom, which is located atop a metal atom of the adsorbent. At room temperature, the molecule is so mobile that no information can be ascertained concerning its motion, but at 5 K, it appears as a

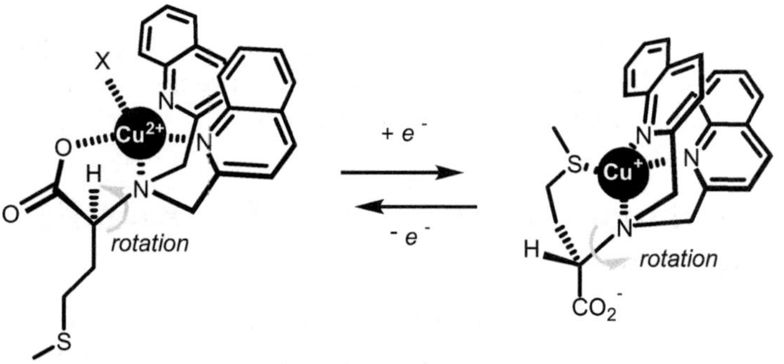

Figure 10.15 A redox chiroptical switch that functions in solution. The redox cycle causes a change in the coordination of the copper ion.

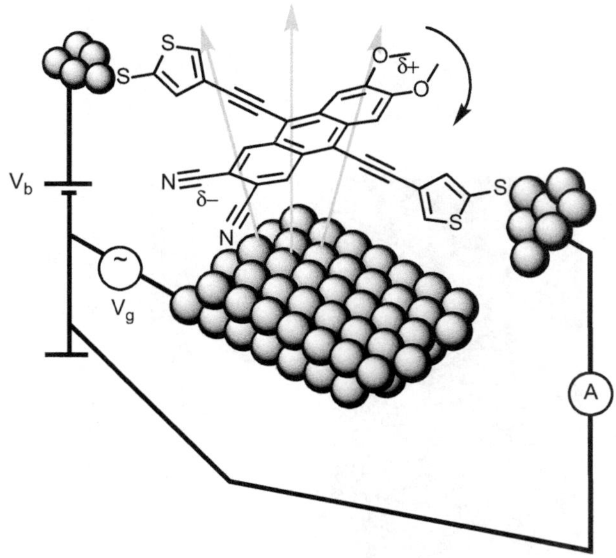

Figure 10.16 A cartoon showing a circuit that might lead to the functioning of a molecular motor, based on the motion of the dipole in the molecule that would be driven by an oscillating gate (g).

six-petalled flower in STM images. The reason for this appearance is that the molecule spins between six defined locations—determined by the substrate symmetry—that interconvert at a rate that is rapid on the STM timescale. The movement of the rotor can actually be monitored by measuring the current at a fixed point of the "flower" as a function of time. Well-defined current levels are observed corresponding to the different positions of the rotor. Directional rotation was claimed using tunnelling electrons from the tip of the microscope, which has a chance chiral point. The tunnelling electrons both excite rotation through interactions with bond vibration[91] as well as monitor the movement of the rotor. This two-terminal device does not have the sophistication of that described earlier for theory, but nonetheless points out subtleties that are all important at a supramolecular level in these mobile systems in very small dimensions.

In a self-assembled system, the application of surface potential has been shown to controllably alter the orientation of molecules. In physisorbed layers at the interface between a solid and a liquid, a reversible transition between an open framework or a stratified bilayer was observed upon changing the potential in the electrochemical STM set-up.[92] The reversible change in organisation of related systems in the electrochemical STM set-up has been interpreted in terms of changes in compression in the cationic adlayer at the interface.[93] In the absence of formal charge, the electric field coming from an STM tip is enough to re-orient a physisorbed chloroaluminium phthalocyanine co-adsorbed with pentacene onto HOPG at low temperature (Figure 10.17).[94] The single molecules—that have a high

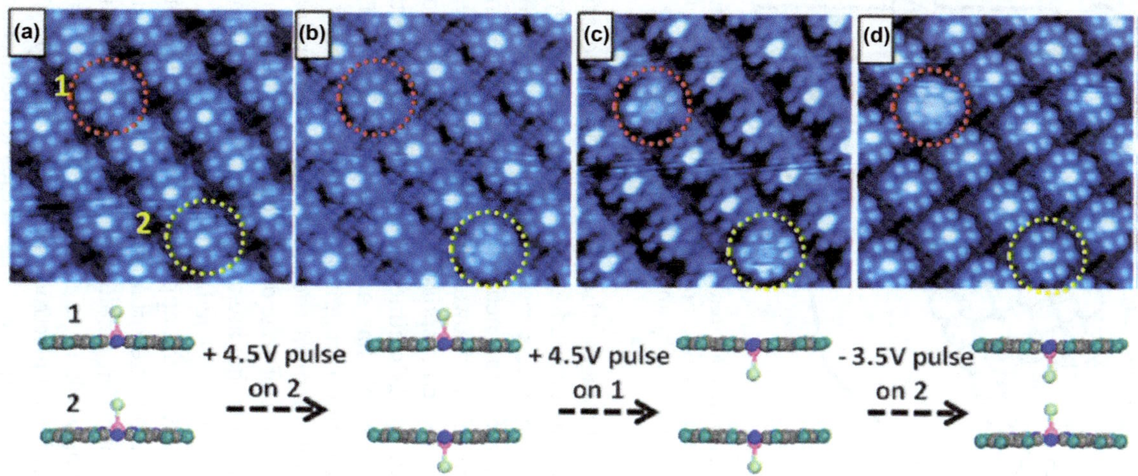

Figure 10.17 STM images and corresponding cartoons showing the STM tip pulse-controlled flipping of chloroaluminium phthalocyanine molecules co-adsorbed onto graphite with pentacene.
Reprinted with permission from ref. 94. Copyright 2014 American Chemical Society.

dipole moment perpendicular to the plane of the interface—flip essentially individually and reversibly, with the surrounding compound unaltered. The result is remarkable given the relatively weak interaction between the surface and adsorbate that must allow the proposed shuttling of the chlorine atom. The system is of interest in that the co-adsorption could aid the spacing of the molecules at the interface, which simultaneously facilitates observation of the orientation as well as ensuring very local switching with the STM tip.

10.5 Supramolecular Chemistry at Surfaces for Biomolecular Applications

The immobilisation of receptors on surfaces immediately opens the door to a wide range of biomedical applications.[95] The adhesion of cells to surfaces is an important process from various viewpoints,[96] be it for tissue repair or for the controlled immobilisation of cells. The adhesion of the protein human serine albumin onto chiral organic gel surfaces has been shown to depend on the chirality of the gelator molecules that is amplified up into the supramolecular structure, giving a chiral interface.[97] The diastereoisomeric amphiphilic gelators incorporating two stereogenic carbon atoms formed gels in mixtures of ethanol and water in which morphologically chiral nanofibres were generated. The chiral materials were deposited onto a quartz crystal microbalance and their interaction with the protein showed that a clear preference for adhesion of the macromolecule onto chiral fibres of one chirality exists. It was claimed that the stereoselectivity in adhesion was significantly amplified for the chiral nanostructures compared with smaller chemical objects. Thus the nanometre-scale chiral surface seems to play a role in the preference for one diastereoisomeric complex compared with the other, a chiral shape phenomenon that is of considerable interest. This aspect of polyelectrolyte multilayers could be of interest, as they are finding increasing attention in the area of antimicrobial coatings because of the importance of these materials in avoiding infection.[98]

Lung surfactants are tremendously important for the organ's function, so any deficiency or compositional change can have dramatic consequences, as in neonatal respiratory distress syndrome. This condition can be treated with replacement surfactants that are effective in about 70% of cases, thus lowering premature infant mortality.[99] A molecular-level understanding explaining the emergence of the condition is lacking though, the presence or absence of cholesterol in treatments being controversial and model systems based on supramolecular effects in monolayers when small quantities of cholesterol are present is a precise way to lead to an interpretation.[100] Indeed, monolayers emulating the lung membrane when exposed to potential environmental pollutants, such as nanoparticles, may provide clues as to their potential toxicity which would be revealed by changes in monolayer structure.[101]

As was pointed out in the introductory chapter of this monograph, biological systems are inspiring in how they function, and the mechanisms of bacteria adhesion are important to understand from the point of view of infection, but also how adhesives might be made to work. AFM has been used to study individual adhesins on living bacteria and in particular to map at the nanoscale on the bacteria.[102] Immobilisation of a sulfated glycoconjugate receptor mimicking heparin onto the tip of an AFM and imaging and force measurements revealed that the adhesin in the bacteria are located in nanodomains and that they require times that are long on the AFM timescale to adhere, meaning that multiple conformational changes and interactions are necessary during the attachment. Supramolecular organisation has been identified as being central to the adhesive function in this and other bacterial systems.[103,104] Learning from these systems and what is known about medical adhesives, for example in the ophthalmic area,[105] will surely lead to improvements in the treatment of lesions.

The physisorption of adsorbates comprising ditopic nucleobases on graphite yields self-assembled nanopatterns thanks to the kind of hydrogen bonding typically taking place in these compounds in natural systems[106] (Figure 10.18). The compounds physisorb at the HOPG–water interface giving striped organisations whose dimensions are determined by the structure of the monomers. These superstructures have been employed to decorate the interface: a monomethyl triethyleneglycol branch gave stable superstructures that inhibited static platelet adhesion at biologically relevant temperatures.[107]

The interface between a liquid and a solid can influence the supramolecular arrangement of molecules in aggregates of guanosine derivatives.[108] A single hydrogen-bonded structure is observed in the bulk solid that rearranges in solution to give a second arrangement. When the solution is placed on a surface and ribbons form by physisorption, STM and AFM indicate that the original solid state packing is generated at the interface, which in solution is clearly not the thermodynamic product. This intriguing situation highlights the delicate balance between self-assembled structures and how it can be tipped at a surface. The phenomenon for biological systems is particularly poignant given the abundance of interfaces in living things.

The use of surface-enhanced infrared spectroscopy has proven to be useful for the characterisation of hydrogen-bonded layers *in situ*.[109] The more complex tip-enhanced Raman spectroscopy has been proven for the detection of nucleic base pairing at an interface.[110] It seems that the development of these and related techniques will be a boon for the understanding of biomolecule-type (and other) self-assembly at interfaces.

A fascinating prospect to understand cell adhesion and to change it at will is to be able to controllably immobilise cells on switchable artificial surfaces. This has been achieved for SAMs on gold, where an azobenzene-containing adsorpt linked to a cell-binding peptide fragment was used as the active

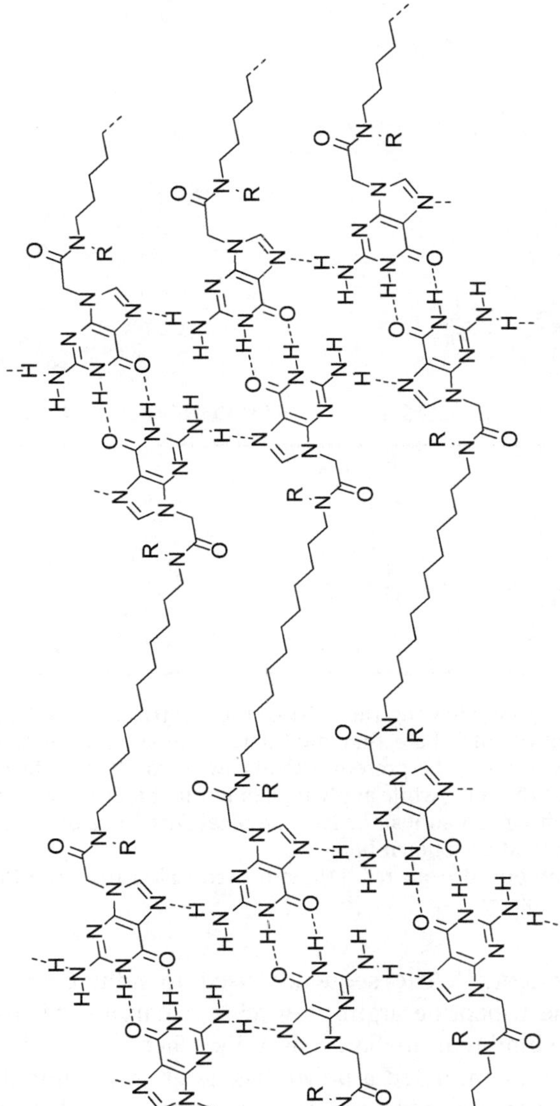

Figure 10.18 A structure representing the SAMs formed by bis-guanine derivatives that form hydrogen-bonded tapes.

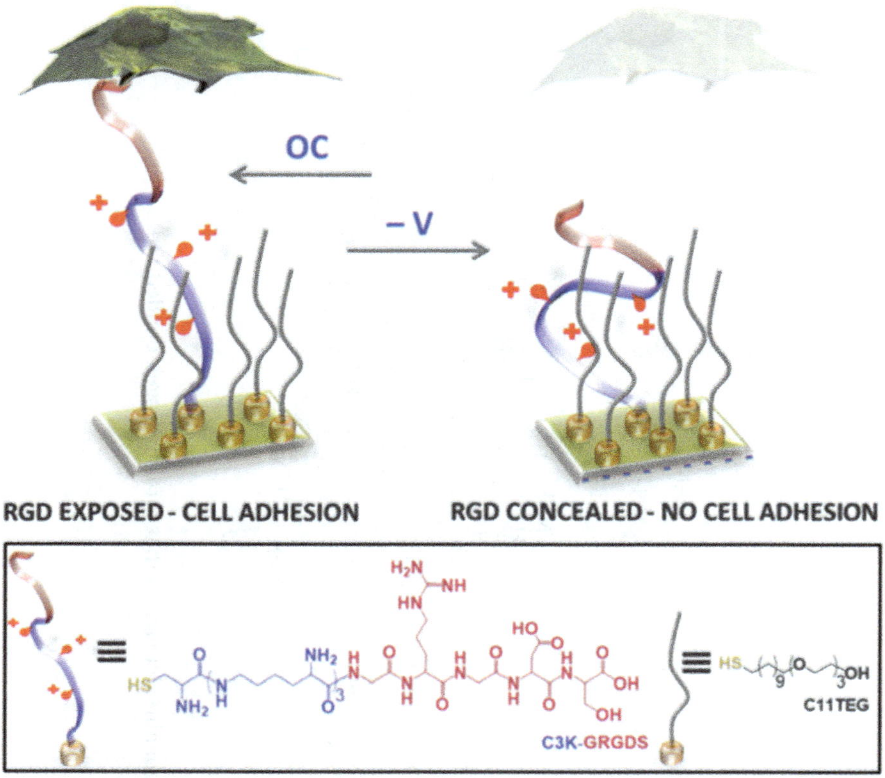

Figure 10.19 A cartoon showing the SAM of the electrically switchable RDG oligopeptide adsorbate that modulates cellular adhesion. With an open circuit (OC), the protein extends away from the surface and interacts with the cells, while applying a potential hides it away in the interface, limiting cell adhesion. The chemical structures of the components of the SAM are shown below.
Reproduced from ref. 115 with permission from the Royal Society of Chemistry.

interface component.[111] The sequence used to adhere the cells at the interface was the tripeptide arginine–glycine–aspartate that had been previously shown to bind mammalian cells at interfaces.[112,113] The system used to switch adhesion comprised a poly(ethylene glycol)-terminated thiol and an activated ester-terminated compound to give a mixed SAM in the first instance that was subsequently reacted with the key tripeptide sequence. Interestingly, extremely low concentrations of the tripeptide in the monolayer proved to be the most effective in the switching experiments. When only 0.1% of all alkanethiolates in the SAMs comprised the tripeptide unit and the azobenzene was in the *E* configuration, the surface supported cell adhesion completely yet resisted adhesion as the photochrome was switched to the *Z* form. When the proportion of tripeptide was increased to 1%, the differentiation between adhesion and non-adhesion was only slight, demonstrating the importance of interface concentration on adhesion

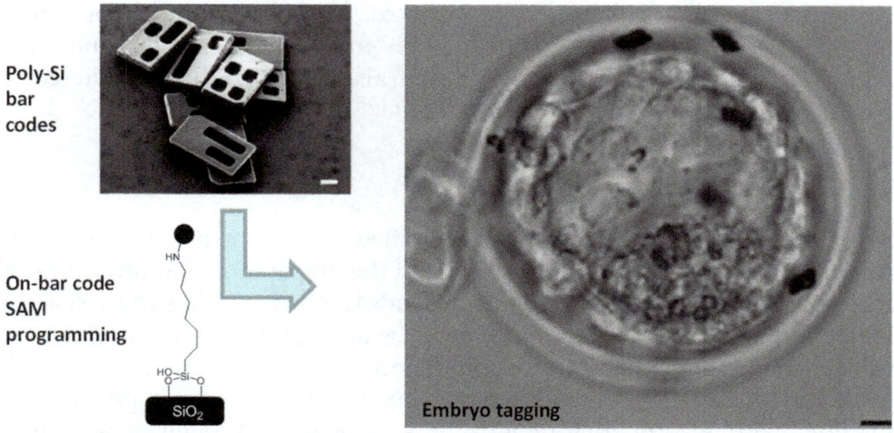

Figure 10.20 Polysilicon bar codes that can be used for embryo tagging when a suitable SAM is attached to their surface.
Images kindly supplied by Lluïsa Pérez-García.

phenomena and particularly their selectivity. The control of cell adhesion through responsive surfaces—photochemically as in this case or electrochemically—will eventually allow their spread from a determined place to screen a drug candidate's cytotoxicity, and responsive surfaces have the potential to switch cell adsorption "on" or "off".[114]

A demonstration of this employing an electrically switchable surface has been achieved for the adhesion of macrophage cells to the monolayer of an arginine–glycine–aspartate (RGD) oligopeptide.[115] The SAM formed by an extended form of this peptide on gold in a mixed system with oligoethylene glycol (to inhibit non-specific adsorption) is responsive to an electric field, and as shown in Figure 10.19, this effect can make the cells adhere, or not, at the interface. It is foreseen that experiments such as these will aid in understanding adhesion signalling in macrophages and allow the development of strategies for extending the lifetime of implantable medical devices and other biomaterials. The interaction of living matter with interfaces will be an increasingly important area of research for a myriad of reasons, from preventing infection, through promoting growth to understanding the fundamental structure and properties of membranes and their interactions with their surroundings.

Perhaps in the future, tiny chips will allow monitoring of biological functions in the body, or at least report on cell activity *in vitro*, and the control of supramolecular chemistry of the surfaces of these chips will determine where they go and what they can do. For polysilicon bar codes—whose identity can be judged by the spaces left in the micro-object—the attachment of specific SAMs improves their ability to attach to cells, for the moment for identification (Figure 10.20).[116] To this end, lectins were attached to the objects though optimisation of the linking chemistry. The orthogonal functionalisation of microchips comprising polysilicon, chromium and gold has been demonstrated.[117] The possibilities for supramolecular control in these systems to enter and go between cells is not only an attractive technological challenge, but also a way to discover the complexities of interfaces in biological material.

10.6 And Beyond

Supramolecular chemistry, whether identified as such or not, is playing an increasingly important role in all aspects of the preparation and properties of molecules at interfaces. Many of the presently trendy topics in surface science are bound to make forward strides larger by taking into full account non-covalent interactions. The use of supramolecular systems that mimic metal-containing enzyme active sites is a case in point.[118] Surfaces can in principle be modified with adsorpts that organise in specific ways through non-covalent interactions so that the interface contains catalytic centres that resemble those of naturally-occurring catalysts, or indeed synthetic ones used in homogeneous media. For studying organic chemistry, the precise single crystal surfaces are bound to yield insight into existing and presently unknown processes.[119] The surface is not only an environment to organise the materials, but the armoury of analytical tools allows new chemistry to be discovered.

One of the challenges for supramolecular chemistry in general is to prepare "intelligent" or "programmed" multicomponent systems that specifically reach a structural outcome, with the aim that this will be usable for some function. Arguably one of the most clever yet inspiringly simple surface-based self-assembling multicomponent systems is that described by Xue and Zimmt where four components sorted themselves into a grating thanks to a design based principally on shape.[120] The system was conceived for adsorbates on HOPG, where alkyl chains align down one of the graphite axes, and where three rigid moieties produce kinks in the structure. These kink shapes, when they coincide, should give a compact structure. But where they are out of register, spaces will result in the monolayer, a situation that is thermodynamically unfavourable. The idea can be appreciated by inspecting Figure 10.21. Take a single component system where a bis(methyleneoxy) anthracene moiety is joined at each heteroatom to a six methylene group chain that in turn is linked to a diyne which has a fourteen carbon chain attached to it. This molecule cannot be packed in a compact manner (without deforming part of it to a state that would be very unfavourable on the HOPG adsorbent). Spaces are left in the layer. However, the situation is different if this molecule is combined with a similar adsorbate with fourteen molecules between the oxygen atom and the diyne and a heptyl terminal chain. Here, like molecules could form alternating lamellae where unlike chains can interdigitate filling space. The shapes of the molecules in the adsorbed layer—determined by their constitution—allow the diyne and anthracene moieties to come into line in the lamellae and where one prochiral face of the anthracene derivative is expected to be adhered in a given domain that will be homochiral. Both the components are C_{2h} symmetric in their ideal conformation, each with its own unique side chain.

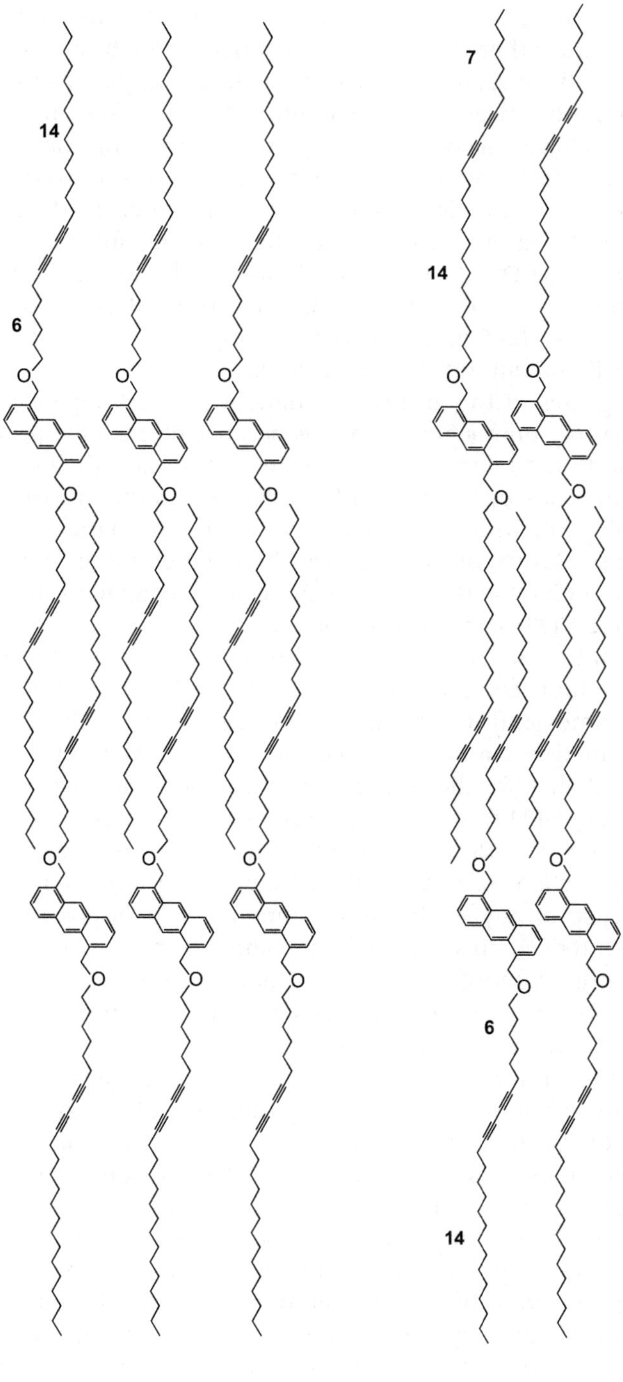

Figure 10.21 Loose packing and tight packing and shape matching. Structures showing the difficulty of close packing and lack of self-complimentarity of an adsorbate for HOPG, on the left. In this structure, the methyleneoxy anthracene moiety is separated by six methylene groups from a diyne that has a fourteen carbon chain attached to it. On the right is a two-component system that would be expected to form a well-packed layer with alternating components perpendicular to the lamellae. In this case, the previous adsorbate alternates with one where the methyleneoxy anthracene moiety is spaced by fourteen methylene groups from the diyne, which has a seven carbon chain attached to its "other end".

Now consider the case where four molecules are combined (they are shown in Figure 10.22); two of them have identical "tails" on the bis(methyleneoxy) anthracene moiety and are C_{2h} symmetric in their ideal conformation while the other two have C_h symmetry and bear two non-identical "tails". The diyne segments are positioned at unique positions of the "tails". This situation means that a combination of all four molecules would be predicted to have only one thermodynamic minimum where space would be filled to full effect, and indeed it is that one that is observed in the ordered regions of the monolayer formed on graphite. STM images (Figure 10.22) of the dry layers formed from phenyloctane show a specific sequence: S1-D1-D2-S2-D2-D1-S1. This was shown by the unique combination of distances between lamellae that results. The unit cell is 23 nm long, and the asymmetric unit half that, and the packing continues at least as far as seven unit cells (about 160 nm), and domains contain at least 2000 molecules. The long axes of the anthracene moieties are all angled in the same direction, meaning homochiral domains because of the adsorption by one of the two enantiotopic faces. The diyne rows (that are distinguished from the anthracene ones by their slightly duller appearance in the STM images) are also aligned, with an opposite tilt to the aromatic moieties. Defects are seen in the layers, at the interfaces between domains and small disordered clusters. Individual defects were shown to rearrange to the more stable situation while STM scanning took place.

The self-sorting into such large ordered structures is remarkable for several reasons: its four components give a unit cell 23 nm long that repeats over seven times and it is formed from one out of 6×10^4 possible six molecule combinations, almost exclusively. The presence of all four blocks is not only sufficient for the assembly, it is also necessary. No ordered layers were observed by STM from the three-component systems when either S1, D1 or D2 were left out. Domains smaller than 35 nm in breadth and width were observed when S2 was omitted. Therefore, the system is disrupted by the absence of any of the four components. It is possible and probable that this kind of chain shape matching and similar strategies based on molecular recognition motifs will lead to more complex and selective systems, and could result in novel functions for the resulting thin films should opto-electronically active units be incorporated.

Another source of complexity is the parallel and non-interfering self-assembly of systems at interfaces for the patterning of materials and possible introduction of further function. In this regard, a particularly conceptually important result is the physisorption of a hexagonal network of perylene-3,4,9,10-tetracarboxylic diimide and melamine to give a hexagonal framework formed by hydrogen bonding whose spaces are used for the chemisorption of thiols on gold.[121] The network (Figure 10.23) was physisorbed onto gold (111) from dimethylformamide such that the melamine units form the nodes and the perylene diimide the struts in the chicken wire-like framework that lies parallel to the metal surface. STM showed that the supramolecular structure is highly regular over the whole of the surface

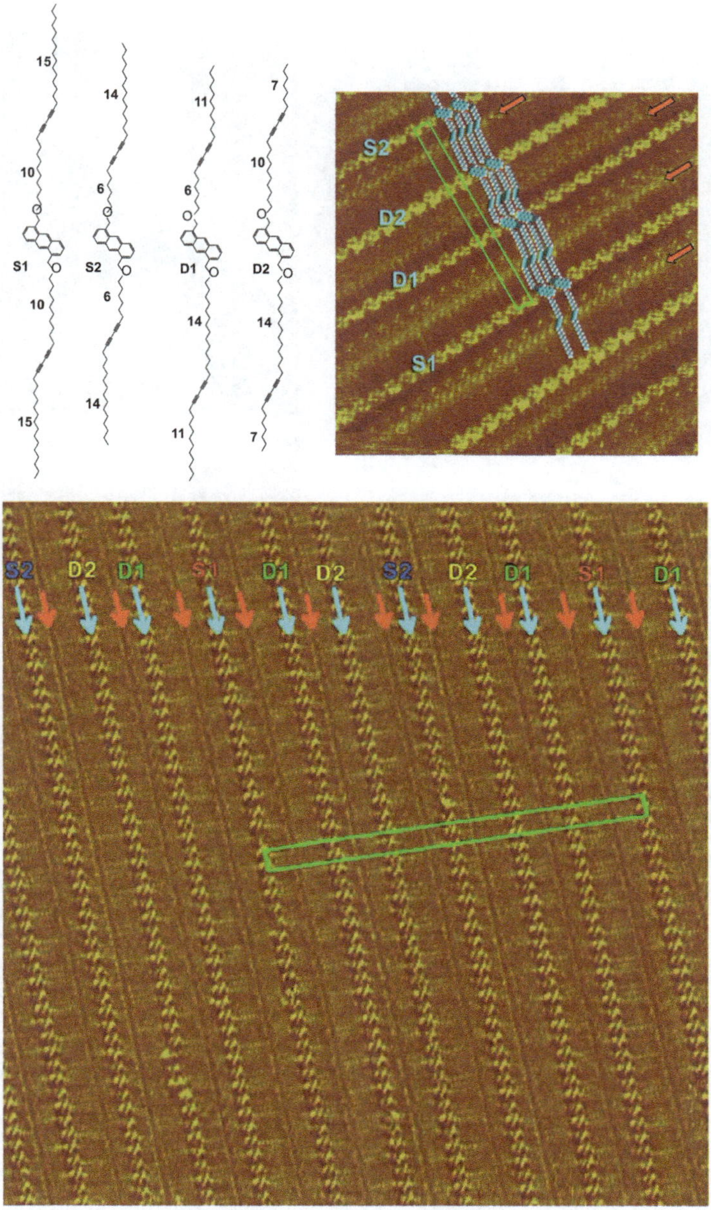

Figure 10.22 The four adsorbates that combine to give exclusively a monolayer on HOPG with the S1-D1-D2-S2-D2-D1-S1 sequence indicated in the STM images that are shown. In both images, the brightest contrast arises from the anthracene moieties, with the diyne bars slightly duller and the methylene chains slightly less intense still. The top right image shows the asymmetric unit in an image measuring 20 nm×20 nm. The lower image is a square 43 nm on each side, with the unit cell indicated in green.
STM images reprinted with permission from ref. 120. Copyright 2012 American Chemical Society.

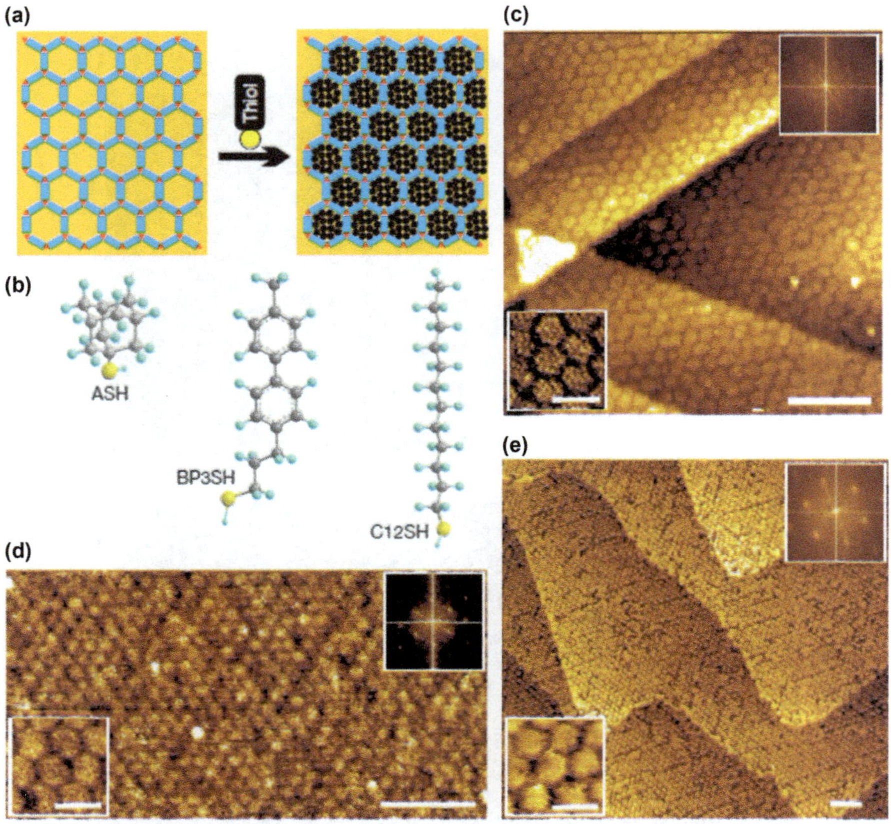

Figure 10.23 The scheme in (a) shows a cartoon of the physisorbed perylene diimide (blue bars)–melamine (red dots) network on gold and its filling by thiols (black dots). Structures of the thiols are shown in (b), where ASH is adamantane thiol, BP3SH is ω-(4′-methylbiphenyl-4-yl)propane thiol and C12SH is dodecanethiol. The STM images of the hybrid structures on Au (111)/mica are shown in (c)–(e) (for spaces filled with ASH (c), C12SH (d) and BP3SH (e), respectively). The lower left and upper right insets in the images show STM of high resolution and Fourier transforms, respectively. The scale bars correspond to 20 nm in the larger images and 5 nm in the insets.
Reprinted by permission from Macmillan Publishers Ltd, *Nature*, Copyright 2008, ref. 121.

(although defects resulting from filling of the voids or missing components, as well as linking defects, were apparent). When this network was exposed to the three thiols shown in Figure 10.23, hybrid self-assembled monolayers were obtained where chemisorption takes place in the spaces of the frames. Introduction of either dodecane thiol or ω-(4′-methylbiphenyl-4-yl)propane thiol leads to chemisorption of the compounds onto the gold that is exposed to the solution for relatively short times, although extended periods lead to the displacement of the physisorbed layer, possibly because of the stability of the thiol SAMs compared to the physisorbed framework that is more reversibly adsorbed. When adamantane thiol was introduced, prolonged exposure could be used to cover the solvent-accessible gold surface, with the physisorbed layer intact, because this adsorbate does not form such stable monolayers on the metal surface. These hybrid layers are stable in liquids that do not dissolve the physisorbed network, so much so that they were used for patterning of copper by electrochemical deposition in the underpotential deposition region. Under these conditions, copper(II) ions are reduced to metallic copper and form a monolayer between the gold and the thiol.[122] The physisorbed network acts as a diffusion barrier, while copper ions can diffuse and be reduced on the gold and keep the hexagonal order. This process in the hybrid monolayer is faster than in complete monolayers of thiols on gold, because of the small island sizes in the spaces in the hexagonal network, so that copper ions approach the gold from the sides of the thiol areas. STM showed that the structure is largely maintained, while the height in the areas of copper insertion (at the early stages of the process) are seen to be thicker. This solution-based approach shows how parallel supramolecular assembly processes can lead to sequential and specific nanometre-scale growth over relatively large areas on surfaces, and it awaits further exploitation.

In systems prepared under UHV, the deposition of components followed by their linkage using covalent chemistry is gaining multifarious routes.[123,124] The variety of C–H bond breaking reactions that take place on copper (110) was used for the preparation of multicomponent molecular networks.[125] The linking of the porphyrin molecules proceeds through the directional formation of organometallic bonds on the surface[126] (see Chapter 5). This directionality leads to the formation of block oligomers that are formed along one axis of the crystalline surface when a different porphyrin is evaporated onto the surface and reacted, and further reaction with pentacene leads to complex covalent structures joined through surface copper–carbon bonds (Figure 10.24). The possibilities for exploring these interface structures and their supramolecular surface chemistry using scanning probe microscopes promises to be a combinatorial way of uncovering new non-covalent interaction pathways. In addition, a hierarchy of interactions from non-covalent interactions to covalent structures can be traced with variable temperature STM, one case being tribromobenzene,[127] and the generation and dynamics of organometallic macrocycles and zig-zag structures on the more symmetric (111) surface of copper show the potential to correlate surface reactions with the formation of interesting new molecules, such as "hyperbenzene".[128,129]

The preparation of high molecular weight components (that might be difficult to deposit or sensitive to the evaporation conditions) of supramolecular systems on surfaces can be exploited for the stepwise preparation of complex supramolecular architectures, as demonstrated for coordination networks comprising a very large bis-terpyridine compound acting as a rod-like linker and tetra-4-pyridylporphyrin that behaves as a node (Figure 10.25).[130] The linear molecule was synthesised on a gold (111)

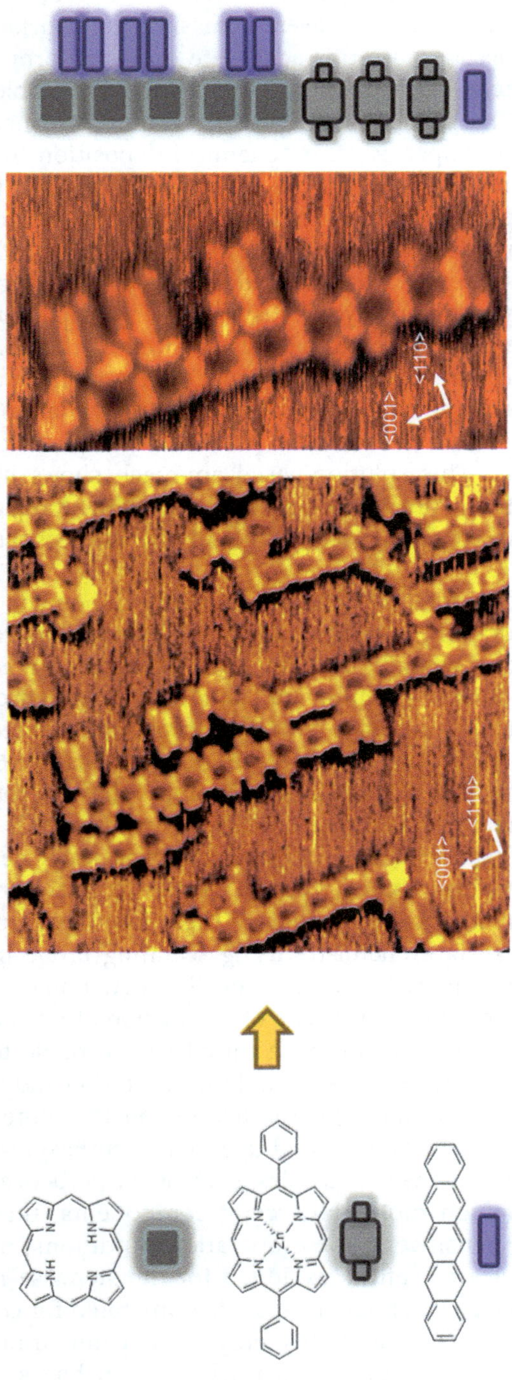

Figure 10.24 STM images of the covalent reaction products of the sequential deposition and annealing of the three molecular components drawn on the left. The cartoon on the right shows a key to the covalent structure indicating the components joined up and an image in the STM image next to it. Original STM images were kindly provided by Sam Haq and Rasmita Raval.

surface by heating the bromophenyl derivative of terpyridine shown in Figure 10.25 to 180 °C on the surface, a temperature at which it is totally mobile. The Ullmann coupling that takes place produces the quaterphenyl-spaced bis-terpyridine compound, which itself formed complexes with iron deposited onto the surface,[131] in this case forming chains.[132] Adsorption of iron and tetra-4-pyridylporphyrin onto the gold with surface-prepared bis-terpyridine resulted in the formation of networks after annealing to 180 °C. These networks were not seen in the absence of the molecule that acts as a four-fold node. STM showed that two types of network are formed, a majority one comprising a distorted square, rhombus-like, frame, and the minority one similar to a Kagomé lattice with a network of hexagonal and triangular pores seen in similar systems.[133] The rhombus network appears with the nodes clearly visible in the STM images, and the terpyridine moieties located precisely at the positions where the pyridyl groups attached to the porphyrin should be directed. Although this topology would be expected to produce a square-like topography, a symmetry mismatch of the Au (111) substrate lattice—that has three-fold symmetry—and the ideal square network results in a distortion to the rhomb shape. The square shape that would be energetically favourable in isolation is not favoured because of the strength of the interaction between the surface and adsorbates and the complicit flexibility of the coordination of the pyridyl ligands to the iron. While the structure is remarkably regular, STM does reveal specific defects that can be attributed either to a missing rod-like connector or to the incorporation of a sporadic monomer unit from the initial covalent coupling on the surface (Figure 10.25(f)). This kind of strategy using the surface to combine covalent and supramolecular steps in hierarchical assembly are bound to flourish as the rules for interface synthesis and complexation become clearer.

The preceding examples comprised pretty rigid components with very well-defined nodes that can be predictably controlled, and they have proven successful partially because of this fact. The control of packing in two dimensions is a lot easier than in three, and here the problem occurs when growing into the third dimension! Apart from non-covalent interactions, shape can have a dramatic effect on the assembly of molecules.[134–136] Barbiturate derivatives of naphthalene compounds bearing long alkyl chains display this effect dramatically (Figure 10.26).[137] One of the regio-isomers forms rod-like aggregates parallel to the surface upon evaporation from methylcyclohexane while the other forms microscopic rings because of the curvature generated in the assembly of the compound. The effect was ascribed to the conformations of the naphthyl rings in relation to the rosette plane, which gives rise to either J- or H-type aggregates when the rosettes pile up. Nanorods of barbiturate derivatives of thiophenes not bearing long alkyl chains have been shown to form nanorods by solution casting or annealing of the thin films that are formed, even with soluble fullerene derivatives present.[138] The absence of long aliphatic chains did not rule out the formation of the rosette rings or stacks, and was actually beneficial for the photovoltaic properties of films made from the compounds compared with

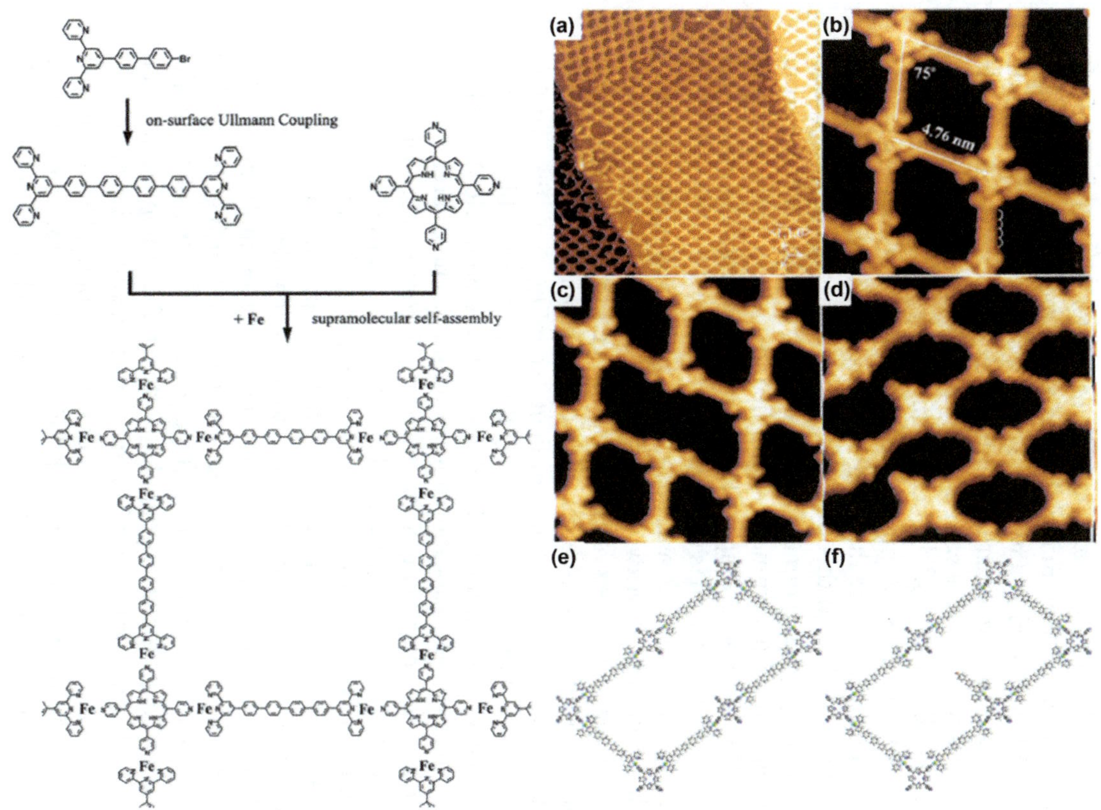

Figure 10.25 A metal–organic framework self-assembled on a surface where one of the components was prepared covalently *in situ*. The scheme on the right shows the two steps, the on-surface Ullmann coupling of two terpyridine-containing where the aryl bromide is reacted to give a bis-terpyridine with a quaterphenyl group between the chelants, and the self-assembly of this ligand with iron atoms and tetra-4-pyridylporphyrin to give the framework. The STM images (a)–(d) show the network on the gold (111) surface at different magnifications ((a) is 100×100 nm^2, (b) 12×12 nm^2, (c) and (d) 15×15 nm^2), the latter two showing defects that are represented as models in (e) and (f).
Reproduced from ref. 130 with permission from the Royal Society of Chemistry.

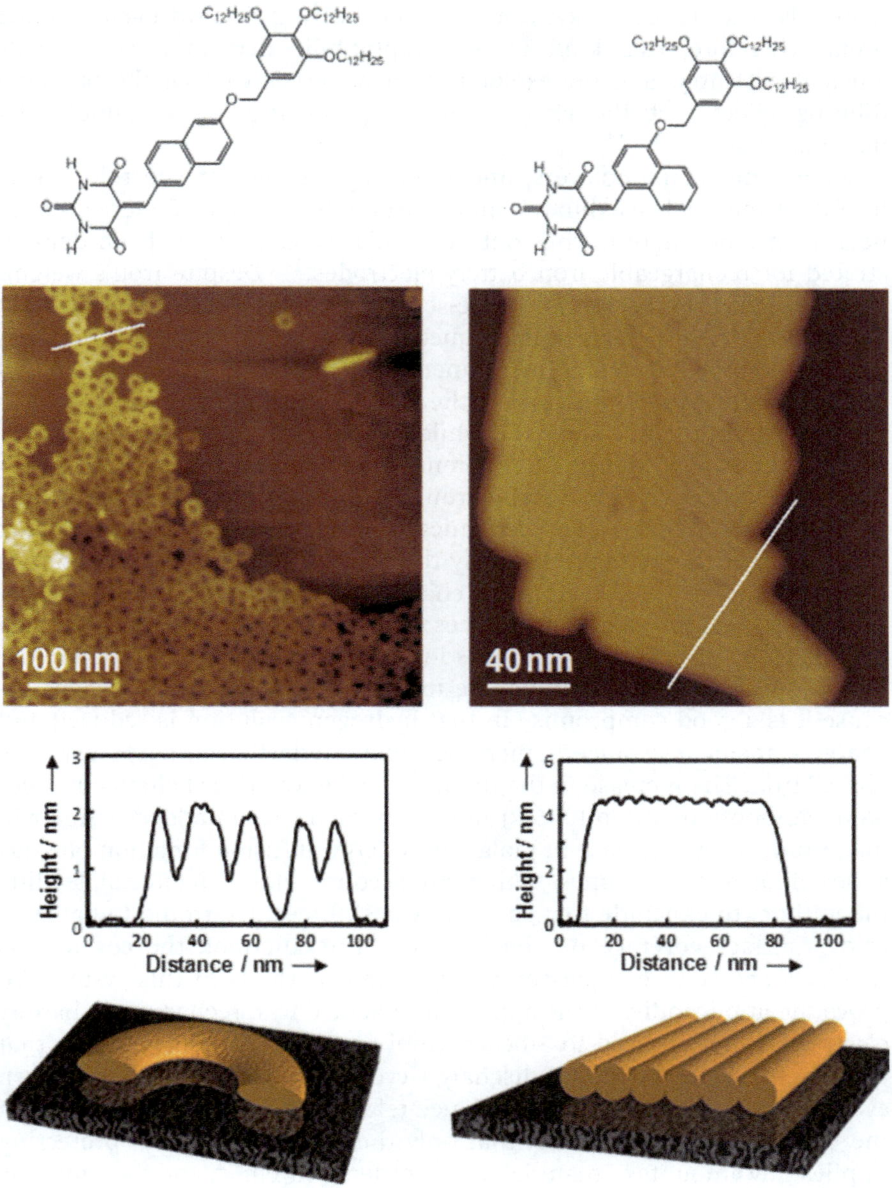

Figure 10.26 Regioisomers of barbiturate derivatives that show completely different assembly at an interface upon evaporation of methylcyclohexane, as seen in the AFM images underneath each compound, along with profiles corresponding to the white lines in the images and impressions of how the cylindrical assemblies appear. The dependence arises from the conformations of the naphthyl rings and their influence on the type of aggregate formed. Gratitude to Shiki Yagai for providing the images and original artwork.

materials bearing long alkyl chains or not bearing the hydrogen-bonding units. Therefore, this kind of non-amphiphilic derivative can provide efficient assembly, and the exploration of materials without the property-diluting effect that the chains have is promising for supramolecular materials research.[139,140]

Self-assembled monolayers, and their supramolecular control, can be useful for materials used in the more sustainable storage of energy, as has been mentioned here in the context of solar cells, and has been demonstrated for rechargeable iron battery electrodes.[141] Despite iron's weight, there is great potential for large-scale energy storage—for grid-scale applications[142]—in rechargeable aqueous alkaline batteries based on this metal, which is abundant, environmentally acceptable and cheap. One of the big challenges in this area is the undesired production of hydrogen taking place in the idle state and while the batteries are being charged.[143] Organic thiols were identified as potential monolayer systems because they form SAMs selectively on metallic iron and not on the oxide or hydroxide-coated material. The electrochemical processes taking place during the cycling of the battery, especially the charging, could be hindered if a totally compact SAM that impedes charge transport were to be employed. Dodecanethiol, for example, reduces hydrogen production considerably but the charge transfer resistance is increased dramatically, and therefore 1-octanethiol or 1-hexanethiol were more attractive as the adsorbent because it is a good compromise in that hydrogen evolution is reduced and charge transfer resistance is increased only modestly compared with the "bare" iron. The increase in the interactions between alkyl chains in these SAMs amounts to around 2.6 kJ mol^{-1} for each carbon added, similar in magnitude to the increase in molar free energy of fusion for a homologous series of long-chain alkanes (taking into account odd–even effects), leading the authors to conclude that van der Waals' forces determine the alkanethiol's surface coverage on the iron. For 1-hexanethiol, the coverage is relatively sparse in comparison to higher alkanethiols in this system. To prove the applicability of the approach, practical iron rechargeable battery electrodes were exposed to 1-hexanethiol and were used in a device that underwent multiple charge–discharge cycles with suppressed hydrogen evolution and high charging efficiency relative to batteries incorporating the untreated metal. The fact that only the interface requires protecting implies tiny amounts of material, of the order of ng cm^{-2}, and therefore the treatment of the iron–aqueous alkaline interface using this strategy promises increased efficiency and improved sustainability in these devices and potentially others.

The preparation of monolayers and multilayers of materials has challenges regarding the preparation of pure phases, and the minimisation of defects, and some examples of strategies to overcome these difficulties have been presented in this monograph. Presently, the state of a surface is summed up nicely in the image and text in Figure 10.27. The opportunities provided by these situations to probe molecules' properties in closely related

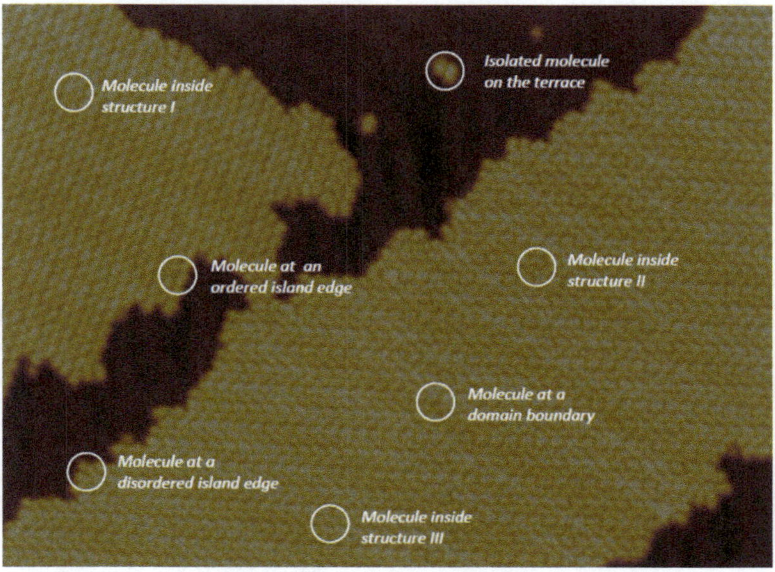

Figure 10.27 An STM image showing the variety of environments that molecules can find themselves in even on a clean surface, where structures I, II and III are polymorphs.
This image was kindly provided by Leonhard Grill.

but subtly different situations is possible using state-of-the-art instrumentation,[144] but the differences also have to be considered when understanding the properties of bulk films.

It is probable and beneficial that supramolecular surface chemistry takes further steps into the exploration of non-covalent assembly in polar media, especially in water. The use of green chemistry concepts[145,146] for the formation of mono- and multi-layers is an intriguing possibility when thinking of potential advantages of supramolecular systems. The formation of the open network of trimesic acid on graphite has been shown from water,[147] as it has for more complex acids.[148] Water is viable for these compounds because of their partial solubility in water owing to their multiple carboxylic acid groups, and the assembly of functional systems that employ aqueous or at least environmentally benign solvents for thin film deposition needs to be made more general.

Looking back at the panel meeting review published in 1987,[149] huge strides have been taken in the patterning of molecules at surfaces, the immobilisation of molecules at metals for sensor applications, or the passivation and control of wettability of surfaces. There are very significant doubts concerning the structure of molecules at surfaces over time. The panel clearly had a sense of humour: in their figure reflecting on this situation (Figure 10.28), the desperate analytical quagmire presented by large molecules at interfaces is still sticky today. However, the drastic improvements in analytical tools—particularly scattering techniques at synchrotrons and

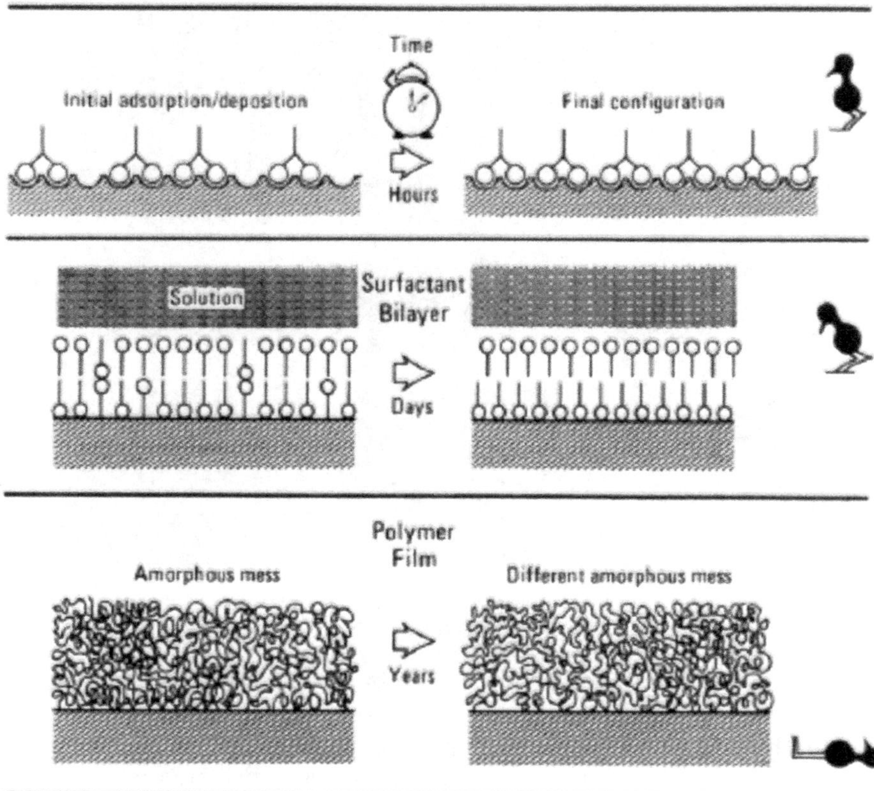

Figure 10.28 A cartoon showing the ageing of molecule-based films and an avian guide to the state of understanding in 1987.
Reprinted with permission from ref. 149. Copyright 1987 American Chemical Society.

powerful optical and electron spectroscopy[150] and microscopy methods—since that report (where they identified this as a challenge) are very likely to facilitate understanding.

Multi- and inter-disciplinary research are increasingly the hallmark of research into supramolecular chemistry at interfaces, as this monograph and a wildly diverse selection of very recent work demonstrate amply.[151–158] The basically flat interfaces allow them to be probed with some of the most powerful analytical and computational tools available in the sciences, and we should take the opportunities available to us to use them to gain the deepest understanding possible of the science there. The concepts that have been applied to assemble molecules at different interfaces will surely be transferred to other systems, particularly colloidal ones. But there are still many supramolecular systems that will continue to amaze and inspire science and technology through systems approaches at interfaces, perhaps providing societal, environmental, or economic impact from where we least expect it.

References

1. W. C. Bigelow, D. L. Pickett and W. A. Zisman, *J. Colloid Sci.*, 1946, **1**, 513.
2. P. M. Dietrich, C. Streeck, S. Glamsch, C. Ehlert, A. Lippitz, A. Nutsch, N. Kulak, B. Beckhoff and W. E. S. Unger, *Anal. Chem.*, 2015, **87**, 10117.
3. J. V. Barth, G. Costantini and K. Kern, *Nature*, 2005, **437**, 671.
4. M.-C. Blüm, E. Ćavar, M. Pivetta, F. Patthey and W.-D. Schneider, *Angew. Chem., Int. Ed.*, 2005, **44**, 5334.
5. M. Bohringer, K. Morgenstern, W. D. Schneider and R. Berndt, *Angew. Chem., Int. Ed.*, 1999, **38**, 821.
6. A. Mourran, B. Tartsch, M. Gallyamov, S. Magonov, D. Lambreva, B. I. Ostrovskii, I. P. Dolbnya, W. H. de Jeu and M. Moeller, *Langmuir*, 2005, **21**, 2308.
7. M. O. Gallyamov, A. Mourran, B. Tartsch, R. A. Vinokur, L. N. Nikitin, A. R. Khokhlov, K. Schaumburg and M. Möller, *Phys. Chem. Chem. Phys.*, 2006, **8**, 2642.
8. E. N. Hoggan, D. Flowers, K. Wang, J. M. DeSimone and R. G. Carbonell, *Ind. Eng. Chem. Res.*, 2004, **43**, 2113.
9. J. Kim, J. P. Rolland, R. G. Carbonell and J. M. DeSimone, *Chem. Mater.*, 2010, **22**, 2411.
10. L. Grill, *J. Phys.: Condens. Matter*, 2010, **22**, 084023.
11. C. Hamann, R. Woltmann, I. P. Hong, N. Hauptmann, S. Karan and R. Berndt, *Rev. Sci. Instrum.*, 2011, **82**, 033903.
12. S. Lorbek, G. Hlawacek and C. Teichert, *Eur. Phys. J.: Appl. Phys.*, 2011, **55**, 23902.
13. A. Robin, P. Iavicoli, K. Wurst, M. S. Dyer, S. Haq, D. B. Amabilino and R. Raval, *Cryst. Growth Des.*, 2010, **10**, 4516.
14. W. J. Saletra, H. Xu, T. Vosch, S. De Feyter and D. B. Amabilino, *CrystEngComm*, 2011, **13**, 5578.
15. J. Seibel, M. Parschau and K.-H. Ernst, *J. Am. Chem. Soc.*, 2015, **137**, 7970.
16. R. Otero, M. Lukas, R. E. A. Kelly, W. Xu, E. Lægsgaard, I. Stensgaard, L. N. Kantorovich and F. Besenbacher, *Science*, 2008, **319**, 312.
17. M. Marschall, J. Reichert, A. Weber-Bargioni, K. Seufert, W. Auwärter, S. Klyatskaya, G. Zoppellaro, M. Ruben and J. V. Barth, *Nat. Chem.*, 2010, **2**, 131.
18. P. Iavicoli, H. Xu, T. Keszthelyi, J. Telegdi, K. Wurst, B. Van Averbekee, W. J. Saletra, A. Minoia, D. Beljonne, R. Lazzaroni, S. De Feyter and D. B. Amabilino, *Chirality*, 2012, **24**, 155.
19. I. Kuzmenko, I. Weissbuch, E. Gurovich, L. Leiserowitz and M. Lahav, *Chirality*, 1998, **10**, 415.
20. I. Weissbuch, R. A. Illos, G. Bolbach and M. Lahav, *Acc. Chem. Res.*, 2009, **42**, 1128.
21. D. Y. Takamoto, E. Aydil, J. A. Zasadzinski, A. T. Ivanova, D. K. Schwartz, T. Yang and P. S. Cremer, *Science*, 2001, **293**, 1292.

22. V. M. Kaganer, H. Möhwald and P. Dutta, *Rev. Mod. Phys.*, 1999, **71**, 779.
23. D. K. Schwartz, *Surf. Sci. Rep.*, 1997, **27**, 241.
24. J. A. Zasadzinski, R. Viswanathan, L. Madsen, J. Garnaes and D. K. Schwartz, *Science*, 1994, **263**, 1726.
25. P. Tippmann-Krayer, R. M. Kenn and H. Mohwald, *Thin Solid Films*, 1992, **210-211**, 577.
26. D. K. Schwartz, J. Garnaes, R. Viswanathan and J. A. Zasadzinski, *Science*, 1992, **257**, 508.
27. R. W. Corkery, *Langmuir*, 1997, **13**, 3591.
28. N. Seiki, Y. Shoji, T. Kajitani, F. Ishiwari, A. Kosaka, T. Hikima, M. Takata, T. Someya and T. Fukushima, *Science*, 2015, **348**, 1122.
29. S.-L. Lee, Z. Yuan, L. Chen, K. S. Mali, K. Müllen and S. De Feyter, *J. Am. Chem. Soc.*, 2014, **136**, 7595.
30. T. J. Roussel, E. Barrena, C. Ocal and J. Faraudo, *Nanoscale*, 2014, **6**, 7991.
31. S. Lee, C.-H. Chen and A. H. Flood, *Nat. Chem.*, 2013, **5**, 704.
32. B. E. Hirsch, S. Lee, B. Qiao, C.-H. Chen, K. P. McDonald, S. L. Tait and A. H. Flood, *Chem. Commun.*, 2014, **50**, 9827.
33. H. N. Tsao and K. Müllen, *Chem. Soc. Rev.*, 2010, **39**, 2372.
34. H. Klauk, *Chem. Soc. Rev.*, 2010, **39**, 2643.
35. W. Shao, H. Dong, L. Jiang and W. Hu, *Chem. Sci.*, 2011, **2**, 590.
36. Y. L. Gao, *Mater. Sci. Eng., R*, 2010, **68**, 39.
37. N. Peor, R. Sfez and S. Yitzchaik, *J. Am. Chem. Soc.*, 2008, **130**, 4158.
38. A. J. Maekinen and G. P. Kushto, *Phys. Rev. B: Condens. Matter Mater. Phys.*, 2011, **83**, 245315.
39. S. J. Schoell, M. Sachsenhauser, A. Oliveros, J. Howgate, M. Stutzmann, M. S. Brandt, C. L. Frewin, S. E. Saddow and I. D. Sharp, *ACS Appl. Mater. Interfaces*, 2013, **5**, 1393.
40. D. Kumaki, S. Ando, S. Shimono, Y. Yamashita, T. Umeda and S. Tokito, *Appl. Phys. Lett.*, 2007, **90**, 053506.
41. A. Della Pia, M. Riello, A. Floris, D. Stassen, T. S. Jones, D. Bonifazi, A. De Vita and G. Costantini, *ACS Nano*, 2014, **8**, 12356.
42. K. Galatsis, K. Wang, Y. Botros, Y. Yang, Y. H. Xie, J. F. Stoddart, R. B. Kaner, C. Ozkan, J. L. Liu, M. Ozkan, C. W. Zhou and K. W. Kim, *IEEE Circuits Devices*, 2006, **22**, 12.
43. A. Facchetti, E. Annoni, L. Beverina, M. Morone, P. Zhu, T. J. Marks and G. A. Pagani, *Nat. Mater.*, 2004, **3**, 910.
44. A. Lv, M. Stolteand and F. Wurthner, *Angew. Chem., Int. Ed.*, 2015, **54**, 10512.
45. N. Fuentes, A. Martin-Lasanta, L. Alvarez de Cienfuegos, M. Ribagorda, A. Parra and J. M. Cuerva, *Nanoscale*, 2011, **3**, 4003.
46. S. J. van der Molen, R. Naaman, E. Scheer, J. B. Neaton, A. Nitzan, D. Natelson, N. J. Tao, H. van der Zant, M. Mayor, M. Ruben, M. Reed and M. Calame, *Nat. Nanotechnol.*, 2013, **8**, 385.

47. T. Bjornholm, T. Hassenkam and N. Reitzel, *J. Mater. Chem.*, 1999, **9**, 1975.
48. W. Wen, Y. Liu, Y. Guo and W. Hu, *Chem. Rev.*, 2011, **111**, 3358.
49. G. Xu, Z. Bao and J. T. Groves, *Langmuir*, 2000, **16**, 1834.
50. C. Suryanarayanan, K. Sankaranarayanan, C. Divakara and N. Somanathan, *RSC Adv.*, 2014, **4**, 38341.
51. A. M. Haruk and J. M. Mativetsky, *Int. J. Mol. Sci.*, 2015, **16**, 13381.
52. D. Bonifaze, O. Enger and F. Diederich, *Chem. Soc. Rev.*, 2007, **36**, 390.
53. T. Jaroch, A. Maranda-Niedbała, M. Góra, J. Mieczkowski, M. Zagórska, M. Salamonczyk, E. Górecka and R. Nowakowski, *Synth. Met.*, 2015, **204**, 133.
54. C.-H. Huang, N. D. McClenaghan, A. Kuhn, G. Bravic and D. Bassani, *Tetrahedron*, 2006, **62**, 2050.
55. T. Aytun, L. Barreda, A. Ruiz-Carretero, J. A. Lehrman and S. I. Stupp, *Chem. Mater.*, 2015, **27**, 1201.
56. J. Yang, D. Yan and T. S. Jones, *Chem. Rev.*, 2015, **115**, 5570.
57. P. Sullivan, T. S. Jones, A. J. Ferguson and S. Heutz, *Appl. Phys. Lett.*, 2007, **91**, 233114.
58. S. Heutz, S. M. Bayliss, R. L. Middleton, G. Rumbles and T. S. Jones, *J. Phys. Chem. B*, 2000, **104**, 7124.
59. S. Nespurek, H. Podlesak and C. Hamann, *Thin Solid Films*, 1994, **249**, 230.
60. S. Heutz, R. Cloots and T. S. Jones, *Appl. Phys. Lett.*, 2000, **77**, 3938.
61. S. Heutz and T. S. Jones, *J. Appl. Phys.*, 2002, **92**, 3039.
62. S. Yim, S. Heutz and T. S. Jones, *Phys. Rev. B: Condens. Matter Mater. Phys.*, 2003, **67**, 165308.
63. T. B. Fleetham, J. P. Mudrick, W. Cao, K. Klimes, J. Xue and J. Li, *ACS Appl. Mater. Interfaces*, 2014, **6**, 7254.
64. J. B. Gilchrist, T. H. Basey-Fisher, S. C'. E. Chang, F. Scheltens, D. W. McComb and S. Heutz, *Adv. Funct. Mater.*, 2014, **24**, 6473.
65. P. Vivo, T. Vuorinen, V. Chukharev, A. Tolkki, K. Kaunisto, P. Ihalainen, J. Peltonen and H. Lemmetyinen, *J. Phys. Chem. C*, 2010, **114**, 8559.
66. T. Vuorinen, K. Kaunisto, N. V. Tkachenko, A. Efimov, A. S. Alekseev, K. Hosomizu, H. Imahori and H. Lemmetyinen, *Langmuir*, 2005, **21**, 5383.
67. A. S. Alekseev, N. V. Tkachenko, A. Y. Tauber, P. H. Hynninen, R. Osterbacka, H. Stubb and H. Lemmetyinen, *Chem. Phys.*, 2002, **275**, 243.
68. M. Smith and J. Michl, *Chem. Rev.*, 2010, **110**, 6891.
69. S. E. Strong and J. D. Eaves, *J. Phys. Chem. Lett.*, 2015, **6**, 1209.
70. M. Castellani, S. Winkler, B. Bröker, M. Baumgarten, K. Müllen and N. Koch, *Appl. Phys. A*, 2014, **114**, 291.
71. T. Kurita, Y. Nishimori, F. Toshimitsu, S. Muratsugu, S. Kume and H. Nishihara, *J. Am. Chem. Soc.*, 2010, **132**, 4524.

72. R. Sakamoto, S. Katagiri, H. Maeda, Y. Nishimori, S. Miyashita and H. Nishihara, *J. Am. Chem. Soc.*, 2015, **137**, 734.
73. Y. Nishimori, K. Kanaizuka, M. Murata and H. Nishihara, *Chem. – Asian J.*, 2007, **2**, 367.
74. R. Davidson, J.-H. Liang, D. Costa-Milan, B.-W. Mao, R. J. Nichols, S. J. Higgins, D. S. Yufit, A. Beeby and P. J. Low, *Inorg. Chem.*, 2015, **54**, 5487.
75. P. T. Bui, T. Nishino, Y. Yamamoto and H. Shiigi, *J. Am. Chem. Soc.*, 2013, **135**, 5238.
76. A. C. Aragonès, N. Darwish, W. J. Saletra, L. Pérez-García, F. Sanz, J. Puigmartí-Luis, D. B. Amabilino and I. Díez-Pérez, *Nano Lett.*, 2014, **14**, 4751.
77. Q. Ferreira, A. M. Bragança, L. Alcácer and J. Morgado, *J. Phys. Chem. C*, 2014, **118**, 7229.
78. Q. Ferreira, L. Alcácer and J. Morgado, *Nanotechnology*, 2011, **22**, 435604.
79. D. S. Turygin, M. Subat, O. A. Raitman, S. L. Selector, V. V. Arslanov, B. König and M. A. Kalinina, *Langmuir*, 2007, **23**, 2517.
80. M. Nakaya, Y. Okawa, C. Joachim, M. Aono and T. Nakayama, *ACS Nano*, 2014, **8**, 12259.
81. N. de Poul and B. Conasson, *ChemElectroChem*, 2015, **2**, 475.
82. S. Zahn and J. W. Canary, *Science*, 2000, **288**, 1404.
83. S. Zahn, D. Das and J. W. Canary, *Inorg. Chem.*, 2006, **45**, 6056.
84. D. Das, Z. Dai, A. E. Holmes and J. W. Canary, *Chirality*, 2008, **20**, 585.
85. H. S. Barcena, A. E. Holmes and J. W. Canary, *Org. Lett.*, 2003, 709.
86. H. S. Barcena, B. Liu, M. V. Mirkin and J. W. Canary, *Inorg. Chem.*, 2005, **44**, 7652.
87. J. S. Seldenthius, F. Prins, J. M. Thijssens and H. S. J. van der Zant, *ACS Nano*, 2010, **4**, 6681.
88. L. Kecke and J. Ankerhold, *Phys. Rev. B: Condens. Matter Mater. Phys.*, 2012, **85**, 245442.
89. A. J. Mota, L. Alvarez de Cienfuegos, S. P. Morcillo, N. Fuentes, D. Miguel, S. Rodrguez-Bolvar, F. M. Gomez-Campos, D. J. Cardenas and J. M. Cuerva, *ChemPhysChem*, 2012, **13**, 3857.
90. H. L. Tierney, C. J. Murphy, A. D. Jewell, A. E. Baber, E. V. Iski, H. Y. Khodaverdian, A. F. McGuire, N. Klebanov and E. C. H. Sykes, *Nat. Nanotechnol.*, 2011, **6**, 625.
91. B. C. Stipe, M. A. Rezaei and W. Ho, *Phys. Rev. Lett.*, 1998, **81**, 1263.
92. K. Cui, K. S. Mali, O. Ivasenko, D. Q. Wu, X. L. Feng, M. Walter, K. Mullen, S. De Feyter and S. F. L. Mertens, *Angew. Chem., Int. Ed.*, 2014, **53**, 12951.
93. K. Cui, O. Ivasenko, K. S. Mali, D. Wu, X. Feng, K. Müllen, S. De Feyter and S. F. L. Mertens, *Chem. Commun.*, 2014, **50**, 10376.
94. J. L. Zhang, J. L. Xu, T. C. Niu, Y. H. Lu, L. Liu and W. Chen, *J. Phys. Chem. C*, 2014, **118**, 1712.

95. S. K. Arya, P. R. Solanki, M. Datta and B. D. Malhotra, *Biosens. Bioelectron.*, 2009, **24**, 2810.
96. G. M. Whitesides, E. Ostuni, S. Takayama, X. Y. Jiang and D. E. Ingber, *Annu. Rev. Biomed. Eng.*, 2001, **3**, 335.
97. K. Lv, L. Zhang, W. Lu and M. Liu, *ACS Appl. Mater. Interfaces*, 2014, **6**, 18878.
98. L. Séon, P. Lavalle, P. Schaaf and F. Boulmedais, *Langmuir*, 2015, **31**, 13385.
99. R. Ramanathan, *J. Perinatol.*, 2006, **26**, S51.
100. K. Kim, S. Q. Choi, Z. A. Zell, T. M. Squires and J. A. Zasadzinski, *Proc. Natl. Acad. Sci. U. S. A.*, 2013, **110**, E3054.
101. S. Tatur and A. Badia, *Langmuir*, 2012, **26**, 628.
102. V. Dupres, F. D. Menozzi, C. Locht, B. H. Clare, N. L. Abbott, S. Cuenot, C. Bompard, D. Raze and Y. F. Dufrene, *Nat. Methods*, 2005, **2**, 515.
103. H. P. Haim, R. M. A. Sullan, P. J. Crowley, S. El-Kirat-Chatel, A. Beaussart, W. X. Tang, R. Besingi, Y. F. Dufrene and L. J. Brady, *J. Biol. Chem.*, 2015, **290**, 9002.
104. R. M. A. Sullan, J. K. Li, P. J. Crowley, L. J. Brady and Y. F. Dufrene, *ACS Nano*, 2015, **9**, 1448.
105. A. M. Oelker and M. W. Grinstaff, *J. Mater. Chem.*, 2008, **18**, 2521.
106. A. M. S. Kumar, S. Sivakova, R. E. Marchant and S. J. Rowan, *Small*, 2007, **3**, 783.
107. A. M. S. Kumar, S. Sivakova, J. D. Fox, J. E. Green, R. E. Marchant and S. J. Rowan, *J. Am. Chem. Soc.*, 2008, **130**, 1466.
108. G. Gottarelli, S. Masiero, E. Mezzina, S. Pieraccini, J. P. Rabe, P. Samori and G. P. Spada, *Chem. - Eur. J.*, 2000, **6**, 3242.
109. Y. Sato, H. Noda, F. Mizutani, A. Yamakata and M. Osawa, *Anal. Chem.*, 2004, **76**, 5564.
110. D. Zhang, K. F. Domke and B. Pettinger, *ChemPhysChem*, 2010, **11**, 1662.
111. D. Liu, Y. Xie, H. Shao and X. Jiang, *Angew. Chem., Int. Ed.*, 2009, **48**, 4406.
112. E. Ruoslahti, *Annu. Rev. Cell Dev. Biol.*, 1996, **12**, 697.
113. D. B. Liu, Y. Y. Xie, H. W. Shao and X. Y. Jiang, *Prog. Chem.*, 2007, **19**, 1965.
114. J. Robertus, W. R. Browne and B. L. Feringa, *Chem. Soc. Rev.*, 2010, **39**, 354.
115. M. Lashkor, F. J. Rawson, A. Stephenson-Brown, J. A. Preece and P. M. Mendes, *Chem. Commun.*, 2014, **50**, 15589.
116. O. Penon, D. Siapkas, S. Novo, S. Durán, G. Oncins, A. Errachid, L. Barrios, C. Nogués, M. Duch, J. A. Plaza and L. Pérez-García, *Colloids Surf., B*, 2014, **116**, 104.
117. S. Durán, M. Duch, T. Patiño, A. Torres, O. Penon, R. Gómez-Martínez, L. Barrios, J. Esteve, C. Nogués, L. Pérez-García and J. A. Plaza, *Sens. Actuators, B*, 2015, **209**, 212.

118. R. Gutzler, S. Stepanow, D. Grumelli, M. Lingenfelder and K. Kern, *Acc. Chem. Res.*, 2015, **48**, 2132.
119. Z. Ma and F. Zaera, *Surf. Sci. Rep.*, 2006, **61**, 229.
120. Y. Xue and M. B. Zimmt, *J. Am. Chem. Soc.*, 2012, **134**, 4513.
121. R. Madueno, M. T. Räisänen, C. Silien and M. Buck, *Nature*, 2008, **454**, 618.
122. C. Silien and M. Buck, *J. Phys. Chem. C*, 2008, **112**, 3881.
123. J. Méndez, M. F. López and J. A. Martín-Gago, *Chem. Soc. Rev.*, 2011, **40**, 4578.
124. R. Lindner and A. Kuehnle, *ChemPhysChem*, 2015, **16**, 1582.
125. S. Haq, F. Hanke, J. Sharp, M. Persson, D. B. Amabilino and R. Raval, *ACS Nano*, 2014, **8**, 8856.
126. F. Hanke, S. Haq, R. Raval and M. Persson, *ACS Nano*, 2011, **5**, 9093.
127. Q. T. Fan, T. Wang, L. M. Liu, J. Zhao, J. F. Zhu and J. M. Gottfried, *J. Chem. Phys.*, 2015, **142**, 101906.
128. Q. T. Fan, C. C. Wang, Y. Han, J. F. Zhu, J. Kuttner, G. Hilt and J. M. Gottfried, *ACS Nano*, 2014, **8**, 709.
129. Q. T. Fan, J. M. Gottfried and J. F. Zhu, *Acc. Chem. Res.*, 2015, **48**, 2484.
130. T. Lin, G. Kuang, X. S. Shang, P. N. Liu and N. Lin, *Chem. Commun.*, 2014, **50**, 15327.
131. T. Bauer, Z. Zheng, A. Renn, R. Enning, A. Stemmer, J. Sakamoto and A. D. Schlüter, *Angew. Chem., Int. Ed.*, 2011, **50**, 7879.
132. W. Wang, Y. Hong, X. Shi, C. Minot, M. A. Van Hove, B. Z. Tang and N. Lin, *J. Phys. Chem. Lett.*, 2010, **1**, 2295.
133. Z. Shi and N. Lin, *J. Am. Chem. Soc.*, 2010, **132**, 10756.
134. Z. Chen, A. Lohr, C. R. Saha-Möller and F. Würthner, *Chem. Soc. Rev.*, 2009, **38**, 564.
135. K. W. Chooi, A. I. Gray, L. Tetley, Y. Fan and I. F. Uchegbu, *Langmuir*, 2010, **26**, 2301.
136. T. Nakanishi, *Chem. Commun.*, 2010, **46**, 3425.
137. S. Yagai, Y. Goto, X. Lin, T. Karatsu, A. Kitamura, D. Kuzuhara, H. Yamada, Y. Kikkawa, A. Saeki and S. Seki, *Angew. Chem., Int. Ed.*, 2012, **51**, 6643.
138. S. Yagai, M. Suzuki, X. Lin, M. Gushiken, T. Noguchi, T. Karatsu, A. Kitamura, A. Saeki, S. Seki, Y. Kikkawa, Y. Tani and K. Nakayama, *Chem. – Eur. J.*, 2014, **20**, 16128.
139. I. Danila, F. Riobé, F. Piron, J. Puigmartí-Luis, J. D. Wallis, M. Linares, H. Ågren, D. Beljonne, D. B. Amabilino and N. Avarvari, *J. Am. Chem. Soc.*, 2011, **133**, 8344.
140. I. Danila, F. Pop, C. Escudero, L. N. Feldborg, J. Puigmartí-Luis, F. Riobé, N. Avarvari and D. B. Amabilino, *Chem. Commun.*, 2012, **48**, 4552.
141. S. Malkhandi, B. Yang, A. K. Manohar, G. K. S. Prakash and S. R. Narayanan, *J. Am. Chem. Soc.*, 2013, **135**, 347.
142. S. R. Narayanan, G. K. Prakash, A. Manohar, B. Yang and S. Malkhandi, *Solid State Ionics*, 2012, **216**, 105.

143. T. S. Balasubramanian, K. Vijayamohanan and A. K. Shukla, *J. Appl. Electrochem.*, 1993, **23**, 947.
144. J. Mielke, S. Selvanathan, M. Peters, J. Schwarz, S. Hecht and L. Grill, *J. Phys.: Condens. Matter*, 2012, **24**, 394013.
145. P. Anastas and N. Eghbali, *Chem. Soc. Rev.*, 2010, **39**, 301.
146. R. M. Izatt, S. R. Izatt, R. L. Bruening, N. E. Izatt and B. A. Moyer, *Chem. Soc. Rev.*, 2014, **43**, 2451.
147. V. V. Korolkov, S. Allen, C. J. Roberts and S. J. B. Tendler, *J. Phys. Chem. C*, 2012, **116**, 11519.
148. I. Cebula, E. F. Smith, M. C. Gimenez-Lopez, S. Yang, M. Schröder, N. R. Champness and P. H. Beton, *J. Phys. Chem. C*, 2013, **117**, 18381.
149. J. D. Swalen, D. L. Allara, J. D. Andrade, E. A. Chandross, S. Garoff, J. Israelachvili, T. J. McCarthy, R. Murray, R. F. Pease, J. F. Rabolt, K. J. Wynne and H. Yu, *Langmuir*, 1987, **3**, 932.
150. A. S. Duwez, *J. Electron Spectrosc. Relat. Phenom.*, 2004, **134**, 97.
151. P. Du, D. Kreher, F. Mathevet, P. Maldivi, F. Charra and A.-Je. Attias, *ChemPhysChem*, 2015, **16**, 3774.
152. S. Bai, C. Sun, H. Yan, X. Sun, H. Zhang, L. Luo, X. Lei, P. Wan and X. Chen, *Small*, 2015, **11**, 5807.
153. A. Fabiano, P. Chetoni and Y. Zambito, *Drug Dev. Ind. Pharm.*, 2015, **41**, 2069.
154. N. Moridi, P. F.-X. Corvini and P. Shahgaldian, *Angew. Chem., Int. Ed.*, 2015, **54**, 14800.
155. E. Sariola-Leikas, Z. Ahmed, P. Vivo, A. Ojanperä, K. Lahtonen, J. Saari, M. Valden, H. Lemmetyinen and A. Efimov, *Chem. – Eur. J.*, 2016, **22**, 1501.
156. N. Bilbao, I. Destoop, S. De Feyter and D. González-Rodríguez, *Angew. Chem., Int. Ed.*, 2016, **55**, 659.
157. V. Singh, Z. Li, X. Zhou, X. Xu, J. Xu, A. Nand, H. Wen, H. Li, J. Zhu and J. Zhang, *RSC Adv.*, 2016, **6**, 3213.
158. O. Roling, L. Stricker, J. Voskuhl, S. Lamping and B. J. Ravoo, *Chem. Commun.*, 2016, **52**, 1964.

Subject Index